I0787946

Chemical Sensors 10 -and- MEMS/NEMS 10

Editors:

A. Simonian
Auburn University
Auburn, Alabama, USA

M. Carter
KWJ Engineering, Inc.
Newark, California, USA

N. Miura
Kyushu University
Kasuga-shi, Japan

Y. Shimizu
Nagasaki University
Nagasaki, Japan

P. Srinivasan
Texas Instruments Incorporated
Dallas, Texas, USA

P. Vanysek
Northern Illinois University
DeKalb, Illinois, USA

P. J. Hesketh
Georgia Institute of Technology
Atlanta, Georgia, USA

B. Chin
Auburn University
Auburn, Alabama, USA

L. Nagahara
National Cancer Institute
Bethesda, Maryland, USA

S. Uchiyama
Saitama Institute of Technology
Saitama, Japan

K. B. Sundaram
University of Central Florida
Orlando, Florida, USA

N. Wu
West Virginia University
Morgantown, West Virginia, USA

Z. Aguilar
Ocean NanoTech
Springdale, Arkansas, USA

G. Hunter
NASA Glenn Research Center
Cleveland, Ohio, USA

M. Sailor
University of California
La Jolla, California, USA

A. Londergan
Qualcomm MEMS Technologies
San Jose, California, USA

O. Tabata
Kyoto University
Kyoto, Japan

Sponsoring Divisions:

 Sensor

 Dielectric Science & Technology

 Electronics and Photonics

 Physical and Analytical Electrochemistry

Published by
The Electrochemical Society
65 South Main Street, Building D
Pennington, NJ 08534-2839, USA
tel   609 737 1902
fax   609 737 2743
www.electrochem.org

ecstransactions ™

Vol. 50, No. 12

Copyright 2012 by The Electrochemical Society.
All rights reserved.

This book has been registered with Copyright Clearance Center.
For further information, please contact the Copyright Clearance Center,
Salem, Massachusetts.

Published by:

The Electrochemical Society
65 South Main Street
Pennington, New Jersey 08534-2839, USA

Telephone 609.737.1902
Fax 609.737.2743
e-mail: ecs@electrochem.org
Web: www.electrochem.org

ISSN 1938-6737 (online)
ISSN 1938-5862 (print)
ISSN 2151-2051 (cd-rom)

ISBN 978-1-62332-011-9 (Hardcover)
ISBN 978-1-60768-360-5 (PDF)

Printed in the United States of America.

Preface

The papers included in this issue of *ECS Transactions* were originally presented in the symposia "Chemical Sensors 10 - Chemical and Biological Sensors and Analytical Systems" and "Microfabricated and Nanofabricated Systems for MEMS/NEMS 10", held during the PRiME 2012 joint international meeting of The Electrochemical Society and The Electrochemical Society of Japan, with the technical cosponsorship of the Japan Society of Applied Physics, the Korean Electrochemical Society, the Electrochemistry Division of the Royal Australian Chemical Institute, and the Chinese Society of Electrochemistry. This meeting was held in Honolulu, Hawaii, from October 7 to 12, 2012.

ECS Transactions, Volume 50, Issue 12
Chemical Sensors 10 -and- MEMS/NEMS 10

Table of Contents

Preface   *iii*

Chapter 1
Nano/Bio Sensors

Phytosensors and Phytoactuators   3
    A. G. Volkov, M. I. Volkova, and V. S. Markin

Miniature Enzymatic Biosensors for Tear Glucose Measurement in Capillary Tubes   13
    B. Peng, Q. Yan, A. Balijepalli, B. Cohan, T. C. Major, and M. E. Meyerhoff

Time Sensors: Circadian Rhythms in Biologically Closed Electrochemical Circuits of   23
Plants
    A. J. Waite, J. D. Wooten, V. S. Markin, and A. G. Volkov

Immobilization of Enzymes and Redox Proteins and Their Electrochemical Biosensor   35
Applications
    V. Mani, B. Devadas, S. Chen, and Y. Li

Signal Optimization for Salmonella Typhimurium Detection on Food Surfaces Using   43
Phage-Based Biosensors
    Y. Chai, S. Li, S. Horikawa, M. Park, V. Vodyanoy, and B. A. Chin

Bio-Inspired Autonomous Sentinel System for Screening Invasive Pathogens   53
    S. Li, H. Wikle III, Y. Chai, M. Park, S. Horikawa, and B. A. Chin

New Multimode Sensors Based on Nanostrucured Materials for Simultaneous   61
Screening of Biological Fluids for Specific Breast Cancer and Hepatitis B Biomarkers
    R. I. Stefan - van Staden and M. Enachescu

Performance of Optimized Phage-Based Magnetoelastic Biosensors for Salmonella   69
Typhimurium Detection on Tomatoes
    M. Park, S. Li, K. Weerakoon, S. Horikawa, Y. Chai, N. Hiremath, V. A. Petrenko,
    and B. A. Chin

Development of Electrochemical Cantilever Sensors for DNA Applications 77
X. Quan, A. Heiskanen, Y. Sun, A. Labuda, A. Wolff, J. Jorge Dulanto, P. Grutter, M. Tenje, and A. Boisen

Fabrication of Minimally-Invasive Patch Type Glucose Sensors 83
M. Yasuzawa, S. Sato, H. Nakanishi, and K. Edagawa

Flexprint Substrate Enzyme Sensor for Determination of Daily Glucose-Profiles of Diabetic 89
Patients
P. D. van der Wal, P. Hadvàry, H. Tschirky, and N. F. de Rooij

Nanodiamond Microelectrode Array with Mesa Structure Fabricated for Bio-Analytical 93
Applications
S. Raina, W. Kang, and N. Ghosh

Electrical Impedance Sensors for Cancer Cell Study 101
L. Yang

Terahertz Chemical Imaging of a Multicomponent Tablet in Pharmaceutical 109
Applications
K. Ajito, Y. Ueno, H. Song, J. Kim, E. Tamechika, N. Kukutsu, W. Limwikrant, K. Yamamoto, and K. Moribe

Chapter 2
Gas and Liquid Phase Chemical Sensors

(Sensor Division Outstanding Achievement Award Presentation) Ceramic Gas Sensors to 119
Oxide Nanostructures: Opportunities and Challenges
S. A. Akbar

Selectivity Enhancement of YSZ-based VOC Sensor Utilizing SnO_2/NiO-SE Via the 129
Application of a Physical Gas-Diffusion Barrier
T. Sato, M. Breedon, and N. Miura

Rapid and Simple Immunoassay Based on Negative Dielectrophoresis with Three- 139
Dimensional Interdigitated Array Electrodes
T. Yasukawa, H. Shiku, T. Matsue, and F. Mizutani

Bismuth-Film Electrodes for Sn^{2+} Sensing: The Roles of Grain Size, Preferred 147
Orientation Ratio, and Surface Roughness
C. H. Lien, C. Hu, Y. Tsai, and D. Wang

vi

High-Throughput Separation Assay for NO Metabolites in Blood Using Microfluidic Electrophoresis   165
  S. Wakida, T. Miyado, K. Shimazu, Y. Shibutani, T. Mizukami, K. Nose,
  and A. Shimouchi

Effects of Surface Modification of Noble-Metal Electrodes with Au on the H_2-Sensing Properties of Diode-Type Gas Sensors   171
  T. Hyodo, T. Yamashita, and Y. Shimizu

Potentiometric YSZ-Based Sensors Using Zn-Ta-O-Based Sensing Electrode for Selective H_2 Detection   179
  S. A. Anggraini, M. Breedon, and N. Miura

Solid Electrolyte Type Ammonia Gas Sensor with High Water Durability   189
  S. Tamura, T. Nagai, and N. Imanaka

Direct Comparison of Anti-Interference Property for Bimetallic PtAu, PtIr, and PtRu Nanoparticle Catalysts in Amperometric Detection for H_2O_2 Based Biosensors   195
  M. Janyasupab, Y. Zhang, C. Liu, and C. Liu

A High-Throughput Assay for Evaluation of Embryoid Bodies Using Local Redox Cycling-Based Electrochemical Chip Device   205
  K. Ino, T. Nishijo, Y. Kanno, H. Shiku, and T. Matsue

Printed Amperometric Gas Sensors   211
  M. T. Carter, J. R. Stetter, M. W. Findlay, and V. Patel

Interaction of Water Vapor with SnO_2 Sensor Materials: A Comparison of DRIFTS and Resistance Measurements   221
  R. G. Pavelko, K. Grossmann, N. Barsan, and K. Shimanoe

Development of Micro Hydrogen Gas Sensor Utilizing Polymerized Gel with Ionic Liquid as a Solvent   231
  T. Yamauchi, T. Matsui, T. Nishiyama, K. Tsunashima, N. Tsubokawa,
  and S. Harada

The Influences of Quenching Times on the Property of Thermal Oxidized Iridium pH Sensor   237
  F. Huang, Y. Jin, L. Wen, D. Mu, and M. Cui

Surface-Enhanced Raman Scattering on Ordered Metal Nanodot Array Obtained Using Anodic Porous Alumina   249
  T. Kondo, K. Nishio, and H. Masuda

vii

Enzyme-Encapsulated Quantum Dot Hydrogels in the Development of Biosensors: 255
A Multifunctional Platform for Both Bio-Catalysis and Fluorescent Probing
J. Yuan, N. Gaponik, and A. Eychmüller

Adaptive Chemical Sampling Device Inspired by Crayfish 259
R. Takemura, K. Takahashi, T. Makishita, and H. Ishida

CO Sensing Properties of Electrochemical Gas Sensors Using an Anion-Conducting 267
Polymer as an Electrolyte
T. Goto, T. Hyodo, K. Kaneyasu, H. Yanagi, and Y. Shimizu

NO_2 Sensing Properties of Porous In_2O_3-Based Powders Prepared by Utilizing 273
Ultrasonic-Spray Pyrolysis Employing PMMA Microsphere Templates: Effects of the
Size of the PMMA Microspheres on Their Gas-Sensing Properties
E. Fujii, T. Hyodo, K. Matsuo, and Y. Shimizu

Redox-Active Alkali Insertion Materials as Inner Contact Layer in All-Solid-State 279
Ion-Selective Electrodes
S. Komaba, C. Suzuki, N. Yabuuchi, T. Akatsuka, S. Kanazawa, and T. Hasegawa

Sensing Characteristics of a Fiber Bragg Grating Hydrogen Gas Sensor Using Sol-Gel 289
Derived Pt/WO_3 Film
S. Okazaki, Y. Maru, and T. Mizutani

Zirconia-Based Electrochemical Oxygen Sensor for Accurately Determining Water Vapor 295
Concentration
R. E. Soltis

Fabrication of Surface Enhanced Raman Scattering (SERS)-Active Substrates by Using 301
Dip-Pen Nanolithography
F. P. Lee, K. Chao, K. H. Hsieh, and K. Ou

Application of Commercial Manufacturing Methods to Mixed-Potential NO_x Sensors 307
C. R. Kreller, P. K. Sekhar, W. Li, P. Palanisamy, E. L. Brosha, R. Mukundan,
and F. H. Garzón

Research on Filter Materials for LP Gas Sensors 315
M. Sai, K. Shinnishi, K. Kaneyasu, T. Suzuki, and M. Takeuchi

Chapter 3
Chemical Sensors Poster Session

Characterization and Electrochemical Response of Sonogel Carbon Electrode Modified with Nanostructured TiO_2 and ZrO_2 Film to Detect Common Neurotransmitters 321
S. K. Lunsford, M. K. Hughes, and P. K. Nguyen

Functionalization of Pyrolyzed Carbon Structures for Bio-Nanoelectronics Platforms 325
M. Hirabayashi, B. Mehta, A. Khosla, and S. Kassegne

Self-Assembled Monolayers of Oligonucleotides as Receptor Layers for Mercury Ion Sensor 333
L. Górski, R. Ziółkowski, and E. Malinowska

Development of Highly-Sensitive Electrochemical Measurement System on Dry Chemistry Using Ionic Liquid 339
S. Arimoto, M. Takahashi, A. Kamei, and T. Yoshioka

Layer-by-Layer Catalytic Interface for Electrochemical Detection of Multiple Substrates Featuring Bio-Functionalized Carbon Nanotubes 345
J. S. Kirsch, X. Yang, and A. L. Simonian

Chalcogenide Glass Chemical Sensor for Cadmium Detection in Industrial Environment 357
M. Milochova, M. Kassem, and E. Bychkov

Electrochemical Pump Consisting of Cu^{2+}-Poly(Acrylic Acid) Gel 363
K. Takada, N. Yamamura, A. Hayashi, T. Yasui, and A. Yuchi

Superoxide Anion Radical Sensor Using GC Electrode Modified with Heparin/PEDOT and Polymerized Iron Porphyrin 369
R. Matsuoka, T. Kondo, and M. Yuasa

High Sensitive Amperometric Detection of Glucose Using Conductive DLC Electrode in Higher Potential Region 377
K. Honda, A. Nakahara, H. Naragino, and K. Yoshinaga

New Application of Produced Pigment from Bacteria to Detect Ammonia in Combination with Flow Injection for Ammonia Analysis 385
Y. Iida and I. Satoh

Electrochemical Immunosensor for Diagnostic of Parasitical Human Diseases 393
C. A. Erdmann, J. Inaba, A. G. Viana, C. A. Pessoa, K. Wohnrath, and J. R. Garcia

Preparation of Fine Implantable Needle Type Biosensors for Blood Vessel Glucose   401
Monitoring
   K. Edagawa and M. Yasuzawa

Chapter 4
Micro/Nanofabrication

Micro-Systems and Nanotechnologies in ELISA and Droplet Generation Applications   409
   C. Yeh and Y. Lin

Wafer Scale Processing of Plasmonic Nanoslit Arrays in 200mm CMOS Fab Environment   413
   K. Malachowski, R. Verbeeck, T. Dupont, C. Chen, Y. Li, S. Musa, T. Stakenborg, D.
   Sabuncuoglu Tezcan, and P. Van Dorpe

Tunable Young's Modulus in Carbon MEMS Using Graphene-Based Stiffeners   423
   C. M. Washburn, T. Lambert, J. Blecke, D. Davis, P. S. Finnegan, B. G. Hance,
   D. R. Wheeler, T. E. Beechem, T. M. Alam, M. T. Brumbach, and J. M. Strong

Residue-Free Dry Etching of Polymer Sacrificial Layer for Microelectromechanical-System   435
Device Fabrication
   K. Takagahara, K. Ono, K. Kuwabara, T. Sakata, H. Ishii, Y. Sato, and Y. Jin

Hydrodynamic Cell Enrichment in Double Spiral Microfluidic Channels   441
   J. Sun, M. Li, C. Liu, G. Hu, and X. Jiang

Chapter 5
Cantilevers and Microdevices

Nano/Micro Patterned Phononic Crystals   449
   B. Kim, J. Nguyen, C. Reinke, M. Ziaei-Moayyed, I. El-Kady, D. Goettler, M. Su,
   Z. C. Leseman, and R. H. Olsson III

Photothermal Cantilever Deflection Spectroscopy   459
   S. Kim, D. Lee, R. Thundat, M. Bagheri, S. Jeon, and T. Thundat

Development of Insulated Conductive AFM Probes for Experiments in Electrochemical   465
Environment
   Y. Wu, T. Akiyama, P. D. van der Wal, S. Gautsch, and N. F. de Rooij

Characterization of Piezoresistive Microcantilever Sensors with Metal Organic   469
Frameworks for the Detection of Volatile Organic Compounds
   I. Ellern, A. Venkatasubramanian, J. Lee, P. J. Hesketh, V. Stavila,
   M. D. Allendorf, and A. L. Robinson

x

Manipulation of Micro Condensed Matter by Direct Peeling Method by Using Atomic   477
Force Microscope Tip
  A. Kawai

A MEMS-Based Platform for Multi-Physics Characterization of Ultra-Thin Freestanding   487
Films
  M. Haque, S. Kumar, and M. Alam

Effects of Adsorbate Surface Diffusion in Focused Electron-Beam-Induced-Deposition   495
  A. Szkudlarek, M. Gabureac, and I. Utke

Electroplating of Micropatterned Nickel Phase Gratings for X-ray Phase Contrast   499
Tomography
  M. Amberger, K. Bade, J. Meiser, D. Kunka, and J. Mohr

Impact of Donor Dopant on Acceptor Solubility in TlBr   507
  S. R. Bishop and H. L. Tuller

Microfabricated Systems to Measure Marine Variables   513
  S. Aravamudhan and S. Bhansali

Chapter 6
MEMS/NEMS Poster Session

Micropatternable, Electrically Conducting Polyaniline Photoresist Blends for MEMS   525
Applications
  C. V. Patel, A. Khosla, and S. Kassegne

Author Index   537

.

Facts about ECS

The Electrochemical Society (ECS) is an international, nonprofit, scientific, educational organization founded for the advancement of the theory and practice of electrochemistry, electronics, and allied subjects. The Society was founded in Philadelphia in 1902 and incorporated in 1930. There are currently over 7,000 scientists and engineers from more than 70 countries who hold individual membership; the Society is also supported by more than 100 corporations through Corporate Memberships.

The technical activities of the Society are carried on by Divisions. Sections of the Society have been organized in a number of cities and regions. Major international meetings of the Society are held in the spring and fall of each year. At these meetings, the Divisions and Groups hold general sessions and sponsor symposia on specialized subjects.

The Society has an active publication program that includes the following:

Journal of The Electrochemical Society — (JES) is the leader in the field of electrochemical science and technology. This peer-reviewed journal publishes an average of 550 pages of 85 articles each month. Articles are published online as soon as possible after undergoing the peer-review process. The online version is considered the final version and is fully citable with articles assigned specific page numbers within specific issues. The date of online publication is the official publication date of record.

Journal of Solid State Science and Technology — (JSS) is one of the newest peer-reviewed journals from ECS launched in 2012. JSS covers fundamental and applied areas of solid state science and technology including experimental and theoretical aspects of the chemistry and physics of materials and devices. Articles are published online as soon as possible after undergoing the peer-review process. The online version is considered the final version and is fully citable with articles assigned specific page numbers within specific issues. The date of online publication is the official publication date of record.

Electrochemistry Letters — (EEL) is one of the newest journals from ECS launched in 2012. It is dedicated to the rapid dissemination of peer-reviewed and concise research reports in fundamental and applied areas of electrochemical science and technology. Articles are published online as soon as possible after undergoing the peer-review process. The online version is considered the final version and is fully citable with articles assigned specific page numbers within specific issues. The date of online publication is the official publication date of record.

Solid State Letters — *(SSL)* is one of the newest journals from ECS launched in 2012. It is dedicated to the rapid dissemination of peer-reviewed and concise research reports in fundamental and applied areas of solid state science and technology. Articles are published online as soon as possible after undergoing the peer-review process. The online version is considered the final version and is fully citable with articles assigned specific page numbers within specific issues. The date of online publication is the official publication date of record.

Electrochemical and Solid-State Letters — (ESL) was the first rapid-publication electronic journal dedicated to covering the leading edge of research and development in the field of solid-state and electrochemical science and technology. ESL was a joint publication of ECS and IEEE Electron Devices Society. Volume 1 began July 1998 and contained six issues, thereafter new volumes began with the January issue and contained 12 issues. The final issue of ESL was Volume 16, Number 6, 2012. Preserved as an archive, ESL has since been replaced by SSL and EEL.

Interface— *Interface* is an authoritative yet accessible publication for those in the field of solid-state and electrochemical science and technology. Published quarterly, this four-color magazine contains technical articles about the latest developments in the field, and presents news and information about and for members of ECS.

ECS Meeting Abstracts— *ECS Meeting Abstracts* contain extended abstracts of the technical papers presented at the ECS biannual meetings and ECS-sponsored meetings. This publication offers a first look into the current research in the field. ECS Meeting Abstracts are freely available to all visitors to the ECS Digital Library.

ECS Transactions— (ECST) is the online database containing full-text content of proceedings from ECS meetings and ECS-sponsored meetings. ECST is a high-quality venue for authors and an excellent resource for researchers. The papers appearing in ECST are reviewed to ensure that submissions meet generally-accepted scientific standards. Each meeting is represented by a volume and each symposium by an issue.

Monograph Volumes — The Society sponsors the publication of hardbound monograph volumes, which provide authoritative accounts of specific topics in electrochemistry, solid-state science, and related disciplines.

For more information on these and other Society activities, visit the ECS website:

www.electrochem.org

CHAPTER 1

NANO/BIO SENSORS

2

Phytosensors and Phytoactuators

A. G. Volkov[a], M. I. Volkova[a], and V. S. Markin[b]

[a] Department of Chemistry, Oakwood University, Huntsville, AL 35896, USA
[b] Department of Neurology, University of Texas, Southwestern Medical Center at Dallas, TX 75390-8813, USA

Plants continuously sense a wide variety of perturbations and produce various responses known as tropisms in plants. It is essential for all plants to have survival sensory mechanisms and actuators responsible for a specific plant response process. Plants are ideal adaptive structures with smart sensing capabilities based on different types of tropisms, such as chemiotropism, geotropism, heliotropism, hydrotropism, magnetotropism, phototropism, thermotropism, electrotropism, thigmotropism, and host tropism. Plants can sense mechanical, electrical and electromagnetic stimuli, gravity, temperature, direction of light, insect attack, chemicals and pollutants, pathogens, water balance, etc. Here we show how plants sense different environmental stresses and stimuli and how phytoactuators response to them. Plants generate various types of intracellular and intercellular electrical signals in response to these environmental changes. This field has both theoretical and practical significance because these phytosensors and phytoactuators employ new principles of stimuli reception and signal transduction and play a very important role in the life of plants.

Introduction

Plants are ideal adaptive structures with smart sensing capabilities based on different types of tropisms, such as chemiotropism, geotropism, heliotropism, hydrotropism, magnetotropism, phototropism, thermotropism, electrotropism, thigmotropism, and host tropism. A phytoactuator is a part of a plant responsible for moving or controlling a specific plant response process. It is operated by a source of electrochemical energy or hydraulic pressure and converts that energy into motion. A phytosensor is defined as a device that can detect, record, and transmit information related to a physiological change/process in a plant. It can also use plant tissue to monitor the presence of various chemicals in a substance. In most successful phytosensors, the principle behind the determination of a chemical or biological molecule is the specific interaction of such an analyte molecule with the plant tissue present in the phytosensor probe device. A variety of plant tissues has been incorporated into various electrochemical transducers to detect and quantify a range of biologically important analytes including drugs, hormones, toxicants, neurotransmitters and amino acids. Plant tissues have received considerable attentions in recent years as alternative biocatalyst for replacing isolated enzymes to construct phytosensors due to their high activity, stability, and low cost. However, they often encounter problems of long response time, low sensitivity, and complex sensor assembly. Based on fast bioelectrochemical signaling events in plants, here we discuss the evidence supporting the foundation for utilizing the entire green plant as a fast phytosensor for monitoring the environmental perturbations in close vicinity of a living plant.

Nerve cells in animals and phloem cells in plants share one fundamental property: they possess excitable membranes through which electrical excitations can propagate in the form of action potentials. Plants generate bioelectrochemical signals that resemble nerve impulses, which are present in plants at all evolutionary levels. Prior to the morphological differentiation of nervous tissues, the inducement of nonexcitability after excitation and the summation of subthreshold irritations were developed in the vegetative and animal kingdoms in protoplasmatic structures (1).

The cells, tissues, and organs of plants transmit electrochemical impulses over short and long distances. It is conceivable that action potentials are the mediators for intercellular and intracellular communication in response to environmental irritants. Action potential is a momentary change in electrical potential on the surface of a cell that takes place when it is stimulated, especially by the transmission of an impulse.

Initially, plants respond to irritants at the site of stimulation; however, excitation waves can be conducted along the membranes throughout the entire plant. Bioelectrical impulses travel from the root to the stem and vice versa. Chemical treatment, intensity of the irritation, mechanical wounding, previous excitations, temperature, and other irritants influence the speed of propagation.

The conduction of bioelectrochemical excitation is a rapid method of long distance signal transmission between plant tissues and organs (2). Plants quickly respond to changes in luminous intensity, osmotic pressure, temperature, cutting, mechanical stimulation, water availability, wounding, and chemical compounds such as herbicides, plant growth stimulants, salts, and water. Once initiated, electrical impulses can propagate to adjacent excitable cells. The change in transmembrane potential creates a wave of depolarization or action potential, which affects the adjoining resting membrane.

The phloem is a sophisticated tissue in the vascular system of plants. Representing a continuum of plasma membranes, the phloem is a potential pathway for transmission of electrical signals. It consists of two types of conducting cells: the characteristic sieve-tube elements, and the companion cells. Sieve-tube elements are elongated cells that have end walls perforated by numerous minute pores through which dissolved materials can pass. Sieve-tube elements are connected in a vertical series known as sieve tubes. The companion cells have nuclei and they are adjacent to the sieve-tube elements. It is hypothesized that they control the process of conduction in the sieve tubes. Thus, when the phloem is stimulated at any point, the action potential can propagate over the entire cell membrane and along the phloem with constant voltage.

Electrical potentials have been measured at the tissue and whole plant level by using the experimental set-up shown in Figure 1. Measurements were taken inside a Faraday cage mounted on a vibration-stabilized table. Nonpolarizable reversible Ag/AgCl electrodes were used to measure the electrical signals. The temperature was held constant since these electrodes are sensitive to the temperature. When electrochemical signals are measured, it is extremely important to take into consideration the *sampling rate* which determines how often the measurement device samples an incoming analog signal (3). According to the sampling theorem, the original analog signal must be adequately sampled in order to be properly represented by the sampled signal. If the sampling rate is too slow, the rapid changes in the original signal between any two consecutive samples cannot be accurately recorded. As a result, higher frequency components of the original signal will be misrepresented as lower frequencies. In signal processing, this problem is known as *aliasing*. According to the Nyquist Criterion, the sampling frequency must be at least twice the bandwidth of the signal to avoid aliasing. Undersampling may result in the mispresentation of the measured signal.

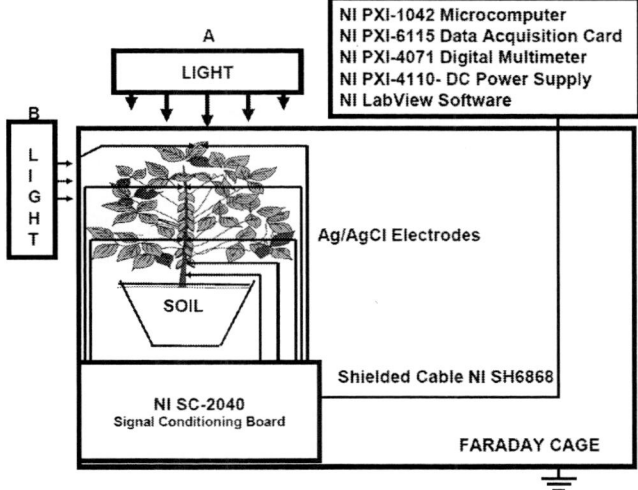

Figure 1. Experimental set-up for measuring electrical signals in green plants.

Plants constantly communicate with the external world in order to maintain homeostasis. Internal biological processes and their concomitant responses to the environment are closely associated with the phenomenon of excitability in plant cells. The extreme sensitivity of the protoplasm to chemical effects is the foundation for excitation. The excitable cells, tissues and organs alter their internal condition and external reactions under the influence of environmental factors, referred to as irritants; this excitability can be monitored. Plants generate different types of extracellular electrical responses in connection to environmental stress. Recent findings have indicated that plants may use a common defense system to respond to various abiotic and biotic stresses, such as heat, cold, drought, flooding, osmotic shock, wounding, high light intensity, UV-radiation, ozone, and pathogens.

Host Tropism: Insect-induced Electrochemical Signals in Plants

Understanding plant-insect interactions is important for ecology and for the development of novel crop protection strategies. Volkov and Haack (4) were the first to create the unique opportunity to investigate the role of electrical signals induced by insects in long-distance communication in plants. Action and resting potentials were measured in potato plants (*Solanum tuberosum* L.) in the presence of leaf-feeding larvae of the Colorado potato beetle (*Leptinotarsa decemlineata* (Say); Coleoptera: Chrysomelidae). When the larvae were allowed to consume upper leaves of the potato plants and action potentials with amplitudes of 40±10 mV were recorded every 2 ±0.5 hours during a two-day test period. The resting potential decreased from 30 mV to a steady state level of 0 ±5 mV. The action potential propagates from plant leaves with Colorado potato beetles down the stem, and to the potato tuber. The speed of propagation of the action potential does not depend on the location of a measuring electrode in the stem of the plant or tuber, or the distance between the measuring and reference electrodes. Therefore, plants can be used as phytosensors for insect attacks.

Phototropism and Heliotropism: Molecular Recognition of the Direction of Light

Plants gain energy from two sources: quantum and thermophysical processes. Photosynthesis is an example of a quantum process, whereas transpiration is a thermophysical one. Plants have evolved sophisticated systems to sense the environment.

Plants can be used as photosensors for the direction of light. Light is important for plant development by influencing nearly all aspects of the life cycle from germination to flowering. Plants perceive light ranging from ultraviolet to far-red light by specific photoreceptors. Natural radiation simultaneously activates more than one photoreceptor in higher plants. These receptors initiate distinct signaling pathways leading to wavelength-specific light responses. Nastic motion causes plants to angle their stems so that their leaves face light sources. The three classes of plant photoreceptors that have been identified at the molecular level are phototropins, cryptochromes, and phytochromes.

Phototropin is a blue light (360-500 nm) flavoprotein photoreceptor responsible for phototropism and chloroplast orientation. The phototropins, such as phot1 and phot2, are a family of flavoproteins that function as the primary photoreceptors in plant phototropism and in intracellular chloroplast movements. Phot1 contains two 12 kD flavin mononucleotide binding domains. LOV1 (light, oxygen, and voltage) and LOV2 are found within its N-terminal region and a C-terminal serine/threonine protein kinase domain. The protein conformation changes in light-activated phototropin. Phot1 and phot2 bind FMN and undergo light-dependent autophosphorylation. Phot2 is localized in the plasma membrane.

Cryptochromes (cry1 and cry2) are flavoproteins in the family of photoreceptors responsible for photomorphogenesis. They perceive (UV-A) light as well as blue light (360 - 500 nm). Although cryptochromes and phototropin share many similarities, they have different transduction pathways. Cry1 plays a significant role in the synthesis of anthocyanin and in the entrainment of circadian rhythms. Cry2 plays a part in the photoperiodic flowering and cotyledon expansion. Cryptochromes were predominantly found in the nucleus. Stomatal opening is also stimulated by blue light and UV irradiation.

Phytochrome (phy) is a protein photoreceptor that regulates many aspects of plant development. Plant phytochromes are also light-modulated protein kinases that process dual ATP-dependent autophosphorylation and protein phosphotransferase activities.

Phototropism is one of the best-known plant tropic responses. A positive phototropic response is characterized by a bending or turning toward the source of light. When plants bend or turn away from the source of light, the phototropic response is considered negative. A phototropic response is a sequence of the four following processes: reception of the directional light signal, signal transduction, transformation of the signal to a physiological response, and the production of directional growth response.

The soybean plant was irradiated inside the Faraday cage in the direction A (Figure 1) with white light for two days with a 12:12 hr light:dark photoperiod prior to the conduction of experiments. Action potentials are not generated when the lights are turned off and on. Changing the direction of irradiation from direction A to direction B generates action potentials in soybean approximately after 1-2 min. The generation of action potentials depend on the wavelength of irradiation light. Irradiation at wavelengths 400-500 nm induces fast action potentials in soybean; conversely, the irradiation of soybean in the direction B at wavelengths between 500 and 630 nm fails to generate action potentials. Irradiation between 500 nm and 700 nm does not induce phototropism (5,6). Irradiation of soybean by blue light induces positive phototropism.

Some voltage-gated ion channels work as plasma membrane nanopotentiostats. A blocker of K^+ ion channels, such as tetraethylammonium chloride (TEACl), stops the propagation of action potentials in soybean induced by blue light and inhibits phototropism (5,6). Voltage-gated ionic channels control the plasma membrane potential and the movement of ions across membranes; thereby, regulating various biological functions.

Green plants sense many parameters in order to adapt to the environment. Figure 2 illustrates the mechanism of bio-signaling in green plants.

Thigmotropism: Mechanosensation in Plants

The concept of mechanosensitive ion channels was primarily developed based on studies of specialized mechanosensory neurons. Due to the applied mechanical force on the cell membrane, these mechanically-gated channels, capable of converting mechanical stress into electrical or biochemical signals, are activated. As a result, they act as molecular transducers and play a vital role in regulating various physiological processes responsible for growth and development in all forms of life, as well as monitoring the surrounding environmental challenges for survival; for example, turgor control in plants. Although the current understanding of the structure and function of mechanosensitive ion channels found in living organisms is limited, significant progress has recently been made in the area of evolutionary origins of mechanosensitive ion channels.

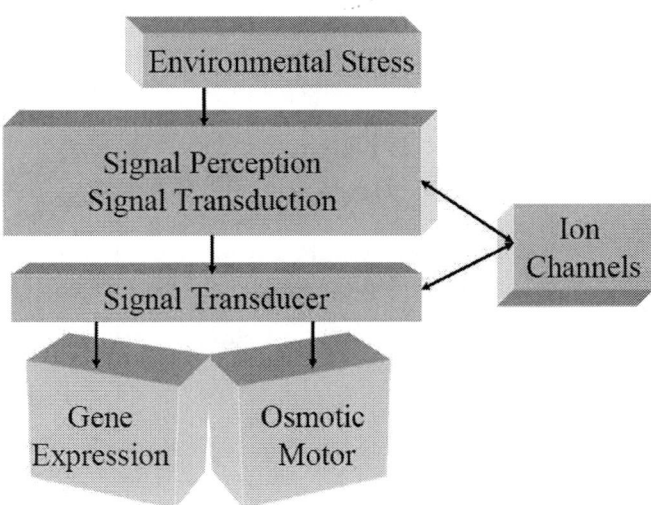

Figure 2. Mechanism of biosignaling in green plants.

The patch-clamp technique has provided the tool for identification of two basic types of mechanosensitive ion channels found in living cells: stretch-activated and stretch-inactivated ion channels. Mechanosensitive ion channels play a vital role in the physiological function of the plant by controlling the ion transport across the plasma membrane, and hence influencing volume and turgor regulation of guard cells.

Plant response to mechanical stimulation has long been known. Perhaps all plants can react in response to the mechanical stimuli, only certain plants with rapid and highly noticeable touch-stimulus response have received much attention; for example, the trap closure of the Venus flytrap. Mechanosensation is considered to have evolved as one of the oldest sensory mechanisms in living organisms.

The Venus flytrap is a marvel of plant electrical, mechanical, and biochemical engineering. The rapid closure of the Venus flytrap upper leaf in about 0.3 s is one of the fastest movements in the plant kingdom. When a prey touches the trigger hairs, these mechanosensors trigger a receptor potential, which generate an electrical action potential. Two stimuli generate two action potentials, which close the trap at room temperature in a fraction of a second. Propagation of action potentials and the trap closing can be blocked by uncouplers, inhibitors of voltage gated channels, and aquaporins (7). We found that the electrical stimulus between a midrib and a lobe can close the Venus flytrap without mechanical stimulation (8-13). It was also shown that the Venus flytrap has a short term electrical memory; using the charge stimulating method we demonstrated that Venus flytrap can accumulate small subthreshold charges, and when the threshold value is reached, the trap closes (12).

In contrast to chemical signals such as hormones, electrical signals are able to transmit information rapidly over long distances. Biologically closed electrical circuits performing these functions operate over large distances in biological tissues. The activation of such circuit can lead to various physiological and biophysical responses. It is often convenient to represent the real electrical and electrochemical properties of biointerfaces with idealized equivalent electrical circuit models consisting of discrete electrical components. We investigated the biologically closed electrical circuits in *Mimosa pudica* L. and in the upper leaf of the Venus flytrap and proposed the equivalent electrical circuits (Figure 3). Stimulus perception (Box 1) can generate electrical signals such as action potentials. Box 2 shows the equivalent electrical circuit for a signal transduction. Electrical signals induce osmotic flow (Box 3) and starts the osmotic motor *M*. Box 4 shows the equivalent electrical circuit for photosynthesis in the *Mimosa pudica*. Signal transduction in *Mimosa pudica* Linn. has been attracting the attention of researchers since the XVI century. It is a sensitive plant in which the leaves and the petiole move in response to intensity of light, mechanical or electrical stimuli, drought, and hot or cold stimuli. It was found by Haberlandt that long distance signal transduction in *Mimosa pudica* occurs through the phloem.

Mechanical movement of *Mimosa pudica* after electrical signal transduction is a defense mechanism and can be the simplest criterion of plant behavior and intelligence. Hooke (14) stated long ago, "that this may be so, it seems with great probability to be argued from the strange phenomena of sensitive plants, wherein Nature seems to perform several animal actions with the same schematism or organization that is common to all vegetables, as may appear by some no less instructive then curious observations that were made by divers eminent members of the Royal Society on some of these kind of plants...".

Mimosa pudica L. is a thigmonastic or seismonastic plant in which the leaves close and the petiole hangs down in response to certain stressors. The unique anatomy of the *Mimosa pudica* contributes to the response mechanism of the plant. The plant contains long slender branches called petioles, which can fall due to mechanical, thermal or electrical stimuli. The petioles contain smaller pinnae, arranged on the midrib of the pinna. The pinnules are the smallest leaflets while the entire leaf contains the petioles, pinnae, and pinnules. A pulvinus is a joint-like thickening at the base of a plant leaf or leaflet that facilitates thigmonastic movements. Primary, secondary, and tertiary pulvini are responsible for the movement of the petiole, pinna, and leaflets, respectively.

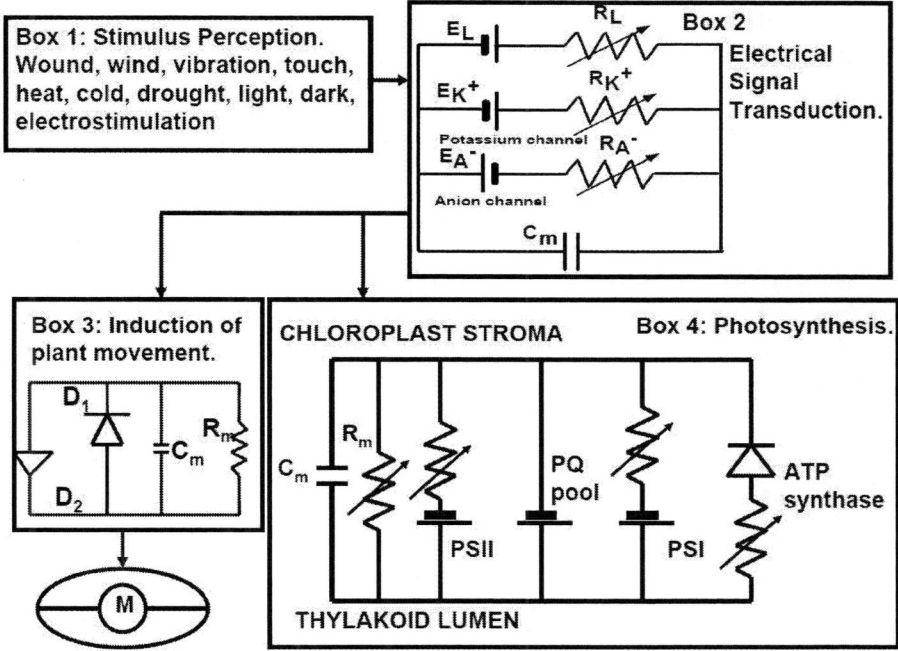

Figure 3. Biologically closed electrical circuits in *Mimosa pudica*. Abbreviations: C – capacitance; D – diode as a model of an ion channel; E - electromotive force; M – osmotic motor; R_m – membrane resistance.

Thigmonastic or seismonastic movements in *Mimosa pudica* appear to be regulated by electrical, hydrodynamical, and chemical signal transduction. The pulvinus of *Mimosa pudica* shows elastic properties, and we found that electrically or mechanically induced movements of the petiole were accompanied by a change of pulvinus shape. As the petiole falls, the volume of the lower part of the pulvinus decreases and the volume of the upper part increases due to the redistribution of water between the upper and lower parts of the pulvinus. This hydroelastic process is reversible. During the relaxation of the petiole, the volume of the lower part of the pulvinus increases and the volume of the upper part decreases. Redistribution of ions between the upper and lower parts of a pulvinus causes fast transport of water through aquaporins and causes a fast change in the volume of the motor cells. The biologically closed electrochemical circuits in electrically and mechanically anisotropic pulvini of *Mimosa pudica* were analyzed using the charge stimulating method for electrostimulation at different voltages. Changing the polarity of electrodes leads to a strong rectification effect in a pulvinus and to different kinetics of a capacitor discharge if the applied initial voltage is 0.5 V or higher.

The pulvinus is comprised of three main parts: a central vascular core and two layers of flexor and extensor cells. The action potential activates voltage gated channels and induces the redistribution of K^+, Cl^-, H^+ and Ca^{2+} ions between extensor and flexor layers which in turn leads to the osmotic movement of water, causing the bending of the pulvinus and movement of the petiole in *Mimosa pudica*.. The potassium concentration in the apoplast of extensor cells of the *Mimosa pudica's* pulvinus increases from 30-70 mM to 100 mM (15). Thus the motor cells shrink and resultantly, the cells from the flexor site also take up K^+ ions from the apoplast and begin to swell. The differential volume changes of flexor and extensor sites result from the transport of ions accompanied by osmotic transport of water. There is a high gradient of osmotic pressure between extensor and flexor cells of about 1 MPa in *Samanea* pulvini (15).

The identification and characterization of bioelectrochemical mechanisms for electrical signal transduction in plants would mark a significant step forward in understanding this under- explored area of plant physiology. Although plant mechanical and chemical sensing and corresponding responses are well known, membrane electrical potential changes in plant cells and the possible involvement of electrophysiology in transduction mediation of these sense-response patterns represents a new dimension of plant tissue and whole organism integrative communication. A short time ago we discovered the bioelectrochemical mechanisms of electrical signal transduction in biologically closed electrical circuits in the pinnae and petioles of *Mimosa pudica* and the plant's responses to electrostimulation (16-18). The studies of the mechanisms of concerted movements in plants from electrical signal transduction to cascades of cellular events will have a potentially broad impact on both fundamental sciences and engineering.

Mechanical movements in *Mimosa pudica* can be induced by electrostimulation if very high applied voltages 200-400 volts were briefly applied between the soil and the primary pulvinus to measure the contractile characteristics of a petiole (19). Jonas (20) used a 0.5 µF capacitor charged by 50, 100 and 150 volts for electrostimulation and found oscillations of leaves and fast petiole movement in *Mimosa pudica* after the application of an electrical shock. The petioles bend downward and the pinnae close after the application of 9 volts to *Mimosa pudica* (21). These high voltages are non-physiological and have a side effect – plant electrolysis. We analyzed, both experimentally and theoretically, the mechanism of mechanical movements in *Mimosa pudica* induced by low voltage electrostimulation of the petiole and pinna (16-18).

Some plants move their leaves upon sudden shaking or touch as seismonastic and thigmonastic movements use osmotic motors, powered by H^+-ATPases. The osmotic motor often resides in specialized leaf organs, midribs or pulvini, at the base of the leaves and leaflets. The osmotic motor transfers water through water channels or aquaporins.

In terms of electrophysiology, these responses in *Mimosa pudica* can be considered in three stages: (1) stimulus perception, (2) signal transmission and (3) induction of response. Action potentials involve effluxes of K^+ and Cl^- and a temporary change of turgor, produced by osmotic motors. Like the action potential, a critical threshold depolarization triggers Ca^{2+} influx, opening of Ca^{2+}-sensitive Cl^- channels and K^+ channels; effluxes last over a short period of time and result in turgor regulation.

Green plants interfaced with a computer through data acquisition systems can be used as fast biosensors for monitoring the environment, detecting effects of pollutants, pesticides, defoliants, predicting and monitoring climate changes, and in agriculture, directing and fast controlling of conditions influencing the harvest. The use of new computerized methods provides opportunities for detection of fast action potentials in green plants in real time. This field has both theoretical and practical significance because discovering these mechanisms would greatly advance our knowledge of natural sensors, principles of their functioning and integration into general system of defense and attack. These systems play a very important role in the life of plants, but their nature is still very poorly understood. Our studies could advance this important scientific field and open new perspectives in technical application of these biological principles.

Acknowledgments

This work was supported by the grant CBET-1064160 from the National Science Foundation.

References

1. A. Goldsworthy, *J. Theor. Biol.*, **103**, 645 (1983).
2. A. G. Volkov, *J. Electroanal. Chem.*, **483**, 150 (2000).
3. E. Jovanov and A. G. Volkov in *Plant Electrophysiology- Methods and Cell Electrophysiology*, A. G. Volkov, Editor, p. 45, Springer, Berlin, (2012).
4. A. G. Volkov and R. A. Haack, *Bioelectrochem. Bioenerg.*, **35**, 55 (1995).
5. A. G. Volkov, T. Dunkley, A. Labady, and C. Brown, *Electrochim. Acta*, **50**, 4241 (2005).
6. A. G. Volkov, T. C.Dunkley, S. A. Morgan, D. Ruff, Y. Boyce, and A. J. Labady, *Bioelectrochem.*, **63**, 91 (2004).
7. A. G. Volkov, K. J. Coopwood, and V. S. Markin, *Plant Sci.*, **175**, 642 (2008).
8. A. G. Volkov, T. Adesina, V. S. Markin, and E. Jovanov, *Plant Signal. Behav.*, **2**, 139 (2007).
9. A. G. Volkov, M. R. Pinnock, D. C. Lowe, M. S. Gay, and V. S. Markin, *J. Plant Physiol.*, **168**, 109 (2011).
10. A. G. Volkov, T. Adesina, V. S. Markin, and E. Jovanov, *Plant Physiol.*, **146**, 694 (2008)
11. A. G. Volkov, H. Carrell, T. Adesina, V. S. Markin, and E. Jovanov, *Plant Signal. Behav.*, **3**, 490 (2008).
12. A. G. Volkov, H. Carrell, A. Baldwin, and V. S. Markin, *Bioelectrochem.*, **75**, 142 (2009).
13. A. G. Volkov, H. Carrell, and V. S. Markin, *Plant Physiol.*, 149, 1661 (2009).
14. R. Hooke, *Micrographia*, The Royal Society, London, (1667).
15. H. L. Gorton, *Plant Physiol.*, 83, 945 (1987).
16. A. G. Volkov, J. C. Foster, T. A. Ashby, R. K. Walker, J. A. Johnson, and V. S. Markin, *Plant Cell Environ.*, **33**, 163 (2010).
17. V. S. Markin and A. G. Volkov, in *Plant Electrophysiology- Signaling and Responses*, A. G. Volkov, Editor, p. 1, Springer, Berlin, (2012).
18. A. G. Volkov, J. C. Foster, and V. S. Markin, *Plant Cell Environ.*, **33**, 816 (2010).
19. R. T. Balmer and J. G. Franks, *Plant Physiol.*, **56**, 464 (1975).
20. H. Jonas, *J. Interdisciplinary Cycle Research*, **1**, 335 (1970).
21. H. Yao, Q. Xu, and M. Yuan, *Plant Signal. Behav.*, **3**, 954 (2008).

12

ECS Transactions, 50 (12) 13-21 (2012)
©The Electrochemical Society

Miniature Enzymatic Biosensors for Tear Glucose Measurement in Capillary Tubes

Bo Peng[a], Qinyi Yan[a], Anant Balijepalli[b], Bruce Cohan[c], Terry Major[d],
Mark E. Meyerhoff[a*]

[a] Department of Chemistry, University of Michigan
[b] Department of Biomedical Engineering, University of Michigan
[c] EyeLab Group, [d] Department of Surgery, University of Michigan, Ann Arbor, MI 48109

> Miniature enzymatic electrochemical sensors were fabricated and applied to detect tear glucose concentrations in anesthetized rabbits. Three µL of tear fluid was sampled from under the lower eyelid via a glass capillary, and the miniature sensor was then inserted into the solution within the capillary for detection. Boluses of insulin were administered to the rabbits to lower the elevated blood glucose concentrations caused from anesthesia. A significant correlation was found between tear and blood glucose levels, suggesting that tear glucose measurements are a potential, minimally invasive alternative for glucose monitoring.

Introduction

According to current statistics from the World Health Organization (WHO), an estimated 346 million people have diabetes mellitus worldwide. Diabetes has been well recognized as one of the major causes of death and disabilities in developed countries, and the prevalence of death from diabetes is expected to double by 2030 (1). Early diagnosis and tight glycemic management are crucial in helping to prevent and control diabetes and its complications, such as cardiovascular disease, kidney failure, and blindness.

Conventional point-of-care glucose monitoring systems usually involve finger pricks to obtain blood drops to be analyzed with an electrochemical strip-based glucometer. A drop of blood is drawn into a test strip loaded with glucose oxidase or glucose dehydrogenase, which reacts with sample glucose and reports a reading of plasma glucose concentrations in a few seconds. For Type 1 diabetics, it is recommended that glucose be monitored up to eight times a day. These repeated finger pricks may result in patient discomfort and, therefore, less frequent blood glucose checks that can lead to further complications.

Hence, a number of new techniques have been examined that might provide a minimally invasive and/or non-invasive approach to blood glucose monitoring to improve patient compliance and glycemic control. These include near-infrared spectroscopy (2, 3), Raman spectroscopy (4), fluorescence affinity sensors (5), photoacoustic probes (6), and a photonic crystal method (7). Unfortunately, none of these techniques have yet achieved the required analytical results necessary to fully substitute for conventional glucometer devices. Other investigations have suggested testing glucose levels in accessible surrogate body fluids, such as interstitial fluids (8, 9), saliva (10), urine (11, 12), and tears (13-16).

Most notably, a commercial automatic glucose monitor called the GlucoWatch® 2 Biographer (Cygnus Inc., Redwood City, CA) received United States Food and Drug Administration (FDA) approval in 2001. It measured glucose continuously from subcutaneous fluid that was brought to the skin surface by reverse iontophoresis via application of a small electric charge. An AutoSensor composed of a replaceable adhesive polymer pad was placed at the back of the device to collect interstitial fluids that further undergo an enzymatic reaction with glucose oxidase at an electrochemical sensor surface, similar to the traditional glucometer. It automatically measured glucose levels through the skin and reported results on a screen every 10 min. However, clinical researchers reported only a 78.4% correlation coefficient ($p<0.05$) between these automated measured glucose values and a finger-prick glucometer (17). Due to poor accuracy, the device required repeated calibration every 5 measurements. Further, according to FDA, at least half of the users experienced skin irritation or itching from the electric discharge, thus the patients had to rotate the device site so as to avoid repeated skin irritation. Sweating was recognized as another factor that led to additional inaccuracy. Due to the lack of reliability and accuracy, the GlucoWatch® G2 Biographer system is no longer produced, effective July 31, 2007.

The use of urine samples for the estimation of blood glucose concentrations has received considerable attention. This method is non-invasive and particularly useful for people who are unable to use a blood glucometer or that reside in rural areas in developing countries. It usually involves the use of a special reagent strip or dipstick that contains color-sensitive reagents to react with glycosuria (glucose in urine). The color developed on the strip provides quantitative information of detected glycosuria levels by referring to a standard color scale. A more recent study, using gold nanoparticles with immoblized glucose oxidase, can measure urine glucose at a lower threshold of 0.56 mM (10 mg/dL), and a color change from red to blue can be differentiated by the naked eye. However, it requires complicated preparations, including gold nanoparticle fabrication and enzyme immobilization (12). Also, glucose will only show up in the urine once it has reached high levels in the blood, i.e. 10 mM (180 mg/dL), which is already a pathological value. Additionally, urine is a complex sample matrix and the correlation between glycosuria and blood sugar is possible only if renal plasma flow and filtration is normal; in case of reduced plasma flow, the renal glucose threshold will be higher. As a result, a glycosuria test is not an ideal alternate to routine blood glucose monitoring and diabetic treatments.

Tear fluid has unique properties among other accessible body fluids and is known to contain glucose (18). Compared to saliva and urine that have a variable dilution effect, tear fluid is maintained at a minute and stable volume (~4 µL). Tear fluid is produced by the lacrimal gland and other accessory glands with a rate of production of 0.5-2.2 µL/min. A thin film of tear fluid (~8 µm thick) keeps the cornea and conjunctiva continuously moist, without any stimulation (19).

The interest in monitoring glucose levels in tear fluid dates back to 1937 when Michail et al. (20) demonstrated an increase of tear glucose concentration in diabetics. Tear glucose measurements provide the possibility of developing a relatively simple and minimally invasive method of monitoring glucose levels, provided that the tear glucose concentrations can be shown to correlate closely to the blood glucose concentrations in individuals. If a good correlation between the two types of samples can be established over a wide glycemic range, tear glucose monitoring could be an attractive alternate for conventional blood glucose measurements within the normal (4-6 mM) as well as hyperglycemic and hypoglycemic ranges. During the measurement, it is important for tear fluid to be collected using a method that will not impact/perturb the eye or cause any damage to the blood capillaries within the eye. Indeed, it is necessary to prevent any capillary glucose leakage into the tear film that would result in false high glucose levels compared to the actual glucose present in the tear fluid. Furthermore, tear fluid must be collected in a manner that will not increase tear production, which would further dilute the glucose concentrations in such samples.

There has been much research focused on the determination of glucose in tears with different methods. For any technique to be effective, it requires a low detection limit, high selectivity over active interferences such as uric acid and ascorbic acid, as well as the ability to quantitatively measure small sample volumes in a short time period. While most of the studies reported to date have found a positive correlation between tear and blood glucose levels, discrepancies still exist in terms of the actual lacrimal glucose concentrations (7-600 µM) and the degree of correlation between blood and tear levels, mainly due to different tear glucose measurement and sampling methods (21-23) (e.g., filter paper and micro-capillaries).

LeBlanc et al. (22) reported a tear glucose concentration analysis system using high-performance liquid chromatography with a pulse amperometric detector (HPLC-PAD). They concluded from 44 paired samples from 5 patients that poor correlation existed between tear and blood glucose in critically ill patients, and that glucose measurement in tear fluid failed to provide an alternative method to blood glucose monitoring. Later, Lane (19) applied a similar chromatographic (LC-PAD) system in 121 diabetic and non-diabetic patients but failed to detect tear glucose at low blood glucose concentrations. In a recent paper, La Belle and co-workers (24) introduced a disposable microfluidics system that measured glucose to levels as low as 43.4 µM using an integrated electrochemical sensor. The authors proposed this device to be a future prototype for a non-invasive glucose testing, but did not provide real sample data. Meanwhile, Baca et al. (13) employed liquid chromatography coupled with electrospray ionization mass spectrometry (LC-ESI-MS) to detect glucose concentrations in samples of 1 µL of tear fluid. They were able to obtain significant correlations between this method and a contact lens-based sensor device. In fact, attempts to continuously monitor tear glucose concentrations using disposable contact lenses have received much research focus as most diabetic patients have complications that need visual correction as well (16). An enzyme immobilized Pt electrode on the surface of a poly-dimethyl silicone (PDMS) was reported to be able to measure tear glucose levels in a rabbit to be 0.12 mmol/L (25); however, no blood samples were collected to validate any relationship between tear and blood glucose levels. In other work, a group of organic boronic acid molecules were investigated as a reversible glucose receptor when immobilized onto a contact lens. By proper design, these fluorescent boronic acids change color as a function of glucose concentration based on a change in the diffraction properties of the lens material as glucose binds to the anchored boronic acid sites (15). One major challenge in the boronic acids doped contact lens system is the complex receptor design and the need for an irradiation to initiate the fluorescent emission. Despite these and other research efforts, discrepancies exist on whether there is a clinically useful correlation between tear glucose and blood glucose, as well as a stable true concentration of tear glucose in normal and diabetic subjects. Consequently, more research is needed to develop more sensitive detection techniques and less invasive sampling methods.

Herein, we propose relatively simple needle-type enzyme electrodes for the detection of tear glucose over a wide range of concentrations (0-800 µM) within a glass capillary that contains 3 µL of tear fluid. The sensors are applied to detect tear glucose in anesthetized rabbits with insulin administration to achieve a wide dynamic range of blood glucose values (2-20 mM). Using the miniature electrochemical devices, a good correlation between the two types of samples is observed; suggesting that tear glucose measurement may provide an attractive alternative method for glucose monitoring.

Experimental

Tear Glucose Sensor Fabrication and Calibration

The amperometric sensor was fabricated in the same way as previously described (26). The device is based on immobilizing glucose oxidase on a Pt/Ir wire and anodically detecting the liberated hydrogen peroxide from the enzymatic reaction (Figure 1). The active length of the exposed Pt wire is 1 mm with an area of 0.40 mm^2. Inner layers of Nafion® and an electropolymerized film of 1,3-diaminobenzene/resorcinol greatly enhance the selectivity for glucose over known electroactive interferents, including ascorbate, urate and acetaminophen. The amperometric sensor design measures steady-state current that is proportional to the concentration of glucose in the samples. With such a sensor configuration, 3 µL of calibrating solution or tear fluid is collected into the capillary (0.58 mm i.d.), and the sensor is capable of achieving a wide dynamic range (10-800 µM) and a very low detection limit of 0.62±0.03 µM, (S/N=3).

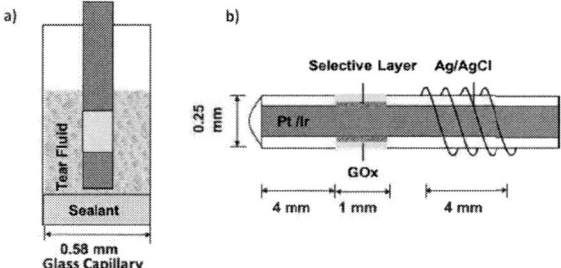

Figure 1. a) Schematic representation of the tear glucose detection in a glass capillary. b) Configuration of the amperometric sensor.

Rabbit Study Protocol

White rabbits (Myrtle's Rabbitry, Thompson's Station, TN) were used in this study to test the correlation between tear and blood glucose levels. A modified anesthesia protocol was applied based on previously reported studies (26) with the additional use of insulin. Human insulin (Sigma, St. Louis, MO) was prepared by dissolving 14.5 mg or 1.45 mg chemical into 10 mL 0.001 N HCl, then further diluted to 100 mL with saline, to make a 4 U/mL or 0.4 U/mL solution. One bolus of insulin (0.4 U/mL in saline, 0.4 mL) was given 1 h after anesthesia commenced, and additional boluses were employed as needed in order to reduce blood glucose levels to less than 4 mM in each rabbit. Every 30 min, 0.5 mL blood was drawn from the rabbit's artery and the glucose level was measured by a Radiometer whole blood analyzer (Radiometer America Inc., Westlake, OH). At the same time, 3 µL of rabbit tear fluid was collected in the capillary and the current from the glucose sensor placed into the tear fluid was recorded 5 minutes after introducing the sensor using a Biostat potentiostat (ESA Biosciences Inc., Chelmsford, MA). The sensor was re-calibrated in capillary tubes with 200 µM glucose at the end of the 7 h experiment, and this calibration point was used to calculate the rabbit tear glucose levels. Statistical data analysis was carried out to examine the correlation between the blood and tear glucose values within a given animal.

Results and Discussion

Tear and Blood Glucose Correlation at High Values

Our group has reported previously on a tear glucose detection system based on a needle-type amperometric enzyme-based electrode coupled with a glass micro-capillary configuration (26). Such miniature sensors are able to be inserted into a micro-capillary that holds microliter volumes of tear samples. In that earlier work, within 5 min, 5 µL of tear fluid was carefully pulled into the capillary tube from the meniscus of the lower eyelids without inflicting any damage to the eye. The results showed a significant correlation between tear and blood glucose levels at the higher end range (7 mM -18 mM) in a rabbit model, with variable correlation coefficients and correlation slopes for each individual animal (see Figure 2 for examples of results from two rabbits). It was concluded from this prior study that with respect to the future clinical application of this technology, after a correlation slope has been established for diabetic patients, the tear glucose levels can be measured by the electrochemical sensor multiple times a day to help monitor blood glucose concentrations without the accumulated pain from blood draws. However, an occasional measurement by a conventional blood glucometer will still be needed to confirm abnormal glycemic levels and determine proper medication or therapy.

It is known that anesthesia will cause an increase of blood glucose concentrations (27), therefore elevating the initial blood glucose levels, that continuously decrease to values still slightly higher than normal (7-8 mM) at the end of the 8 h test period. The use of insulin in the new animal model will enable us to examine whether a similar correlation exists in this animal model at more normal and even hypoglycemic blood glucose levels.

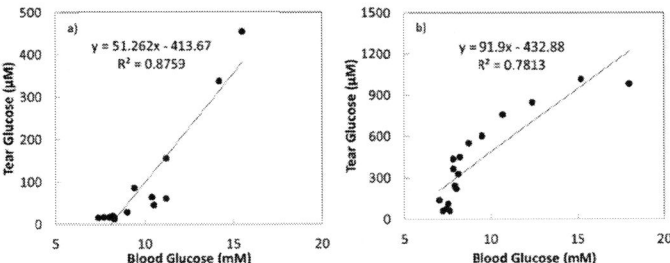

Figure 2. Tear and blood glucose level correlation results using amperometric glucose sensor in capillary configuration from two individual rabbits a) and b), with no addition of insulin to the animals.

Tear and Blood Glucose Correlation at Low Values

With a newly modified animal protocol, insulin was administrated into the rabbits to lower the elevated blood glucose levels 1 h after anesthesia was initiated. Figure 3a shows a pilot experiment result with continuous infusion of insulin (4 U/mL in saline) into the rabbit ear vein for 75 min. The blood glucose dropped from an initial value of 10.4 mM (187.2 mg/dL) to 7.6 mM (136.8 mg/dL) in 30 min, and then gradually decreased to 0.6 mM (10.8 mg/dL) at the end of 7 h experiment. The highly sensitive amperometric sensor was able to detect the very low corresponding glucose levels in tear fluid and shared the same decreasing trend as the blood glucose, indicating that blood and tear glucose levels do indeed change in tandem over this wide concentration range. A linear least-square regression was obtained for the measured glucose levels in tears and blood samples for just such pilot experiment. A strong Pearson's correlation was observed with a determined r^2=0.9309, p<0.05.

Since it is difficult for the animal to survive the surgery at very low glucose levels (< 3 mM) for a long time period, the initial insulin protocol was modified to giving an insulin bolus into the rabbit vein instead of a continuous infusion. The concentration of insulin was decreased to one-tenth of the original value, to 0.4 U/mL as well. In this way, the blood glucose drops from initial values in the range of 12-18 mM (216-324 mg/dL) to the range of 5-9 mM (90-162 mg/dL) in 30 min, and then gradually decrease to 2-3 mM (36-54 mg/dL) at the end of a 7 h experiment. A representative result from one rabbit using this optimal protocol is shown in Figure 3b. The rabbit was first left under anesthesia without insulin so as to collect samples at high glucose levels for one hour. Then, one shot (0.4 mL) of the insulin solution was administrated, and the blood glucose dropped from 9.0 mM to 5.1 mM within 30 min, followed by the lowest point at 2.2 mM after 90 min. The Pearson's correlation analysis reveals a significant relationship between tear and blood glucose concentrations (r^2=0.8956, p<0.05) and a linear regression shows a good fit of the data. Besides the strong correlation across a wide glucose concentration range, the measured tear glucose levels (20-200 µM) are consistent with most recent reports in the literature (25). Separate correlation coefficients are still observed in the modified animal protocol for different rabbits subjects, indicating the necessity of pre-calibrating the tear-blood correlation for each individual in potential real-world applications.

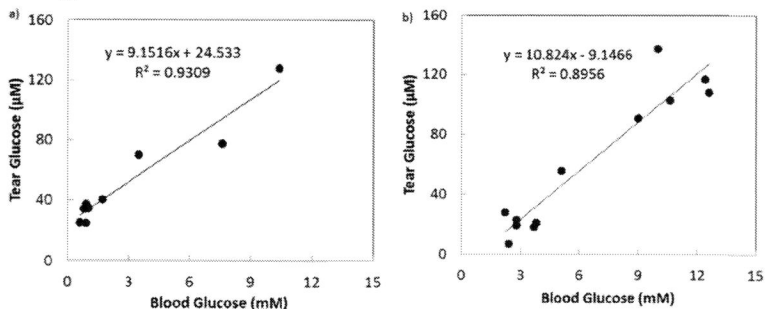

Figure 3. Tear and blood glucose level correlation results from two individual rabbits a) and b) under a modified wide glucose range protocol.

Coulometric Sensor Design

An alternate coulometric glucose sensor design was also developed in order to detect very low levels of glucose in tear fluid. Besides the excellent sensitivity of such a sensor, the coulometry technique provides a more convenient analytical method for future applications, potentially without need for calibration. Compared to the amperometric glucose sensor, there are two major differences in the new coulometric sensor design. First, there is a larger sensing area on the Pt working electrode (3.81 mm^2). Second, the device employs a thinner enzyme layer so that there is better efficiency in collecting all the peroxide produced from glucose. Ten millimeters Pt from a Teflon® coated wire (80 mm total) was exposed to the sample as the working electrode. Ten layers of glucose oxidase (5% in Bovine Serum Albumin) crossed-linked with another 5 layers of glutaraldehyde (2.5%) was dip coated onto the Pt surface. Such a method provides a thinner enzyme layer and, therefore, enhances the diffusion of glucose and peroxide to the electrode surface. Overall, the larger surface area of the sensor increases the surface to volume ratio, which facilitates consumption of the glucose in the 3 microliter sample of glucose in the capillary.

For tear glucose testing, it would be important to complete a glucose measurement in a reasonably short time period (1-10 min). A useful way of reducing response time is to accelerate the overall kinetics of the enzymatic reaction and diffusion by increasing temperature. During the coulometric sensor calibration and real sample measurements, the temperature was raised to 50°C. At higher temperature, 50% glucose was depleted within 5 min, compared to 20 min at 25°C. The typical calibration curve for the coulometric tear glucose sensor in capillary tubes under this higher temperature condition is shown in Figure 4. The coulometric sensor will be employed in future rabbit studies to monitor tear fluid glucose levels, and the measured tear glucose concentrations will be compared with the amperometric sensors so that two orthogonal methods can prove the validity of the tear and blood glucose relationship.

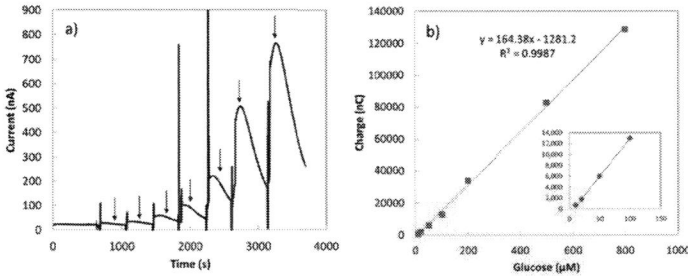

Figure 4. Typical calibration of coulometric tear glucose sensor using 3 µL solution in capillary at elevated temperature, 50 °C: a) Amperometric signals from solutions in the order of 10 µM, 20 µM, 50 µM, 100 µM, 200 µM, 500 µM, 800 µM glucose. b) Resulting calibration curve of tear glucose sensor for 5 min data (current) integration. Inset: Low-end glucose calibration (10-100 µM).

Conclusions

We have developed simple, minimally invasive electrochemical tear glucose biosensors based on amperometric or coulometric measurements coupled with a fluid collection glass micro-capillary configuration. The electrochemical sensor was employed in anesthetized rabbit experiments to detect tear glucose levels while a commercial blood analyzer was concurrently employed to measure corresponding blood glucose levels. With the optimized administration of insulin, a reasonable correlation between tear and blood glucose levels was established over a wide range (from hypoglycemic to hyperglycemic), however, the exact correlation coefficient varies from animal to animal. Therefore, in using tear fluid as an alternate sample to evaluate blood glucose for human subjects, it will likely require a pre-calibration step to obtain the exact correlation ratio between blood and tears for each individual. Once this correlation is established, the proper algorithm can be employed for the reported tear glucose results in order to reflect a corresponding value of blood glucose. Ongoing research on the new coulometric sensor that can measure the total amount of charge that occurs due to near depletion of all glucose in the capillary sample, will likely enable a more sensitive method, and may not require any sensor pre-calibration, provided a reproducible and constant volume of tear fluid can be collected and that all the glucose present is consumed in the time course of the measurement. Efforts to utilize the new coulometric design in the same rabbit experiments as the amperometric sensor to measure the concentrations of glucose in tear fluid are currently in progress.

Acknowledgments

We gratefully acknowledge EyeLab Group for financial support of this research.

References

1. http://www.who.int/mediacentre/factsheets/fs312/en/index.html.
2. Sämann, A., et al., *Exp Clin Endocrinol Diabetes.* **108**, 406 (2000).
3. Abookasis, D. and J.J. Workman, *J. Biomed. Opt.* **16**, 027001 (2011).
4. Dingari, N., et al., *Anal. Bioanal. Chem.* **400**, 2871 (2011).
5. Ballerstadt, R., et al., *Diab. Tech. Ther.* **8**, 296 (2006).
6. Weiss, R., et al., *Diab. Tech. Ther.* **9**, 68 (2007).
7. Muscatello, M.M.W., L.E. Stunja, and S.A. Asher, *Anal. Chem.* **81**, 4978 (2009).
8. Woderer, S., et al., *Anal. Chim. Acta.* **581**, 7 (2007).
9. Cengiz, E. and W.V. Tamborlane, *Diab. Tech. Ther.* **11**, S (2009).
10. Yamaguchi, M., M. Mitsumori, and Y. Kano, *IEEE Eng. Med. Biol. Mag.* **17**, 59 (1998).
11. Lu, J., et al., *Diab. Res. Clin. Pract.* **93**, 179 (2011).
12. Radhakumary, C. and K. Sreenivasan, *Anal. Chem.* **83**, 2829 (2011).
13. Baca, J.T., et al., *Clin. Chem.* **53**, 1370 (2007).
14. Lewis, J.G., *Br. Med. J.* **1**, 585 (1957).
15. Zhang, J., et al., *J. Diab. Sci. Techn.* **5**, 166 (2011).
16. Baca, J.T., D.N. Finegold, and S.A. Asher, *Ocul. Surf.* **5**, 280 (2007).
17. Rhee, S.Y., et al., *J Korean Med Sci.* **22**, 70 (2007).
18. Iwata, S., *Int. Ophthalmol. Clin.* **13**, 29 (1973).

19. Lane, J.D., et al., *Curr. Eye Res.* **31**, 895 (2006).
20. Michail, D. and N. Zolog, *C. R. Soc. Biol., Paris.* **126**, 1042 (1937).
21. Chen, R., Z. Jin, and L.A. Colon, *J. Cap. Elec.* **3**, 243 (1996).
22. LeBlanc, J., et al., *Intens. Care Med.* **31**, 1442 (2005).
23. Sen, D.K. and G.S. Sarin, *Br. J. Ophthalomol.* **64**, 693 (1980).
24. LaBelle, J.T., et al., *J. Diab. Sci. Techn.* **4**, 307 (2010).
25. Chu, M.X., et al., *Biomed. Microdevices.* **13**, 603 (2011).
26. Yan, Q., et al., *Anal. Chem.* **83**, 8341 (2011).
27. Saha, J.K., et al., *Exp. Biol. Med.* **230**, 777 (2005).

22

ECS Transactions, 50 (12) 23-33 (2012)
©The Electrochemical Society

Time Sensors: Circadian Rhythms in Biologically Closed Electrochemical Circuits of Plants

A. J. Waite[a], J. D. Wooten[a], V. S. Markin[b], A. G. Volkov[a]

[a] Department of Chemistry, Oakwood University, Huntsville, Alabama 35896, USA
[b] Department of Neurology, University of Texas, Southwestern Medical Center, Dallas, Texas 75390-8813, USA

> The circadian clock regulates a wide range of electrophysiological and developmental processes in plants. Here, we discuss the direct influence of a circadian clock on biologically closed electrochemical circuits *in vivo*. These circuits in the leaves of *C. miniata* (Kaffir lily), *Aloe vera* and *Mimosa pudica*, which regulate their physiology, were analyzed using the charge stimulation method. The electrostimulation was provided with different voltages and electrical charges. Phytosensors memorize daytime and nighttime. Even at continuous light or darkness, plants recognize nighttime or daytime and change the electrical input resistance. The circadian clock can be maintained endogenously and has electrochemical oscillators, which can activate voltage gated ion channels in biologically closed electrochemical circuits. The activation of voltage-gated channels depends on the applied voltage, electrical charge, and the speed of transmission of electrical energy from the electrostimulator to plants.

Introduction

The biological clock regulates a wide range of physiological and developmental processes in plants. In plants, circadian rhythms are linked to the light–dark cycle. Many of the circadian rhythmic responses to day and night continue in constant light or dark, at least for a period of time. The biological clock in a plant is an endogenous oscillator with a period of approximately 24 hours. The circadian clock in plants is sensitive to light, which resets the phase of the rhythm. Molecular mechanism underlying circadian clock function is poorly understood, although it is now widely accepted for both plants and animals. The circadian clock was discovered in 1729 by De Mairan (1) in his first attempt to resolve experimentally the origin of rhythm in the leaf movements of *Mimosa pudica*. This rhythm continued even when the *Mimosa pudica* plant was maintained under continuous darkness. We investigated the electrical activity of *Mimosa pudica* in the day light, at night, and in darkness the following day. The response in darkness the following day was similar to a typical daily response.

Mimosa pudica is a nyctinastic plant that closes its leaves in the evening; the pinnules fold together and the whole leaf droops downward temporarily until sunrise. The leaves open in the morning due to a circadian rhythm, which is regulated by a biological clock with a cycle of about 24 hours. During photonastic movement in *Mimosa pudica*, leaves recover their daytime position. During a scotonastic period, the primary pulvini straighten up and pairs of pinnules fold together about the tertiary pulvini. The closing of pinnae depends upon the presence of phytochrome in the far-red absorbing form.

23

A monocotyledon *Clivia miniata* (Lindl.) Regel, is a vascular plant with dark green, strap shaped leaves with somewhat swollen leaf-bases which arise from fleshy roots. The name *Clivia* was given to this ornamental genus of plants by John Lindley to compliment Lady Clive, the Duchess of Northumberland, who first cultivated this plant in England.

Aloe vera (L.) is a member of the Asphodelaceae (Liliaceae) family with crassulacean acid metabolism (CAM). In *Aloe vera*, stomata are open at night and closed during the day. CO_2 acquired by *Aloe vera* at night is temporarily stored as malic and other organic acids, and is decarboxylated the following day to provide CO_2 for fixation in the Benson-Calvin cycle behind closed stomata. *Aloe vera* is a model for the study of plant electrophysiology with crassulacean acid metabolism.

Electrical phenomena in plants have attracted researchers since the eighteenth century (2-6). The cells of many biological organs generate electric potentials that result in the flow of electrical currents. Electrical impulses may also arise as a result of stimulation. Once initiated, these impulses can propagate to adjacent excitable cells. The change in transmembrane potential creates a wave of depolarization, which affects the adjoining, resting membrane. Electrical signals can propagate along the plasma membrane on short distances in plasmodesmata and on long distances in conductive bundles. Light-sensitive movements of the leaflets of *Mimosa pudica* depend on ion fluxes across the plasma membrane of extensor and flexor cells in the pulvinus, which create changes in turgor.

Isolated pulvinar protoplasts are responsive to light signals *in vitro* (7-9). In the dark period, the closed inward-directed K^+ channels of extensor cells are opened within 3 minutes by blue light. Conversely, the inward-directed K^+ channels of flexor cells, which are open in the darkness, are closed by blue light. In the light period, however, the situation is more complex. Premature darkness alone is sufficient to close the open channels of extensor protoplasts, but both darkness and a preceding pulse of red light are required to open the closed channels in the flexor protoplasts.

The Charge Stimulation Method (CSM) was used to evaluate the electrical equivalent schemes of biological tissue in plants and to estimate the amount of electrical energy necessary to induce a plant response. This DC method permits direct *in vivo* evaluation of the simplest electrical circuits in a cluster of cells or in a single cell (10-16). Equivalent electrical schemes of biologically closed electrical circuits were then evaluated inside the pulvinus of *Mimosa pudica*, leaves of *Aloe vera* and *Clivia miniata* during the day, night and darkness period during next day.

Materials and Methods

All measurements were conducted in the laboratory at 21°C inside a Faraday cage mounted on a vibration-stabilized table (Figure 1). Ag/AgCl electrodes were prepared in the dark from Teflon coated silver wires (A-M Systems, Inc., Sequim, WA, USA) by electrolysis in 0.05 M KCl aqueous solution. The response time of Ag/AgCl electrodes was less than 0.1 µs. Plants were allowed to rest after electrode insertion.

To perform the electrostimulation experiments we used the PXI machine (PCI extensions for Instrumentation), a rugged PC-based platform, as a high-performance measurement and automation system. A NI-PXI-4071 digital multimeter (National Instruments, Austin, TX, USA), connected to 0.2 mm thick nonpolarizable reversible Ag/AgCl electrodes, was used to record the digital data. A NI PXI-4110 DC Power

Supply (National Instruments Austin, TX, USA) was also used to provide a voltage source for capacitor charging.

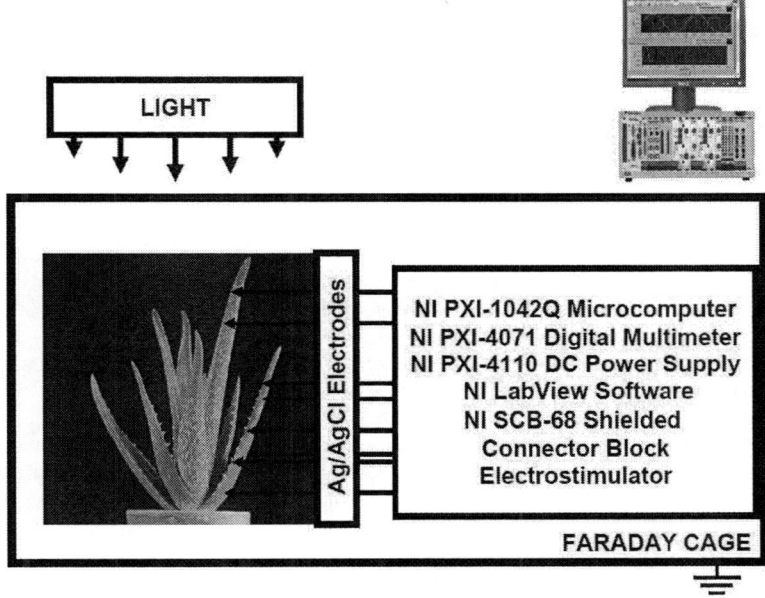

Figure 1. Experimental setup.

Experimentation with electrical stimulation of plants requires precise control of plant stimulation. Our primary objective was to precisely determine the voltage or the amount of charge that generates a given biological effect. The CSM method was used to estimate, with high precision, the amount of electrical energy necessary to induce a response. We implemented two types of electrostimulation: a manual switch and a custom made application specific controller. Manual stimulation is convenient for a single stimulation, as it does not require additional equipment. It was implemented using a double pole double throw (DPDT) switch to connect the 10 µF capacitor to the NI PXI-4110 DC Power Supply during charging, and then to the plant during plant stimulation. By changing the switch position, we could instantaneously connect the charged capacitor to the plant and induce a response.

We designed and implemented the plant stimulator to allow multiple stimulations with precise timing, voltage, and charge during stimulations (17). The plant stimulator is a battery powered portable device controlled by a low-power microcontroller MSP430F1611 (Texas Instruments, Texas, USA). A specialized PC program allows flexible configuration of the controller and communicates with the controller through an optically isolated USB interface. During each stimulation cycle, the controller charges a capacitor with a predefined voltage using an integrated digital to analog (DA) converter of the microcontroller. Each pulse can be controlled with microsecond resolution. A dual integrated SPDT analog switch is controlled by the microcontroller and connects the capacitor to the DA converter during charging and to the plant during stimulation.

Plants were exposed to a 12:12 hour light/dark photoperiod at 21 °C. Volume of soil was 3.9 L. *C. miniata* plants had 25 - 55 cm leaves. The average humidity was 40%. Irradiance was 500-700 µmol photons $m^{-2}s^{-1}$. All experiments were performed on healthy adult specimens.

Results and Discussion

Circadian Rhythms in Electrical Circuits of *Clivia miniata.*

We investigated electrical responses of *C. miniata* to electrical stimulation during the day in daylight, darkness at night, and the following day in darkness with different timing and voltages. *C. miniata* is a model for the study of circadian rhythms in plants. Circadian variation of the *Clivia miniata, Aloe vera* and *Mimosa pudica* are sensitive to electrical stimulation. The biologically closed electrochemical circuits in the leaves of *C. miniata* (Kaffir lily), which regulate its physiology, were analyzed *in vivo* using the charge stimulation method. The electrostimulation was provided with different voltages and electrical charges. Resistance between Ag/AgCl electrodes in the leaf of *C. miniata* was higher at night than during the day or the following day in the darkness. The biologically closed electrical circuits with voltage gated ion channels in *C. miniata* are activated the next day, even in darkness. *C. miniata* memorizes daytime and night time. At continuous light, *C. miniata* recognizes nighttime and increases the input resistance to the nighttime value even under light. These results show that the circadian clock can be maintained endogenously and has electrochemical oscillators, which can activate voltage gated ion channels in biologically closed electrochemical circuits. The activation of voltage-gated channels depends on the applied voltage, electrical charge and speed of transmission of electrical energy from the electrostimulator to the *C. miniata* leaves.

As it is well known, if a capacitor of capacitance C with initial voltage U_0 is discharged through a resistor R (Figure 2a), the voltage decreases with time t as

$$U(t) = U_0 \cdot e^{-t/\tau} \tag{1}$$

where

$$\tau = RC \tag{2}$$

denotes the time constant. Equation (1) in the logarithmic form reads:

$$\ln(U(t)/U_0) = -t/\tau \tag{3}$$

The time constant, τ, can be determined from the slope of this linear function. The circuit time constant τ governs the discharging process. As the capacitance or resistance increases, the time of the capacitor discharge increases according to equation (1). The resistance of the linear circuit can be easily found from eq. (1).

However, if the function (3) is not linear, then one can find the "input resistance" at any moment of the time:

$$R_{input} = -\frac{U}{C\ dU\ /\ dt} \tag{4}$$

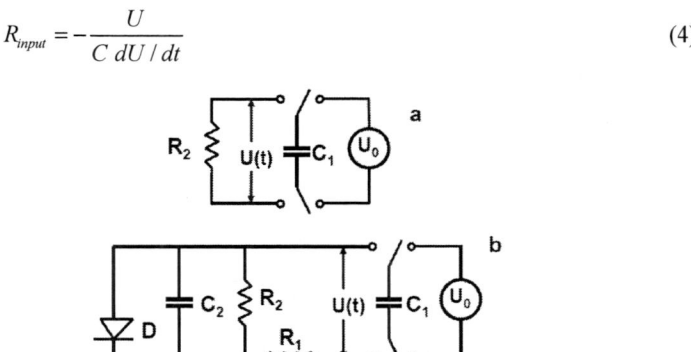

Figure 2. Electrical equivalent schemes of a capacitor discharge in plant tissue. Abbreviations: C_1 – charged capacitor from voltage source U_0; C_2 – capacitance of plant tissue; R – resistance, D – a diode as a model of voltage gated ion channels.

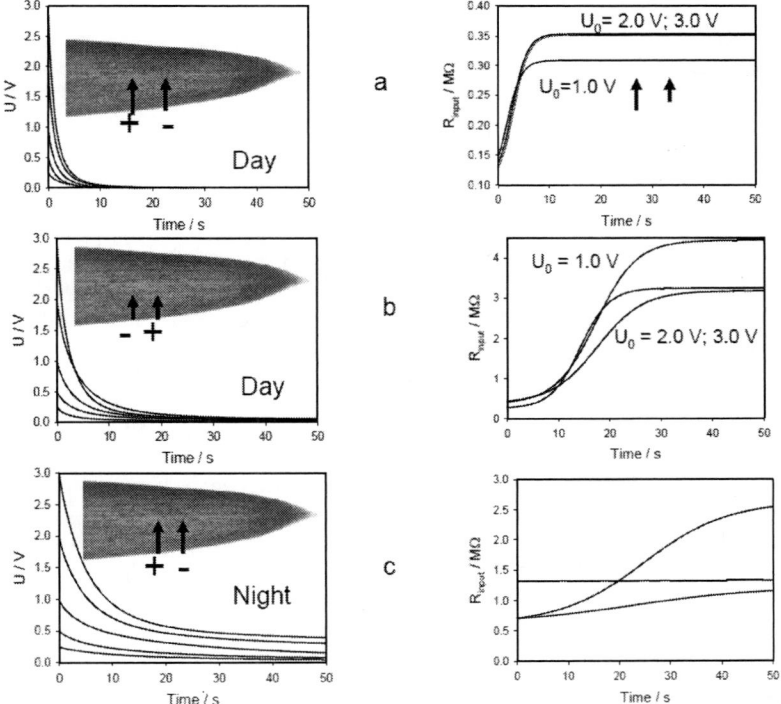

Figure 3. Time dependencies of electrical discharge and input resistance in the *C. miniata* leaf between two Ag/AgCl electrodes connected to the 10 μF charged capacitor during a day (a,b) and at night (c). *U* is the capacitor voltage and U_0 is the initial voltage in volts.

We studied electrical discharge of 10 μF capacitor between two Ag/AgCl electrodes in the leaf of *C. miniata* parallel to the conductive bundles. The results obtained in the daylight during the first day are presented in Fig. 3. The difference between the two experiments is the polarity of the electrodes: the positive pole is closer to the base of a leaf in Fig. 3a and the positive pole is closer to the apex in Fig. 3b. The dependence of resistivity on polarity can be explained by the opening of voltage gated ion channels, and can be modeled as diodes in Fig. 2b. Opening and closing of voltage gated channels results in the effect of electrical rectification. Similar rectification effects were found in *Aloe vera*, the Venus flytrap and *Mimosa pudica*. While using a silicon rectifier Schotky diode NTE583 as a model of a voltage gated channel, we reproduced experimental dependencies of the capacitor discharge in plant tissue.

The same electrical discharge in darkness during the nighttime is shown in Fig. 3c. The kinetics of the night discharge is significantly slower (Fig. 3c). This means that leaf resistance strongly increases at night (Fig. 3c).

The biological clock in *C. miniata* recognizes the daytime, even in darkness. The discharge during the following day in darkness is very similar to the first day. Input resistance in the initial moment of the capacitor discharge was the same as during the day light. During the third day when the lights are on, the results are the same as shown in Fig. 3, which were reproduced on different leaves of *C. miniata* plants.

In all three examples, kinetics of a capacitor discharge depends on polarity of electrodes in a leaf of *C. miniata* due to electrical anisotropy of the leaf. Dependence of the discharge kinetics on polarity of electrodes shows the rectification effect. This can be caused by the opening or closing of voltage-gated ion channels and decreasing or increasing of the resistance in plant tissue, correspondingly.

Fig. 4 illustrates the plant memory of a "sunset". Normally, we switch off the lights at 5:00 pm, however in this experiment we did not switch off the light at that time. Any time from the morning and during a day until 4:00 pm, the time dependencies of a 10 μF capacitor discharge coincide. At 5:00 pm resistance in leaves starts to increase and at 7:00 pm it reaches the same parameters as at night even at continuous light. Fig. 4b presents the normalized voltage U/U_0 in semi-logarithmic coordinates. As one can see from Fig. 4, the capacitor discharge is fast during daytime, but speed of the capacitor discharge decreases after 4:00 pm even under continuous light and reaches minimal value at 7:00 pm as in the dark during nighttime. Darwin (18) found that leaves in *Clivia* move periodically: "A long glass filament was fixed to a leaf, and the angle formed by it with the horizon was measured occasionally during three successive days. It fell each morning until between 3 and 4 p.m., and rose at night. The smallest angle at any time above the horizon was 48°, and the largest 50°; so that it rose only 2° at night; but as this was observed each day, and as similar observations were nightly made on another leaf of a distinct plant, there can be no doubt that the leaves move periodically. The position of the apex when it stood highest was 0.8 of an inch above its lowest point." The periodical movement of leaves in *Clivia* has the electrophysiological component as it is shown in Fig. 4. Two hours before the light turning, the speed of electrical discharge decreases due to increasing of a resistance in leaves, which is probably related to the closing of voltage gated channels. This result is very impressive: the biological clock in *C. miniata* recognizes the approaching of darkness after 4 p.m. even under constant light from 6.00 a.m. to 6.00 p.m.

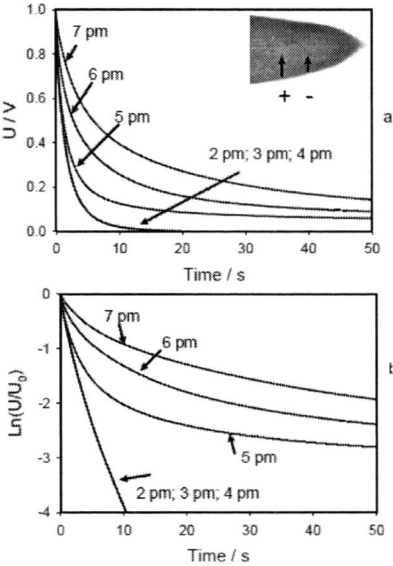

Figure 4. Time dependence of the capacitor discharge in the *C. miniata* leaf afternoon and at evening during continuous light. U is the capacitor voltage and U_0 is the initial voltage in volts.

Circadian oscillators are components of the biological clocks that regulate the activities of plants in relation to environmental cycles and provide an internal temporal framework. The circadian rhythm regulates a wide range of electrophysiological and developmental processes in plants such as floral development, flowers opening in the morning and closing towards evening, seeds germination, leaf growth and movement, stem elongation chloroplast movements, photosynthetic capacity, stomata conductance, cell division.

Plant tissues have biologically closed electrochemical circuits that are involved in these regulations. Voltage gated ion channels are devices in the engineering sense: they have signal inputs, power supplies, and signal outputs. We found periodic activation and deactivation of these circuits in the leaves of *C. miniata* controlled by internal clock, rather than environmental clues. We tested characteristics of capacitor discharge during the day, night, and next day in the darkness. Response of circuits in *C. miniata* during the day was significantly faster than at night and these variations were steadily repeated day after day.

While in darkness the following day, the plant remembers the time and the rate of discharge drastically increases approaching the rate of the first day. However, the input resistance is the same as during day light in the beginning of the capacitor discharge and increases to night values during the discharge process. That means that the internal clock does change electrical conductance, but alone, without environmental clues (light), cannot ideally generate the same values of day and night conductance. The plant needs additional environmental information, and then the properties of electrical circuits will be restored to the same conditions.

These results demonstrate that the circadian clock can be maintained endogenously, probably involving electrochemical oscillators, which can activate or deactivate ion channels in biologically closed electrochemical circuits. This circadian rhythm can be related to the differences found in the membrane potentials during the day and nighttime, which were found in different plants. The expression of many ion transporters in plants is regulated by the circadian rhythm.

Circadian Rhythms in Electrical Circuits of *Aloe vera* and *Mimosa pudica.*

The difference in kinetics of discharge between night and day in the *Aloe vera* leaf, when the response during the night is subtracted from response during the day, is shown in Figure 5a. Discharge of the capacitor at night was faster than during the day. This means that leaf resistance decreases at night. The difference in discharge kinetics between day light and the next day in the darkness is shown in Figure 5b. The difference in kinetics discharge during the day and during the following day in the darkness is about two times less than the difference in kinetics of discharge between the day and night. The biological clock in *Aloe vera* recognizes the day time, even in darkness, but in the absence of light voltage gated ion channels have some electrical responses about two times less than at night.

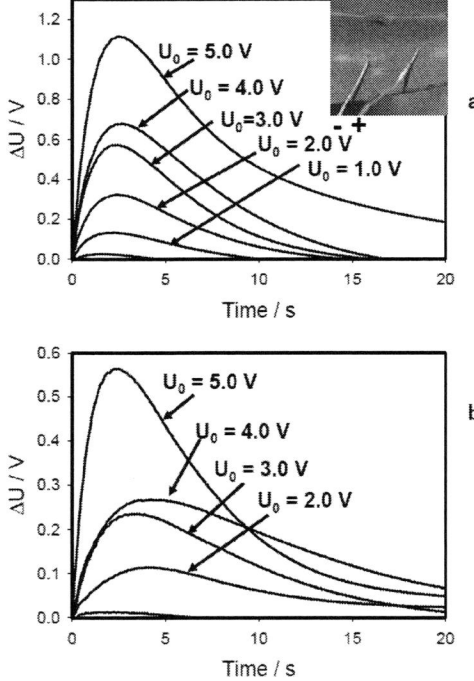

Figure 5. (a): Difference in time dependencies of electrical discharge in the *Aloe vera* leaf at night and at day time (response during the night subtracted from the response during the day). (b): Difference in time dependencies of electrical discharge in the *Aloe vera* leaf at day time light and darkness during next day (response during the darkness subtracted from the response during the day light).

We analyzed kinetics of a capacitor discharge during a day in *Mimosa pudica* and found the input resistance between electrodes inserted across the pulvinus. Here we studied the difference in the discharge kinetics between day and night time (Figure 6). Initial difference in the speed of the response (faster during the day), can be explained by activation of ion channels, equivalent to the high rectification effect. This effect depends on the applied stimulation voltage.

Figure 7a shows the kinetics of a capacitor discharge in a pulvinus during the day time, but in the dark. *Mimosa pudica* was exposed to a 12:12 hour light/dark photoperiod during a 2 month period, but in the morning of the experiment, the light source was not switched on for the day. Figure 7b shows the difference in time dependence of electrical discharge in the dark in the day time. The biologically closed electrical circuits with voltage gated ion channels in *Mimosa pudica* are activated the next day even in the darkness. This phenomenon can be caused by biological clock in *Mimosa pudica*. The nonlinear effect of activation of electrical circuits during the day time is stronger than during the next day in the darkness.

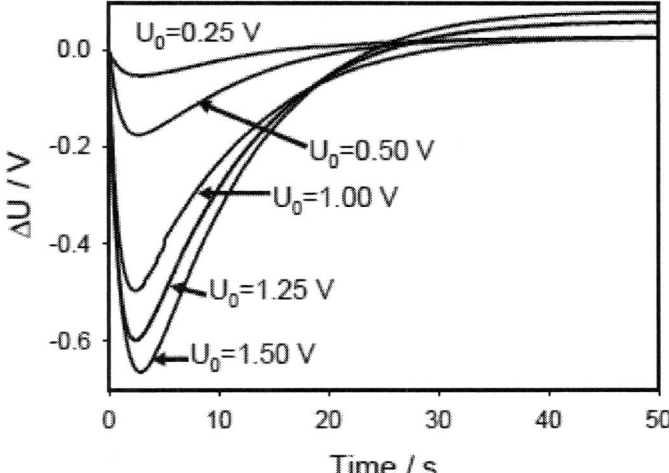

Figure 6. Difference in time dependencies of electrical discharge in a pulvinus of *Mimosa pudica* at night and at daytime (night – day response).

Circadian oscillators are components of the biological clocks that regulate the activities of plants in relation to environmental cycles and provide an internal temporal framework. The circadian clock regulates a wide range of electrophysiological and developmental processes in plants. Plant tissues have biologically closed electrochemical circuits that are involved in these regulations. We found periodic activation and deactivation of these circuits in the leaves of *Aloe vera*, *C. miniata* and *Mimosa pudica* controlled by internal clock, rather than environmental clues. We tested characteristics of capacitor discharge. Response of circuits in *Aloe vera* at night was faster than during the day and these variations steadily repeat day after day. This plant response is probably related to crassulacean acid metabolism in *Aloe vera* with stomata closed during the day and open at night. In contrast to *Aloe vera* the discharge of the capacitor in a pulvinus of *Mimosa pudica* was faster during the day, because *Aloe vera* is a CAM plant.

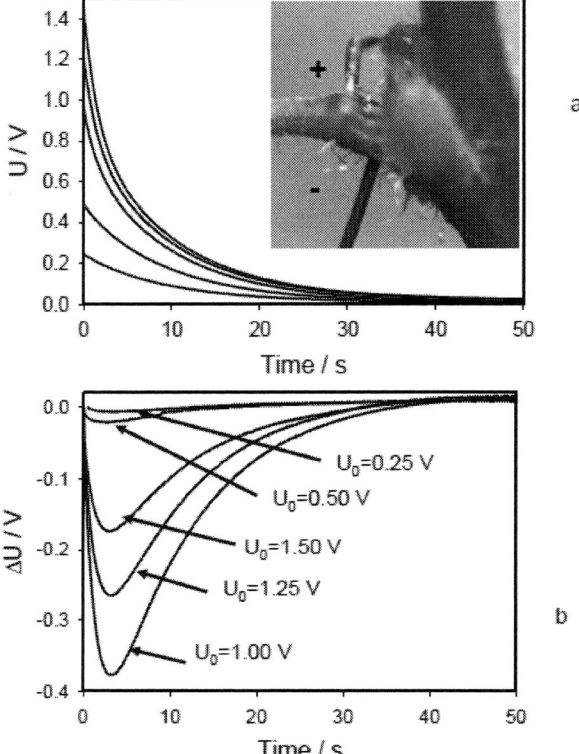

Figure 7. (a): Time dependence of electrical discharge in a pulvinus of *Mimosa pudica* in the dark during day time with positive stimulation on the flexor side of the pulvinus; (b): Difference in time dependencies of electrical discharge at night and in the dark at day time. U is the capacitor voltage and U_0 is the initial voltage in volts.

This circadian rhythm can be related to the difference in the membrane potentials during the day and night time, which was found in pulvini of different plants (19, 20). During the day in darkness, there is still a rectification effect in the pulvinus (Figure 7b), however, in the presence of light during the daytime, this rectification effect and the resistance decrease in the pulvinus are two times higher. These results demonstrate that the circadian clock can be maintained endogenously, probably involving electrochemical oscillators, which can activate or deactivate ion channels in biologically closed electrochemical circuits.

Acknowledgments

This work was supported in part by the grant CBET-1064160 from the National Science Foundation and by the U. S. Army Research Laboratory and the U. S. Army Research Office under contract/grant number W911NF-11-1-0132.

References

1. M. De Mairan, *Histoire de l'Academie Royale de Sciences,* p. 35, Paris (1729).
2. M. Bertholon, *De L'electricite Des Vegetaux: Ouvrage Dans Lequel on Traite De L'electricite De L'atmosphere Sur Les Plantes, De Ses Effets Sur Leconomie Des Vegetaux, De Leurs Vertus Medico,* P.F. Didot Jeune, Paris (1783).
3. J. C. Bose, *The Nervous Mechanism of Plants,* Longmans Green, London (1926).
4. A. G. Volkov, Editor, *Plant Electrophysiology- Methods and Cell Electrophysiology,* Springer, Berlin (2012).
5. A. G. Volkov, Editor, *Plant Electrophysiology- Signaling and Responses,* Springer, Berlin (2012).
6. A. G. Volkov, Editor, *Plant Electrophysiology,* Springer, Berlin (2006).
7. G.G. Coté, *Plant Physiol.,* **109**, 729 (1995).
8. H. Y. Kim, G. G. Coté and R. C. Crain, *Science,* **260**, 960 (1993).
9. H. Y. Kim, G. G. Coté and R. C. Crain, *Plant Physiol.,* **99**, 1532 (1992).
10. A. G. Volkov, J. C. Foster, T. A. Ashby, R. K. Walker, J. A. Johnson and V. S. Markin*, Plant Cell Environm.,* **33**, 163 (2010).
11. A. G. Volkov, T. Adesina, V. S. Markin and E. Jovanov, *Plant Physiol.,* **146**, 694 (2008).
12. A. G. Volkov, H. Carrell, A. Baldwin and V. S. Markin, *Bioelectrochem.,* **75**, 142 (2009).
13. A. G. Volkov, H. Carrell and V.S. Markin, *Plant Physiol.,* **149**, 1661 (2009).
14. A. G. Volkov, J .C. Foster and V. S. Markin, *Plant Cell Environm.,* **33**, 816 (2010).
15. A. G. Volkov, J. C. Foster, E. Jovanov and V. S. Markin, *Bioelectrochem.,* **81**, 39 (2011).
16. A. G. Volkov, J. D. Wooten, A. J. Waite, C. R. Brown and V. S. Markin*, J. Plant Physiol.,* **168**, 1753 (2011).
17. A. G. Volkov, K. Baker, J. C. Foster, J. Clemmons, E. Jovanov, and V. S. Markin, *Bioelectrochem.,* **81**, 39 (2011).
18. C. Darwin, *The Power of Movements in Plants,* John Murray, London (1880).
19. R. Racusen and R. L. Satter, *Nature,* **255**, 408 (1975).
20. B. I. H. Scott and H. F. Gulline, *Nature,* **254**, 69 (1975).

ECS Transactions, 50 (12) 35-41 (2012)
©The Electrochemical Society

Immobilization of enzymes and redox proteins and their electrochemical biosensor applications

Veerappan Mani, Balamurugan Devadas, Shen Ming Chen[*], Ying Li

Department of Chemical Engineering and Biotechnology, National Taipei University of Technology, No. 1, Section 3, Chung-Hsiao East Road, Taipei 106, Taiwan (ROC)

Abstract

Immobilization and direct electrochemistry of proteins and enzymes can provide a good model for mechanistic studies of their electron transfer activity in biological systems. Moreover, achieving direct electron exchange between redox proteins or enzymes and electrodes simplifies third generation biosensors, enzymatic bioreactors and biomedical devices by removing the requirement of chemical mediators, and thus has a great significance. We have immobilized enzymes and redox proteins, such as glucose oxidase (GOx), alcohol dehydrogenase (ADH), horseradish peroxidase (HRP), and catalase (CAT) on the surface of the modified electrode. The matrixes which we prepared for the enzyme immobilization are gelatin-Multiwalled carbon nanotube (GCNT), Bismuth oxide nanoparticles-MWCNT composite, didodecyldimethylammonium bromide (DDAB) present on nafion dispersed multiwalled carbon nanotubes (MWCNTs-NF) and biocomposite of Toluidine blue O with adsorbed alcohol dehydrogenase (ADH). The surface morphology of the enzyme immobilized modified electrode were characterized by Scanning electron microscopy and atomic force microscopy. GOx/GCNT, NF/HRP/Bi_2O_3–MWCNT, MWCNTs-NF-(DDAB/CAT) and ADH/TBO/NF films exhibited a wide linear response from 6.30 to 20.09 mM (glucose), 8.34–28.88 mM (H_2O_2), 0.5 to 1.2 mM (H_2O_2) and 283–856 mM ethanol respectively. These films showed sensitivity of 2.47 μAmM^{-1} cm^{-2}, 26.54 $\mu AmM^{-1}cm^{-2}$, 101.74 $\mu AmM^{-1}cm^{-2}$ and 7.91 $\mu AM^{-1}cm^{-2}$ respectively. All the developed sensors show good stability with appreciable sensitivity, selectivity and wide linear range.

* Corresponding author. Fax: +886 2270 25238; Tel: +886 2270 17147, E-mail:

smchen78@ms15.hinet.net

Introduction

Glucose oxidase (GOx) based enzyme sensors have been widely employed to monitor the blood glucose level [1]. However, one of the most challenging aspects in the development of highly sensitive glucose sensor is the selection of a suitable matrix for GOx immobilization. With high mechanical strength, excellent conductivity and antifouling properties, multiwalled carbon nanotubes (MWCNTs) have been immensely used in glucose sensors for GOx immobilization [2]. Horseradish peroxidase (HRP) is an important heme containing redox enzyme which belongs to class III of the plant peroxidase super family [3]. HRP contains ferriprotoporphyrin IX prosthetic group at its active centre. It has been majorly used in amperometric H_2O_2 biosensors due to its unique ability to catalyze H_2O_2 reduction process at a specific low potential via direct electron transfer [4].

Catalase (CAT) is an active enzyme, which catalyzes the disproportionation process of H_2O_2. This catalytic ability of CAT has been applied in the development of enzyme based H_2O_2 sensors. However; the rapid response of such CAT based sensors relies on the rate of electron transfer between CAT and the electrode. Earlier, few attempts were made to achieve direct electron transfer between CAT and GCE and graphite electrodes using electrode modification methods[5-7].Alcohol dehydrogenase (ADH) based ethanol biosensors have received much interest, Because irreversible oxidation of ethanol occurs in presence of both ADH and its cofactor nicotinamide adenine dinucleotide (NAD^+)at their close proximity[8]. Attempts have also been made to design a disposable ethanol sensor using polystyrene modified screen printed carbon electrode as the ADH immobilizing matrix.

Experimental

Materials and methods

Multiwalled carbon nanotube, Glucose oxidase, Gelatin, Peroxidase from horseradish typeVI-A, Bismuth (III) nitrate pentahydrate, CAT from bovine liver, ADH from Saccharomyces cerevisiae (332 U mg−1protein) and NAD^+ from yeast and D (+) glucose were purchased from sigma Aldrich and used as received. All the solutions were prepared by using double distilled demonized water. The 0.05 M phosphate buffer solution (pH 5 and pH 7) used for all electrochemical experiments. Nitrogen gas passed for all electrochemical experiments. All the cyclic voltammetry experiments were carried out by using CHI 1205A instruments. CHI-750 potentiostat was used for amperometric i–t curve studies. For all electrochemical instruments employed by using single compartment of three electrode system, standard Ag/AgCl as reference electrode, platinum wire with 0.5 mm diameter as counter electrode and Glassy carbon electrode with surface area of 0.079 cm^{-2} was used as a working electrode. Surface morphology of all prepared film was characterized by using Hitachi S-3000 H scanning electron microscopy (SEM).

Preparation of modified electrode

Before preparation of modified electrode, the glassy carbon electrode (GCE) was polished with alumina slurry (0.05μm) and washed and ultrasonicated with double distilled deionized water for few minutes. The pre-cleaned GCE has been used for the further modification with the respective composite. Moreover, the enzymes GOx, HRP, CAT, and ADH have been immobilized onto the modified electrode by simple adsorption strategy.

Result and Discussion

Surface morphology of modified electrode

The surface morphology of enzyme immobilized modified GCE was examined using SEM. Fig. 1(a) shows the surface of bare GCE. Fig. 1(b) shows the SEM images of only GOx film. Small spherical voids surrounded by small bead like structures are seen on the film surface. Fig (c) SEM images of uniform GCNT/GOx/GAD composite film closely packed larger bead like structures are seen. This validates higher GOx loading at GCNT matrix and it could be attributed to the larger surface area of GCNT. Fig 1(d) shows the SEM image of MWCNTs, where several coiled MWCNT bundles were found along with few agglomerated MWCNT bundles. Fig. 1(a') shows the SEM image of Bi_2O_3–MWCNT with bright Bi_2O_3 nanoparticles coated MWCNT networks. Whereas, (b') shows the needle shaped DDAB coated on the ITO. The arrow in (b') indicates a needle shaped structure of DDAB. Fig (c') shows the DDAB/CAT, MWCNTs-NF-(DDAB/CAT) films where the bud like structures of CAT incorporate with needle shaped structures of DDAB. Fig (d') morphology of ADH/TBO/NF films it possesses more uniform surface morphology.

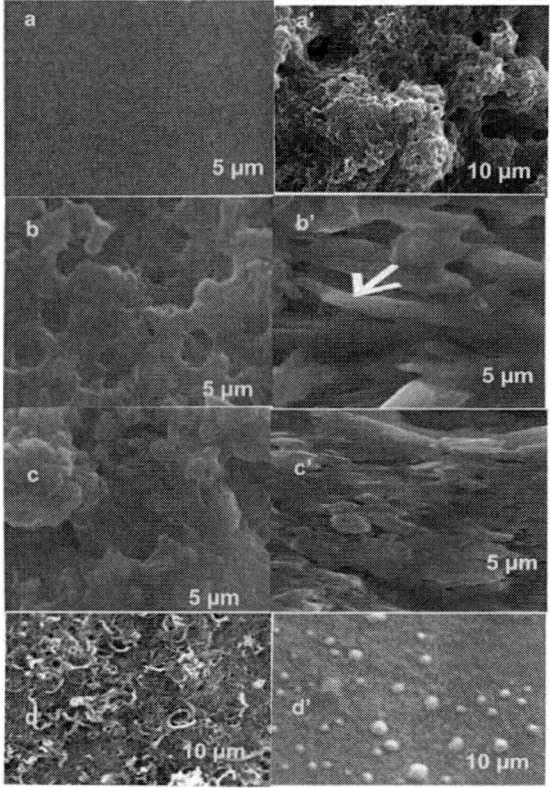

Figure 1.SEM images (a) Bare GCE, (b) GOx, (c) GCNT/GOx/GAD (d) MWCNT, (a')
NF/HRP/Bi$_2$O$_3$-MWCNT, (b') CAT, (c') MWCNTs-NF-(DDAB/CAT), (d') ADH/TBO/NF

Electrochemical biosensor:

The figure 2 (A) shows the amperometric response of glucose oxidase immobilized
modified GCE with determination glucose in the applied potential of -0.44 V in stirred
oxygenated PBS solution. Small shoulder peaks appeared immediately for the successive
addition of minimum glucose concentrations (0.01 and 0.02 mM).These shoulder peaks appeared
as a result of oxygen reduction. With further increase in glucose addition, the oxygen
consumption increases and thus the catalytic current decreases gradually and finally levels off for
very high glucose concentration (25.84 mM) (see Fig. 2(A) upper inset). This GOx/GCNT/GAD
film exhibits the linear range towards glucose was 6.30 to 20.09 mM.

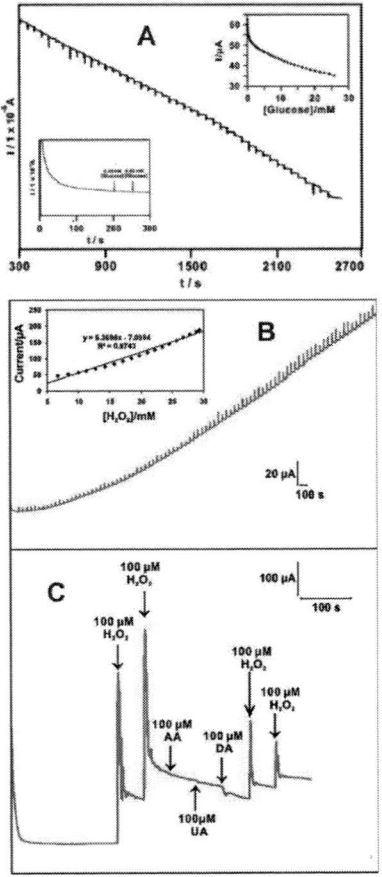

Fig. 2 (A) Amperometric i–t response at GCNT/GOx/GAD composite film modified rotating disc GCE for the addition of 0.01 to 25.84 mM glucose in − 0.44 V. Fig 2. (B) Amperometric i–t responses at NF/HRP/Bi_2O_3–MWCNT film modified rotating disc GCE upon successive additions of 6.63–30.12 mM H_2O_2. The inset is the plot of linear response current vs. [H_2O_2] = 8.34–28.88 mM. Figure (C) Amperometric i–t responses at NF/HRP/Bi_2O_3–MWCNT film modified rotating GCE for the successive 100 μM of H_2O_2, AA, UA, and DA concentration additions. Applied potential: −0.3 V; Supporting electrolyte: continuously stirred N_2 saturated 0.05 M PBS (pH 7).

The amperometric response of Horseradish peroxidase (HRP) immobilized on Bi_2O_3-MWCNT composite modified GCE with determination of H_2O_2 in the applied potential of -0.3 V Fig. 2(B).This enzyme film shows the good response towards H_2O_2 biosensor. Depends on the

H_2O_2 concentration increases the amperometric current increased linearly. HRP/ Bi_2O_3-MWCNT film shows the linear range of 8.34 to 28.88 mM towards the detection of H_2O_2. Fig 2(C) depicts the amperometric response of 100 μM H_2O_2 with interfering species of 100 μM of Ascorbic acid, Uric acid and Dopamine. No amperometric signal was observed with AA, UA and DA. This indicates that HRP immobilized biosensor is highly selective film for the H_2O_2 biosensor.

Stability studies:

The storage stability of GOx, HRP, ADH and Lactase immobilized on modified glassy carbon electrode biosensor employed by using CV. GOx/GCNT/GAD, HRP/ Bi_2O_3-MWCNT, MWCNTs-NF-(DDAB/CAT) and ADH/TBO/NF film modified GCE was stored in 0.05 M PBS (pH 7) at 4°C. The peak currents have been measured for the each biosensor every day. The peak current is slightly decreased on every consecutive days and biosensor retained its 95% of current after its one month of its storage. Throughout the one week storage the enzyme immobilized film exhibits well defined catalytic biosensor response. By employing MWCNT as a electrode material, we achieved "direct electrochemistry" which is one of the primary evidence that enzyme activity is not denature by incorporating enzymes in MWCNT. Further, we performed stability test for the entire enzyme modified electrode and the results showed that the enzyme immobilized MWCNT modified electrode retained its 95% of its response even after one month. This also confirmed that enzyme activity is not denatured when it got adsorbed onto the surface of MWCNT.

Acknowledgments

This work was supported by National Science Council (NSC) of Taiwan (ROC).

References

1.   M. Musameh, J.Wang, A.Merko ci, Y. Lin, Low-potential stable NADH detection at carbon nanotube modified glassy carbon electrodes, *Electrochem.Commun.*, 4 (2002) 743–746.

2.   R.T. Kachoosangi, M.M. Musameh, I.A. Yousef, J.M. Youse f, S.M. Kan an, L. Xiao, S.G.Davies, A. Russell, R.G. Compton, Carbon nanotube-ionic liquid composite sensors and biosensors, *Anal. Chem.*, 81 (2009) 435 – 442.

3.   N.C. Veitch, Horscradish peroxidase: A modern view of a classic enzyme, *Phytochemistry*, 65 (2004) 249–259.

4.   S.V. Dzyadevych, V.N. Arkhypova, A.P. Soldatkin, A.V. Elskaya, C. Martelet, N. Jaffrezic-Renault, Amperometric enzyme biosensors: Past, present and future, *ITBM–RBM*, 29 (2008) 171.

5.   M.E. Lai, A. Bergel, Direct electrochemistry of catalase on glassy carbon electrodes, *Bioelectrochemistry*, 55 (2002) 157.

6.   E. Horozova, Z. Jordanowa, V. Bogdanovskaya, Enzymatic and electrochemical reactions of catalase immobilized on carbon materials, *Z. Naturforsch.* 50 (1995) 499.

7.   A.M. Azevedo, D.M.F. Prazeres, J.M.S. Cabral, L.P. Fonseca, Ethanol biosensors based on alcohol oxidase, *Biosens. Bioelectron.*, 21 (2005) 235–247.

8.   Y.V. Rodionov, O.I. Keppen, M.V. Suckhacheva, A Photometric Assay for Ethanol, *Appl. Biochem. Microbiol.*, 38 (2002) 395–396.

Signal Optimization for *Salmonella* Typhimurium detection on food surfaces using phage-based biosensors

Yating Chai[a], Suiqiong Li[a], Shin Horikawa[a], Mi-kyung Park[a], Vitaly Vodyanoy[b], Bryan A.Chin[a]

[a] Materials Research and Education Center, Auburn University, Auburn, AL 36849, USA

[b] Department of Anatomy, Physiology and Pharmacology, Auburn University, Auburn, AL 36849, USA

This paper presents the direct detection of Salmonella typhimurium on egg shells using a phage-based magnetoelastic (ME) biosensor. The ME biosensor consists of an ME resonator as the sensor platform and E2 phage as the bio-recognition element. The bio-recognition element is genetically engineered to bind specifically with Salmonella typhimurium. The ME biosensor, a wireless sensor, vibrates with a characteristic resonant frequency under an externally applied magnetic field. Multiple sensors can easily be remotely monitored. Multiple measurement and control sensors were placed on egg shells contaminated by Salmonella typhimurium solutions with different known concentrations. The resonant frequency of sensors before and after the exposure to the spiked egg shells was measured. The frequency shift of the measurement sensors was significantly different than the control sensors indicating Salmonella contamination. Scanning electron microscopy was used to confirm binding of Salmonella to the sensor surface and the resulting frequency shift results.

Introduction

The current microbiological detection techniques, including enzyme-linked immunosorbent assay (ELISA)[3], polymerase chain reaction (PCR)[4], fourier transform infrared spectroscopy (FTIR)[5], and the recently developed desorption electro-spray ionization (DESI)[6], are time-consuming and manpower-intensive. Compared with these methods, the phage-based magnetoelastic (ME) biosensor used in this investigation is rapid, label-free, easy to use and inexpensive[7]. In addition, biomolecular recognition elements composed of phage-based structures are more robust environmentally than antibody-based structures, making our biosensor more stable in different detection situations and easier to store.

This paper presents an investigation into the direct detection of *Salmonella typhimurium* on egg shells using a phage-based magnetoelastic (ME) biosensor. The ME biosensor is composed of a magnetoelastic resonator platform coated with a phage bio-recognition layer. The ME resonator is made up of a ferromagnetic, amorphous material that elongates and contracts along the direction of a modulated applied magnetic field[8]. The ME resonator vibrates at a characteristic resonant frequency under an externally applied time varying magnetic field. The resonant frequency is related to the ME sensor's dimensions and material properties. The resonator investigated in this paper was a freestanding, strip-shaped platform, the thickness of which is much smaller than the length and width. The fundamental resonant frequency of the resonator under longitudinal vibration is expressed as[9]:

$$f = \frac{1}{2L}\sqrt{\frac{E}{\rho(1-v)}} \qquad (1)$$

where E, ρ, and v are the elastic modulus, density, and Poisson's ratio of the material. When the ME sensor comes in contact with target pathogens, the pathogen will be captured and bound to the sensor's surface through the bio-molecular recognition element (E2 phage). The capture of the pathogens causes an increase in the mass of the ME biosensor (Δm) which results in a decrease in the sensor's resonant frequency (Δf). When this additional non-magnetoelastic mass (Δm) is much smaller than the mass of the sensor (m), the change in the resonant frequency Δf is approximated by[10]:

$$\Delta f \approx -\frac{f}{2}\frac{\Delta m}{M} \qquad (2)$$

This equation reflects that the frequency change depends on the additional mass added to the sensor surface. This additional mass is equivalent to the number of bacterial cells bound to the ME sensor surface.

This ME sensor has been successfully applied in both chemical and biological sensing[11,12]. Because the ME sensor is a wireless device, without a connection between the sensor and detection system, multiple sensors can be easily monitored. This study focuses on *Salmonella typhimurium* detection directly on the egg shell surface which eliminates the sample collection and preparation steps.

Experiment

Sensor fabrication and metal deposition

METGLAS 2826MB alloy ribbon, obtained from Honeywell International, was used to fabricate the magnetoelastic resonator platforms for the biosensors. The ribbon was diced into strip-shaped platforms 0.028 mm × 0.2 mm × 1 mm by a computer-controlled micro-dicing saw.The platforms were cleaned with acetone and ethanol and annealed at 220 °C in vacuum (10^{-3} Torr) for 2 hr. Annealing is necessary to remove residual stresses resulting from the dicing process. Two metal layers (Cr and Au) were then sputter-deposited onto all surfaces of the platforms. The layer of Cr serves as an adhesive interface between the platform and the gold layer. The Au layer provides corrosion resistance and a compatible surface for bio-probe immobilization[13].

E2 phage immobilization

The filamentous E2 phage which has highly specific and selective properties towards *S. typhimrium*, was derived from a landscape f8/8 phage library[14] by Dr. Valery Petrenko of Auburn University. Dr. James M. Barbaree's lab in the Department of Biological Sciences at Auburn University prepared and provided the E2 phage for this research. The concentration of the E2 phage solution was 5×10^{11} vir/ml in 1×Tris-Bufferd Saline (TBS). The resonator platforms were immersed in the phage solution for 1 hr. The immobilization of the E2 phage on the platform surface is based on physical adsorption. After phage immobilization, the biosensors were washed with deionized water twice in order to remove unbound phage and salts from TBS. In order to compensate for the environmental effects and non-specific binding, control sensors were prepared in the same manner but without E2 phage immobilization.

Salmonella detection on food surface

S. typhimrium (ATCC 13311) culture with the concentration of 5×10^8 CFU/ml was provided by Dr.James M. Barbaree's lab. The culture solution was stored in a refrigerator at 4 °C and equilibrated to room temperature before testing. In this research, the original *S. typhimrium* solution was diluted to lower concentrations (5×10^7 CFU/ml and 5×10^6 CFU/ml) with deionized water. The eggs were purchased from the local Kroger supermarket, and the shell egg surfaces were cleaned with deionized water.

The *S. typhimrium* solutions of three different concentrations (from 5×10^6 to 5×10^8 CFU/ml) with the same volume (2.5×10^{-4} ml) were pipetted onto the shell surface. After the solutions dried for 20 min, both measurement and control sensors were placed on the contaminated shell surface. The egg shells with the sensors were then stored in 100% humidity for 20 min. Seven measurement sensors and three control sensors were used for each solution concentration. The sensor resonant frequencies were measured with an HP 8751A network analyzer before and after binding with *S. typhimrium*.

The network analyzer was connected to a solenoid coil wound around a glass tube. The network analyzer is responsible for the generation of a time-varying external magnetic field and signal detection. The attachment of *S. typhimrium* cells can be detected by the sensor's resonant frequency shift. After measuring the resonant frequencies, all sensors and egg shells were exposed to osmium tetroxide (OsO_4) vapor for 45 min.

Results and Discussion

The improvement of signal performance.

For the sensor's vibration signal, the resonant frequency change determines the quantity of the bacteria on sensor surface. In order to detect the accurate resonant frequency, the amplitude and Q value of the signal peak are important. There are several different ways to improve the amplitude of the signal peak. The relationship of magnetostriction and magnetic field is shown in Figure 1. At different H points: H1, H2 and H3, with the same magnetic field change, the magnetostriction changes are different. This means that the static magnetic field can be used to adjust the H point and find the largest change of magnetostriction. Table 1 shows the signal amplitude for different static magnetic fields. The static magnetic field is created using a permanentmagnetic bar. The distance between the magnetic bar and solenoid coil influences the strength of the magnetic field. Table 1 displays the signal amplitude increase when the distance between the magnetic bar and solenoid coil is changed from 4.5 inches to 3.5 inches. The signal decreased when the static magnetic field was stronger. When the static magnetic density equaled 9.35 Gauss, the signal amplitude is at itshighest point.

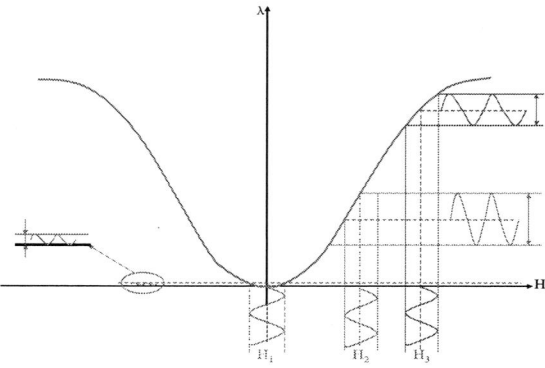

Figure 1: Response of a ME material under different DC bias

Table 1 the signal amplitude with different static magnetic density

distance(inch)	static magnetic density(Gauss)	amplitude(mU)
4.5	5.02	16.115
4	6.8	25.834
3.5	9.35	35.194
3	14.1	15.936
2.5	22.45	3.0138
0	0	4.7905

The solenoid coil creates the harmonic magnetic field to excite the sensor's vibration. Meanwhile, the solenoid coil also picks up the signal of the sensor's vibration and transmits it back to the network analyzer. Therefore, the size of the solenoid coil affects the signal performance. The harmonic magnetic field is stronger with a larger size coil. However, the noise is also increases with larger coils. Table 2 shows the amplitude of the signal versus the length of the coil. For detection of a

1 mm long sensor, the table shows the signal amplitude for lengths of coil (3mm, 2mm and 1mm). Compared with the signal amplitude of five samples, the coil with the length of 1 mm had the largest signal amplitude. This result shows that the signal is largest when the coil is approximately the same size as the sensor.

Table 2: The relationship between the amplitude(mU) and the length of coil:

# of sample	3mm	2mm	1mm
1	1.6347	1.6762	3.7704
2	1.1194	1.1333	3.2045
3	1.8139	1.8781	2.8558
4	1.4622	1.9984	3.1028
5	1.1261	1.7763	2.2663

The effect of the egg's surface curvature on the contact area between the sensors and *Salmonella*.

Figure 2 shows a representative image of the egg's outline which has non-uniform curvatures. Hence, the surface curvature needs to be evaluated to estimate the contact between a 1 mm-long sensor and *Salmonella* on egg shell. Assuming the egg is axisymmetric around the axis OP, only the top half of the egg outline (i.e.,OEP) is considered. The curvature of $\overset{\frown}{OEP}$ decreases from point O, which has the largest curvature, to point E and increases from point E to point P. In the environment,*Salmonella* cells move and assemble with cilia, creating a dynamic, changing layer on the egg shell surface. Based on the curvature calculations with the estimated *Salmonella* layer thickness of larger than 2 μm, 100% contact between the sensor and *Salmonella* can be achieved on $\overset{\frown}{MEP}$, where point M is located between points O and E. The left of Figure 2 shows 100% contact at point M, while less than 100% contact at point O. As the area with larger curvatures ($\overset{\frown}{MON}$) is less than 5% of the whole surface, 1 mm-long sensors are able to detect the contamination on over 95% of the egg shell surface.

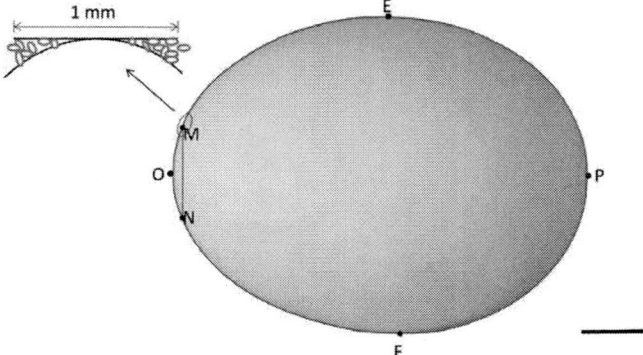

Figure 2: Curvature of the egg surface. (Scale bar: 1cm)

The effect of reaction time on the S. typhimrium cells distribution at the egg surface.

The egg shell surface contaminated with 5×10^8 CFU/ml S. *typhimrium* for 0 min, 30 min, 4 hr, shown in Fig. 3(a)–(c), Fig, 3(d) is the clean egg shell surface without *Salmonella* exposure. Compared with the fresh surface (Fig. 3d), the *Salmonella* cells lay on the egg shell in high concentration and covered the entire surface at 0 min. (Fig. 3a). After 30 min, the shell egg surface had a similar number of bacterial cells which nearly covered the whole drop area. Hence, if the ME biosensor test time is 20 min after exposure, the detection result is similar to what would occur with the total number of bacterial cells in one drop. After 4 hrs, the surface coverage of the *Salmonella* cells is about 10% due to the cells entering the egg. (Fig. 3c) These SEM images show that the number of S. *typhimrium* cells on the egg shell surface decreased with the increasing time. The *Salmonella* cells, attracted to the nutrients inside the egg, penetrate the shell. Holes in the egg shell provide the passage for the cell penetration.

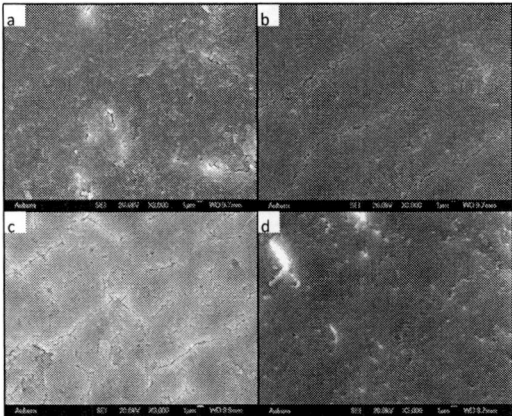

Figure 3. The contamination of egg shell surface for different exposure times. (a) 0 min, (b) 30 min, (c) 4 hr, and (d) fresh shell surface.

The frequency changes with S. typhimrium detection

The resonant frequency changes for ME measurement and control sensors are shown in Fig. 3. In the figure, the black curve is the frequency of ME sensors before binding with *S. typhimrium* cells; the red curve is the frequency after *S. typhimrium* attachment. The resonant frequency shifts as a result of the bacterial cells bound to the ME sensor surface.The control sensor has a small resonant frequency change due to measurement errors and non-specific binding. However, the frequency change of the control sensor (820 Hz) is negligible compared with measurement sensor (5900Hz).

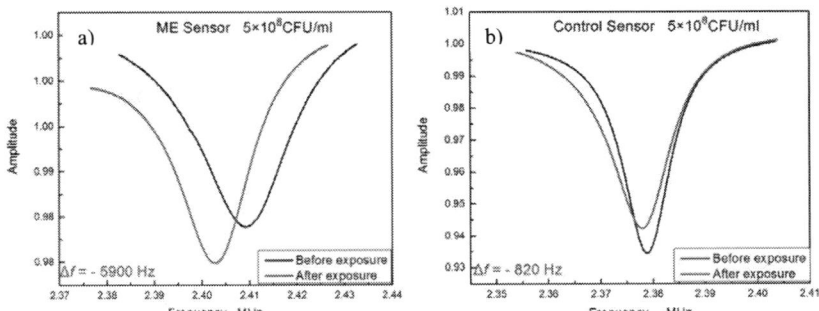

Figure 4.The frequency curves before and after exposure to spiked egg shells for ME measurement and control sensors: a) measurement sensor; b) control sensor

The bacterial cells on the surface are not typically well-distributed, especially when the bacterial concentration is low. This is because the cells can migrate and move freely on the humid surface, and they prefer to aggregate in areas with nutrition and water. Based on this situation, the test result for each sensor highly depends on where the ME biosensors are placed on the egg shell surface. In other words, a single sensor will likely be unable to provide contamination information of the whole egg shell when the bacterial cells are non-uniformly distributed on the shell surface. Therefore, multiple biosensors can help to solve this problem. Seven measurement sensors and three control sensors were used for each of the *S. typhimrium* concentrations (5×10^6 to 5×10^8 CFU/ml). The sensors were placed on different areas of egg surface. Figure 5 plots the resonant frequency changes for both measurement (blue diamonds) and control biosensors (red rectangles). The resonant frequency changes for the measurement biosensors highly depend on the concentration of the bacterial solution. The average frequency changes decreased with decreasing *Salmonella* concentration as anticipated. Furthermore, the resonant frequency changes for the control sensors are much smaller and are not related to the bacterial solution concentration.

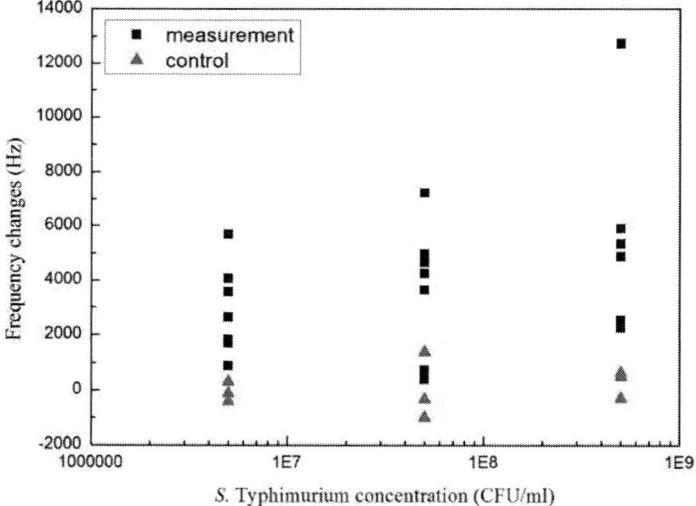

Figure 5. The resonant frequency changes for measurement and control sensors exposed to different concentrations of *Salmonella* on egg shell surfaces.

Conclusions

This research demonstrates that remote, direct detection of *Salmonella typhimurium* on egg shell surfaces is possible with the wireless ME biosensor. The phage-based ME biosensor can detect *S. typhimrium* on egg shells based on the resonant frequency changes after the attachment of bacterial cells on the sensor surface. The amplitude of the signal can be improved by increasing the static magnetic field and adjusting the solenoid coil shape. The resonant frequency changes were proportional to the quantity of *Salmonella*on the shell eggs. The control sensors allow the compensation for non-specific binding and environmental conditions. The humidity is important for binding reaction between the phage and *Salmonella*. Meanwhile, the 3-dimensional structure of the protein will change without adequate water. The pVIII proteinwill lose the capability of specific binding to bacteria cells in low humidity environments. The detection methodology is rapid, requiring only 20 min for the reaction between the biosensor and *S. typhimrium* cells to be detected. The movement of bacterial cells on egg shell surfaces was demonstrated at a high humidity. The non-uniform distribution of S. *typhimrium* cells on egg shell surfaces is believed to be due to the nutrition competition among cells. For lower bacterial concentrations, greater non-uniformity of*Salmonella* cell distribution on the egg shell surface was observed.

References

1. Schoeni, J. L., Glass, K. A., MoDermott, J. L. and Wong, A. C.L., "Growth an penetration of *Salmonella enteritidis, Salmonella heidelberg* and *Salmonella typhimurim* in eggs," Food Microbiology 24, 385-396 (1995).
2. Messens, W., Duboccage, L., Grijspeerdt, K., Heyndrickx, M. and Herman, L., "Growth of *Samonella serovars* in hens' egg albumen as affected by storage prior to inoculation," Food Microbiology 21(1), 25-32 (2004).
3. Lequin, R.M., "Enzyme immunoassay (EIA)/enzyme-linked immunosorbent assay (ELISA)," Clinical Chemistry 51 (12), 2415–2418 (2005).
4. Kim, Y., Flynn, T.R., Donoff, R.B., Wong, D.W., Todd, R., "Then gene: the polymerase chain reaction and its clinical application," Journal of Oral and Maxillofacial Surgery 60 (7), 808–815 (2002).
5. Yang, H., Irudayaraj, J., "Rapid detection of foodborne microorganisms on food surface using Fourier transform Raman spectroscopy," Journal of Molecular Structure 646 (1–3), 35–43 (2003).
6. Wilson, S., A., Jourdain, R., P.J., Zhang, Q., Dorey, R., A., Bowen, C., R., etc., "New materials for micro-scale sensors and actuators an engineering review," Materials Science and Engineering R 56, 1-129(2007).
7. Li, Q., Li, Y., Chen, H., Horikawa, S., Shen, W., Simonian, A. and Chin, B., A., "Direct detection of *Salmonellatyphimurium* on fresh produce using phage-based magnetoelastic biosensors," Biosensor and Bioelectronic, 26(4), 1313-1319(2010).
8. Lakshmanan, R., S., Guntupalli, R., Hu, J., Kim, D., Petrenko, V., A., Barbaree, J., M. and Chin, B., A., "Phage immobilized magnetoelastic sensor for the detection of *Salmonellatyphimurium*," Journal of Microbiological Methods 71, 55-60(2007).

9. Liang, C., Morshed, S. and Prorok, B. C. "Correction for longitudinal mode vibration in thinslender beams," Applied Physics Letters, 90, 221912 (2007).
10. Stoyanov, P. G.; Grimes, C. A. "A remote query magnetostrictive viscosity sensor," Sensors and Actuators, 80, 8-14 (2000).
11. Johnson, M., L., Wan, J., Huang, S., Cheng, Z., Petreko, V., A., Kim, D., Chen, I., Barbaree, J., M., Hong, J., W. and Chin, B., A., "A wireless biosensor using microfacbricated phage-interface magnetioelastic particles," Sensor and Actuators A 144, 38-47(2008).
12. Chai, Y., Li, S., Horikawa, S., Park, M., Vodyanoy, V., Chin, B., A., "Rapid and sensitive detection of *Salmonella* Typhimurium on eggshells by using wireless biosensors," Journal of food protection, 75(4), 631-636(2012).
13. Huang, S., Hu, J., Wan, J., Johnson, M. L., Shu, H. and Chin, B., A., "The effect of annealing and gold deposition on the performance of magnetoelastic biosensors," Materials Science and Engineering C, 28, 380-386 (2007).
14. Petrenko, V.A. and Sorokulova, I.B., "Detection of biological threats. A challenge for directed molecular evolution," Journal of Microbiological Methods 58 (2), 147–168 (2004).

Bio-inspired Autonomous Sentinel System for Screening Invasive Pathogens

S. Li[a], H. C. Wikle III[a], Y. Chai[a], M.-K. Park[a], S. Horikawa[a], and B. A. Chin[a]

[a] Materials Research & Education Center, Auburn University, Auburn, AL 36849, USA

> This paper presents the results of a preliminary investigation to develop a bio-inspired autonomous pathogen screening system that mimics the function of natural immune systems, such as white blood cells. This screening system is comprised of multiple magnetoelastic (ME) biosentinels. The sentinels are composed of a wireless magnetoelastic resonator immobilized with a bio-probe layer that captures specific target pathogens. The biosentinels are actuated, monitored, and controlled wirelessly by external magnetic fields. The autonomous sentinel system is envisioned to have the capability of seeking out invasive pathogens in liquid environments, detecting and capturing them. In the paper, the proof-in-principal of the concept of autonomous sentinels is demonstrated by comparing the response of the sentinels in dynamic and static *Salmonella* analytes.

Introduction

Infectious pathogens pose an imminent threat to public health and result in great economic loss, thus much attention and effort have been devoted to develop a system that allows the real-time tracking of small amounts of infectious pathogens. For centuries, humankind has attempted to mimic the designs of Nature to develop new engineering materials and systems. The human immune system is one of Nature's amazing creations that inspires us to develop this magnetoelastic biosentinel system for pathogen screening. As part of the immune system, white blood cells are the main defensive mechanism against pathogenic invaders. There is a variety of white blood cell types (neutrophil, eosinophil, lymphocytes, etc.) that target different pathogens. When health threatening pathogens invade our body, the white blood cells will seek out, capture and kill the target pathogens. This capability serves as the model for a bio-inspired system of autonomous sentinels for the detection and capture of invasive pathogens described in this paper (Figure 1).

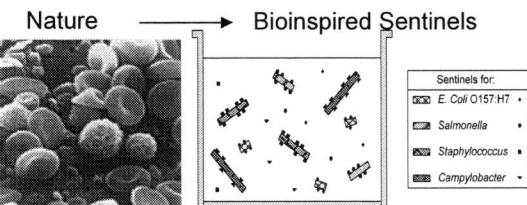

Figure 1. Bio-inspired sentinels will target different types of bacteria (*E. coli*, *Salmonella* Typhimurium, etc.) mimicking white blood cells that target different invasive pathogens (1)

This screening system is comprised of multiple magnetoelastic (ME) biosentinels. A ME biosentinel is constructed of a freestanding ME resonator (transducer platform) coated with a biorecognition layer that specifically binds the target pathogen. The freestanding sentinels require no on-board power for motion or to signal detection of a target pathogen. Due to their magnetic nature, the ME biosentinels can be wirelessly driven to move through an analyte using a magnetic field. Upon contact with the target pathogen, the sentinel binds with the target cell, increasing the sentinel's mass, resulting in an instantaneous decrease of the sentinel's resonant frequency. By wirelessly monitoring the resonant frequency, a sentinel can signal the presence and concentration of target pathogens. Therefore, in this screening system, magnetoelastic (ME) biosentinels can seek out, capture, and signal intelligence of target pathogens in liquid environments.

The objective of this paper is to demonstrate proof-in-principal of the concept of autonomous ME biosentinels as a pathogen screening system. In this study, E2 phage (specifically binding with *Salmonella*) coated ME sentinels were fabricated. In order to mimic the movement of the ME sentinels in the analytes, the ME sentinels were place in flowing analytes containing different *Salmonella* concentrations. The response of the ME sentinels in flowing anaytes was compared with the response of the ME sentinels placed in static anaytes. The results show that the ME sentinels in flowing analytes gave higher response due to the capture of more pathogen cells. This demonstrates proof-in-principal of autonomous ME sentinels.

Theory of ME Sentinel

A ME biosentinel is comprised of a free-standing sensor platform made of a strip-shaped magnetoelastic resonator and a biomolecular recognition element (antibody, phage, enzymes, etc.) immobilized on the sensor platform surface. Figure 2 shows the detection principle of the ME biosentinels. Under an external magnetic field, magnetostrictive materials undergo a change in shape. If the magnetic field is varied at the proper frequency and aligned along the length direction of the resonator, the structure can achieve resonance. The ME sentinel is actuated into resonance by the application of an alternating magnetic field, meanwhile the magnetic flux emitted from the sentinels can be measured by a pick-up coil. Upon contact with the specific target bacteria, the bio-molecular recognition element on the sentinel's surface captures the target bacterial cells, causing the overall sensor mass to increase. This mass increase results in a decrease in the resonant frequency. The resonant frequency change is remotely and wirelessly measured using a pick-up coil. Interrogated through magnetic signals, no physical contact or onboard power is required by a sentinel. For a strip-shaped ME sentinel, the sensitivity (2) can be expressed as follows (3, 4):

$$S_m = \frac{\Delta f}{\Delta m} = -\frac{1}{L^2 wt} \sqrt{\frac{E}{\rho^3 (1-\upsilon)}} \qquad [1]$$

where L, w, and t are the length, width and the thickness of the sentinel. E, ρ, and v are the Young's modulus, density, and Poisson ratio of the material respectively. The negative sign in the equation shows that with a small mass loading Δm on the sensor, the resonant frequency (Δf) decreases. The equation indicates that the sensitivity of the sentinel exponentially increases with the decrease of the sentinel dimensions. ME sentinels in the scale of 100 μm have the capability of detection one single bacteria cell.

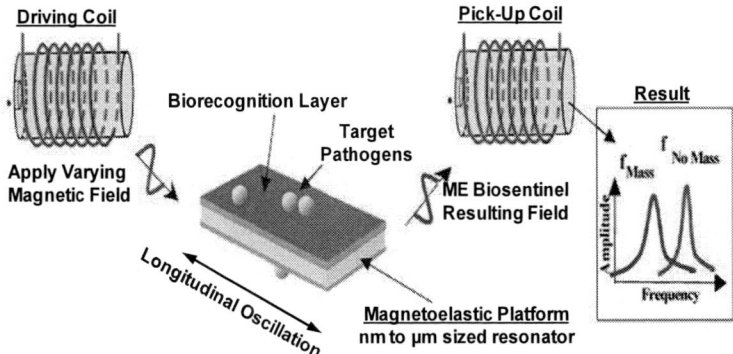

Figure 2. Detection principle of a magnetoelastic (ME) biosentinel. A driving coil generates a modulated magnetic field that drives the ME resonator into vibrational resonance. Binding of the target bacteria to the bio-molecular recognition layer immobilized onto the resonator increases the mass of the sensor resulting in a decrease in resonant frequency.

The unique advantages of ME sentinels enable them to mimic the function of nature immune system. The sentinels are wireless devices, enabling in-situ remote detection of multiple target pathogens (5-8). For any bio-detection device, the detection can only happen when the target pathogen makes contact with the device. At low bacterial concentration analytes, the first contact may take very long time. For ME sentinels, the odds of detection are improved either by increasing the number of ME sentinels deployed or by exposing the sentinels to a dynamic environment. For detection in liquid media, dynamic exposure can be achieved by flowing the media past the immobilized sentinels or by moving the sentinels around within the media. Since ME sentinels are remote/wireless detectable and manipulative, a simpler approach to achieve greater exposure is to harness the magnetic field that is currently used only for pathogen detection to provide the forces for sentinel motion. A nonuniform magnetic field can be used to propel and steer the sentinels.

Due to its wireless nature, a large number of sentinels can be deployed simultaneously, which significantly enhances the probability of binding with a target pathogen. More importantly, the binding of target pathogens on only one out of many sentinels can be easily detected. By taking advantage of these properties and capabilities of phage-coated ME resonators, a system of sentinels that mimics the functions of white blood cells can be built and deployed for enhanced medical diagnostics, food safety, or water quality applications (Figure 3).

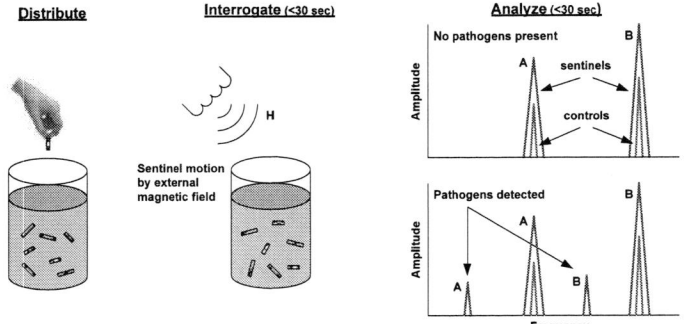

Figure 3. A large number of sentinels targeting different pathogens may be mixed together and interrogated simultaneously for pathogen detection. Different pathogens may be detected simultaneously since the sentinels for different pathogens are designed to operate in different frequency ranges.

Experimental

ME Sentinel Fabrication

Magnetoelastic strip-shaped resonator platforms of in the size of 28 x 200 x 1000 μm were fabricated from METGLAS® 2826MB alloy, obtained from Honeywell International. The as-received alloy was diced into rectangular strips of the desired size. The sensor platforms were ultrasonically cleaned, first in acetone, and then in methanol, followed by annealing at 220 °C for 2 hours in a vacuum (10^3 Torr) to remove residual stresses. After annealing, two layers of thin films (Cr & Au) were sputtered onto both sides of the sensor platforms. The Cr layer was deposited first in order to improve the adhesion between the ME platform and the Au layer. The Au layer provides a corrosion protection layer for the sensor platform, and a biological compatible surface for the phage immobilization.

To form functional sentinels, E2 phage was immobilized on the ME sentinel surface by direct physical adsorption. E2 phage, which was provided by Dr. V.A. Petrenko and propagated by Dr. Barbaree, is an Auburn University patented, genetically engineered, filamentous phage that binds specifically with *Salmonella* bacteria (9, 10). Each ME sentinel platform was placed in a vial containing 300 μL of phage E2 suspension (5×10^{11} vir/mL in 1 x Tris-Buffered Saline (TBS)). These vials were then rotated on a rotor (running at 8 rpm) for 1 hour. After the immobilization process, the sentinels were washed three times with sterile distilled water in order to remove salt and any unbound or loosely bound phage.

Control sentinels are used to calibrate for environmental changes and nonspecific binding that may influence the measured resonant frequencies. The control sentinel is identical to the measurement sentinel except it lacks the E2 phage coating. Both control and E2-coated sentinels were immersed into 1 mg/ml BSA solution for one hour, followed by a distilled water rinse. After BSA treatment, measurement and control sentinels are ready to use.

Detection of *Salmonella* in Flowing and Static Analytes

The resonant frequency of the sentinels was measured using an HP network analyzer 8751A with S-parameter test set. The ME sentinels were placed in a tube, outside of which a pick up coil was wound and connected to the network analyzer. The analyzer scanned, measured and recorded the resonant frequency spectrum of the ME sentinel.

The *S.* Typhimurium (ATCC13311) culture used in this investigation was provided by Dr. James M. Barbaree's lab in the Department of Biological Sciences at Auburn University, Auburn, AL. The cultures obtained from Dr. Barbaree's lab were provided in the form of a suspension at a concentration of 5×10^8 cfu/ml. The suspensions were serially diluted in water to prepare bacterial suspensions with the concentrations ranging from 5×10^1 to 5×10^7 cfu/ml. All test solutions were prepared on the same day as the biosensor testing. The test solutions were stored at 4°C (during transfer and storage) and equilibrated to room temperature in a water bath prior to the experiments. The sentinels were exposed to *Salmonella* analytes with different concentrations (5×10^1 to 5×10^8 cfu/ml). The resonant frequency of sentinels was measured using the method described above. The sentinels were then exposed to *Salmonella* solutions with different concentrations dynamically or statically. After exposure, the sentinels were gently washed by distilled water once, dried in air, and the resonant frequency was measured again. For each concentration, the responses of sentinels in dynamic and static analytes were compared.

In the dynamic system, a peristaltic pump is used to drive the analyte from the container through a horizontal glass tube. For each concentration, three ME measurement and three control sentinels were placed in the glass tube. A flow rate of 50 µl/min was used in order to ensure laminar flow and the sentinel not moving with the flow. The internal diameter of the tube is 600 µm, therefore the flow velocity is 44.2 mm/min. As the analyte flows past the sentinels, it is equivalent to moving the sentinel through the analyte with the surface parallel to the moving direction at a velocity of 44.2 mm/min. For each test analyte, the analyte was flowed for 20 minutes, thus a total 1 mL of solution flowed past the sentinels. The resonant frequency of each sentinel was measured before and after the sentinel was placed in the flowing analyte, and the shift in resonant frequency was recorded.

In order to detect the *Salmonella* in static analyte, 1 mL of the test analyte was placed in a 1.5 mL PCR tube and the measurement sentinel was placed at the bottom of the PCR tube for 20 minutes. Next, the resonant frequency of the sentinel was measured and compared with the resonant frequency before the sentinel was placed in the test analyte.

A JEOL-7000F scanning electron microscope (SEM) was used to confirm the binding of S. typhimurium on the phage-coated ME biosensor. After the detection, the ME sensors were exposed to osmium tetroxide (OsO4) vapour for 45 minutes. The sensors were then mounted onto aluminum stubs and examined using the SEM.

Results and Discussions

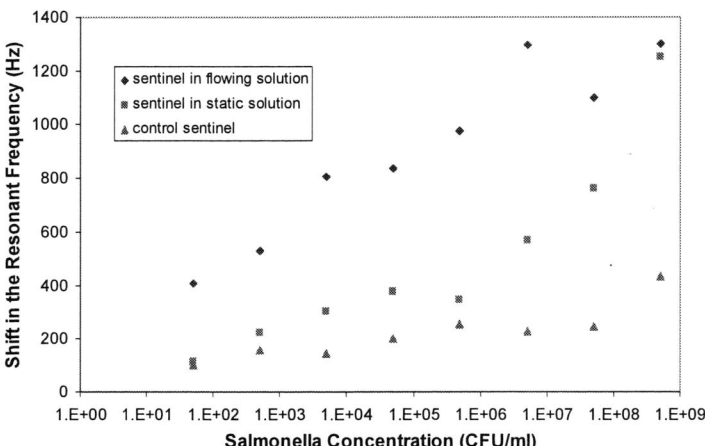

Figure 4. The response of ME sentinels as the sentinels detect *Salmonella* in dynamic (flowing) and static analytes. In dynamic detection, the *Salmonella* solution flowed past the sentinel for 20 minutes at a flow rate of 50 μL/min. For static detection, the sentinels were exposed to 1 mL of static solution for 20 minutes. The results of control sentinels were obtained from dynamic detection.

Figure 4 shows the resonant frequency shift of ME sentinels after the sentinels were placed in dynamic and static analytes. The response of control sentinels is also shown in Figure 4. As shown in Figure 4, the response of sentinels increases as the concentration of the *Salmonella* solution increases for both dynamic and static experiments, while the control sentinels show negligible shifts. However, at lower concentrations ($<5\times10^6$ CFU/mL), the average resonant frequency shift in dynamic sentinels is at least 50% higher than the response of static sentinels. When the sentinels measure *Salmonella* in static analytes, the chance of the *Salmonella* cells making contact with the sentinel depends on the diffusion of the bacterial cell in the solution. When the concentration of cells is low, the cells take a longer time to encounter a sentinel surface. Obtaining more binding (and higher response) requires longer time. On the other hand, as the analyte flows past the sentinel surface or the sentinel moves in the analyte, more cells come into contact with the sentinel surface, reducing the diffusion depletion zone around the sentinel and enhancing the binding of cells to the sentinel. In high concentration analytes, the diffusion depletion zone at the sentinel surface is much smaller, and more cells are available for the binding reaction. Therefore, the improvement of the detection rate due to dynamic flow is much less significant. These results demonstrate the proof-in-principle of the concept of autonomous sentinels actively seeking out and detecting invasive pathogens.

Conclusions

This paper demonstrates proof-in-principle of the concept of autonomous sentinels for the detection of invasive pathogens. Magnetoelastic strip-shaped resonators coated with a bio-molecular recognition layer can be moved through a liquid using a non-uniform magnetic field and then measured remotely and wirelessly to detect the binding and capture of specific pathogenic bacteria. In this paper, detection of *Salmonella* in flowing analytes by ME sentinels is used to simulate the sentinel moving through the solution and signaling the presence of the target pathogens. In the dynamic test system, the sentinels show higher response than the sentinels detecting *Salmonella* in static analytes, demonstrating the capability of ME sentinels to seek out and detect target pathogens.

References

1. B. Wetzel and H. Schaefer, *National Cancer Institute* (1982).
2. V. A. Petrenko and G. P. Smith, *Protein Engineering*, **13**, 589 (2000).
3. C. Liang, S. Morshed and B. C. Prorok, *Applied Physics Letters*, **90** (2007).
4. L. D. Landau and E. M. Lifshitz, Pergamon Press. (1986).
5. S. Huang, H. Yang, R. S. Lakshmanan, M. L. Johnson, J. Wan, I. H. Chen, H. C. Wikle, V. A. Petrenko, J. M. Barbaree and B. A. Chin, *Biosensors & Bioelectronics*, **24**, 1730 (2009).
6. M. L. Johnson, J. H. Wan, S. C. Huang, Z. Y. Cheng, V. A. Petrenko, D. J. Kim, I. H. Chen, J. M. Barbaree, J. W. Hong and B. A. Chin, *Sensors and Actuators a-Physical*, **144**, 38 (2008).
7. R. S. Lakshmanan, R. Guntupalli, J. Hu, V. A. Petrenko, J. M. Barbaree and B. A. Chin, *Sensors and Actuators B-Chemical*, **126**, 544 (2007).
8. W. Shen, R. S. Lakshmanan, L. C. Mathison, V. A. Petrenko and B. A. Chin, *Sensors and Actuators B-Chemical*, **137**, 501 (2009).
9. J. R. Brigati and V. A. Petrenko, *Analytical and Bioanalytical Chemistry*, **382**, 1346 (2005).
10. V. A. Petrenko, *Expert Opinion on Drug Delivery*, **5**, 825 (2008).

60

New Multimode Sensors based on Nanostrucured Materials for Simultaneous Screening of Biological Fluids for Specific Breast Cancer and Hepatitis B Biomarkers

R.I. Stefan-van Staden[a] and M. Enachescu[b]

[a] Laboratory of Electrochemistry and PATLAB Bucharest, National Institute of Research for Electrochemistry and Condensed Matter, Str. Splaiul Independentei Nr. 202, Bucharest, 060021, Romania, e-mail: iustinavanstaden@yahoo.com
[b] Center for Surface Science and Nanotechnology, University "Politehnica" of Bucharest Splaiul Independentei, Nr. 313, Bucharest-060042, Romania, e-mail: marius.enachescu@upb.ro

New materials (e.g., dextrins) were proposed for the design of multimode sensors for screening of biological fluids for specific biomarkers for breast cancer antigen and hepatitis B. The selection of the materials was based on its nanostructure (for stochastic mode), and its catalytic activity (for amperometric mode). The qualitative analysis was performed using the stochastic mode, and it was followed by the quantitative analysis of the antigens using stochastic mode and amperometric mode. The biomarkers were determined directly from the biological fluids (e.g., whole blood). The response characteristics of the new multimode sensors for CA15-3, and hepatitis B biomarkers (HBV antigen and D-pipecolic acid) as well as their recovery in the biological fluid of the patients are shown.

Introduction

Hepatitis B is a potentially life-threatening liver infection caused by the hepatitis B virus. Therefore, it is a major global health problem, being in the same time the most serious type of viral hepatitis. Accordingly with WHO, worldwide, an estimated two billion people have been infected with hepatitis B virus (HBV), and more than 350 million have chronic (long-term) liver infections, part of them acquired from the transfusions for which the limits of quantification was not low enough to detect the presence of the HBV antigen in the blood sample of the donor. Researchers found a link between the hepatitis B virus and formation of cancer centers in the body, centers which will later on grow uncontrollably producing the cancer cells. Viruses, typically initiate cancer development by suppressing the host's immune system, causing inflammation over a long period of time, or by altering host genes. 41% of the breast cancer patients are carrying HBV (1). Therefore it is a high needs for tools that can simultaneous detect specific biomarker(s) for HBV as well as specific biomarker(s) for breast cancer.

Multimode sensors represent a new concept in sensors' technology. They can function for the same analyte, as stochastic sensors (to perform the qualitative and quantitative analysis), amperometric sensors (to perform using, e.g., differential pulse voltammetry (DPV), the quantitative analysis), or potentiometric sensors (to perform the quantitative analysis based on chronopotentiometric analysis). Therefore, one need nanostructured materials presenting channels or pores with diameters on micro or nano levels (for stochastic measurements), that can be in the same time good electrocatalysts (to be used for DPV measurements) and if it is possible to have a favorable architecture for selectively binding of the analyte(s) of interest.

The stochastic mode can be used reliable for the qualitative analysis of analytes from complex mixtures, the principle being based on channel/pore conductivity (2-6). After identification of the analytes of interest from the mixture using the stochastic mode, they can be quantified using the same diagrams used for qualitative analysis and switching to any of DPV or chronopotentiometric modes.

Maltodextrins with dextrose equivalence 4-7, nanostructured materials with native channels due to its helix structure were selected for the design of a diamond paste based multimode sensor which can reliable be used for simultaneous qualitative and quantitative analysis of breast cancer biomarker (CA15-3), hepatitis B virus biomarkers (HBVAg and D-pipecolic acid (D-PA)). The stochastic mode was combined with DPV mode (due to its electrocatalytic activity) to give the multimode behavior.

Experimental

Materials and Reagents

Natural monocrystallin diamond powder having particle size of $1\mu m$ (99.9%), maltodextrin (dextrose equivalent 4-7), CA15-3, HBV antigen (HBVAg), and D-pipecolic acid were purchased from Aldrich (Milwaukee, USA); paraffin oil was purchased from Fluka (Buchs, Switzerland). 0.1mol/L Phosphate buffer solution (pH=7.4), and NaN_3 were purchased from Merck.

Deionized water was obtained using a Millipore Direct-Q 3 System (Molsheim, France). The solutions of biomarkers were prepared using deionized water and 0.1mol/L phosphate buffered (pH=7.4) containing 0.1% NaN_3 in a ratio water:buffer solution=1:1 (v/v). Serial dilution technique was used for preparation of solutions of different concentrations – needed for the evaluation of the multimode sensor.

Design of Multimode Sensor

The multimode sensor was prepared by mixing 200 mg of natural monocrystalline diamond powder with 20 μL paraffin oil to form a diamond paste. 100μL from the solution of maltodextrin with dextrose equivalent 4-7 (10^{-3}mol/L) was added to the diamond paste. The modified paste was placed into a plastic tube. The diameter of the sensor was about 300μm. Electric contact was obtained by inserting an Ag/AgCl wire into the modified diamond paste. The surface of the sensor was wetted with deionized water and polished with alumina paper (polishing strips 30144-001, Orion) before using. When not in use, the sensor was stored in a dry state at room temperature.

Apparatus

A PGSTAT 12 and software Nova were used for all chronoamperometric and DPV measurements. A Pt electrode and an Ag/AgCl electrode served as the counter and reference electrodes in the cell. A Cyberscan PCD 6500 pH/mV-meter from Eutech

Instruments was employed for all pH measurements. Atomic force microscopy (AFM) investigations were carried out with an Agilent 5500 SPM system, described by PicoSPM controlled by a MAC Mode module and interfaced with a PicoScan controller from Agilent Technologies, Tempe, AZ, USA (formally Molecular Imaging). A multipurpose large scanner and Point Probe Plus Silicon SPM Sensor cantilevers (PPP-FM cantilevers), n^+- silicon material with no coating, of about 227 μm length, 1.8 N m^{-1} spring constants, with the tips oscillated near their resonant frequencies in air, of about 64 kHz were used for all measurements. All AFM measurements were done by scanning the surface at a rate of 0.8-1.2 lines per second and were done at room temperature in the tapping mode.

Proposed Modes

Two modes are proposed to be used with the maltodextrin based sensor: stochastic mode (chronoamperometry) and differential pulse voltammetry (DPV) mode. For the stochastic mode, a chronoamperometric technique was used for the measurements of t_{on} and t_{off} at 125mV. The electrodes were dipped into a cell containing solutions of biomarker of different concentrations. No stirring was applied during measurements. Equations of calibration $1/t_{on}$ = f(Conc.) for each of the biomarkers (CA15-3, HBVAg, and D-PA) are determined using statistics. The unknown concentrations of biomarkers in whole blood samples were determined from the calibration equations. For the DPV mode - the applied pulse potential was 25mV/sec. vs. Ag/AgCl. The multimode sensor, together with the reference and counter electrodes were dipped into the cell containing the buffered analyte. The peak heights were measured at the potentials given in Table I (vs Ag/AgCl) and were plotted against the concentrations of each biomarker; no stirring was applied during measurements. The unknown concentrations of the biomarker in whole blood samples were obtained from the calibration graphs. Nova software allowed the switching between the modes.

TABLE I. Response characteristics of the multimode sensor based on maltodextrin, in DPV mode.

Biomarker	E (mV)	Linear concentration range	Sensitivity	Equation of calibration
CA15-3	780	$5 - 500$ [a]	5pA U/mL	H= 0.3 + 5.0xC;[a,d] R= 0.9382
HBVAg	526	$10^{-12}- 5x10^{-9}$ [b]	2.9nA g/mL	H= 3.1 + 2.9xC;[b,e] R= 0.9976
D-PA	350	$10^{-11}-10^{-8}$ [c]	155.4pA mol/L	H= 0.2 + 0.16xC;[c,e] R= 0.9915

[a]U/mL; [b]g/mL; [c]mol/L; [d]<H>=pA; [e]<H>=nA

AFM measurements

High-resolution imaging using the atomic force microscope (AFM) has been applied to observe the differences of surface structures between unmodified diamond and maltodextrin modified diamond paste. The topography of the surfaces was investigated by AFM in tapping mode in a range of scan lengths from 50 μm to 5 μm. The images were processed by first order flattening in order to remove the background slope and the contrast and brightness were adjusted. The original images for the samples, the 3D topographical images and section analysis over the unmodified or modified diamond pastes were performed using the PicoView SPM Software, version 1.6.2, Molecular Imaging.

Results and Discussion

Characterization of the active surface of the multimode sensor

Figure 1 shows the AFM topographical images of plane and modified diamond paste based sensor. The surface roughness analysis was performed for both modified and unmodified diamond pastes. Because the roughness can be defined using S_q(root mean square height) and S_a(arithmetical mean height) value, we determine these parameters using the PicoImage tool and accordingly with ISO25178, these values were automatically calculated. The values obtained for the unmodified diamond paste were S_q=13.6nm, and S_a=10.8nm, while for the diamond paste modified with maltodextrin were S_q=364nm, and S_a=311nm. The significant difference between roughness of plane and modified pastes as well as of images obtained proved that the maltodextrin was immobilized into the diamond paste. The presence of the round channel in the modified paste justified the stochastic behavior of the sensor when a constant value of potential (125mVvs Ag/AgCl) was applied and the current value vs time was recorded.

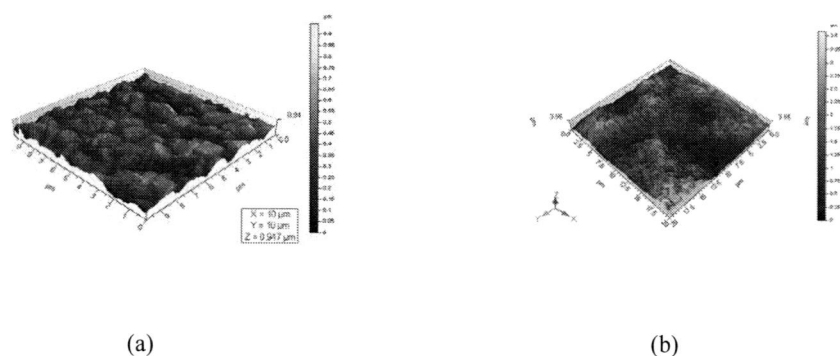

(a)   (b)

Figure 1. AFM topographical images for the surfaces of (a) unmodified diamond paste and (b) maltodextrin based diamond paste.

Response characteristics of the multimode sensor

For the stochastic mode, the signature of each biomarker (CA15-3, HBVAg, D-PA) were determined using the value of t_{off} shown in Table 2. These values can be used for the qualitative analysis of the biomarkers in any of the biological matrices whole blood, serum, saliva. For the quantitative analysis in stochastic mode, the measurements were performed for 6 months in order to determine the stability as well as the sensitivity of the sensor. The equations of calibration in the stochastic mode were determined using the values of t_{on} - defined as the binding time (2), measured in sec. obtained for different concentrations of the biomarkers (Table II).

TABLE II. Response characteristics of the multimode sensor based on maltodextrin, in stochastic mode.

Biomarker	t_{off} (s)	Linear concentration range	Sensitivity	Equation of calibration[d]
CA15-3	3.3	$5-500$ [a]	44mU s/mL	$1/t_{on}= 20.11 + 0.044 \times C;$[a] R= 0.9608
HBVAg	1.9	$10^{-10}\text{-}10^{-8}$ [b]	3.9×10^5 g s/mL	$1/t_{on}= 0.05 + 3.9 \times 10^5 \times C;$[b] R= 0.9997
D-PA	4.9	$10^{-12}\text{-}10^{-10}$ [c]	3.0×10^8 mol s/L	$1/t_{on}= 0.02 + 3.0 \times 10^8 \times C;$[c] R= 0.9488

[a]U/mL; [b]g/mL; [c]mol/L; [d]$<t_{on}>$ = s.

The very low limits of quantification: 5U/mL for CA15-3, 10^{-10}g/mL for HBVAg, and 10^{-12}mol/L for D-PA will make possible early detection of the biomarkers in the biological fluids. It is a significant difference between the signatures of these biomarkers which will make possible their simultaneous assay without any need of their separation from the biological fluid. The relative standard deviation (RSD%) per biomarker was less than 0.09% for the response/sensitivity of the sensor, values being recorded for the 6 months.

For the DPV mode (Table I), each biomarker (CA15-3, HBVAg, D-PA) can be assayed at a different potential making possible the accurate quantification. For the quantitative analysis in DPV mode, the measurements were performed for 6 months in order to determine the stability as well as the sensitivity of the sensor. The lower levels of quantification using this mode: 5U/mL for CA15-3, 5×10^{-12}g/mL for HBVAg, and 10pmol/L for D-PA made possible their quantification at an early stage DPV competing in this regard with the stochastic mode, and in the same time favoring the detection of the biomarkers without any concentration or dilution of the sample of biological fluid. A RSD (%) less than 0.10% was recorded when the multimode sensor was used for the assay of the three biomarkers over a period of 6 months.

Analytical Applications

Qualitative analysis of the biomarkers in whole blood samples

Qualitative analyses of CA15-3, HBVAg, and D-PA were performed using the stochastic mode. Their identification in the whole blood was done accordingly with their signatures given in Table II.

TABLE III. Qualitative analysis of biomarkers using the multimode sensor, based on maltodextrin in stochastic mode, and standard method.

Biomarker	Sample 1		Sample 2		Sample 3	
	Stochastic mode	Standard method	Stochastic mode	Standard method	Stochastic mode	Standard method
CA15-3	+	+	+	+	+	+
HBVAg	+	+	+	+	+	+
D-PA	+	+	+	+	+	+

Three control samples were obtained from the specialized clinical laboratory which performed the analysis of these biomarkers using ELISA (for CA15-3, and HBVAg) and capillary zone electrophoresis (CZE) (for enantioanalysis of D-PA). The results are presented in Table III and are shown a good correlation between the proposed method and standard methods used in specialized clinical laboratories.

Quantitative analysis of the biomarkers in whole blood samples

Quantitative analyses of CA15-3, HBVAg, and D-PA were performed using both stochastic and DPV modes (Table IV). The results obtained using the proposed methods were in the agreement with those obtained using the standard methods used in clinical laboratories of analysis. The slightly higher values obtained when stochastic or DPV modes are used are due to the fact that ELISA and CZE are using serum for analysis and any sampling process performed on the biological fluid may decrease the concentration of the analyte.

TABLE IV. Quantitative analysis of biomarkers using the multimode sensor based on maltodextrin, in stochastic and DPV modes and standard method.

Biomarker	Sample 1			Sample 2			Sample 3		
	Mode1	Mode2	Std.	Mode1	Mode2	Std.	Mode1	Mode2	Std.
CA15-3[a]	10.2	10.8	9.6	43.7	42.9	42.0	87.2	87.9	86.0
HBVAg[b]	420	419	417	327	327	324	529	528	524
D-PA[c]	325	326	320	219	216	211	625	627	620

[a]U/mL; [b]ng/mL; [c]pmol/L; Mode1 = Stochastic mode; Mode2=DPV mode; Std.=standard method.

The results obtained using the proposed multimode sensor based on maltodextrin for simultaneous analysis of biomarkers CA15-3, HBVAg, and D-PA in whole blood samples proved that the sensor is suitable for clinical analysis, and that it can be reliable used for early diagnosis of breast cancer and hepatitis B.

Conclusions

A multimode sensor based on maltodextrin was proposed for qualitative and quantitative simultaneous assay of CA15-3 (biomarker for breast cancer), HBVAg and D-PA(biomarkers for hepatitis B virus). The modes used were stochastic mode (for qualitative and quantitative analysis), and DPV mode (for quantitative analysis). The low limits of quantification obtained in both modes proved that the multimode sensor can be used for early diagnosis of breast cancer and hepatitis B. The recovery tests performed using controlled whole blood samples obtained from specialized clinical laboratories proved that it is a good agreement between the proposed method and well known standard methods (e.g., ELISA and CZE), and that the multimode sensor can be used for the simultaneous assay of CA15-3, HBVAg, and D-PA for whole blood samples. The method is simple, fast, low cost and reliable, making it a good choice for clinicians.

Acknowledgments

This work was supported by PNII Program Capacity, 2012-2014, Contract nr. 3ERC-like/2012, program Ideas 2011-2014, Contract nr. 123/05.10.2011 and by UEFISCDI, ERANET contract nr. 4_001/7.04.2011, project acronym NANODIATER under the frame of EuroNanoMed.

References

1. M.S. Dai, R.Y. Shyu, J.J. Lu and T.Y. Chao, *Liver Int.*, **24**, 540 (2004).
2. H. Bayley and P.S. Cremer, *Nature*, **413**, 226 (2001).
3. R.I. Stefan-van Staden, J.F. van Staden and S.C. Balasoiu, *Anal.Lett.*, **44**, 2280 (2011).
4. L.Q. Gu, O. Braha, S. Conlan, S. Cheley and H. Bayley, *Nature*, **398**, 686 (1999).
5. A. Liu, Q. Zhao and X. Guan, *Anal.Chim.Acta*, **675**, 106 (2010).
6. R.I. Stefan-van Staden, I. Stefanescu, J.F. van Staden and M. Enachescu, *Analyst*, 00, 000 (2012).

68

Performance of Optimized Phage-based Magnetoelastic Biosensors for *Salmonella* Typhimurium Detection on Tomatoes

M-K. Park[a], S. Li[a], K. Weerakoon[a], S. Horikawa[a], Y. Chai[a], N. Hiremath[a], V.A. Petrenko[b], and B.A. Chin[a]

[a] Materials Research and Education Center, Auburn University, Auburn, Alabama 36849, USA
[b] Department of Pathobiology, Auburn University, Auburn, Alabama 36849, USA

> An optimized, phage-based magnetoelastic (ME) biosensor method
> was performed to detect *Salmonella* Typhimurium contaminates on
> the surface of tomatoes and to determine the detection limit. After
> serial inoculations of *S.* Typhimurium on the surface of the
> tomatoes, the *S.* Typhimurium on the surfaces was analyzed using
> an optimized ME biosensor method. Detection limit and
> sensitivity of the ME biosensor for detection of *S.* Typhimurium on
> tomatoes were determined. The resonant frequency shifts of the
> ME measurement sensors increased linearly with an increase in the
> concentration of *S.* Typhimurium. Contrastingly, the control
> sensors showed relatively constant and low magnitude frequency
> shifts. Detection limit and sensitivity of the ME biosensor method
> was determined to be (1.78 ± 0.17) log CFU/cm^2 and 922.7 Hz.

Introduction

Public interest in eating healthy has led to an increase in the consumption of fresh fruits and vegetables (1-3). Associated with this increase in consumption of fruits and vegetables is the potential for an increase in foodborne illness outbreaks. Among foodborne pathogens, *Salmonella* is one of the greatest concerns associated with fresh fruits and vegetables due to the high potential for growth prior to consumption and the low infectious dose necessary to cause illness (4).

In an effort to address the recent outbreaks of *Salmonella* associated with fresh produce, phage-based magnetoelastic (ME) biosensors have been developed as a novel and practical device for *Salmonella* detection (5-8). The phage-based ME biosensor is composed of a ME resonator platform that is immobilized with genetically engineered filamentous phages (E2 phage) for the specific recognition of *S.* Typhimurium (9). The ME resonator platform is made from a magnetostrictive material that elongates or contracts when subjected to an applied external magnetic field. Under an applied alternating magnetic field, the ME resonator platform undergoes a corresponding oscillating shape change. The resonant frequency is remotely measured using a standard pick-up coil in the ME biosensor system (Figure. 1). The characteristic resonant frequency of the ME resonator depends on the material properties, shape, and dimensions of the sensor platform as well as any mass bound to the surface of the resonator (9). ME biosensors have recently been shown to detect *Salmonella* on the surfaces of fresh produce without requiring the usual preparation steps such as stomaching or purification of the sample (10-11). This is in contrast to the time consuming protocols of traditional methods such as culturing, expensive and highly technical polymerase chain reaction (PCR), or enzyme-linked immunosorbent assay (ELISA) methods.

The direct detection method using ME biosensors has been found to be dependent on: 1) how many bacteria the sensors come into contact with when placed on the tomato surface; 2) the number of *S.* Typhimurium distributed on the surface of fresh produce and 3) the binding of *S.* Typhimurium with E2 phage during incubation (9-10). Hence, the incubation temperature and time were optimized in order to enhance the binding of bacteria with E2 phage on the surface of fresh produce (12-13). The purpose of this study is to directly detect *S.* Typhimurium on the surface of tomatoes using an optimized ME biosensor method. Detection limit and sensitivity of the ME biosensor method for *S.* Typhimurium detection on the tomato were determined for later comparison with those of other fresh produce.

Experimental Procedures

Preparation of magnetoelastic strip-shaped resonator platforms and E2 phage immobilization

Magnetoelastic (ME) strip-shaped resonator platforms, 0.028 mm × 0.2 mm × 1 mm in size, were fabricated following the procedures described in a previous study (12-13). Filamentous E2 phage (clone E2 - displaying foreign peptide VTPP-TQHQ developed by Dr. Valery Petrenko was used in this work) was propagated by Dr. Barbaree's Laboratory [1.0 × 10^{12} vir/ml in Tris-Buffered Saline (TBS, pH 7.4)]. The E2 phage suspension was mixed with equal amounts of TBS buffer to order to adjust the concentration of E2 phage. Each ME resonator platform was placed in an Eppendorf tube containing 300 μl of diluted E2 phage suspension and incubated on a Barnstead LabQuake tube rotator (Fisher Scientific, Inc., PA, USA) at 8 rpm for 1 h. After washing with the TBS buffer, any unbound area of the ME resonator platform was blocked with 300 μl of 1% bovine serum albumin (BSA, Sigma-Aldrich Co., St. Louis, MO, USA) at 22 °C for 1 h. Finally, the ME resonator platform was washed three times with sterilized water and allowed to air dry. The ME resonator platform was then ready for use as a measurement sensor. Control sensors were prepared by the same procedure, with the exception that the E2 phage immobilization step was eliminated, in order to compensate for the effects of environmental changes. The resonant frequency of each sensor were measured with an HP 8751A network analyzer (HP/Agilent 8751A, Agilent Technologies Inc., CA, USA) combined with an *S*-parameter test set prior to placement of both measurement and control sensors on the tomato surface.

Direct inoculation of *S.* Typhimurium on the tomato surface

Figure 1 shows the process of direct detection of *S.* Typhimurium on the tomato surface. Red, ripe tomatoes (*Lycopersicon esculentum var. esculentum*) were purchased from a local grocery store (Auburn, AL, USA). Only tomatoes free of visible defects such as bruises, cuts, or abrasions were used. All tomatoes were hand washed with sterilized distilled water (DW) five times to remove all possible contaminants. One cm diameter circles were drawn on the surface with fine-tip permanent marker using a sterile paper template for spot-inoculation.

The culture of *S.* Typhimurium (ATCC 13311) was provided by Dr. Barbaree's Laboratory in the Department of Biological Sciences at Auburn University, Auburn, AL. The culture of *S.* Typhimurium culture (5×10^8 CFU/ml) was serially diluted with 9 ml of sterilized distilled water prior to inoculation. Twenty μl of *S.* Typhimurium was inoculated on each of three tomato surfaces within a 1 cm diameter circle. The inoculated tomato were placed under a laminar flow hood at 22 °C for 90 min to allow the *S.* Typhimurium to attach to the surface. After attachment, the inoculated samples were placed in individual glass containers equilibrated with approximately 100% RH.

Placement of base-lined sensors on the tomato surface inoculated with *S.* Typhimurium

The previously base-lined sensors (measurement and control sensors with measured resonant frequencies) were then placed on the inoculated surfaces of tomato. Each tomato was placed into a humidity controlled container equilibrated to approximately 100% RH and incubated at 30 °C for 30 min. Finally, the resonant frequencies of the sensors were measured 10 times with the network analyzer which provided an average resonant frequency. This procedure was repeated at least 3 times per sensor (yielding 3 data points from 30 measurements). The resonant frequencies of the sensors before and after placement on the tomato were compared.

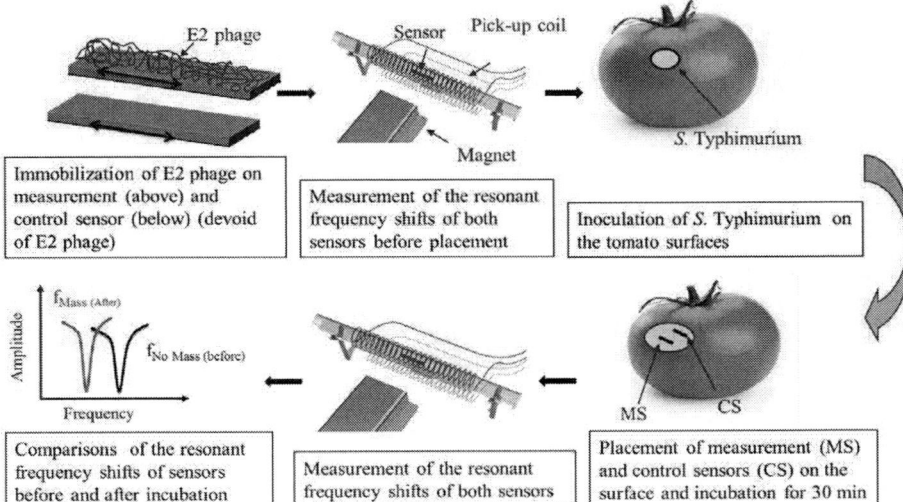

Figure 1. Scheme used for the direct detection of *S.* Typhimurium on a tomato surface using an optimized phage-based ME biosensor method (13).

Microscopic analysis

For the confirmation of *S.* Typhimurium binding to the sensors, the sensors were exposed to OsO4 vapor for at least 45 min. Finally, the tomato and spinach surfaces and sensors were mounted on aluminum stubs to be examined using a JEOL-7000F scanning electron microscope (SEM).

Statistical analysis

Comparisons between various treatments and/or groups were conducted using the one-way analysis of variance (ANOVA) and the Tukey-Kramer multiple comparisons test. Differences were considered to be statistically significant if the "P value" was less than 0.05. Statistical analysis was performed using GraphPad and InStat v.5 (GraphPad, San Diego, CA).

Results and Discussion

Fresh produce has a diverse surface morphology (1-2, 14). The previous SEM microscopic images (9-10) showed that a tomato is composed of ridges and depressions. In addition, the attachment and distribution of bacteria on the tomato surface was not uniform, especially when the concentration of *S.* Typhimurium was low. Although the distribution of *S.* Typhimurium over the tomato surface became more uniform with increasing bacteria concentration, these non-uniform distributions of bacteria affected the performance of the ME biosensors on the contaminated surface. Hence, three sets of sensors were placed on the tomato surface in different locations in order to minimize experimental errors stemming from the non-uniform distribution of bacteria on the produce surface and where the sensor contacts the produce.

Figure 2 shows the response of multiple measurement and control sensors after the exposure of sensors to *S.* Typhimurium on a tomato surface. The solid blue circles represent the resonant frequency shifts of the measurement sensors and the open black circles represent the resonant frequency shifts of the control sensors. As the concentration of *S.* Typhimurium inoculant increased, the resonant frequency shift also increased proportionally. The measurement sensors showed much larger resonant frequency shifts for the larger concentrations of *S.* Typhimurium inoculated. However, there were no significant differences in the resonant frequency shifts of the control sensors despite the increase in *S.* Typhimurium concentration ($P > 0.05$).

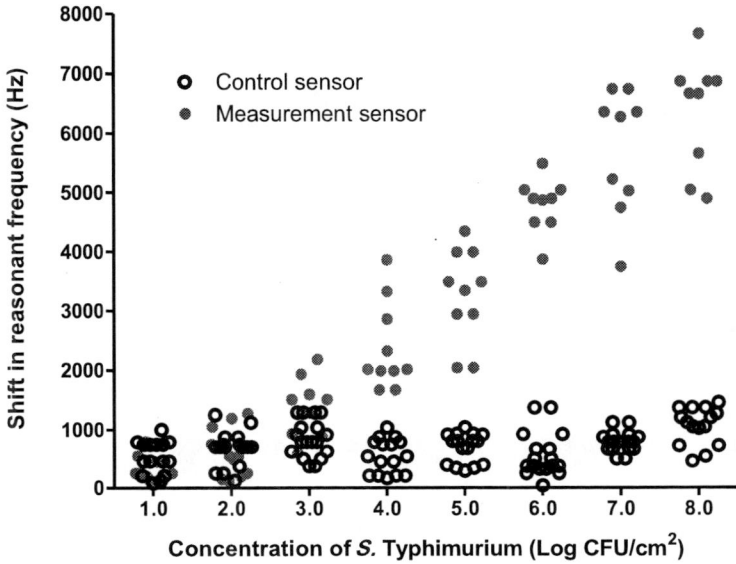

Figure 2. Resonant frequency shifts of ME measurement and control sensors as a function of concentration of *Salmonella* Typhimurium inoculated on tomato surfaces.

The SEM images of control (Figure 3A) and measurement (Figure 3B-3D) sensors confirmed that the resonant frequency shifts of the measurement sensors were caused by the specific binding of *S.* Typhimurium on the measurement sensors. Contrastingly, the SEM images of the control sensor showed no or very little binding with *S.* Typhimurium except for a few nonspecific bindings from salt debris from the phage buffer solution and/or unknown debris from tomato surface (Figure 3A). SEM images of the measurement sensors (Figure 3B-3D) also confirmed that the increasing resonant frequency shifts were due to an increased number of *S.* Typhimurium cells bound to the measurement sensors.

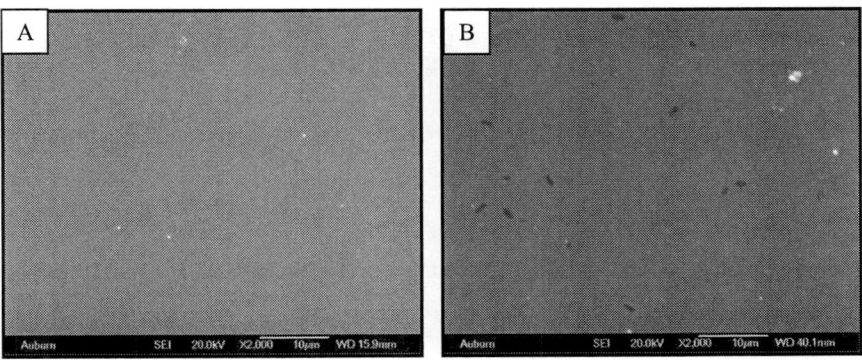

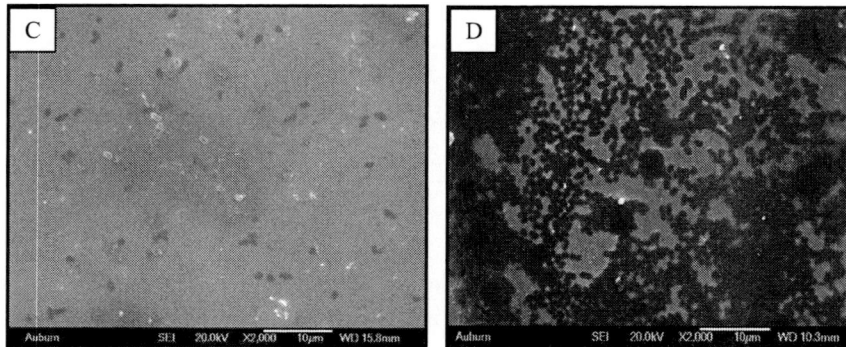

Figure 3. SEM images of *S*. Typhimurium on the tomato surfaces with inoculation of 3 log CFU/cm^2 on (A) control sensor and (B) measurement sensor, (C) 5 log CFU/cm^2 on measurement sensor, and (D) 8 log CFU/cm^2 on measurement sensor.

Figure 4 shows the measured resonant frequency shift versus inoculation concentration curve for the measurements taken on the surface of tomato. The tomato surfaces showed the linear relationship with a correlation coefficient (R^2) of 0.90 and a generated slope of 922.7 ± 35.12. The detection limit is newly defined as the point of intersection of two linear lines for the measurement and control sensors in this study instead of using the student's t-test. Therefore, the detection limit for the performance of optimized ME biosensors on a tomato surface was determined to be (1.78 ± 0.17) log CFU/cm^2. The sensitivity, defined as the slope of the standard curve (15), was calculated to be 922.7 Hz/decade.

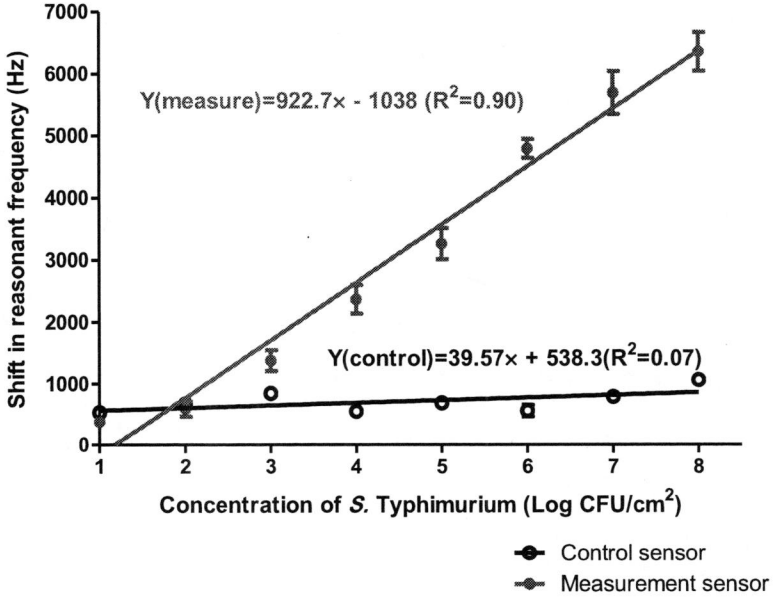

Figure 4. Standard curve for a 10-fold dilution series of *S*. Typhimurium on tomato surface after the performance of optimized ME biosensor method.

Conclusions

This study demonstrates the direct detection of *S*. Typhimurium on the surface of tomato using an optimized ME biosensor method. ME biosensor measurements increased linearly on a semi logarithmetic plot as the concentration of *S*. Typhimurium on the surface of tomato increased. Contrastingly, control sensors exhibited only very small frequency shifts due to no or negligible binding of *S*. Typhimurium on the control sensors. Detection limit and sensitivity were determined to be (1.78 ± 0.17) log CFU/cm^2 and 922.7 Hz using the linear equation obtained from the standard curve. The detection limit will be compared later with that of various fresh produce with different surface morphologies. The comparison will be useful in evaluating how the different surface morphologies of fresh produce affect the performance of the ME biosensor method.

Acknowledgments

This research was supported by the Auburn University Detection and Food Safety (AUDFS) Center with funding from the USDA-CSREES grant (204327 130851 2000).

References

1. L. R. Beuchat, L.R., *Microbes and Infection.*, **4**, 413 (2002).
2. J. C. Heaton, K. Jones, *J. Appl. Microbiol.*, **104**, 613 (2008).
3. M-K. Park, J-H. Oh, *J. Food Sci.*, **77**(2), M127 (2012)
4. J. W. Buck, R. R. Walcott, L. R. Beuchat, http://www.apsnet.org/publications/apsnetfeatures/Documents/2003/Microbiologic alSafety.pdf (2003).
5. R. Guntupalli, R. S. Lakshmanan, J. Hu, T. S. Huang, J. M. Barbaree, V. Vodyanoy, B. A. Chin, *J. Microbiol. Meth.*, **70**(1), 112 (2007).
6. S. Huang, H. Yang, R. S. Lakshmanan, M. L. Johnson, J. Wan, I. H. Chen, H. C.
7. Wikle, V. A. Petrenko, J. M. Barbaree, B. A. Chin, *Biosens. Bioelectron.*, **24** (6), 1730 (2009).
8. R. S. Lakshmanan, R. Guntupalli, J. Hu, V. A. Petrenko, J. M. Barbaree, B. A. Chin, *Sens. Actuators B*, **126** (2), 544 (2007).
9. M-K. Park, S. Li, B. A. Chin, *Food Bioprocess Tech.*, DOI:10.1007/s11947-011-0708-2 (2011).
10. Li, Y. Li, H. Chen, S. Horikawa, W. Shen, A. Simonian, B. A. Chin, *Biosens. Bioelectron.*, **26** (4), 1313 (2010).
11. Y. Chai, S. Li, S. Horikawa, M-K. Park, V. Vodyanoy, B. A. Chin, *J. Food Protect.*, **75**(4), 631 (2011).
12. M-K. Park, J-H. OH, B. A. Chin, *Sensor Actuators B*, **160**, 1427 (2011).
13. M-K. Park, H. C. Wikle, Y. Chai, S. Horikawa, W. Shen, B. A. Chin, *Food Control*, **26**, 539 (2012).
14. R. Kroupitski, R. Pinto, M. T. Brandl, E. Belausov, S. Sela, *J. Appl. Microbiol.* **106**(6), 1876 (2009).
15. W. Shen, R. S. Lakshmanan, L. C. Mathison, V. A. Petrenko, B. A. Chin, *Sens. Actuators B*, **137**, 501 (2009).

Development of Electrochemical Cantilever Sensors for DNA Applications

Xueling Quan[a], Arto Heiskanen[a], Yi Sun[a], Aleks Labuda[b], Anders Wolff[a], Jorge Dulanto[b], Peter Grutter[b], Maria Tenje[c], Anja Boisen[a]

[a] Department of Micro- and Nanotechnology Technical University of Denmark, Lyngby, Denmark

[b] Department of Physics, McGill University, Montreal, Quebec, Canada

[c] Department of Measurement Technology and Industrial Electrical Engineering, Lund University, Lund, Sweden

> In this work, we develop a generic DNA based sensing platform used for characterizing surface functionalization and detecting DNA hybridization. Silicon nitride cantilever sensors are fabricated with an integrated three-electrode system and integrated in a microfluidic chip. Cantilevers with gold electrodes are functionalized with thiol-modified single stranded DNA (ssDNA) probes to detect target DNA. During functionalization and hybridization, information related to nanomechanical changes on the surface are obtained by optical measurements of changes in cantilever deflection. Simultaneously, the process is monitored electrochemically. The results clearly indicate that the electrochemical cantilever sensor is very sensitive for detecting DNA hybridization at the cantilever surface.

Introduction

Microcantilevers have been widely investigated for label free high-sensitivity detection of molecular interactions. Monitoring of DNA hybridization is one of the many applications in which cantilever sensing has been used (1-2). Electrochemical methods for detection of DNA hybridization have also gained an intensive focus in research (3). Combining electrochemical techniques with cantilever devices is an efficient way to improve sensor performance. Combined measurements, where the cantilever also serves as a working electrode have been investigated over the past decades (4). In the past five years the electrochemical cantilever sensors have been significantly developed by integrating a three-electrode system together with cantilevers onto a single chip (5). Here we develop a generic DNA based sensing platform used for characterizing surface functionalization and detecting DNA hybridization. We show the fabrication of electrochemical cantilever chip (ECC) by conventional microfabrication techniques and demonstrate the capabilities of combined measurements. The electrochemical cantilever sensor described in this study shows a great potential for (bio)chemical applications.

Experimental

The ECC chip presented here is an optimized design based on our previous study (5). The chip was designed to simultaneously function as a cantilever array and to provide a three-electrode system. The requirements that: i) the reference electrode should have a higher and stable resistance and ii)the counter electrode should have a low resistance in comparison to the working electrodes were fulfilled by regulating the dimensions of the designed electrodes. As shown in Figure 1, the chip is 12 mm × 11.3 mm and it is fabricated on a 375 ìm thick Si substrate. Four cantilevers are placed in the middle of the chip, each having the dimensions 100 ìm (width) x 550 nm (thickness) x 300 ìm or 400 ìm (length). Each cantilever is capable of functioning as an independent working electrode (WE). The reference electrode (100 ìm × 200 ìm) is placed on the right side of cantilevers and the counter electrode (2700 ìm × 400 ìm) on the left side. The spring constant of the 300 ìm and 400 ìm long cantilevers is 6 mN/m and 5 mN/m, respectively.

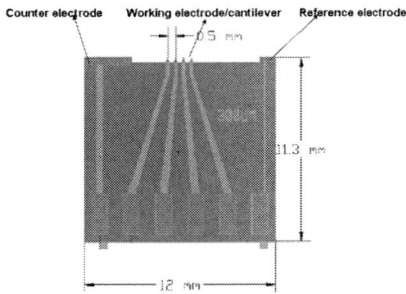

Figure 1. A schematic illustration of the electrochemical cantilever sensor.

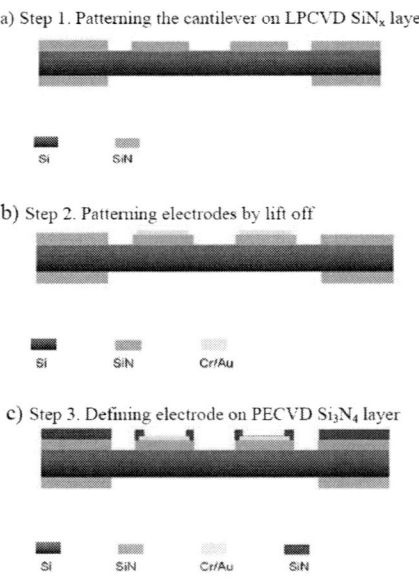

Figure 2. Schematic diagram of the key steps in the fabrication process of the electrochemical cantilever chip.

The main steps of the ECC fabrication process are shown in Figure 2. First, a 400 nm thick SiN_x layer, to be used as cantilever device layer as well as protective masking layer when defining the chip body, is deposited on Si wafers using Tempress LPCVD nitride furnace. Double-sided UV photolithography and subsequent reactive ion etch (RIE) (Figure 2 a) is used to define cantilevers (front side of wafer) and windows for releasing the cantilevers (front and back side). Secondly, a 2.2ìm thick photoresist layer is spun onto the front side SiN_x layer to pattern the electrodes. The wafers with the patterned photo-resist are mounted in a Wordentec QCL800 e-beam evaporation system for metal layer deposition. 3 nm Cr and 30 nm Au is deposited on the wafers, followed by a lift-off process using ultrasonic bath with acetone, ethanol and de-ionized water, successively, to reveal the electrode patterns (Figure 2 b). Then, another 100 nm thick Si_3N_4 layer is deposited on top of the metal by PECVD and the electrode window is etched by RIE (Figure 2 c). Finally, a 3h bath in potassium hydroxide (KOH, 80 °C) is used to etch through the Si wafer from both front and back side to release the cantilevers.

The chips are cleaned before use to remove any contamination. The cleaning method should provide a standard cleanliness level of electrode surfaces, which is very important for improving the reproducibility of results. Cantilever chips are immersed in a piranha solution for 6 min, and then rinsed thoroughly in MilliQ-water and ethanol (6). Finally, the chips are blown dry with nitrogen gas and mounted in a microfluidic chamber.

As shown in Figure 3, the microfluidic chamber encapsulates the ECC biosensor and creates a three-dimensional flow path leading into the chip and through the channel past the cantilevers. The fluid enters from reference electrode side and exits from the counter electrode side. An optical window and electrical contacts are both integrated into the microfluidic chamber. An additional option is to use a standard reference electrode (e.g. Ag/AgCl reference electrode) in the microfluidic chamber.

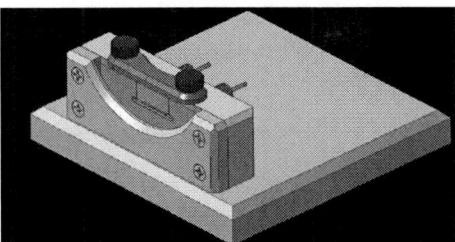

Figure 3. Schematic diagram of electrochemical cantilever fluidic setup.

An optical beam technique is used to monitor the microcantilevers' deflections. The cantilever deflection is proportional to the output voltage signal (V_{Def}) of a position sensing photodetector (PSD). A commercial potentiostat (CH Instruments 1030A electrochemical workstation) is used to conduct the electrochemical measurements. The fluidic operations are handled using a syringe pump in the reverse mode.

The cantilevers are repeatedly characterized by cyclic voltammetry between -0.3 V and 0.5 V (versus Ag/AgCl) at a scan rate of 20 mV/s using 1 mM hexacyanoferrate (II) and hexacyanoferrate (III) ($[Fe(CN)_6]^{3-/4-}$) (Fluka) in 10mM Tris-HCl and 200 mM NaCl electrolyte. The characterization is done directly after the cleaning step, after functionalization with 25-mer thiol-modified single stranded DNA (ssDNA) (2-h incubation after introduction of 2.5 ìM ssDNA sample) and after hybridization with target

DNA (2-h incubation after introduction of 650-nM DNA solution). Prior to each characterization step, a flow of $[Fe(CN)_6]^{3-/4-}$ solution is maintained for 5 min and then cyclic voltammogram (CV) and deflection are recorded under stop-flow conditions. Direct and complementary information related to nanomechanical changes on the cantilever surfaces are obtained by optical measurements of the cantilever deflection after functionalization and DNA hybridization.

.

Results and Discussion

Figure 4 shows CVs recorded at a cantilever electrode after immobilization (ssDNA) and after hybridization (dsDNA). The CV is changed significantly as ssDNA is bound to the electrode surface. The ssDNA-modified Au cantilever generates clearly lower anodic and cathodic peak currents than the bare Au cantilever. Due to the even stronger passivation of the electrode surface upon DNA hybridization, relatively small peak currents are observed in comparison with the ones obtained with the ssDNA-modified gold cantilever. As shown in Figure 5, the deflection voltage change is recorded simultaneously during the potential scanning (6 cycles from 0.5 V to -0.3 V) for each step in the process. Both the current and deflection voltage show the same variation trend. The deflection voltage curve obtained in buffer with 1mM $[Fe(CN)_6]^{3-/4-}$ indicates that the motion of the cantilever is the result of a potential controlled change combined with the charge accumulation corresponding to the faradic current. Due to the ssDNA coverage on the surface, the influence of the charge accumulation decreases significantly and the stress caused by the ssDNA immobilization may result in a movement that opposes the potential-driven deflection. After hybridization, the deflection decreases even more strongly. This is probably caused by combined stress due to different factors, such as potential-controlled, electrostatic, steric, and hydrophobic interactions.

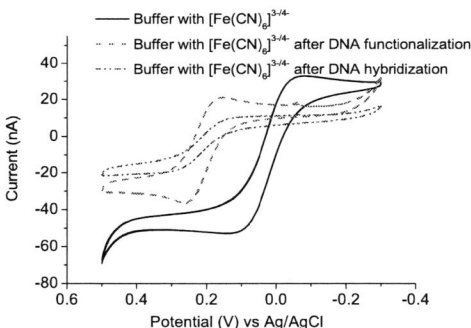

Figure 4. Cyclic voltammograms of 1mM of $[Fe(CN)_6]^{3-/4-}$ redox couple in 10 mM Tris-HCl and 200 mM NaCl electrolyte at a bare cantilever electrode as well as the same cantilever after ssDNA functionalization and hybridization (dsDNA).

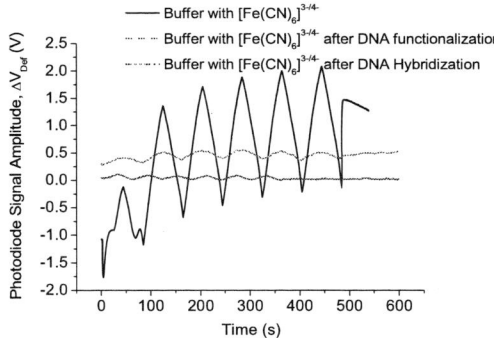

Figure 5. Deflection voltage changes versus time of a cantilever undergoing DNA functionalization and hybridization recorded during potential scanning between 0.5 V and -0.3 V in 10 mM Tris-HCl and 200 mM NaCl electrolyte containing 1 mM $Fe[(CN)_6]^{3-/4-}$ redox couple.

Conclusions

The aim of this work was to combine cantilever-based detection and electrochemical detection to investigate *in situ* the process of DNA hybridization. Monitoring of the entire process from ssDNA functionalization to DNA hybridization could be realized in a microfluidic chamber allowing simultaneous electrochemical and cantilever-based detection. This concept will be explored further as a sensitive and selective sensing platform where we have the possibility of speeding up the functionalisation and hybridization by controlling the cantilever potential.

Acknowledgments

This research was funded through the Villum Kann Rasmussen centre of Excellence 'NAMEC'. We thank the staff at DTU Danchip for their efforts in microfabrication, Hadi Izadi and Yoshihiko Nagai in McGill University for helping in DNA sample preparation and scientific discussions.

References

1. J.Fritz, M.K. Baller, H. P. Lang, H. Rothuizen, P. Vettiger, E. Meyer, H.-J. Guntherodt, Ch. Gerber, J. K. Gimzewski, *Science*, 288, 316 (2000).
2. K.M. Hansen, H.F. Ji, G.H. Wu. R. Datar, R. Cote, A. Majumdar, T. Thundat, *Anal. Chem.*, 73, 1567 (2001).
3. E. Hvastkovs and D. A. Buttry, *Anal.*, 135, 1817 (2010).
4. R.A. Fredlein, A. Damjanov, J. O. Bockris, *Surf. Sci.*, 25, 261 (1971).
5. L. M. Fischer, P. Christoffer, E. Karl, N. Nadine-Nicole, D. Søren, A. Boisen and T. Maria, *Sens. Actuators B.*,157(1), 321 (2011).
6. X.L. Quan, Y. Sun, A. Heiskanen, A. Wolff, P. Grutter, A. Boisen, *Proceedings of the 9th Nanomechanical Sensing workshop NMC*, 95 (2012).

Fabrication of Minimally-Invasive Patch Type Glucose Sensors

M. Yasuzawa, S. Sato, H. Nakanishi, K. Edagawa

Department of Chemical Science and Technology,
The University of Tokushima, Tokushima 770-8506, JAPAN

A fine tube type glucose sensor, which has a sensing region at the inside wall of fine tube tip, was prepared and applied for minimally invasive glucose monitoring. The enzyme glucose oxidase was immobilized on the platinum surface located in the tube by the combination of electrodeposition and electro-polymerization of *o*-phenylenediamine. Amperometric responses of the prepared electrodes to glucose were measured at a potential of 0.60 V (vs Ag/AgCl). The tube type electrode showed good response with good linear relationship up to the glucose concentration of 22.4 mM, which was actually higher than normal sensor without outer film for permeability restriction. This may be due to its unique sensor structure.

Introduction

The number of diabetic patients has rapidly increased around the world and self-monitoring of blood glucose (SMBG) using blood glucose degree meters are commonly used for their health management, since the recognition of glucose level and appropriate treatment is essential for the elimination of serious diabetes complications. Dangerous diabetes complications occur not only at hyperglycemia, but also at hypoglycemia. Many diabetic patients feel fear on the occurrence of hypoglycemia, since continuous hypoglycemic state of less than 30 mg/dl may present damage of brain and occasionally lose one's life. Moreover, hyperglycemia occurs frequently at sleeping time (after the midnight), when the SMBG actions are difficult and stressful. Therefore, it is helpful and stress reducing, if the blood glucose can be recognized without SMBG performance. Recently, several continuous glucose monitoring system (CGMS) are released in the market and it is getting popular as an effective equipment that lower physical and mental load of diabetes patient on glucose measurement. However, glucose sensor probes of longer than 1 cm in length, must be implanted in the tissue. Therefore, the development of lower invasive CGMS is longed to improve the quality of life of the diabetic patients.

In this study, a unique fine needle tube type glucose sensor, which has sensing region at the tip inside wall of a fine tube, was proposed. The schematic illustration of the tip of such glucose sensor is shown in Figure 1. Since the sensing region is at the inside of fine needle tube, it has space to immobilize certain amount of enzyme, while it require only the tip of sensor probe to be implanted in the tissue for glucose monitoring. CGMSs in the market are designed to measure glucose concentration in subcutaneous tissue, while the proposed sensor has the possibility to be inserted in dermis and measure its glucose concentration, which is significantly shallow from the skin surface. Combination of electrodeposition and electropolymerization techniques was employed for a specific precise enzyme immobilization. The potential of the prepared sensor as a minimally invasive glucose sensor probe was investigated.

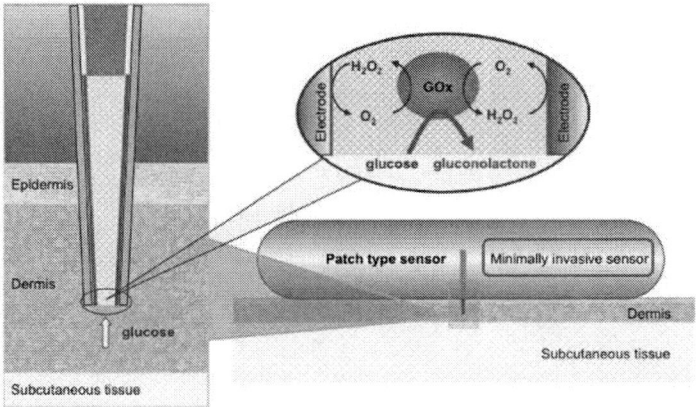

Figure 1. Schematic illustration of fine needle tube type glucose sensor implanted inside the skin.

Experimental

Sensor Preparation

Platinum-iridium (Pt 90%) tube with the outer diameter of 0.5 mm and inner diameter of 0.3 mm was employed as a starting material. One end of Pt-Ir tube was filled with solvent soluble polymer for protection and length of 10 mm from the end was polished in order to obtain tapered structure. The outside surface of tapered area was coated with perhydropolysilazane solution (Aquamica, AZ Electronic Materials) and amorphous fluoropolymer solution (CYTOP, Asahi Glass), and heated at 150°C for 90 min to obtain dielectric film. Cyclic voltammetry of the tube electrode was carried out in 1.0 mM potassium ferricyanide aqueous solution containing 0.1 M potassium chloride, to confirm the formation of dielectric film.

Polymer filled at one end was removed and glucose oxidase (GOx) (244 U/mg, purified from Aspergillus niger, Biozyme laboratories) was immobilized inside the tapered tube using the combination of electrodeposition and electropolymerization, according to the procedure of Matsumoto et al. (1) That is, a 10 mg/mL GOx 0.05 M phosphate buffer solution (pH 7.0) containing 0.8 mM Triton X-100 was poured inside the tapered Pt-Ir tube and a potential of 1.3 V (vs Ag/AgCl) was applied for 1 h to form GOx layer on the surface. GOx immobilized Pt-Ir tube was then putted into a degassed 50 mM o-phenylenediamine 0.1 M phosphate buffer solution (pH 7.4) and a potential of 0.7 V (vs Ag/AgCl) was applied for 15 min to induce the electropolymerization reaction.

Glucose Sensor Response Measurement

The amperometric responses of the prepared electrodes to glucose were examined at 25°C in a 0.1 M phosphate buffer solution (pH 7.4) containing 0.1 M NaCl, by measuring the electrooxidation current at a potential of 0.6 V (*vs.* Ag/AgCl) for hydrogen peroxide detection. Amperometric measurements were performed with a Potentiostat Model 3104 (Pinnacle Technology Inc.).

The calibration of the sensor was carried out by adding increasing amounts of glucose to the measuring solution. The current was measured at the plateau (steady-state response), and was related to the concentration of the analyte.

Results and Discussion

Figure 1 shows cyclic voltammograms of Pt-Ir tube electrode in potassium ferricyanide solution before and after Silica/CYTOP insulating film formation. Redox current of ferricyanide obtained on Pt-Ir tube electrode was not observed on the electrode after Silica/CYTOP film formation. This indicates that insulating film was successfully prepared on the surface of Pt-Ir tube electrode.

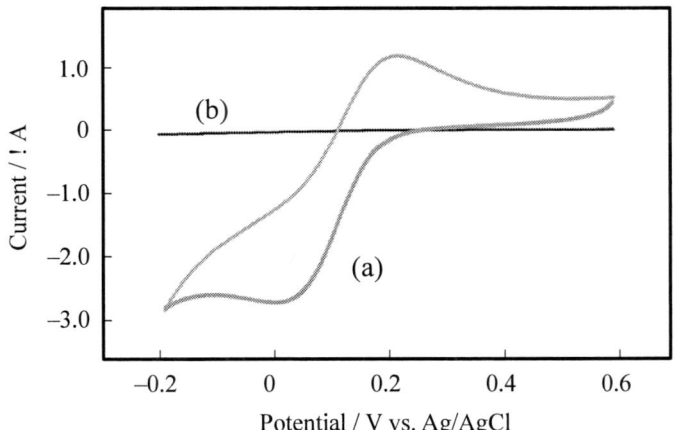

Figure 1.   Cyclic voltammograms of Pt-Ir tube electrode without (a), and with Silica/CYTOP insulating film (b). The sweep rate was 0.05 V s^{-1}.

Figure 2 shows typical calibration curves of prepared tube electrode. The response current increased with increasing concentration of glucose. Good linear relationship was obtained up to the glucose concentration of 22.4 mM, which was actually higher than normal sensor without outer film for permeability restriction. This may due to high concentration of oxygen obtained by sufficient reproduction and maintenance oxygen, since both electrochemical and enzymatic reactions occur in the tube and the release of oxygen can be restricted.

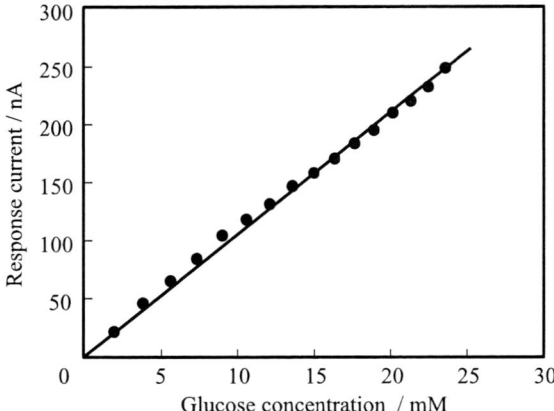

Figure 2. Typical calibration curve of Pt-Ir tube type glucose sensor having sensing region inside the tube tip in a 0.1 M phosphate buffer solution (pH 7.4) containing 0.1 M NaCl at 25°C.

Figure 3 shows the variation of the relative response in time on the Pt-Ir tube type glucose sensor having sensing region inside the tube tip. The response of the electrode was tested in 5.6 mM glucose at different intervals at 40°C and stored in phosphate buffer at 4°C when not in use. The response current of 5.6 mM glucose obtained on the first day was defined as the 100% relative response. After two days of use, the response of the sensor was relatively constant for few days, while still the sensitivity remained unstable and was lowered down to 30 to 50% of the initial sensor response. However, sufficient

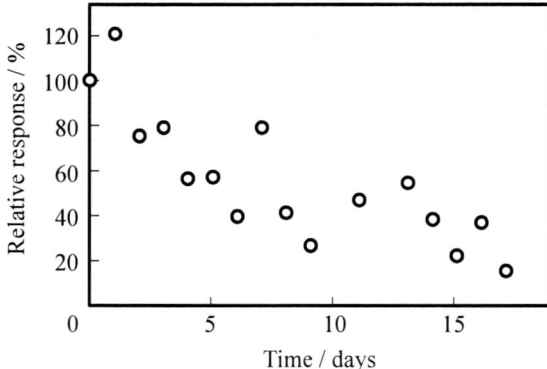

Figure 3. Typical long term time dependence of the sensor response of Pt-Ir tube type glucose sensor having sensing region inside the tube tip in a 0.1 M phosphate buffer solution (pH 7.4) containing 0.1 M NaCl at 25°C.

glucose sensor response was still obtained even after 17 days from the day of sensor preparation.

Conclusion

A fine tube type glucose sensor having sensing region inside the fine tube tip, was prepared and its sensor properties were investigated in a phosphate buffer solution. The combination of electrodeposition and electropolymerization of *o*-phenylenediamine were applied for the precise localization of glucose oxidase on the platinum surface located inside the tip of the tube. The obtained tube glucose sensor showed good response with good linear relationship up to the glucose concentration of 22.4 M. Although the sensitivity of the obtained sensor was not stable, sufficient response current was observed even after two weeks of use. The abilities of the prepared sensor as a minimally invasive glucose sensor was not proved in this study, however, its possibility was expanded.

References

1. N. Matsumoto, X. Chen and G. S. Wilson, *Anal Chem*, **74**, 368 (**2002**).

Flexprint Substrate Enzyme Sensor For Determination Of Daily Glucose-profiles Of Diabetic Patients

P.D. van der Wal[a], P. Hadvary[b], H-J Tschirky[b], and N.F. de Rooij[a]

[a] Ecole Polytechnique de Lausanne, Sensors Actuators and Microsystems laboratory (EPFL-IMT-SAMLAB), 2002 Neuchâtel, Switzerland
[b] PharmaSens AG, Kägenstrasse 17A, 4153 Reinach, Switzerland

> The goal of our project is to develop a miniaturized system allowing continuous measurement of glucose concentration profiles over 24 hours for several days. The unique feature of the PharmaSens / IMT approach lies in the insertion mechanism and the needle size to assure optimum user-friendliness and patient comfort. The measurement method is based on classical amperometric detection using an enzymatic membrane. The sensor substrate is made on a flexprint substrate and consists of a 3-electrode cell on a needle-shaped tip of ~100 μm width and is covered with polymeric membranes. The sensor is integrated in a device that combines the insertion mechanism and electronics, not needing a separate insertion tool.

Introduction

There is a lot of interest in developing glucose sensors for continuous monitoring. The ultimate goal of most of these efforts is to develop a system that combines continuous monitoring of glucose with an insulin pump. However, such a development is quite a challenge, especially with respect to safe operation and the long term behavior of the sensor. During the last few years, the emerging application of continuous glucose monitoring focused on hospital use and the profiling of daily glucose levels of diabetes patients during 1-7 days for a subset of highly motivated type I diabetic patients as a help in adjusting self-administered insulin dosing and for behavioral education, supported by health care professionals. As summarized in reference (1) and (2), in 2010 there were 4 continuous direct glucose sensors on the market: Abbott Freestyle navigator™, Minimed Paradigm® REAL-Time System, Minimed Guardian® RT and Dexcom™ Seven®Plus. Both Medtronic Minimed systems use the same sensor. Recently Medtronic introduced the Enlite sensor system (3) with a smaller version of the sensor.

Typically, these devices consist of a sensing unit, an insertion unit, a transmitter and a receiver. For use, the sensor is loaded into the insertion unit, inserted by the patient, separated from the insertion unit and connected to the electronics unit which contains the potentiostat for sensor-measurement and the transmitter. Sensor probe dimensions for the Abbott, Medtronic and Dexcom device are 21, 23, and 26 gauge (0.82, 0.64, and 0.46 mm), respectively, as given in (2). The Medtronic Enlite sensor is smaller but the insertion needle has approximately the same width as before.

The use of a separate insertion and measuring device as well as the size of the insertion guide needle makes these systems complicated to use and insertion into the skin can be painful. Therefore, only highly-motivated Type I diabetics use these systems and a user-friendly, painless, simple and cost-effective diagnostic glucose profiling device giving the health care professional valuable information and reaching also the Type II diabetic patient population remains a high and unmet medical need.

PharmaSens is combining the sensor, sensor insertion unit and electronics / transmitter into one single device, thus making the handling extremely easy and safe. The dimensions of the sensor and insertion needle will be such that the patient will have no sensation of pain when the sensor is implanted thus optimizing user-friendliness and patient comfort. In this contribution we focus on the development of the sensor. A prerequisite was that the dimensions for implantation into the skin needed to be less than < 250 μm overall diameter of the implanted part for implanting without any sensation.

Sensor Development

Sensing Principle

The sensing principle is the indirect measurement via hydrogen peroxide (H_2O_2) that is generated by the enzyme glucose oxidase. The sensor setup is a classical 3 electrode cell. The enzyme glucose oxidase (GOx) is immobilized on the platinum working electrode. Upon contact with glucose, the enzyme reactions are:

$$glucose + O_2 \rightarrow gluconolactone + H_2O_2 \qquad [1]$$

The hydrogen peroxide is then oxidized at the working electrode, the current being proportional to the amount of hydrogen peroxide produced and therefore a measure of the glucose concentration. At lower concentrations there is a linear relationship between the current and the glucose concentration. However, as can be seen in the reaction scheme [1], oxygen is also needed for the enzyme reaction. At higher glucose concentrations not enough oxygen is available and the signal saturates, see also Figure 1. This is especially true in subcutaneous conditions where the oxygen concentration is lower than in oxygenated blood. Without further precaution, the linear range will end at only 2-4 mM of glucose. The desired measuring range of an in-vivo sensor is between 2-25 mM. With a response curve as presented in Figure 1 (left), the requested range is not covered.

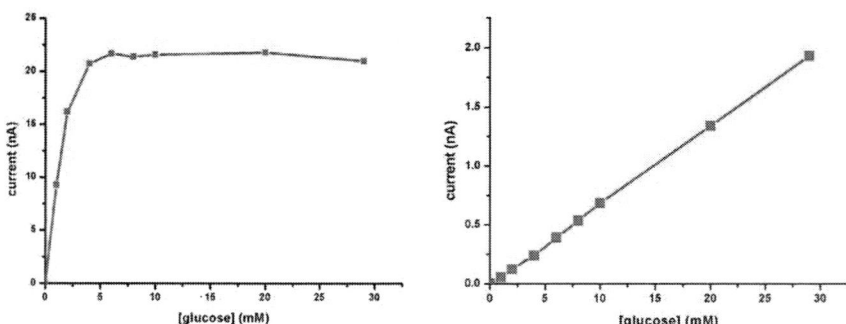

Figure 1. Calibration curve of glucose sensor without diffusion limiting membrane (left) and with diffusion limiting membrane (right).

The use of a diffusion limiting membrane (DLM) that partially limits glucose diffusion towards the enzyme, while not or only slightly limiting oxygen diffusion will reduce the sensor sensitivity but increases the linear range substantially. In Figure 1 (right), a calibration curve is shown of a sensor with a DLM resulting in a linear range up to at least 30 mM of glucose.

Sensor Substrate

Different sensor substrates were studied. Besides biocompatibility and the above mentioned total diameter of the implantable part of less than 250 μm, prerequisites included robustness to resist implantation and explantation, and possibilities of cost-effective mass-production of the sensor. The solution was found in flexprint-based substrates, to be implanted with the help of a guiding needle. Flexprint technology is well-established in electronics manufacturing and is suitable for cost-effective mass production. The challenges of the adaptation of the technology to our needs were in the field of miniaturization at the cutting-edge of available technologies and the high demand for the electrochemical characteristics of the electrodes. In several optimization cycles, electrodes with adequate characteristics could be obtained. Besides the dimensions, copper corrosion, a material that is standardly used in this technology, were the main difficulties that needed to be overcome. The resulting double sided ~100 μm wide needle shaped sensor substrate with 3 electrodes (working electrode on one side, counter and reference electrode on the other side) is shown in Figure 2. Although complete bending and forcing should be avoided, these flexprint substrates are sufficiently robust to resist normal and repetitive bending. Therefore it is expected they will survive normal use without damage to the electrical contact.

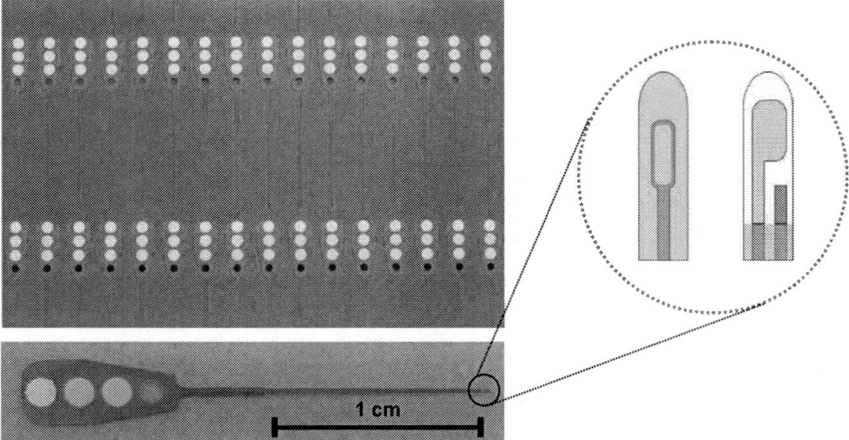

Figure 2. Flexprint substrate; top left: as produced in frame, bottom left: single substrate, right: drawing of the layout of the electrodes on the tip (doubles sided).

The flexprint based sensors are removed from the frame followed by coating of the working electrode with the enzyme membrane and diffusion limiting layer. The response of the sensor placed in a flow-through cell is shown in Figure 3. Measurements were done in phosphate buffered saline solutions (PBS) with glucose added to it. The platinum working electrode was polarized at 0.65 V vs the Ag/AgCl reference electrode.

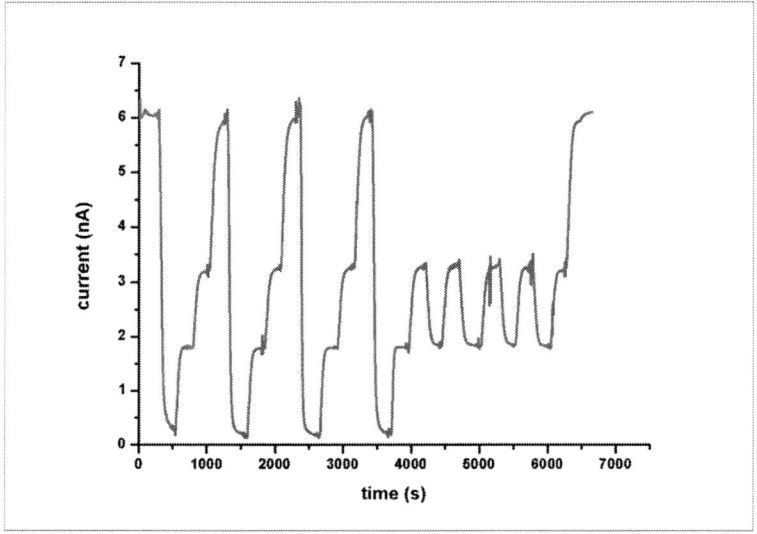

Figure 3. Response of a flexprint based glucose sensor placed in a flow cell to solutions of 0, 5, 10, and 20 mM glucose in PBS.

In summary, we could build a glucose sensor using sensor substrates made with flexprint technology. This relatively simple technology enables us to fabricate 3 electrode sensors on needle shaped sensor prints with a width of ~100 µm. With an enzyme membrane and diffusion layer a good linearity up to 30 mM of glucose was shown.

Acknowledgments

CTI grant 9008.1 PFLS-LS is gratefully acknowledged

References

1. A. Heller, B. Feldman, *Chem. Rev.*, **108**(7), 2482 (2008)
2. http://www.childrenwithdiabetes.com/continuous.htm
3. http://www.medtronic-diabetes.co.uk/product-information/index/enlite-sensor.html

Nanodiamond microelectrode array with mesa structure fabricated for bio-analytical applications

S. Raina[a*], W. P. Kang[a], N. Ghosh[a]

[a] Department of Electrical Engineering and Computer Science, Vanderbilt University, Nashville, TN 37235 USA

*Corresponding author: supil.raina@vanderbilt.edu

Ultra-microelectrode arrays (UMEAs) are widely used because of advantages such as low capacitance and low iR losses, which provides a high signal/noise ratio and allows use in highly resistive media. They also can be used in applications requiring high temporal and spatial resolution. Hemispherical diffusion enhances the flux density at the microelectrode surface to provide greater sensitivity than macroelectrodes. Catecholamine neurotransmitters, such as dopamine, play important roles in the mammalian central nervous system and require constant monitoring and measurement. A nitrogen incorporated nanodiamond UMEA was fabricated using silicon microfabrication technology and microwave PECVD process by using a 'top-down' approach. This process allows greater control over the UMEA dimensions and nanodiamond microstructure. The array consists of 2500 UME 'mesas' projecting above the surrounding insulating matrix of spin-on-dielectric. Cyclic voltammetry was used to characterize the UMEA for detection of dopamine.

Introduction

Ultra-microelectrodes (UMEs) have several advantages over macroelectrodes, such as enhanced rates of mass transport of electroactive species due to radial diffusion to the UME surface allowing rapid attainment of steady state currents and thereby providing excellent temporal resolution. The double-layer capacitance is also lowered which results in a smaller RC time constant in the electrochemical cells. Lower capacitance also improves the S/N ratio of the UMEs. The currents observed at UMEs typically lie in the pA to nA range. Due to this, the ohmic drop (iR) values are also reduced. Lower iR drop minimizes cyclic voltammogram distortion allowing their application in less conductive media. The small size of UMEs provides excellent spatial resolution allowing exploration of microscopic domains. This also allows their use for electroanalysis in small volume samples (1-3). Despite having multiple advantages, UMEs do have a major drawback. Due to the small size, the limiting currents are very small which limits the range of measurement. This may require additional instrumentation for signal amplification. However, by using an array of such UMEs (i.e. UMEA), with optimum size and spacing, we can amplify the current without using external signal amplifiers.

Nanodiamond is an excellent electrode material for electrochemical detection of redox-active bio-analytes. It is being widely recognized for its properties, also observed in boron-doped diamond, such as a wide working potential window which enables study of a wider range of analytes with more extreme oxidation-reduction potentials, not possible previously with conventional electrode materials. It also provides a low and stable background current, high resistance to deactivation due to surface adsorption, excellent biocompatibility, high sensitivity and selectivity for detection of target analytes (4-5). Excellent control over the nanodiamond film thickness, growth rates, microstructure and electrical conductivity has become possible with advances in CVD processes such as microwave plasma enhanced (MPECVD) and hot filament (HF) chemical vapor deposition (6). These favorable characteristics of nanodiamond have made it the material of choice in this work.

Dopamine (DA) is a catecholamine neurotransmitter secreted by dopaminergic neurons and relays messages to target cells via dopamine receptors (7-8). It is derived in the biosynthetic pathway from tyrosine, which after a hydroxylation step forms L-DOPA followed by a decarboxylation step, resulting in DA. In addition to being a precursor of norepinephrine (NE), DA has biological activity in the periphery, most particularly in the kidney, and plays an active role in several important pathways in the mammalian central nervous system (CNS) (9-13). It is a central player in the brain "reward" system and is also involved in motor and cognitive functions (9-13). Dysfunction of the dopaminergic system in the CNS has been related to neurological disorders such as schizophrenia and Parkinson's disease (9-13). Electrochemical techniques can be employed to detect DA as it is electrochemically active and can be oxidized to form DA-o-quinone (DAQ) along with transfer of two electrons and protons. In addition to the ease and flexibility, electrochemical techniques offer better spatial and temporal resolution and require simpler and less expensive analytical systems (7,14).

Nitrogen incorporated thin-film nanodiamond electrodes, in a macro form and as UMEAs, have been used successfully to detect DA (15-17). The UMEAs were characterized under steady state conditions at slower scan rates, when hemispherical diffusion of the analyte to the microelectrode surface is dominant. Under these conditions, sigmoidal shaped cyclic voltammograms are obtained. The nanodiamond UMEAs characterized previously were fabricated using a 'bottom-up' approach. However, this fabrication process provided limited control over the microelectrode dimensions and the nanodiamond microstructure. Consistency in UMEAs is critical in the development of a successful device which can meet the strict quality controls as well as provide reliable electroanalytical results. In this study we describe the fabrication and characterization of a nanodiamond UMEA using a 'top-down' approach, which we believe can solve the issues of inconsistent size and microstructure, while providing excellent sensitivity for DA detection.

Experimental

The nanodiamond UMEA's (ND-UMEAs) fabricated with both the 'bottom-up' and 'top-down' approaches used conventional silicon microfabrication techniques and MPECVD nanodiamond growth processes. The 'bottom-up' technique has been described in detail previously (16-17). In this work, we focus on the alternative process.

A highly doped n-type silicon substrate was used to grow a ~2μm thick nanodiamond film using a MPECVD process in a 1.5kW, 2.45GHz ASTEX machine with an input gas mixture of H_2, CH_4 and N_2, where methane was the carbon source gas. A thin-film of aluminum was deposited using DC sputtering technique in a Cressington 308R system. After spin-coating photoresist, conventional UV photolithography was used to pattern and develop the desired array layout. The array was designed to optimize the UME dimensions and spacing, so that the individual diffusion layers do not overlap during electrochemical analysis. Wet-etch of aluminum was followed by anisotropic nanodiamond etch, RIE (reactive ion etch), using an oxygen plasma in a STS AOE machine. This produced an array of nanodiamond *'mesas'* with the protective aluminum layer on the top. A spin-on-dielectric was applied on this substrate and after the curing

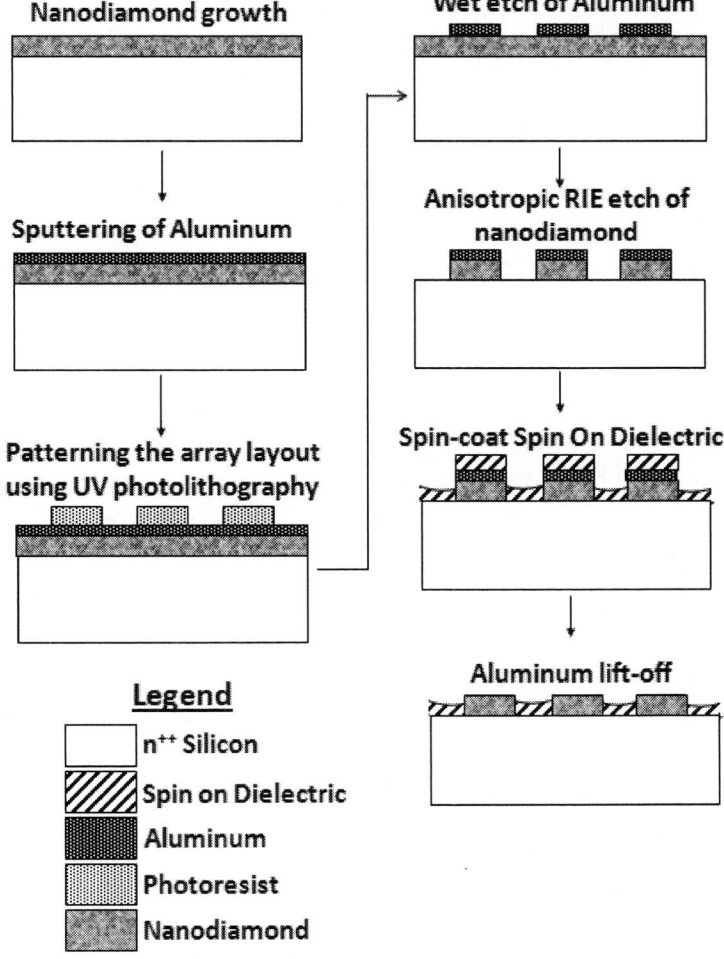

Figure 1. Flow chart for the fabrication of the nanodiamond UMEA with a *'top-down'* approach.

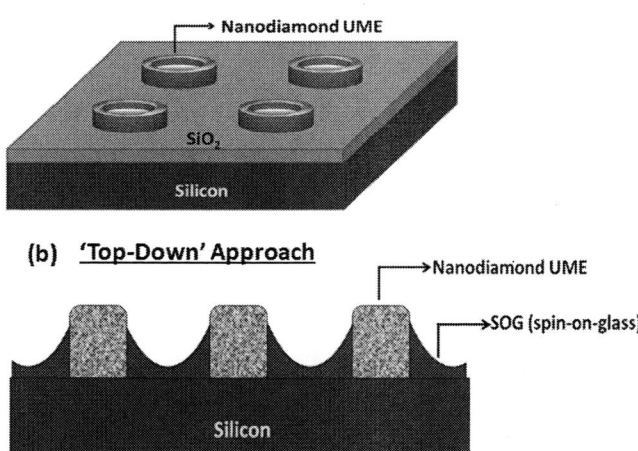

Figure 2. (a) A perspective representation of the final structure of a ND-UMEA fabricated using the *'bottom-up'* approach; (b) A cross-sectional representation of the ND-UMEA with a *mesa* structure fabricated by the new *'top-down'* approach.

process the aluminum layer was lifted-off as the final step. The fabrication process flow can be seen in figure 1. A three dimensional perspective of the ND-UMEA fabricated using the *'bottom-up'* approach can be seen in figure 2(a), which can be compared to the cross-sectional representation of the final structure from the newer *'top-down'* approach in figure 2(b).

The ND-UMEA structure and surface morphology was studied using a Hitachi S-4200 scanning electron microscope. Electrochemical characterization was performed using a glass flat cell in a 3 electrode configuration with a platinum wire as the counter electrode and a Ag/AgCl (3M KCl) reference electrode. The ND-UMEA working electrode is sealed in by an o-ring on the top and the electrical connection is made from the back side of the Si substrate using a copper tape. Potassium ferrocyanide and potassium chloride were obtained from Fisher Scientific, Dopamine from Alfa Aesar and a premixed powder of PBS from EMD Chemicals. All solutions were prepared with de-ionized water (18 MΩ-cm). Cyclic voltammetry experiments were performed using a CHI 660C (CH Instruments) work station and the accompanying data acquisition software.

Results and Discussions

SEM Characterization

The SEM micrographs of the ND-UMEA can be seen in figure 3. The mesa structure formed after the nanodiamond RIE step can be seen in figure 3(a). The aluminum layer, which acts as the mask during RIE, is still visible on top of the nanodiamond microelectrode, which itself stands on the silicon substrate. An individual nanodiamond microelectrode in its final form can be seen in figure 3(b). The spin-on-dielectric forms the insulating layer between the microelectrodes. After aluminum lift-off, the nanodiamond microstructure is exposed and a disk-shaped microelectrode can be observed. A low magnification SEM micrograph in figure 3(c) shows a tilt view of a section of the ND-UMEA, the bright spots correspond to the individual microelectrodes.

The array consists of 2,500 microelectrodes arranged in a square array, each with a diameter of ~9µm and 100µm spacing. The UMEs project above the surrounding dielectric matrix. The microstructure of the nanodiamond film can be seen in the high resolution image of figure 3(d), consisting of nanoplatelets and nanocrystallites on the side-walls. This unique structure is formed due to the competing processes: nanodiamond nucleation & growth and etching due to the plasma.

Based on the fabrication process and the SEM images, the strengths of this *'top-down'* approach can be observed. This process ensures that the nanodiamond microstructure, which forms the starting point, is consistent every time, suitable for batch fabrication. The height of the *'mesa'* depends on the thickness of the nanodiamond film and can be adjusted by varying the growth time. In the present case, the microelectrodes exposed at the top form a two-dimensional microdisk. However, this can be easily modified by varying the spin-on-dielectric coating process, such as the spin-coating speed, etc. An array with a three-dimensional microelectrode surface can therefore be easily developed. Additionally, the lateral dimensions of the microelectrodes and the overall array that are defined during the photolithography process can be maintained throughout the fabrication process. Such control was not possible previously, where the nanodiamond growth rate was greater at the periphery resulting in a 'donut' like microelectrode geometry, as seen in figure 2(a). The final dimensions would always exceed those defined in the photo-mask.

Additional material characterization using Raman spectroscopy and X-ray Photoelectron Spectroscopy (XPS) of the nanodiamond film has been performed and reported previously (18-19).

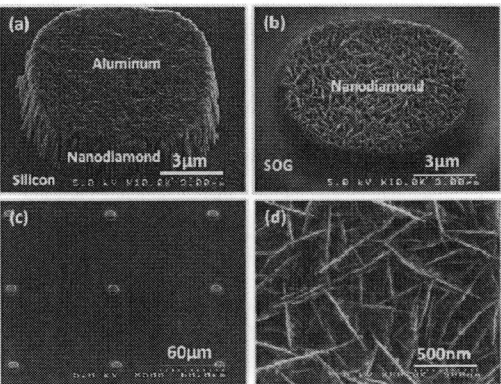

Figure 3. SEM micrographs of the ND-UMEA: (a) SEM image of a stand-alone nanodiamond microelectrode with a layer of the aluminum mask after the anisotropic RIE etch process; (b) Final structure of an individual nanodiamond microelectrode after the application of the spin-on-dielectric layer and aluminum lift-off step; (c) Tilt view of a section of the ND-UMEA; (d) High resolution SEM image of the nanodiamond microstructure with the characteristic nanoplatelets and nanocrystallites on the side-walls.

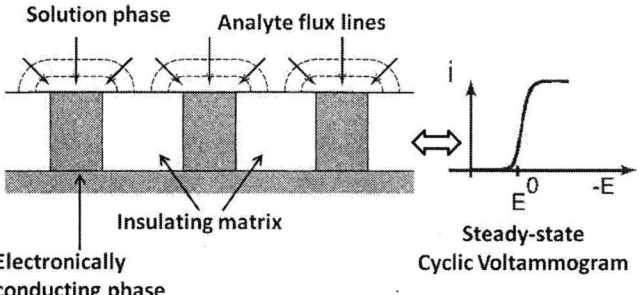

Figure 4. Left: Hemispherical diffusion at the UMEA surface, where individual diffusion layers do not overlap; Right: Sigmoidal shaped cyclic voltammogram under steady state conditions.

Electrochemical Characterization

The response of a UMEA working at the same potential depends on the dimensions of the individual microelectrodes and the thickness of the diffusion layer with respect to the microelectrode (1). Diffusion layer is the region where a concentration gradient exists between the electrode surface and the bulk solution (1). At very short time scale, as in fast scan cyclic voltammetry (FSCV), the diffusion layer thickness is relatively smaller than the radii of the microelectrodes and semi-infinite linear diffusion (planar diffusion) is dominant. At larger time scales, the individual diffusion layers overlap creating pseudo planar diffusion conditions. Under these conditions, the array behaves like a planar electrode where the current is proportional to the entire array area, including the insulating regions in between the microelectrodes. In between the two time scales, the diffusion layer thickness grows larger than the microelectrode dimension (r) at which point, hemispherical diffusion is present and steady state conditions are observed, as seen in figure 4. Under these conditions we obtain a sigmoidal shaped cyclic voltammogram.

Initial electrochemical characterization of the ND-UMEA was performed using the ferri/ferrocyanide redox couple in 0.1 M KCl as the supporting electrolyte. Cyclic voltammograms (CVs) were recorded at 100mV/s for different ferrocyanide concentrations of 1mM, 2mM, 3.9mM, 5.9mM and 8mM. As predicted, sigmoidal shaped CVs were obtained, consistent with steady state conditions. The CVs plotted as J (mA/cm^2) vs. Potential (V) vs. Ag/AgCl can be seen in figure 5(a). A plot of the limiting currents versus the ferrocyanide concentration is called the calibration plot and the slope of this curve is used to determine the sensitivity of the UMEA. A linear calibration plot can be seen in figure 5(b) which gave a sensitivity value of ~2mA/cm^2.mM for ferrocyanide detection in the range of 1mM-8mM. Closer analysis of the curves using the Tomeš criteria ($E_{3/4}-E_{1/4}$), which is the difference between the three-quarter wave potential and quarter wave potential, suggests a quasi-reversible response (1).

Electrochemical techniques provide excellent means to monitor and measure DA concentrations since it is an electroactive bio-analyte. It can be oxidized to form dopamine-o-quinone along with the transfer of 2 electrons and 2 protons, as shown in figure 6. Cyclic voltammograms were recorded at 100mV/s with 0.1M phosphate buffered saline (PBS) as the supporting electrolyte at the physiologic pH7.4. The CVs for different concentrations of DA: 100µM, 200µM, 400µM, 600µM, 800µM and 1000µM; have been overlaid and can be seen in figure 7(a). We obtained sigmoidal shaped CVs consistent with steady state conditions. On plotting the limiting currents with DA concentration, we obtain a linear calibration plot seen in figure 7(b), with a linear dynamic range of 100µM-1mM. The slope of the calibration plot, which is also the sensitivity value of the UMEA, was calculated to be ~2.7µA/cm^2.µM. This represents a **10x** increase in sensitivity for DA detection as compared to a nanodiamond macroelectrode characterized previously (16).

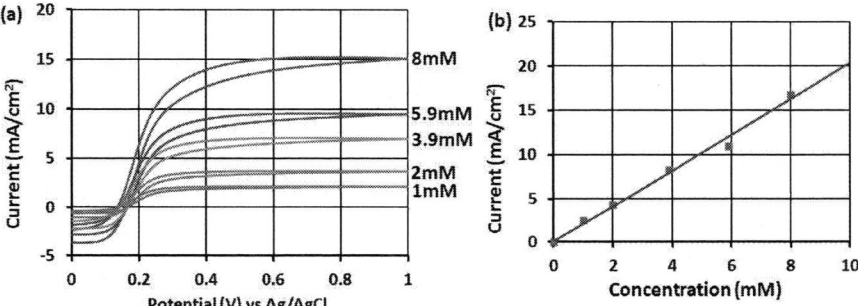

Figure 5. (a) Cyclic voltammograms recorded at 100mV/s for different concentrations of ferrocyanide in 0.1M KCl as the supporting electrolyte; (b) Linear calibration plot.

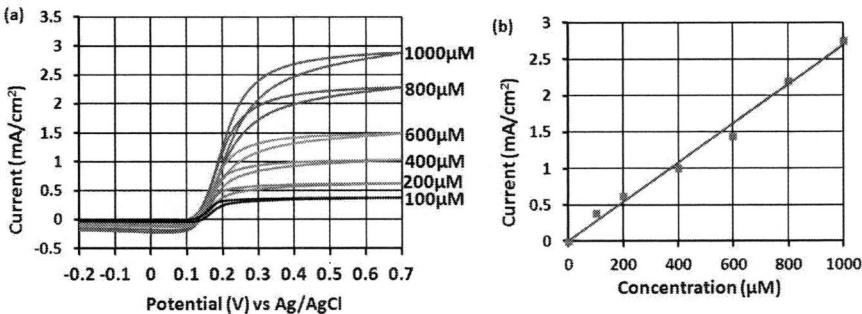

Figure 6. DA is oxidized to form DA-o-quinone and involves transfer of 2 e$^-$ and 2 H$^+$.

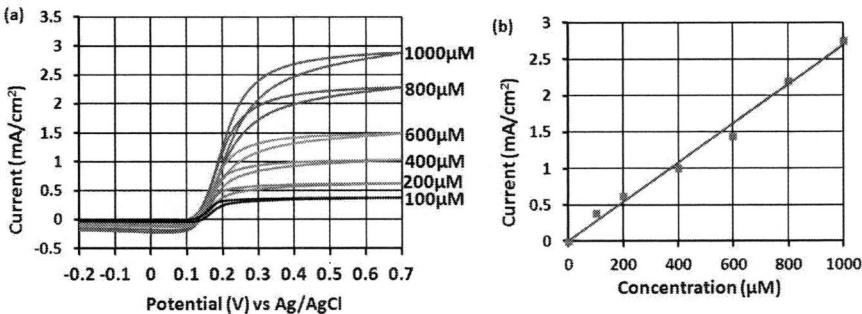

Figure 7. (a) Cyclic voltammograms recorded at 100mV/s for different concentrations of dopamine in 0.1M PBS at pH7.4 as the supporting electrolyte; (b) Linear calibration plot.

Conclusions

A nitrogen incorporated nanodiamond UMEA was successfully fabricated by using a *'top-down'* approach consisting of conventional silicon microfabrication processes and MPECVD technique for nanodiamond deposition. This approach provides excellent control over the lateral dimensions of the UMEA as well as the nanodiamond film microstructure. It also provides flexibility in terms of creating a 2D or 3D microelectrode surface by varying the spin-on-dielectric application parameters. The ND-UMEA supported quasi-reversible electrochemical response and was successfully able to detect dopamine with nearly *10x* greater sensitivity than a nanodiamond macroelectrode.

References

1. A. J. Bard and L. R. Faulkner, *Electrochemical Methods: Fundamentals and Applications*, p. 156, Wiley, New York (2001).
2. R. J. Forster and Tia E. Keyes, in *Handbook of Electrochemistry,* C. G. Zoski, Editor, p. 155, Elsevier, Amsterdam (2007).
3. J. Wang, *Analytical Electrochemistry*, 100, Wiley-VCH, New York (2001).
4. S. Raina, W. P. Kang and J. L. Davidson, *Diamond Relat. Mater.*, **18**, 718 (2009).
5. J. Park, Y. Show, V. Quaiserova, J. J. Galligan, G. D. Fink and G. M. Swain, *J. Electroanal. Chem.*, **583**, 56 (2005).
6. M. Hupert, A. Muck, J. Wang, J. Stotter, Z. Cvackova, S. Haymond, Y. Show and G. M. Swain, *Diamond Relat. Mater.*, **12**, 1940 (2003).
7. D. L. Robinson, A. Hermans, A. T. Seipel, and R. M. Wightman, *Chem. Rev.*, **108**, 2554 (2008).
8. T. Kondo, Y. Niwano, A. Tamura, J. Imai, K. Honda, Y. Einaga, D. A. Tryk, A. Fujishima and T Kawai, *Electrochim. Acta*, **54**, 2312 (2009).
9. M. J. Kuhar, K. Minneman and E. C. Mully, in *Basic Neurochemistry: Molecular, Cellular, and Medical Aspects*, G. J Siegel, R. W. Albers, S. T. Brady and D. L. Price, Editors, p. 211, Elsevier Academic Press, Canada (2006).
10. B. J. Venton and R. M. Wightman,. *Anal. Chem.*, **75**, 414A (2003).
11. G. Eisenhofer, T. Huynh, M. Hiroi, and K. Pacak, *Rev. Endocr. Metab. Disord.*, **2**, 297 (2001).
12. D. L. Robinson, A. Hermans, A. T. Seipel, and R. M. Wightman, *Chem. Rev.*, **108,** 2554 (2008).
13. G. C. Wild and E. C. Benzel, *Essentials of Neurochemistry,* p. 58, Jones and Bartlett Publishers, Boston (1994).
14. S. Threlfell and S. J. Cragg, in *Electrochemical Methods for Neuroscience*, A. C. Michael and L. M. Borland, Editors, p. 125, CRC Press, Boca Raton, Florida (2007).
15. S. Raina, W. P. Kang and J. L. Davidson, *Sensors,* 1780 (2009).
16. S. Raina, W. P. Kang and J. L. Davidson, *Diamond Relat. Mater.*, **19**, 256 (2010).
17. S. Raina, W. P. Kang, J. L. Davidson and J. H. Huang, *ECS Trans.,* **28**, 21 (2010).
18. S. Raina, W. P. Kang and J. L. Davidson, *Diamond Relat. Mater.*, **17**, 790 (2008).
19. Supil Raina, X. C. LeQaun, W. P. Kang and J. L. Davidson, *ECS Trans.*, **19**, 23 (2009).

ECS Transactions, 50 (12) 101-108 (2012)
©The Electrochemical Society

Electrical Impedance Sensors for Cancer Cell Study

L. Yang*

Department of Pharmaceutical Sciences, Biomanufacturing Research Institute and
Technology Enterprises (BRITE), North Carolina Central University, Durham, NC 27707

> This research used the label-free impedance technique to analyze
> the cellular activities oral squamous cell carcinoma (OSCC) cells
> and non-cancer oral epithelial cells. Three major research activities
> were included: 1) to study various cellular activities of OSCC cells,
> including cell spreading, attachment, proliferation, drug-induced
> apoptosis and inhibition of apoptosis; 2) to distinguish oral cancer
> cells and normal cancer in a label-free manner; 3) The resistance
> and capacitance components induced by the two types of cells on
> the microelectrodes were obtained by fitting appropriate equivalent
> circuits. This study demonstrated that the impedance-based method
> has the potential to be a useful analytical approach for cancer
> research.

Introduction

Cellular analysis has been an effective approach to understand many biological and
biomedical problems. It is also important for developing rapid and simple methods to
study the characteristics of different cell types and their interactions with drugs,
especially in anti-cancer drug discovery and screening. Current approaches such as
cytometry and microscopic imaging can provide insights into the physiological function
of any particular cell or of pathological changes that may have occurred. However, they
usually require fluorescent or radioactive labeling steps which often involve destruction
of cells. The labeling processes may thus lead to the loss of very important biological
information about live cells. Label-free and non-invasive analysis of cells and their
functions can provide real-time and kinetic cellular activities of live cells, which could be
of great use in many biomedical applications. Among a number of label-free technologies,
electrical impedance spectroscopy (EIS) has been recognized as a powerful
electrochemical technique that can monitor live cell behavior in real-time. By culturing
biological cells on an electrode surface, EIS can directly sense detailed information about
cellular activities occurring on an electrode or substrate's surface by measuring the
induced capacitance and/or resistance changes, eliminating multiple labeling and
amplification steps typically used in many other cell-based methods, and allowing label-
free and non-invasive study of cellular properties (1-3) and monitoring drug induced
cellular activities for drug discovery (4-9). In addition, impedance technique is amenable
to miniaturized electronic systems to meet the growing needs of microdevices for point-
of-care analysis.

In this study, electrical impedance technique was used for studying the behaviors of
oral cancer cells and normal oral epithelial cells on microelectrodes in a label free and
real-time manner. Oral cancer is one of the most common cancers worldwide, and oral
squamous cell carcinoma (OSCC) occurs with the highest frequency. In the US, OSCC

101

represents 2%-4% of the annually diagnosed malignancies, accounting for 8,000 deaths every year (10,11).

The research has been focused on three major activities: 1) the use of impedance technique to study various cellular activities of OSCC cells, including cell spreading, attachment, proliferation, drug-induced apoptosis and inhibition of apoptosis; 2) distinguishing oral cancer cells and normal cancer in a label-free manner by impedance measurement; 3) determining the resistance and capacitance components induced by the two types of cells on the microelectrodes by fitting appropriate equivalent circuits.

Materials and Methods

Cell culture

The OSCC cell line, CAL 27 (ATCC # CRL-2095), and the noncancer-derived esophageal epithelial cell line, Het-1A (ATCC# CRL-2692), purchased from the American Types Culture Collection (ATCC), were used in the experiments. CAL 27 cell line was cultured in Dulbecco's modified eagle's medium (DMEM) (Sigma) supplemented with 10% of fetal bovine serum, 100 IU/ml penicillin and 100 µg/ml streptomycin. The Het-1A cell line was culture in bronchial epithelial cell basal medium (BEBM) supplemental with a bronchial epithelial growth media (BEGM) kit which contains supplements and growth factors including BPE, hydrocortisone, hEGF, epinephrine, insulin, triiodothyronine, transferrin, gentamicin/amphotericin-B and retinoic acid). For Het-1A cells, the flasks were pre-coated with a mixture of 0.01 mg/ml fibronectin, 0.03 mg/ml bovine collagen type I and 0.01 mg/ml bovine serum albumin (BSA) dissolved in culture medium, as recommended by ATCC. Cells were cultured in 75 cm^2 flask and were incubated at 37 °C in an atmosphere of 5% CO_2 in air. The medium was renewed every 2-3 days. When confluent, cells were washed with Dulbecco's phosphate buffered solution (DPBS) and detached from the flask using 0.25% trypsin with 0.53 mM EDTA solution. Cells were centrifuged to remove trypsin and resuspended with fresh DMEM or BEBM medium. Cell number in the suspension was determined using the Vi-cell XR cell counting system (Beckman Coulter, Miami, FL). Desired cell concentrations were obtained by diluting the cell suspension with fresh DMEM or BEGM for further experiments.

Real-time impedance measurement of cellular activities

The real-time impedance-based measurement of cellular activities was performed using the RT-CES system (ACEA Biosciences Inc. San Diego, CA). The system consists of an electronic sensor analyzer station and three 16-well E-plates for culturing cells. In the 16-well plate, each well is equipped with an array of circle-on-line gold microelectrodes on the bottom of the well. Detailed information about the components and the principle of the system was reported in previous studies (4,12,13).

For measurements, 50 µl of medium was added to the wells for taking background readings first, then 100 µl of CAL 27 or Het-1A cells in medium with desired concentration was added to the wells. Impedance-based cell index was measured at designed period of time, varying from 5 min to 30 min. Based on measured impedance, cell index (CI), was derived and recorded as a function of time (12,13).

Electrical/Electrochemical impedance spectroscopy (EIS)

EIS measurements were carried out using an IM-6 Impedance Analyzer (Zahner-elektrik CMBH & Co. KG, Germany) with the IM-6/THALES software. The IME chip device was assembled by a gold IME on a glass substrate (ABTECH Scientific, Inc. Richarmond, VA) and a poly dimethylsiloxane (PDMS) micro-chamber right above the electrode area. The IME chips were 1.0 cm wide x 2.0 cm long x 0.05 cm thick. The IMEs possessed 50 pairs of finger electrodes with 15 μm width and space and 5 mm length. The IMEs were cleaned with acetone, 2-propanol and then washed profusely with DI water. The chips were then cleaned with a solution of 1:1:5 (v/v) $H_2O_2(30\%)/NH_4OH(30\%)/H_2O$ for 5 s, rinsed with DI water, and dried with air, before they were assembled with the PDMS chamber. The IMEs were connected using a two-electrode configuration wherein one set of the array electrodes was connected to both the test and sense probes and the other set was connected to both the reference and counter electrode probes on the impedance analyzer.

EIS measurements were carried out in cell growth medium. For measurement, 40 μl of cell sample with desired number of CAL 27 cells in DMEM medium were seeded on the IMEs and incubated in at 37 °C. At desired time, the chips with cells were taken out from the incubator and cooled down to room temperature, and EIS measurements were performed over the frequency range 10 Hz to 1 MHz with an amplitude of ±50 mV at room temperature. After each measurement, the chip was placed back to the incubator until the subsequent measurement.

Bode impedance spectra (impedance and phase vs. frequency) were recorded. Simulations were performed using the SIM program of THALES software. To perform the simulation, proper equivalent circuits were proposed, then 100 data points from each measured spectrum were automatically selected by the software as inputs to the equivalent circuit to generate a fitting spectrum. The agreement between the measured spectrum and the fitting spectrum indicated the feasibility of the equivalent circuit for representing the electrochemical behavior of the IME system.

Imaging

Fluorescence images of cells on the electrodes were taken on the Nikon Laser-Scanning Confocal Microscope. Cells were cultured on the IMEs, after 24 h, they were stained with two fluorescent dyes: the blue Hoechst dye 33342 (Invitrogen, Carlsbad, CA) and the red celltracker (Invitrogen, Carlsbad, CA).

The bright field microscopic images of cells on IMEs were taken on a Nikon ECLIPSE E600FN microscope (Japan) with a Coolsnap HQ camera (Roper Scientific, Inc.—Photometric, Tucson, AZ) using reflective bright field mode.

Results and Discussion

Real time impedance monitoring of cell adhesion, spreading, and proliferation

Fig. 1A shows the growth curve of OSCC cells seeded on microelectrodes. The cell index was derived from impedance measurement (12). As shown in the figure, the impedance-based cell index increased slowly due to the increasing number of cells settled down on the electrode surfaces in the first 1-2 h. In the next stage, the cell spreading caused the most significant and rapid increase in the cell index, followed by the cell proliferation which generated slower increases in the cell index. The rate of cell index increase due to cell spreading was about 5 times the rate measured during initial cell adhesion or cell proliferation. The result indicated that impedance-based measurement is able to monitor various cellular activities of OSCC cells, including cell adhesion, spreading, and proliferation.

If different cell numbers of OSCC cells were seeded on the electrodes, at each time point, the higher cell number has higher cell index. As the cell size, morphology, the extent of cell spreading, and the rate of proliferation are relatively constant in the same type of cells, the impedance—based cell index should be indicative of the cell number. Fig. 1B shows the relationship between the cell number and the impedance-based cell index measured at 15 h during the cell proliferation stage. It shows that the cell index is proportional to the cell number in the range of cell number between 3,500 and 35,000 cells/well, with the linear regression equation of I (cell index) $= 8x \ 10^{-5} \ N$ (cells/well) $-$ 0.1124. The impedance cell index is related to the cell coverage on electrode surface. Estimated from several images taken at different locations on each electrode, the average cell coverage generated by 1,000, 5,000, 10,000, and 20,000 cells/well were approximately 2.4%, 12.0%, 38.0%, 72.0%, respectively. The results indicated that a ~10% cell coverage on the electrode surface is required to generate a detectable impedance-based cell index. The sample with 1,000 cells did not generate a cell index that is significantly different from the blank. The results here demonstrated that this impedance-based method could offer a label-free and non-invasive quantitative method to detect OSCC cells. This can be useful in clinic diagnostics, such as detection of the presence of cancer cells in body as a potential indicator of prognosis and diagnosis in oncology.

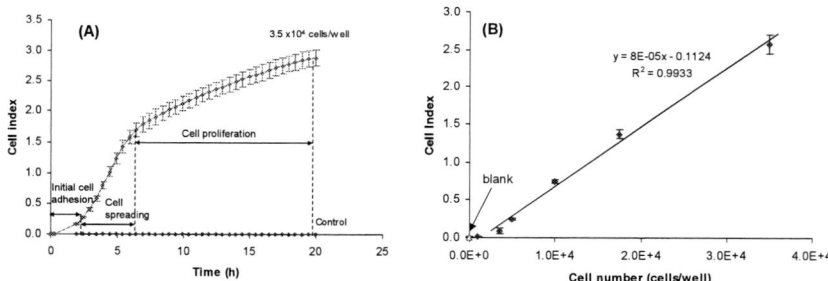

Figure 1. (A) Impedance-based cell index curve of the growth of OSCC cells on the microelectrodes, showing the impedance responses in different stages including the initial cell adhesion, cell spreading and cell proliferation stages; (B) The linear relationship between the impedance-based cell index signal and the cell number seeded on the electrodes.

Real-time impedance monitoring of cellular responses to drug treatments

Fig. 2A shows the impedance-based cell index curves of OSCC cells in response to a well-known anti-cancer drug, cisplatin, and the inhibition of cisplatin-induced apoptosis by nicotine. While the control (untreated) cells resulted in a growth curve with impedance –based cell index continuously increasing with time, the cisplatin treated cells resulted in growth curves with decreasing cell index after the treatment. The magnitude in decreased cell index was related to the cisplatin concentration. The results suggest that the impedance-based cell index can be used as reliable measure for cellular response to drug treatments (12). In addition, this method was also able to monitor cellular response to combination drug treatments. Fig. 2B shows the impedance-based cell index curves of OSCC cells in response to the combination treatment of cisplatin and nicotine. The cell index curve of OSCC cells with 10 μM nicotine treatment is similar to that of the control, indicating that nicotine only did not affect the OSCC cells' growth. The cells treated with 20 μM cisplatin shows reduced cell index compared with the control, indicating the cisplatin-induced apoptosis. The cells co-treated with 1 μM nicotine and 20 μM cisplatin presents higher cell indexes than that with 20 μM cisplatin treatment, indicating that 1 μM nicotine has an inhibitory effect on the cisplatin-induced cell apoptosis. We also found that 0.1 μM nicotine did not cause significant inhibition on the cisplatin-induced apoptosis. Treatment with 0.5 μM nicotine exhibited significant inhibition on the cisplatin-induced apoptosis. However, when the nicotine concentration increased to 5 μM, the cell indexes are lower than those with 1 μM and 0.5 μM nicotine treatments, but still higher than those of cisplatin treatment only, which suggest that the 5 μM nicotine inhibitory effect on the cisplatin-induced apoptosis is not as strong as that by 1 μM and 0.5 μM nicotine (Figure 2C). The results are consistent with several other studies on the effect of nicotine on head and neck cancer cells (14-16). For example, Xu et al. (14) reported that nicotine inhibited cisplatin-induced apoptosis in human oral cancer cell line Tca8113. Others reported that nicotine affected the signaling of the death pathway and resulted in a decreased cytotoxicity of various anticancer agents such as cisplatin and gamma-radiation in head and neck cancer cell lines UMSCC 10b and UMSCC5 (16).

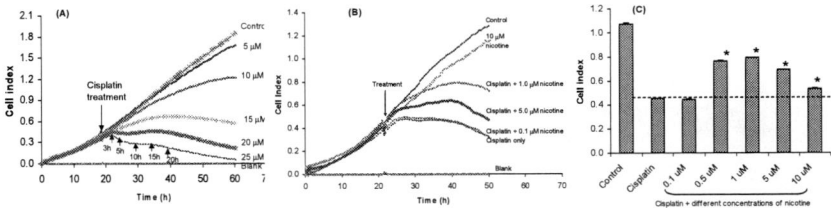

Figure 2. (A) The impedance-based cell index curves of OSCC cells in response to treatments of cisplatin at concentrations ranging from 5μM to 25 μM; (B) The impedance-based cell index curves of OSCC cells in response to combination treatment of 20 μM cisplatin and different concentrations of nicotine ranging from 0.1 μM to 10 μM; (C) The changes in cell index due to the combination treatments of cisplatin and nicotine.

Distinguishing cancer cells and non-cancer cells by impedance-based measurement

Impedance measurement also enabled us to distinguish the OSCC cells (CAL-27 cells) and non-cancer oral epithelial cells (HET-1A cells) based on their distinct behaviors on the microelectrodes (13). Fig. 3 shows the cell index curves of the two types of cells. Different characteristics of the two types of cells included: (1) the cell index curves showed a difference in the overall pattern for CAL-27 cells and Het-1A cells (Figure 3A); (2) the kinetics of cell index change in cell spreading stage were different for the two cell types (Figure 3B); (3) At a given time, CAL-27 cells showed a higher cell index than Het-1A cells (Figure 3C). Figure 3D and 3E show the representative confocal microscopic images of CAL 27 cells and Het-1A cells after 2.0 x 10^4 cells were seeded in 96-well plate for 48 h. It is clearly seen that CAL 27 cells tend to be in clusters, while Het-1A cells tend to be individually. CAL 27 cells seem to interact more closely to each other than Het-1A cells. Figure 3E shows the percentage Z-profile as a function of scanning steps for the red channel (cytoplasm). If we took 10% of the fluorescence intensity as the threshold, for Het-1A cells, fluorescence was detected between steps 15 to 39, total of 24 steps (0.5 µm/step), by which the thickness of the cell layer was estimated as 12 µm of thickness. For CAL 27 cells, fluorescence was detected between steps 22 to 52, total of 30 steps, which was estimated as 15 µm of thickness. The results suggest that the CAL 27 cells form a thicker layer on the electrode surface than the Het-1A cells. A number of factors could account for the difference in the impedance based cell indexes for the two cell types, possibly including the overall cell-electrode interactions, cell-cell interactions, and the thickness of the cell layer. However, the results indicated that the impedance-based cell index measurement could provide a potential approach to distinguish the two cell types in a real time and label free manner.

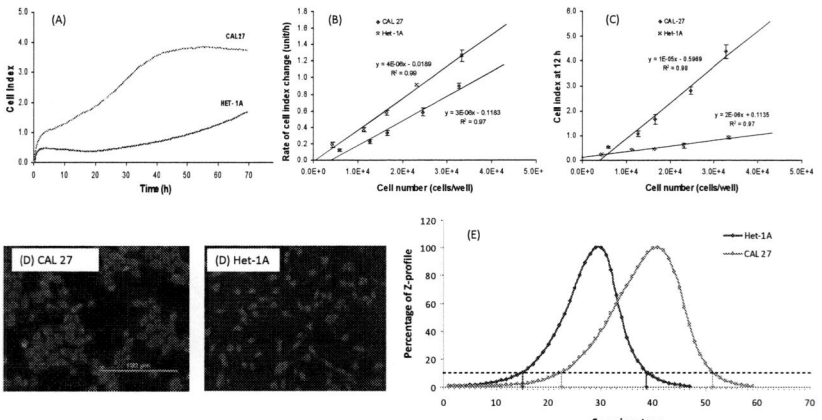

Figure 3. (A) the impedance-based cell index curves of OSCC cells and Het-1A cells on the microelectrodes; (B) The rate of cell indexes for OSCC cell and Het-1A cells during the cell spreading stage in relation to cell numbers; (C) The cell indexes of OSCC cells and Het-1A cells measured at 15 h in relation to cell numbers; (D) and (E) Confocal microscopic images of CAL-27 and Het-1A cells; (E) Z profiles of CAL 27 and Het 1A cells measured in the confocal images.

Table 1 shows the values of different impedance components induced by CAL 27 cells and Het-1A cells on the IME electrodes measured by the impedance spectroscopy and simulated using the proper equivalent circuit (17). It is understood that the attachment of cells on the electrode surface is the cause of increase in the resistance component, R_{cell}, because of the highly insulating cell membrane and the increased cell coverage on the IME. When looked at these components over time, the slight increase in R_{sol} with time is possibly because of the change in the solution microenvironment in the immediate neighborhood of the electrodes due to cell adhesion, spreading and proliferation. The initial cell adhesion on the electrode surface induced a capacitance. This capacitance decreased during cell spreading and decreased further during cell proliferation on the electrode surface, and the most significant decrease in C_{cell} was during the cell spreading stage. As compared to the percentage change of R_{cell}, the results here also suggested that the decrease in capacitance due to cell spreading and proliferation contributed to the increase in impedance, as capacitance is inversely correlated to impedance.

TABLE I. Simulated values of electrical element in the equivalent circuits for the controls and the cells samples of CAL 27 cells (1.40×10^6, 40 µl) and Het-1A cells (1.99×10^6, 40 µl) on the IMEs.

	C_{dl} (nF)	R_{sol} (Ω)	C_{cell} (nF)	R_{cell} (Ω)
Control	564.6	55.85	-	-
CAL 27	698.4	63.09	184.4	99.08
Percentage change	23.7%	13.0%	-	-
Control	489.4	27.49	-	-
Het-1A	504.2	31.2	640.8	25.46
Percentage change	3.0%	13.5%	-	-

Simulation impedance error (mean): 1.6% for Cal-27 cells and 1.2% for Het-1A cells.

Conclusions

This study have demonstrated that impedance-based measurements are capable of measuring various cellular activities of cancer cells and non-cancers on electrodes, and distinguishing oral cancer cells and non-cancer cells in a real time and label free manner. This study suggested that the impedance-based method be a useful analytical approach for cancer research.

Acknowledgments

This research was supported by National Science Foundation (CBET#0916138).

References

1. I. Giaever and C.R. Keese, *Proc. Natl. Acad. Sci. USA*, **81**, 3761 (1984).
2. I. Giaever and C.R. Keese, *Proc. Natl. Acad. Sci. USA*, **88**, 7896 (1991).
3. I. Giaever and C.R. Keese, *Nature*, **366**, 591 (1993).

4. K. Solly, X. Wang, X. Xu, B. Strulovici and W. Zheng, *Assay Drug Dev. Technol.,* **2**, 363 (2004).
5. P. Linderholm, J. Vannod, Y. Barrandon and P. Renaud, *Biosens. Bioelectron.,* **22**, 789 (2007).
6. R. McGuinness, *Curr. Opin. Pharmacol.* **7**, 535 (2007).
7. B.D. Klo, R. Kurz, H.G. Jahnke, M. Fischer, S. Rothermel, U. Anderegg, J.C. Simon and A.A. Robitzki, *Biosens. Bioelectron.,* **23**, 1473 (2008).
8. Y. Chen, J. Zhang,Y. Wang, L. Zhang, R. Julien, K.Tang and N. Balasubramanian, *Biosens. Bioelectron.,* **23**, 1390 (2008).
9. Q. Liu, J. Yu, L. Xiao, J.C.O. Tang, Y. Zhang, P. Wang and M. Yang, *Biosens. Bioelectron.,* **24,** 1305 (2009).
10. CDC (Centers for disease Control and Prevention). MMWR, **47(RR-14)**, 1 (1998).
11. USDHHS (U.S. Department of Health and Human Services), Oral health in America: a report of the surgeon general---executive summary. Rockville (MD): US Department of Health and Human Service, National Institute of Dental and Craniofacial Research, National Institutes of Health, p.1-13, (2000).
12. L.R. Arias, C.A. Perry and L. Yang, *Biosensors and Bioelectronics*, **25(10)**, 2225 (2010).
13. L. Yang, L.R. Arias, T.S. Lane, M.D. Yancey and J. Mamouni, *Anal. Bioanal. Chem.*, **339**, 1823 (2011).
14. J. Xu, H. Huang, C. Pan, B. Zhang, X. Liu and L. Zhang, *Int. J. Oral Maxillofac. Surg.,* **36**, 739 (2007).
15. J. Arredondo, A.L. Chernyavsky and S.A. Grando, *Cancer Biol. Ther.,* **5**, 511 (2006).
16. N. Onoda, A. Nehmi, D. Weiner, S. Mujumdar, R. Christen and G. Los, *Head Neck,* **23**, 860 (2001).
17. J. Mamouni and L. Yang, *Biomed. Microdevices*, **13**, 1075 (2011).

Terahertz Chemical Imaging of a Multicomponent Tablet in Pharmaceutical Applications

Katsuhiro Ajito [a,*], Yuko Ueno [a], Ho-Jin Song [a], Jae-Young Kim [a],
Emi Tamechika [a], Naoya Kukutsu [a],
Waree Limwikrant [b,c], Keiji Yamamoto [b], Kunikazu Moribe [b]
[a] NTT Microsystem Integration Laboratories
NTT Corporation, 3-1, Morinosato-Wakamiya, Atsugi, Kanagawa, 243-0198 Japan
[b] Graduate School of Pharmaceutical Sciences
Chiba University, 1-8-1, Inohana, Chuo-ku,
Chiba 260-8675, Japan
[c] Present Address: Faculty of Pharmacy, Mahidol University
447 Sri Ayudhya Road, Ratchatewi, Bangkok 10400, Thailand
* Corresponding author, Email: ajito.katsuhiro@lab.ntt.co.jp

Molecular networks based on hydrogen bonds have resonant
frequencies in the terahertz (THz) region. THz spectroscopy is a
powerful tool for identifying molecular networks created by inter-
molecular or intra-molecular hydrogen bonds in pharmaceutical
and biomedical samples such as drugs, proteins, and cancer cells.
A THz chemical imaging (TCI) system was developed by
combining a THz time-domain spectrometer with a translational
stage to obtain two-dimensional distributions of molecular
networks in tablet samples. Since THz spectral peaks of
pharmaceuticals are broad at room temperature, multicomponent
chemical analysis is limited using the TCI system. In this paper, we
describe multicomponent chemical analysis of pharmaceuticals
using a sample chamber cooled by a cryostat. TCI measurement at
low temperature sharpens spectral peaks and/or shifts peak
frequencies, enabling us to determine the distribution of several
kinds of pharmaceutical chemicals within a tablet.

Introduction

Molecular networks are created by inter-molecular or intra-molecular hydrogen bonds,
which have resonant frequencies in the terahertz (10^{12} Hz, THz) region. THz
spectroscopy is a powerful tool for identifying such inter-molecular and intra-molecular
hydrogen bonds in pharmaceutical and biomedical samples, such as organic acids (1,2),
amino acids (3, 4), sugars (5), pharmaceuticals (6), proteins (7), and cancer cells (8, 9).
THz spectra of amino acids and pharmaceuticals have several peaks due to the presence
of hydrogen bonds, making quantitative analyses possible (3, 10). Applications such as
pathological examination of tissues or identification of drugs or explosives in postal
packages have received attention (11, 12). THz chemical imaging (TCI) based on THz
spectroscopy has the potential to reveal not only molecular distributions but also
molecular networks, which could lead to new medical diagnostic and drug evaluation
techniques. THz waves penetrate pharmaceutical tablets, enabling us to inspect the
homogeneity of their coatings (13) and to identify polymorphic forms of crystals (14, 15).

However, spectral peaks of chemical compounds in the THz frequency region are broad, so the number of chemical species that can be separated is limited.

This paper describes temperature-dependent THz imaging of pharmaceutical crystals using a TCI system for multicomponent chemical analysis. The system provides frequency-dependent THz images of a tablet at low temperature and enables us to determine the distribution of several kinds of pharmaceuticals within a tablet.

Experimental

Figure 1 is a diagram of the TCI system composed of a THz time-domain spectroscope (THz-TDS) with the vacuum chamber mounted on a 3-dimensional translational stage. For multicomponent chemical analysis, a cryostat was added to the vacuum chamber in the TCI system. The sample holder inserted into the chamber has a quartz plate bottom, and can be cooled down to a temperature of 120 K by the cryostat using liquid nitrogen and an electric heater. The stability of the cryostat is less than 2 K. The THz–TDS consists of a 9-fs near-infrared pulse laser (Integral Pro, Femtolasers), two gallium arsenide photoconductive antennas (AISPEC), a mechanical stage delay line, and mirrors. One photoconductive antenna is an emitter, and the other is a detector. A 13-fs near-infrared pulse laser (Fusion, Femtolasers) is also used for TDS measurement. The delay line is used to obtain a time-domain waveform, which is converted to a frequency-domain spectrum by Fourier transformation. The 3-dimensional translational stage with 0.1-mm-step resolution and the THz-TDS are controlled by a personal computer to obtain a THz time-domain spectrum at each point in a sample tablet. The acquisition time for obtaining a 12 × 12 mm^2 image is about 7 hours. The image consists of 60 × 60 pixels in 200-μm increments. The sample measurement spatial resolutions in the horizontal and vertical directions are about 0.5 and 1 mm, respectively. The number of accumulations for obtaining time-domain waveforms was 32, and no accumulations were acquired at each pixel position during THz imaging.

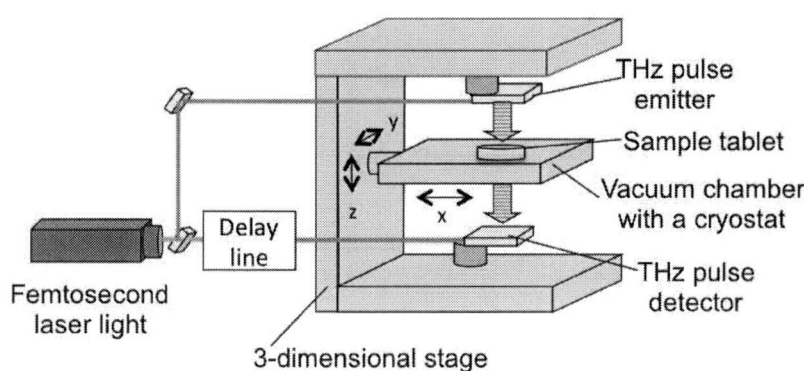

Figure 1: Diagram of TCI system composed of a THz-TDS, vacuum chamber with a cryostat, and a 3-dimensional translational stage.

Sample tablets studied were made of famotidine, which is a histamine H2-receptor antagonist for the prevention and treatment of stomach and intestinal ulcers. Two polymorphic crystalline forms of famotidine, form A and form B, were obtained by recrystallization in hot water and hot methanol aqueous solution, respectively. Original famotidine reagent was purchased commercially (ICN Pharmaceuticals). The polymorphic form was determined by differential scanning calorimetry (SSIC-5200, Seiko Instruments), measuring heat capacity at the melting point for each form. D-mannitol of analytical grade (Sigma-Aldrich) was used without further purification. D-mannitol is sometimes used in famotidine pills to control solubility in the body. The crystals were crushed into a fine powder and then diluted with polyethylene powder (Sigma-Aldrich). Tablets were then formed with a mechanical compress machine. The diameter and thickness of each 100 mg tablet was about 10 mm and 1.5 mm, respectively. For the test tablets used in imaging measurements, a piece of each tablet was cut and multiple tablet pieces were compressed within additional polyethylene powder to form a tablet 10 mm in diameter and about 2 mm thick.

Results and Discussion

Figure 2 shows the concentration dependent spectra of the pharmaceuticals studied, i.e., famotidine form A, famotidine form B, and D-mannitol. The form A spectrum shows peaks at 0.90 and 1.62 THz, and form B has peaks at 1.05, 1.19, and 1.34 THz in the range of 0.3 to 2 THz as shown in Figs. 2A and 2B. Those peaks less than 2 THz are indicated by the symbol '*'. The number of famotidine peaks over 2 THz is greater than those less than 2 THz, and broad spectral peaks over 2 THz overlap; therefore, for the use of the TCI technique, peaks less than 2 THz are very useful for determining various kinds of pharmaceuticals. Both form A and form B crystals are composed of the same molecule; however they have different crystal forms depending on the synthesis method used. Since THz spectroscopy detects mainly hydrogen bonds between molecules in crystals, both forms of famotidine show different spectral peak positions. The peaks of famotidine form A and form B were assigned to vibrational modes using the CONFLEX software (16). D-mannitol shows peaks at 1.10 and 1.48 THz in the range of 0.3 to 2 THz as shown in Fig. 2C.

A set of plots showing the average optical density at the peaks 0.90 and 1.62 THz in the famotidine form A spectrum against its concentration are shown in Fig. 3A. Linear fits for all plots are shown as solid lines. Similarly, two sets of plots showing the average optical density at the famotidine form B and D-mannitol spectral peaks are also given in Figs. 3B and 3C, respectively. The optical density, or absorbance, is proportional to the concentration of famotidine form A and B at the peaks in the 0.3 to 2 THz region; however, it is not proportional to the concentration of D-mannitol due to the baseline of the D-mannitol spectrum, which increases with increasing THz frequency. The origin of the baseline is probably related to scattering off the powder and is still under investigation. Furthermore, the spectral peaks of the three pharmaceuticals are too close together to determine their distribution in an image at 1.05 THz for famotidine form B or 1.10 THz for D-mannitol. In order to resolve this problem a cryostat was added to the vacuum chamber in the TCI system.

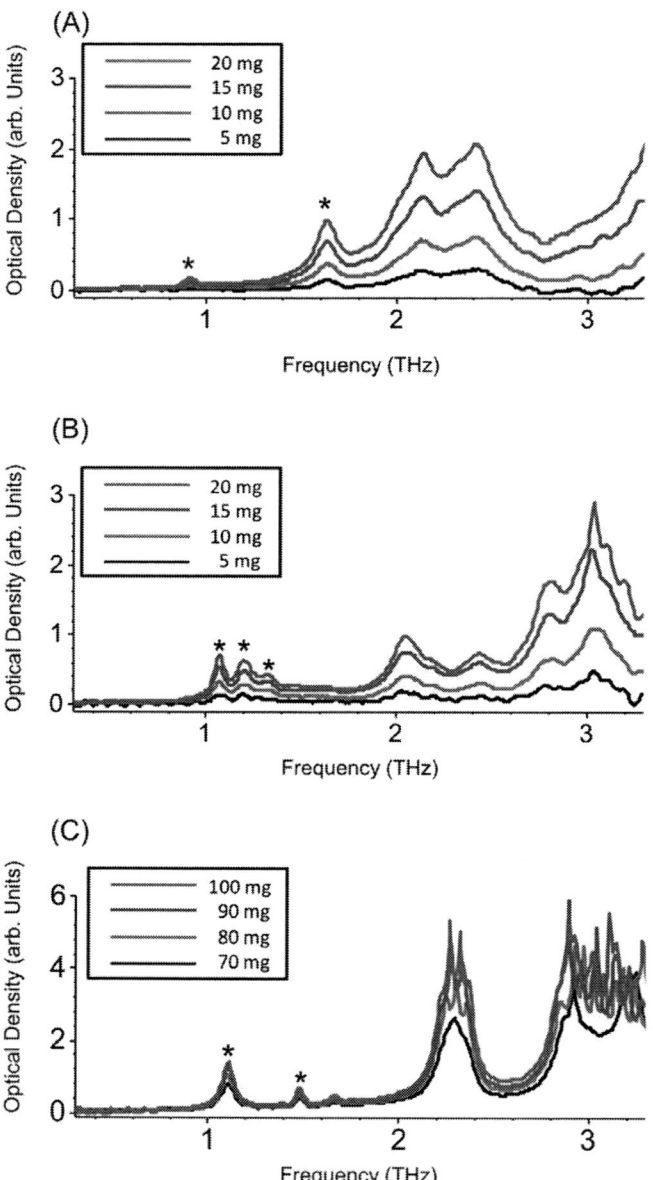

Figure 2: Concentration dependent THz spectra of famotidine (A) form A and (B) form B, and (C) D-mannitol. The peaks at less than 2 THz indicated by the symbol '*' are used in the following calibration process.

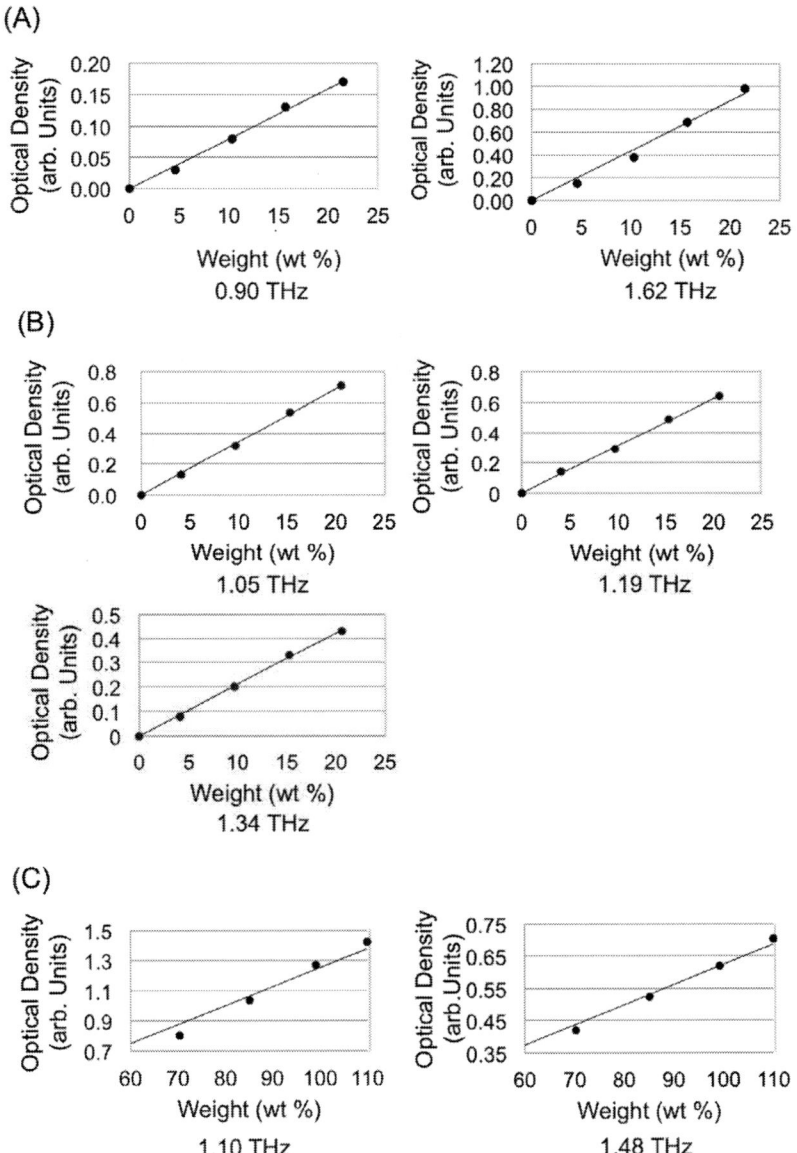

Figure 3: Plots of average optical density of famotidine (A) form A, (B) form B, and (C) D-mannitol against their concentration. Linear fit for each plot is shown as solid line.

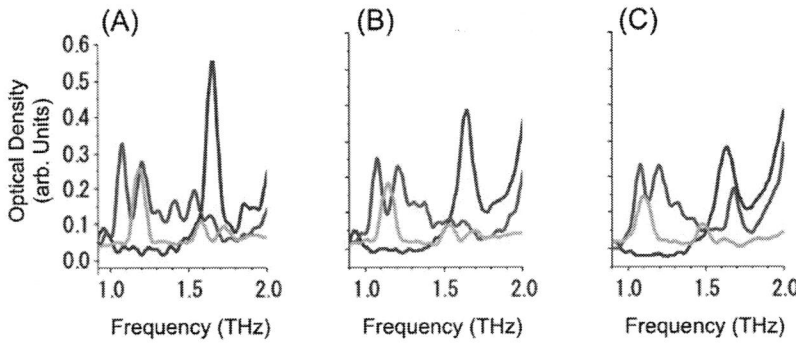

Figure 4: Temperature-dependent THz spectra of famotidine form A (blue line) and form B (red line), and D-mannitol (green line). The measurement temperatures were (A) 120 K, (B) 220 K, and (C) 298 K. The concentrations of all pharmaceutical tablets were 10 wt%.

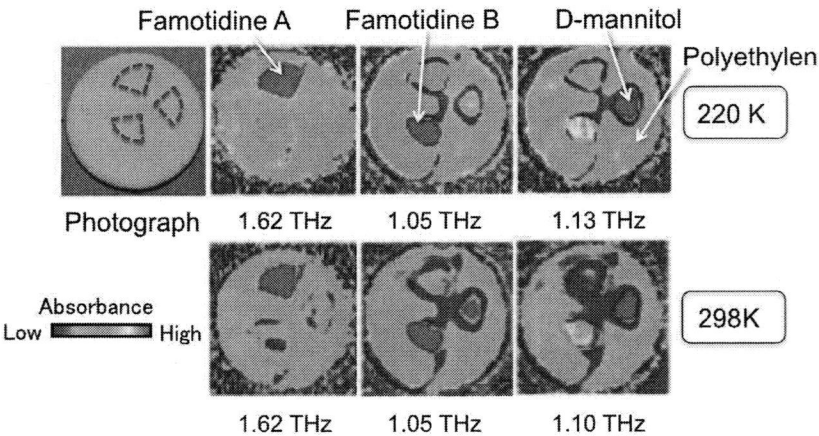

Figure 5: Temperature-dependent THz chemical images of a compressed pellet containing pieces of famotidine polymorphic forms A and B and D-mannitol crystals, measured at 220 K (upper) and 298 K (lower). The pellet is 10 mm in diameter and 2-mm thick.

Figure 4 shows a set of temperature-dependent THz spectra of famotidine forms A and B, and D-mannitol. The concentrations of all pharmaceutical tablets are 10 wt%. Peaks in the spectra of both famotidine forms are sharpened at low temperature and their frequencies are shifted very little. However, peaks in the spectrum of D-mannitol are greatly shifted, indicating that the neighboring peaks of famotidine form B and D-mannitol can be separated at low temperature.

Figure 5 shows temperature-dependent THz chemical images of a compressed pellet containing pieces of famotidine polymorphic forms A and B, and D-mannitol crystals measured at 220 K and 298 K. The peak frequencies of the famotidine B and D-mannitol spectral peaks are close to each other at 298 K, or room temperature. Therefore, the 298K image at 1.05 THz, which is one of the peak positions of famotidine form B, shows both famotidine form B and D-mannitol pieces. However, only famotidine form B distribution is observed at 1.05 THz in the image measured at 220 K. As a result, the two-dimensional distribution of three pharmaceuticals is almost completely determined at 220 K. This result indicates multicomponent chemical analysis of pharmaceuticals is possible when a cryostat is added to the vacuum chamber in a TCI system.

Conclusion

A TCI system was developed by combining a THz time-domain spectrometer with a translational stage to obtain two-dimensional distributions of pharmaceutical chemicals within a tablet. For multicomponent chemical analysis, a cryostat was added to the vacuum chamber in the TCI system. The use of the system at low temperature enables us to identify three pharmaceutical chemicals, e.g. famotidine polymorphic forms A and B, and mannitol crystals in a compressed tablet. Even though the frequencies of the THz peaks of famotidine B and mannitol are close to each other at room temperature, their two-dimensional distribution was obtained at a lower temperature, where the peaks were sharpened and/or shifted in frequency. This result indicates the potential utilization of multicomponent chemical analysis using a TCI system as a pharmaceutical evaluation technique, and could lead to new pharmaceutical and biomedical chemical imaging analyses.

Acknowledgements

We thank Ms. Maro Yamaguchi of WDB and Ms. Danielle Charron of University of Waterloo for their kind assistance with the THz measurements, Ms. Miho Kanazawa of NTT-AT for helping to purify the famotidine, and Dr. Kimihisa Aihara of NTT for their encouragement.

References

1. Y. Ueno and K. Ajito, *Anal. Sci.,* **23**, 803 (2007).
2. H. Hoshina, Y. Morisawa, H. Sato, A. Kamiya, I. Noda, Y. Ozaki, C. Otani, *Appl. Phys. Lett.,* **96**, 101904 (2010).
3. Y. Ueno, R. Rungsawang, I. Tomita, and K. Ajito, *Anal. Chem.,* **78**, 5424 (2006).

4. R. Rungsawang, Y. Ueno, I. Tomita, and K. Ajito, *J. Phys. Chem.* B, **110**, 21259 (2006).

5. R. Rungsawang, Y. Ueno, I. Tomita, and K. Ajito, *Opt. Express,* **14**, 5765 (2006).

6. K. Ajito and Y. Ueno, IEEE Trans. *THz Sci. Technol.*, **1**, 293 (2011).

7. S. Ebbinghaus, S. J. Kim, M. Heyden, X. Yu., M. Gruebele, D. M. Leitner, and M. Havenith, *J. Am. Chem. Soc.,* **130**, 2374 (2008).

8. S. J. Oh, J. Kang, I. Maeng, J.-S. Suh, Y. -M. Huh, S. Haam, and J. -H. Son., *Opt. Express,* **17**, 3469 (2009).

9. P. C. Ashworth, E. Pickwell-MacPherson, E. Provenzano, S. E. Pinder, A. D. Purushotham, M. Pepper, and V. P. Wallace, *Opt. Express,* **17**, 12444 (2009).

10. K. L. Nguyen, T. Friscic, G. M.Day, L. F. Gladden, and W. Jones, *Nature Materials,* **6**, 206 (2007).

11. H. Hoshina, Y. Sasaki, A. Hayashi, C. Otani, and K. Kawase, *Appl. Spectrosc.,* **63**, 81 (2009).

12. M. R. Leahy-Hoppa, M. J. Fitch, X. Zheng, L. M. Hayden, and R. Osiander, *Chem. Phys. Lett.,* **434**, 227 (2007).

13. J. A. Zeitler, Y. Shen, C. Baker, P. F. Taday, M. Pepper, and T. Rades, *J. Pharm. Sci.,* **96**, 330 (2007).

14. G. M. Day, J. A. Zeitler, W. Jones, W. T, Rades, and P. F. Taday, *J. Phys. Chem. B,* **110**, 447 (2006).

15. K. Ajito, Y. Ueno, H. -J. Song, E. Tamechika, and N. Kukutsu, *Mol. Cryst. Liq. Cryst.,* **538**, 33 (2011).

16. K. Ajito, Y. Ueno, H.-J. Song, E. Tamechika, and N. Kukutsu, *ECS Trans.*, **35**, 157, (2011).

CHAPTER 2

GAS AND LIQUID PHASE CHEMICAL SENSORS

Ceramic Gas Sensors to Oxide Nanostructures: Opportunities and Challenges

S. A. Akbar

Department of Materials Science and Engineering, Ohio State University, Columbus, Ohio 43210, USA

This article summarizes R&D efforts in the author's laboratory starting with the development of ceramic-based gas sensors to the fabrication of ordered/oriented oxide nanostructures exploiting intrinsic material properties. Over the past twenty years, our focus has been on the development of a series of high-temperature gas sensors specifically for combustion processes. We have developed both the resistive and electrochemical sensors and the underlying theme of our work has been the use of materials science and chemistry to promote high-temperature performance with selectivity. Our recent work has led to the development of surface modification techniques for the fabrication of oxide nanostructures that are inexpensive, highly scalable and do not require use of lithography. These nano-structures can be used as platforms for chemical sensing, photocatalysis, electroemission and biomedical applications. This article is concluded with preliminary results on chemical sensing along with future directions.

Introduction

Chemical sensors are widely used for health and safety (e.g., air quality monitoring, detection of toxic, flammable, and explosive gases, and medical diagnostics), energy efficiency, and emission control in combustion processes, and industrial process control for improved productivity. There is a continuing need for the development of low-cost sensors for applications in automotive, aerospace, food-processing, heat treating, metal processing and casting, glass, ceramic, pulp and paper, utility and power, and chemical and petrochemical processing industries. Besides being sensitive, many applications require these sensors to be selective and fast-responding. Our work at Ohio State University (OSU) has specifically focused on the development of a series of high-temperature gas sensors for combustion processes [1-10]. The underlying theme in our sensor development has been the use of materials science and chemistry to promote high-temperature performance with selectivity. We have developed both the resistive as well as the electrochemical sensors [11, 12]. In a resistive sensor one detects a change in the resistance due to the interaction of the target gas with surface adsorbed oxygen on the sensor film that involves charge transfer. An electrochemical sensor, on the other hand, uses a Galvanic cell that detects the difference in the chemical potential/activity of the target gas between the sensing and the reference electrodes. The sensor response in terms of emf signal is created by the electrochemical reactions that occur at the triple phase boundary (TPB) of the electrolyte, the electrode and the gas phase.

The sensor work at OSU began with a very modest project funded by Orton Ceramic Foundation in 1990. The initial focus was on ceramic oxide-based bulk sensors that eventually led to extensive work on thick-film sensors funded by the Edison Materials Technology Center (EMTEC) of the State of Ohio. The first breakthrough came by winning a Collegiate Inventor's Award in 1993 for the development of a TiO_2-based CO and H_2 sensor. Gradually, the research became more focused on materials chemistry and sensing mechanism leading to a collaborative effort involving chemists, physicists and other engineering disciplines. This led to the establishment of the Center for Industrial Sensors and Measurements (CISM) at OSU in 1996 that was funded in two phases through 2004. CISM was co-funded by the National Science Foundation (NSF), the State of Ohio and a consortium of industries. The focus at CISM has primarily been on the development of high-temperature gas sensors for the detection of combustion gases with major success in CO, CO_2, O_2 and NO_x sensors. CISM-developed sensors, thus far, has earned 3 R&D 100 awards, 1 NASA TGIR award, 8 patents with 2 being licensed, more than 100 peer-reviewed journal articles, and completion of more than 20 graduate degrees. More recently, the focus has expanded to thin-film sensors as well as sensors based on nano-structured oxides with emphasis on low-temperature applications.

This article is based on the presentation as part of the ECS Sensor Division Outstanding Achievement Award. This article presents synopsis of key results produced at OSU over the past 20 years in the author's laboratory starting with bulk and thick-film sensors to thin-film and nano-structure sensors. Hence, it represents an overview article summarizing highlights of results published previously and the readers may refer to the original papers for details.

Specific Gas Sensors

Oxide-based CO Sensor

Change of electrical conductance of a metal oxide in presence of a gas is a relatively simple method of detection. ZnO, SnO_2, TiO_2 and Ga_2O_3 have been extensively studied for gas sensing applications [13-15]. Addition of dopants to modify sensitivity, selectivity and stability has also been exhaustively studied. Nonetheless, three major drawbacks of the metal oxide sensors at high temperatures (400-800°C)

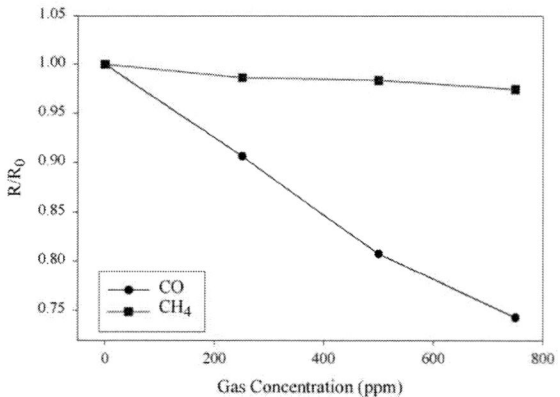

Figure 1: Change in resistance of an ALC sample at 600°C upon exposure to CH_4 and CO. [adapted from ref.16]

remain and they are: (i) cross-sensitivity (interference from other gases), (ii) drift of the sensor response and (iii) poor stability and reproducibility.

To achieve better stability and reproducibility for CO sensing at temperatures above 600°C, we have taken the strategy of doping TiO_2 (anatase) with lanthanum oxide, which provides microstructural, crystallographic and electrical stability during long-term operation at high temperatures, thus minimizing drift. To the La stabilized anantase, we then add CuO (labeled ALC) for improving sensitivity [16]. Figure 1 shows the results of changes in the electrical resistance of an ALC sample upon exposure to CH_4 and CO at 600°C. The sensor response is defined as the ratio of steady-state resistance (R) of the sensor at a given concentration of the target gas to that in the absence of the target gas (R_o). As evident from Figure 1, the sensor response to CH_4 is practically nonexistent, making it selective to CO detection.

Oxide-based O_2 Sensor

For high temperature applications, potentiometric tube-type yttria stabilized zirconia (YSZ) sensors are most commonly used [17, 18]. However, such sensors require a source of reference oxygen (usually air) which implies the need for plumbing to get the air in or placement close to an air source. Isolating and sealing an internal oxygen reference would make it possible to eliminate the need for an air reference, thus allowing for unrestricted placement of the sensors and significant miniaturization. It is known that a metal/metal oxide mixture encapsulated within a ceramic superstructure and placed in intimate contact with a Pt electrode would generate a stable oxygen pressure [5]. These sensors have not been commercialized because ceramic seals that contain intermediary bonding agents strain against mismatched thermal expansions in the course of long-term high temperature operation and fail.

We have used grain boundary sliding for sealing, which involves heating the materials under a load [19, 20]. The procedure to make the oxygen sensor package using grain boundary sliding is demonstrated in Figure 2. The electrolyte, ring and bottom wafer were made of 3 mol% yttria-stabilized tetragonal zirconia polycrystals (YTZP).

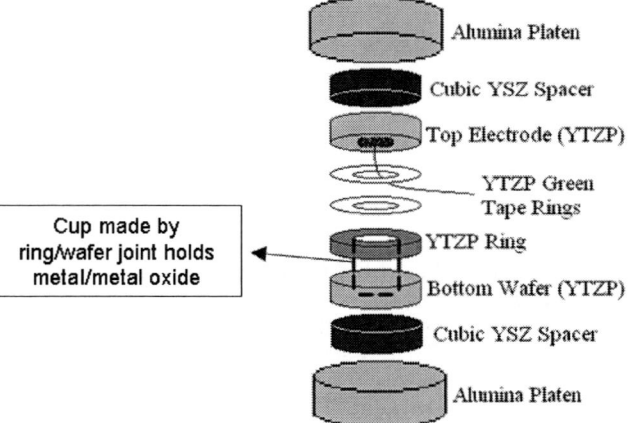

Figure 2: Components of the sensor package. The cubic YSZ spacers were necessary as bonding occurred between YTZP and alumina during initial joining. [adapted from ref. 19]

The devices were constructed by sealing 100% metal oxide in the reference chamber (PdO, RuO_2, NiO).

To assemble the sensor, the "sandwich" in Figure 2 was compressed in an argon atmosphere at temperatures ranging from 1250°C to 1290°C in a universal testing machine (Instron, Model 1125) at crosshead speeds of 0.01-0.02 mm/min and a strain rate of 4×10^{-5} s^{-1}. Upon reaching the target temperature, the system was left under a 5 N load for 30 min to attain thermal equilibrium. In order to complete the sensor fabrication, a glass plug was applied to the region of the sensor package where the Pt wire breached the inner-to-outer environment. Figure 3 shows the sensing of a Pd/PdO-containing reference electrode to changes in external oxygen concentration (3, 5, 7, 10, 14, 21 % O_2) over 24 h at 700°C. Over an eight-day test cycle, there was no baseline drift, no loss of sensitivity and the sensor exhibited near-Nernstian behavior.

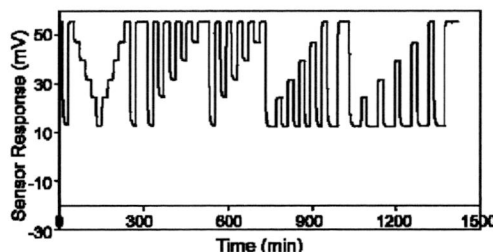

Figure 3: Performance of the Pd/PdO-based sensor at 700°C to changing external oxygen concentrations. [adapted from ref. 19].

Oxide-based CO_2 Sensor

There are several types of commercially available CO_2 gas sensors and most of them are based on Non-Dispersed Infra-Red (NDIR) and electrochemical methods. NDIR type CO_2 sensors allow highly specific detection via the absorption of CO_2 in the infrared region [21]. However, because of bulk size, limited operation temperature range (<328 K) and high cost, their applications are not widespread.

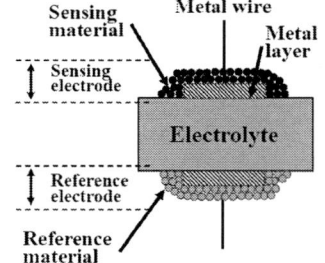

A solid-state electrochemical CO_2 sensor with Li_3PO_4 electrolyte was developed in our laboratory in early 2000 [9]. For the sensor structure, we adopted the open reference system with a bi-phase mixture of Li_2TiO_3 and TiO_2 as the reference electrode and Li_2CO_3 as the sensing electrode, as shown in Figure 4. While the interaction of the CO_2 gas on the sensing electrode allows its detection, the reference electrode fixes Li+ activity as illustrated by the electrochemical reactions.

Figure 4: Schematic of the sensor with open reference electrode. [adapted from ref. 12.]

$$2Li^+ + CO_2(s) + \frac{1}{2}O_2(g) + 2e^- \leftrightarrow Li_2CO_3(s) \qquad \text{(sensing electrode)}$$

$$2Li^+ + TiO_2(s) + \frac{1}{2}O_2(g) + 2e^- \leftrightarrow Li_2TiO_3(s) \qquad \text{(reference electrode)}$$

The use of the bi-phase mixture as the reference electrode is a unique feature of this design allowing a simpler device than the closed reference system and associated gas-tight sealing challenges, particularly for high-temperature applications.

This sensor showed good sensitivity, selectivity and linear response (Fig.5) in the temperature range between 823 K and 873 K [9], though its response systematically deviated from ideal Nerstian values, especially at high temperatures and low CO_2 concentrations. Lee et al. [10] investigated two possibilities for the non-Nernstian behavior: (1) lack of reversibility of electrode reaction and (2) partial electronic conductivity of the electrolyte. Based on Electrochemical Impedance Spectroscopy (EIS) study, the reaction on $Li_2TiO_3 + TiO_2$ mixture was found to be sluggish, while on Li_2CO_3 was relatively fast. To solve the kinetic problem, addition of gold powder or porous sputtered gold electrode were investigated that improved the sensitivity closer to the Nernstian value as shown in Figure 5. However, since this could not explain the temperature dependence, mixed ionic-electronic conduction of Li_3PO_4 electrolyte was suspected. Based on EMF and Hebb-Wagner (HB) polarization measurements, an n-type electronic conduction of Li_3PO_4 electrolyte was confirmed, particularly in low CO_2 concentration ranges (~ppm levels) and at higher temperatures [10].

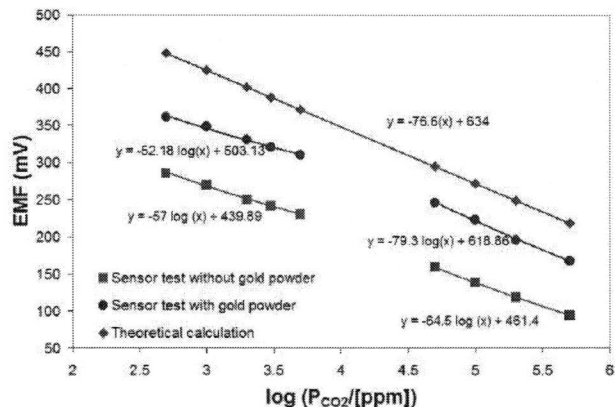

Figure 5: Comparison of EMF between sensors with and without gold powder at 773 K along with theoretical Nernstian values. [adapted from ref.12]

This sensor also showed humidity interference and Lee et al. [22] used Li_2CO_3-$BaCO_3$ bi-phase as the sensing electrode to eliminate the problem. Unlike other binary carbonate studies, Li_2CO_3 was coated by $Ba(NO_3)_2$ via a wet chemical method and then $BaCO_3$ outer-layer was formed on Li_2CO_3 in the presence of CO_2. The heat-treatment allowed eutectic reaction promoting adhesion of the sensing electrode as well. The sensitivity of the sensor for 5-20 % CO_2 was nearly Nernstian at 773 K and the humidity interference was practically eliminated [22].

While this sensor works well at high temperatures (above 673 K), for many applications there is a need for a low temperature sensor. For such a sensor, the selection of the electrolyte and the electrode materials is critical. Lithium-Lanthanum-Titanate ($Li_{3x}La_{(2/3)-x}TiO_3$, LLTO) is known as the highest lithium (Li)-ion conducting material at room temperature [23]. A sensor based on this electrolyte is currently being investigated for low temperature applications (<473 K). A remaining challenge for such a sensor is to overcome the slow CO_2 reaction kinetics at low temperatures.

Oxide-based NO_x Sensor

The two main components of nitrogen oxides in combustion environments are NO and NO_2, which generate opposite signals in an electrochemical sensor. Many NO_x sensors focus on NO since it is the major component of NO_x at high temperatures [24]. However, in lean-burn conditions, NO_2 is also present in significant amounts. Thus, sensors that can discriminate between the two gases or provide total NO_x (NO + NO_2) are required. Moreover, selectivity needs to be considered since typical lean-burn engine exhausts contain 1 to 10 ppm NO_x along with 20% CO_2, 10% H_2O, 3%O_2, 10 ppm NH_3, 1000 ppm hydrocarbons and 2000 ppm CO [25].

A typical electrochemical NO_x sensor design involves the use of two electrodes on an oxygen-ion conducting ceramic such as YSZ, as shown in Figure 6a. Prof. Dutta's group at OSU has extended our earlier work [8] and has obtained optimal results with a Pt electrode covered with Pt containing zeolite Y (PtY) as the reference electrode and WO_3 as the sensing electrode [26-28].

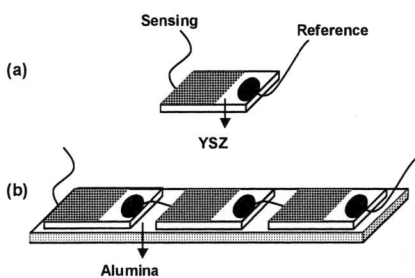

These electrodes were identified by temperature programmed desorption of NO from NO_x/O_2-exposed PtY and WO_3, and their ability to equilibrate a mixture of NO and O_2. Significant reactivity differences were found between the PtY

Figure 6: Potentiometric sensors composed of YSZ, WO_3 sensing electrodes, and PtY/Pt reference electrodes. (a) single sensor (b) 3-sensor array on an alumina substrate. [adapted from ref. 12]

and WO_3, with the latter being largely inactive toward NO_x equilibration. Thus, with PtY as the reference electrode, NO and NO_2 reaches equilibrium ($2NO + O_2 \leftrightarrow 2NO_2$) upon passing through the PtY before reaching the TPB and thus not contribute to an electrochemical signal. On the other hand, because of the poor chemical reactivity on WO_3, NO_x species reach the TPB chemically unmodified and undergo the electrochemical reaction, $2NO + 2O^{2-} \leftrightarrow 2NO_2 + 4e^-$, making this electrode primarily responsible for the sensor signal.

In order to obtain a total NO_x sensor ($NO+NO_2$), a second optimization step of passing the gases through the PtY filter prior to the sensor is necessary. It is required to maintain a temperature difference between the filter (typically at 400°C) and the sensor at 600°C, to obtain a signal. When NO or NO_2 passes through the PtY filter in the presence of oxygen, an equilibrium mixture of NO and NO_2 is formed. The NO/NO_2 ratio depends only on the filter temperature when the oxygen level is fixed, e.g. in 3% oxygen, NO_2 is 37.7% of total NO_x at 400°C and 5.3% at 600°C. Thus, a NO/NO_2 equilibrated mixture emerging from the PtY filter at 400°C will, upon contact with a sensor at 600°C, generate a new equilibrium (NO_2 converting to NO), and is the basis for the total NO_x sensing.

This filter/sensor combination also has an added advantage of reducing the interference to CO, CO_2, NH_3, propane, O_2, and H_2O. There are two strategies to increase sensitivity in this design. The first is to increase the temperature difference between the sensor and the filter [28]. A second method is by connecting sensors in series (Figure 6b), where the EMF is additive. It has been demonstrated that a 10-sensor array can detect NO concentrations as low as a few ppb [29, 30].

Sensors Based on Nano-structured Oxides

Recent work in the author's laboratory has led to the development of novel and inexpensive techniques to fabricate oriented and self-assembled oxide nano-structures without the use of lithography. All one needs is a high-temperature furnace allowing controlled atmosphere heat treatment by flowing reactive gases from gas cylinders. One such process creates crystallographically oriented nanofiber arrays of single crystal TiO_2 by hydrogen containing gas phase (H_2/N_2) reaction [31]. H_2/N_2 heat treatment was also used to grow nanofibers on polycrystalline SnO2 in regions of the sample coated with gold, showing directional growth on grains with crystal facets [32]. We have also developed a process to create nanofibers of TiO2 on Ti metal and Ti alloys via oxidation under a limited supply of oxygen [33, 34]. Lately, we have succeeded in converting the 1-D TiO2 nano-fiber grown by thermal oxidation to nano-dendritic titanates by hydrothermal treatment (Figure 7) [35]. We have developed yet another unique nano-structure during thermal annealing of an oxide such as Gd-doped CeO_2 (GDC) on top of another oxide substrate such as YSZ that self-assembles along the softest elastic direction of the substrate [36, 37]. A distinct characteristic of these methods is that they exploit intrinsic material properties to fabricate oriented and self-ordered nanostructures. These methods provide an economical way to mass-produce high surface area ceramic nano-structures that are attached to a

Figure 7: SEM micrograph of the dendritic barium titanate formed by hydrothermal conversion of TiO_2 nanowire grown on Ti foil by thermal oxidation in Ar (containing 10s of ppm of oxygen) bubbled through water.

substrate. This makes them ideal platforms with catalytic, gas-sensing, electronic and antimicrobial functions for a variety of chemical manufacturing, environmental, transportation, and biomedical applications. Here we report some preliminary gas sensing results using 1-D nano-structures fabricated by gas-phase reaction. The author's group has recently published a comprehensive review article on chemical sensing using 1-D nano-structured oxides [38].

To inspect the possibility of TiO_2 nanofibers as gas sensors, a sensing test of TiO_2 disks containing nanofibers was performed in H_2 gas at 400°C and the results were published earlier [39]. For comparison, a TiO_2 disk with no nanofibers (just sintered disks) was also tested. The resistance values of the TiO_2 nanofiber specimens exhibited a significant decrease upon exposure to increasing concentrations of hydrogen gas, which indicated an n-type behavior. Compared with a sintered titania, which showed practically no response, the nanofiber-based sensor exhibited good response because of increased surface area.

Development of mixed oxide nanostructure is another approach to investigate the performance of a TiO_2-based sensor. Carney et al. [40] synthesized $Ti_{0.9}Si_{0.1}O_2$ nanofibers by the nanocarving process developed by Yoo et al. [31]. Due to the difference in the sintering temperature both solid solution (at 1450 °C) and spinodally decomposed (at 1200 °C) $Ti_{0.9}Si_{0.1}O_2$ samples were obtained, which upon H_2-etching produced nanofiber and nano-lamelar structure, respectively. Both these samples showed good sensitivity toward H_2 gas with a response of ~1.3 for 2% H_2 at 400 °C. The response time and recovery time was 1–2 min and 5–7 min, respectively.

Finally, it is worth discussing some potential applications of TiO_2 nanofibers grown on Ti alloy particles via thermal oxidation. Extraordinary merit of these types of nano-structures for sensing has recently been demonstrated in SnO_2 by the author through his collaboration with Prof. Jong-Heun Lee's group [41, 42] and the plan is to extend similar measurements to TiO_2-based sensors. The sensor fabrication process is illustrated in Figure 8.

Another exciting possibility is to use two different Ti alloys, one which makes an n-type oxide and one which makes a p-type oxide after heat treatment. This would create an n/p junction where two fibers connect effectively making a diode. This diode would only allow the flow of electrons in one direction which could also aid in increasing the response and sensitivity of the sensor. Moreover, the dendritic titanate

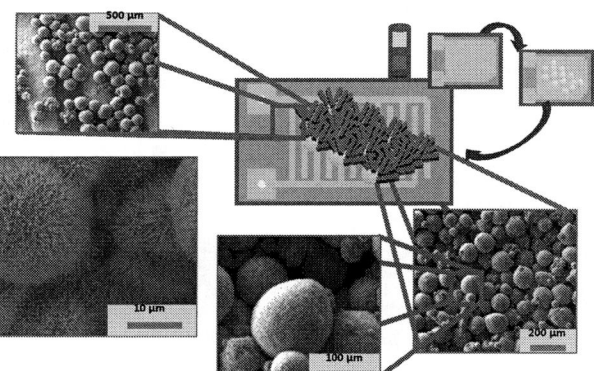

Figure 8: Schematic depicting the process for making a thick film sensor from Ti alloy particles. First, the particle paste is deposited on the interdigitated substrate via screen printing. Next, the sensor is heat treated to produce nanostructures.

(shown in Figure 7), being a ferroelectric, presents opportunities for selective detection of polar molecules such as NH_3, SO_2, H_2S and acetone.

Challenges and Opportunities

In spite of the recent advancements in gas sensors, further research and development are needed for their commercialization. Ceramic gas sensors based on stable oxides are attractive because of their low cost, simple structure, ease of fabrication and compatibility with electronic systems. However, for commercial success, major advances in these sensors are required in terms of selectivity, long-term stability and miniaturization. Moreover, the operation temperature of the sensor should cover a wider range and clearly there are challenges specifically for low-temperature ($< 200°C$) electrochemical sensors. The enhancement of electrochemical reactions at the TPBs particularly under dry condition will require multidisciplinary effort in the development of heterogeneous catalysts, new electrode and electrolyte materials, optimized electrode morphology, as well as study of electrochemical reaction kinetics. Additionally, novel fabrication processes should be developed for mass-scale production of devices. Traditional ceramic processing and wet chemical methods may not be practical for maintaining quality control from one device to another. Like the semiconductor industry, thin film technology can be a promising solution. The development of new processes will obviously open opportunities and challenges for the synthesis and characterization of novel materials, as well as fundamental studies on a variety of surface and interface problems.

The surface modification techniques described in this article represent an innovation in nano-processing without requiring lithography, making them cost-effective, that will assist in the proliferation of miniaturized devices such as nano-sensors. Moreover, there are opportunities for multidisciplinary studies involving characterization of surface/interface structures and characteristics of gas-solid interaction on these structures. These studies combined with computer modelling and simulation would aid in the fundamental understanding of the mechanisms that would allow creation of a wide-range of nanostructures and extend the technique to other technologically important ceramics. In terms of broader impact, these techniques should provide a new avenue of process pathways for making nano-structures on bulk, thin-film and particles with potential applications in chemical sensing including bio-sensing, micro- and nano-electronics, photo-catalysis and bio-medical devices.

Acknowledgements

This article is written based on Ph.D. dissertations of former students, Drs. Sehoon Yoo, Nancy Savage, John Spirig, Chong-hoon Lee, Carmen Carney, J.-C. Yang, Michael Rausher, Inhee Lee, Huyong Lee and Benjamin Dinan, and current student, Haris Ansari. Contributions of my collaborators, Profs. Prabir Dutta, Kenneth Sandhage and Suliman Dregia, are greatly acknowledged.

References

1. L. D. Birkefeld, A. M. Azad and S. A. Akbar, *J. Am. Ceram. Soc.*, **75**, 2694 (1992).
2. N. Savage, B. Chwieroth, A. Ginwalla, B. R. Patton, S. A. Akbar and P. K. Dutta, *Sens. Actuator B*, **79**, 17 (2001).
3. N. F. Szabo, H. Du, S. A. Akbar, A. A. Soliman and P. K. Dutta, *Sens. Actuator B*, **82**, 142 (2002).
4. A. Kohli, C.C. Wang and S. A. Akbar, *Sens. Actuator B*, **56**, 121 (1999).
5. A. K. M. S. Chowdhury, S. A. Akbar, S. Kapileshwar and J. R. Schorr, *J. Electrochem. Soc.*, **148**, G91 (2001).
6. B. Narayanan, S. A. Akbar and P. K. Dutta, *Sens. Actuator B*, **87**, 480 (2002).
7. C. Reddy, P. K. Dutta and S. A. Akbar, *Sens. Actuator B*, **92**, 351 (2003).
8. N. Szabo, H. Du, S. A. Akbar, A. Soliman and P. K. Dutta, *Sens. and Actuator B*, **82**, 142 (2002).
9. C. Lee, S.A. Akbar, and C.O. Park, *Sens. Actuator B*, **80**, 234 (2001).
10. C. Lee, P. K. Dutta, R. Ramamoorthy and S. A. Akbar, *J. Electrochem. Soc.*,**153**, H4 (2006).
11. C. O. Park and S. A. Akbar, *J. Matls. Sci.*, **38**, 4611 (2003).
12. S. A. Akbar and P. K. Dutta, in *Encyclopedia of Electrochemistry*, Savinell and C. C. Liu, Editors, Electrochemistry (2012).
13. U. Lampe, M. Fleischer, N. Reitmeier, H. Meixner, J.B. McMonagle, and A. Marsh, in *Sensors Update,* W. Göpel and J. Hesse, Editors, p. 1, VCH, New York (1996).
14. N. Yamazoe, G. Sakai and K. Shimanoe, Catal. Surv. Asia, **7**, 63 (2003).
15. G. Korotcenkov, *Sens. Actuator B*, **107**, 209 (2005).
16. N. Savage, S. A. Akbar and P. K. Dutta, *Sens. Actuator B*, **72**, 239 (2001).
17. R. Ramamoorthy, P. K. Dutta and S. A. Akbar, *J. Mater. Sci.*, **38**, 4271 (2003).
18. W. C. Maskell, B. C. H. Steele , *J. Appl. Electrochem.*, **16**, 475 (1986).
19. J. V. Spirig, R. Ramamoorthy, S. A. Akbar, J. L. Routbort, D. Singh and P. K. Dutta, *Sens. Actuator B*, **124**, 192 (2007).
20. J. V. Spirig, J. L Routbort, D. Singh, G. King, P. M Woodward and P. K. Dutta, *Solid State Ionics*, **179**, 550 (2008).
21. D. L. Auble and T. P. Meyers, *Boundary-Layer Meterology*, **59**, 243 (1992).

22. I. Lee, S. A. Akbar and P.K. Dutta, *Sens. and Actuator B*, **142**, 337 (2009).
23. Y. Inaguma, et al., *Solid State Communications*, **86**, 689 (1993).
24. N. Miura, G. Lu and N. Yamazoe, *Solid State Ionics*, **136**, 533 (2000).
25. F. Menil, V. Coillard and C. Lucat, *Sens. Actuator B*, **67**, 1 (2000).
26. J.-C. Yang and P. K Dutta, *Journal of Physical Chemistry C*, **111**, 8307 (2007).
27. J.-C.Yang and P. K. Dutta, *Sens. and Actuator B*, **136**, 523 (2009).
28. J.-C. Yang and P. K. Dutta, *Sens. and Actuator B*, **125**, 30 (2007).
29. G. W. Hunter, J. C. Xu, A. M. Biaggi-Labiosa, D. Laskowski, P. K. Dutta, S. P. Mondal, B. J. Ward, D. B. Makel, C. C. Liu, C. W. Chang and R. A. Dweik, *J. Breath Res.*, **5**, 037111 (2011).
30. S. P. Mondal, P. K. Dutta, G. W. Hunter, B. J. Ward, D. Laskowski and R. A. Dweik, *Sens. and Actuator B*, **158**, 292 (2011).
31. S. Yoo, S. A. Akbar and K. H. Sandhage, *Advanced Materials*, **16**, 260 (2004).
32. C. Carney, Y. Cai, S. Yoo, K. H. Sandhage and S. A. Akbar, *Journal of Materials Research*, **23**, 2639 (2008).
33. B. Dinan and S. A. Akbar, *Functional Nanomaterials Letters*, **2**, 87, (2009).
34. H. Lee, S. Dregia, S. Akbar and M. Alhoshan, *Journal of Nanomaterials*, **vol. 2010**, Article ID 503186, 7 pages, doi:10.1155/2010/503186 (2010).
35. B. Dinan, Ph.D. Thesis, The Ohio State University, Columbus, OH (2012).
36. M. Rauscher, S. A. Dregia, A. Boyne and S. A. Akbar, *Advanced Materials*, **20**, 1699 (2008).
37. H. Ansari and S. A. Akbar, *Sci. Adv. Mater.*, **3**, 821 (2011).
38. M. Arafat, B. Dinan, S. A. Akbar and A. S. M. A. Haseeb, *Sensors*, **12**, 7207 (2012).
39. S. Yoo, S. A. Akbar and K. H. Sandhage, *Ceramics International*, **30**, 1121 (2004).
40. C. M. Carney, S. Yoo and S. A. Akbar, *Sens. Actuators B*, **108**, 29 (2005).
41. H. -R, Kim, K. -I. Choi, J. -H. Lee, and S. A. Akbar, *Sens. and Actuator B*, **136**, 138 (2009).
42. J. -H. Lee, *Sens. Actuator B*, **140**, 319 (2009).

Selectivity Enhancement of YSZ-based VOC Sensor Utilizing SnO₂/NiO-SE via the Application of a Physical Gas-diffusion Barrier

T. Sato[a,b], M. Breedon[b,c] and N. Miura[c]

[a] Interdisciplinary Graduate School of Engineering Sciences, Kyushu University,
Kasuga-shi, Fukuoka 816-8580, Japan
[b] Japan Society for the Promotion of Science, Chiyoda-ku, Tokyo 102-8471, Japan
[c] Art, Science and Technology Center for Cooperative Research, Kyushu University,
Kasuga-shi, Fukuoka 816-8580, Japan

The sensing characteristics of an yttria-stabilized zirconia (YSZ)-based gas sensor utilizing an $SnO_2/NiO(+Al_2O_3)$ sensing-electrode (SE) were evaluated, with aspirations of selective ppb level detection of indoor volatile organic compounds (VOCs). The fabricated sensor gave a preferential response towards 50 ppb C_7H_8 (toluene), while exhibiting negligible responses towards interfering gases (C_3H_6, H_2, CO, NO_2, C_2H_5OH (ethanol)) at their average atmospheric concentrations. However, it was observed that toluene detection was strongly affected by high concentration ethanol, which can peak to approximately 10 times higher in concentration in indoor environments, owing to the ubiquitous use of alcohol in beverages, cooking and cleaning products. To overcome this limitation, a physical gas-diffusion barrier which was comprised of nano-Al_2O_3 particles was formed on the exposed edges of the NiO-SE, which assisted in the effective oxidation of ethanol in the SnO_2 catalytic layer, by avoiding direct penetration of ethanol into NiO-SE. The developed YSZ-based sensor utilizing an $SnO_2/Al_2O_3/NiO(+Al_2O_3)$-SE was found to be capable of selectively detecting aromatic VOCs lower than the indoor guideline concentrations, with low interference from high ethanol concentrations.

Introduction

Since the gas sensor was commercially developed in 1964 in Japan, various types of gas sensors utilizing different gas sensitive materials and transducer pathways have been developed including: semiconductor, quartz crystal microbalance, solid-electrolyte, optical fiber and field-effect-transistor have been reported. Among these sensors, solid-electrolyte (i.e. yttria-stabilized zirconia (YSZ))-based gas sensors which utilize electrochemical reactions of an analyte remain an attractive candidate for the development of high-performance sensors, due to YSZ's distinguished chemical, thermal and mechanical stability. Solid-electrolyte gas sensors can be roughly classified into 3 types with different sensing modes; potentiometric (1), amperometric (2) and impedancemetric (3).

Recently, we reported that a potentiometric sensor utilizing a YSZ electrolyte and a NiO sensing electrode (SE) was capable of detecting ppb levels of C_7H_8 (toluene) (4,5) which belongs to the class of volatile organic compounds (VOCs), and is one of the causative agents of indoor sick building syndrome (SBS) (6). For VOCs detection, Sasahara *et al.* reported a prototype highly sensitive adsorption/combustion-type VOCs sensor utilizing Pd/γ-Al$_2$O$_3$ sensing material fabricated by MEMS technique (7). While Kanda *et al.* reported a sensitive and selective semiconductor-type VOCs sensor utilizing WO$_3$/Pd material (8). Both sensors were reported to be capable of detecting very low concentration VOCs, as low as tens ppb levels which are lower than the Japanese guideline values, although ethanol which is not toxic to human body at low concentrations, has no established guideline value was considered to be one of the detectable VOCs, suggesting that the selective sensor only for toxic VOC detection can be more advantageous.

In our previous paper (9), sensing characteristics of the NiO-SE towards ethanol was also evaluated; unfortunately a 6-times larger preferential response towards ethanol, rather than the desired toluene was recorded. Gas selectivity is one of the most important sensing characteristics, which has attracted considerable attention in the sensor field. Our group reported several methodologies to improve the selectivity of YSZ-based sensors: the cancellation of electrical responses between ZnCr$_2$O$_4$-SE and ZnCr$_2$O$_4$(+Au)-SE for exhaust CO sensor (10), the addition of Pt nano particles into a In$_2$O$_3$-based SE, or the lamination of a ZnO catalytic layer on an SnO$_2$-SE for atmospheric C$_3$H$_6$ sensing (11,12) and the sputtering of nano Au onto NiO-SE for exhaust NO$_2$ sensing (13). Quite recently, we reported that the toluene selectivity of the YSZ-based sensor utilizing an NiO(+Al$_2$O$_3$)-SE was improved by the application of an in-line SnO$_2$ catalytic filter which oxidized interfering gases (especially ethanol) (9).

In this study, the YSZ-based sensor was fabricated by laminating the SnO$_2$ on the NiO(+Al$_2$O$_3$)-SE as a catalytic layer, aiming at the selective detection of indoor ppb level toluene. In addition, a new approach for selectivity enhancement was proposed by the application of a physical gas-diffusion barrier which consisted of nano-Al$_2$O$_3$ particles.

Experimental

Fabrication of the Sensing Device

First, 5 wt.% ethyl cellulose (Kanto Chemical, Japan) was dissolved into α-terpineol (Wako) in an ultrasonic bath operated for 12 h to achieve the appropriate viscosity of the α-terpineol mixture, later used as an organic binder. Commercial nano-Al$_2$O$_3$ powder (Sigma, Japan) mixed in different weight ratios (0, 5, 10 and 20 wt.%) was added to NiO powder (Kishida Chemical, Japan), and the resulting mixed oxides were thoroughly blended with the α-terpineol, with a weight ratio of 1:1. The relevant NiO(+Al$_2$O$_3$) pastes as well as a commercial Pt paste (TR-7601, Tanaka Kikinzoku, Japan) were respectively painted on the outer and inner surface of a hemi-spherically terminated YSZ-tube (8 mol.% Y$_2$O$_3$ doped ZrO$_2$, length: 300 mm, inside diameter: 5 mm, outside diameter 8 mm, Nikkato, Japan). The coated-YSZ was then thermally treated in air at 1000°C for 2 h, forming the outer NiO sensing electrode (SE) and inner Pt reference-electrode (RE).

For the fabrication of a laminated SE, pre-sintered SnO$_2$ (Kojundo Chemical Lab., Japan) powder was mixed with α-terpineol and was applied onto the NiO-SE which had already been sintered, while taking care to avoid contact between the SnO$_2$ and YSZ, which might alter sensing performance. Finally, a gas-diffusion barrier consisting of nano-Al$_2$O$_3$ particles (+α-terpineol) was formed at the edges of the SnO$_2$/NiO-SE, and subsequently sintering at 450°C for 12 h.

130

Evaluation of Sensing Characteristics

A digital electrometer (R8240, Advantest, Japan) was connected to a personal computer which was used to continuously record the electromotive force (*emf*) between SE and RE as the sensing signal. Here, the RE was always exposed to the atmospheric environment, while the SE was placed in flowing base gas (21 vol.% O_2 + N_2 balance) or a sample gas (50 ppb C_3H_6, 500 ppb H_2, 100 ppb CO, 40 ppb NO_2, 10-300 ppb C_7H_8, 80-480 ppb C_2H_5OH) which were diluted by the base gas. The total gas flow-rate was fixed at 100 cm^3 min^{-1} which contained 1.35 vol.% water vapor (RH $\approx$ 32%, at 25°C) and 400 ppm CO_2, in order to replicate a realistic atmospheric environment. The operational temperature of the fabricated sensor was fixed at 450°C for all measurements.

The polarization (I−V) curves of the sensors in the base gas and 50 ppb toluene were measured with a potentiostat (HZ-3000, Hokuto Denko, Japan) by applying electrical potentials across SE and RE. The scan rate was fixed at 1 mV min^{-1}.

SEM Observation

The SnO_2/NiO(+ 20 wt.% Al_2O_3)-SE edged with nano-Al_2O_3 layer was formed on a YSZ plate following the previously described procedure. The morphology of the SnO_2 catalytic layer, NiO(+Al_2O_3)-SE and nano-Al_2O_3 gas-diffusion barrier was observed by a field-emission scanning electron microscope (FE-SEM; JSM-6340F, JEOL, Japan) at 10 kV.

Results and Discussion

Effect of Al_2O_3 Addition on Sensing Performance

Different amounts of nano-Al_2O_3 powder were added into the SE paste to physically stabilize the NiO-SE, which must withstand future lamination of an SnO_2 catalytic layer on NiO-SE. Figure 1 shows the response transients towards different gases for the YSZ-based sensor utilizing (a) NiO-SE, (b) NiO(+ 5 wt.% Al_2O_3)-SE, (c) NiO(+ 10 wt.% Al_2O_3)-SE and (d) NiO(+ 20 wt.% Al_2O_3)-SE at an operational temperature of 450°C under humidified (RH $\approx$ 32%) and carbonized (400 ppm CO_2) conditions. Here, the sensitivity (Δemf) was defined as follows:

$$\Delta emf = emf_{(sample\ gas)} - emf_{(base\ gas)} \qquad [1]$$

It can be clearly seen from Fig. 1 that the sensitivity, selectivity as well as the response and recovery rates, even the base *emf* values for these sensors were similar, despite varying the Al_2O_3 additions. Since Al_2O_3 is a well-known inert material which has extremely low catalytic activity to gas-phase oxidation reaction, Al_2O_3 additions into NiO-SE will have no effect on the gas-phase reaction of sample gases in the SE layer. However, for the solid-electrolyte type gas sensor, the electrochemical catalytic activity at the SE/YSZ interface also plays an important role. The modified polarization (I-V) curves of the NiO(+Al_2O_3)-SEs operated at 450°C are presented in Fig. 2, which were measured in base gas (21 vol.% O_2) as well as in 50 ppb toluene, diluted with base gas under humidified (RH $\approx$ 32%) and carbonized (400 ppm CO_2) conditions. The anodic polarization curve (shaded markers) was obtained by subtracting current values in the base gas from those in 50 ppb toluene, while the cathodic polarization curve (unshaded markers) was estimated by reversing the sign of current values in the base gas (14-17). The anodic and the cathodic reactions are as follows:

Oxidation of C_7H_8 $(1/9)C_7H_8 + 2O^{2-} \rightarrow (7/9)CO_2 + (4/9)H_2O + 4e^-$ [2]

Reduction of O_2 $O_2 + 4e^- \rightarrow 2O^{2-}$ [3]

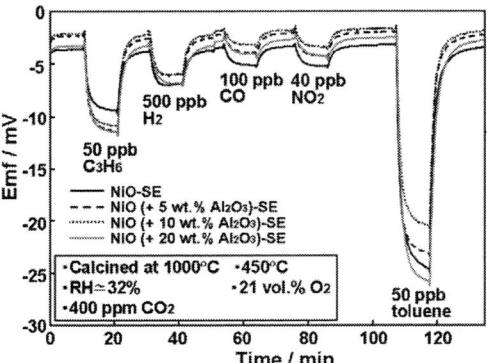

Figure 1. Response transients towards different gases for YSZ-based sensors utilizing NiO(+Al$_2$O$_3$)-SEs with different additions of Al$_2$O$_3$, at an operational temperature of 450°C under humidified and carbonized conditions (RH≈32%, 400 ppm CO$_2$).

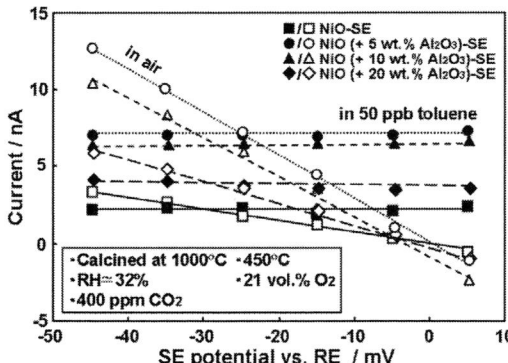

Figure 2. Anodic and cathodic polarization curves of YSZ-based sensors utilizing NiO(+Al$_2$O$_3$)-SEs with different additions of Al$_2$O$_3$, at an operational temperature of 450°C under humidified and carbonized conditions (RH≈32%, 400 ppm CO$_2$).

It can be seen from Fig. 2 that 5 wt.% Al$_2$O$_3$ addition into NiO-SE greatly increased both the anodic and cathodic currents as well as the gradient of the cathodic slope, which is speculated to be due to improved particle packing within the electrode and adhesion to the YSZ substrate caused by the addition of Al$_2$O$_3$. However, polarization current decreased after the further addition of Al$_2$O$_3$ (10 and 20 wt.%) into the SE. This behavior can be explained by the increase in the number of insulating Al$_2$O$_3$ particles which are largely electrochemically-inactive, decreasing the number of active electrochemical reaction-sites at an SE/YSZ interface. With respect to the anodic polarization curves, the obtained current was constant against the varying potentials for each SE, indicating that the toluene diffusion in SE layer is the rate-determining step rather than the electrochemical reaction of toluene at the triple phase boundary (TPB), which easily occurs at low concentration levels (18). Thus, among the SEs tested, NiO(+ 20 wt.% Al_2O_3) was selected as the base SE for the direct lamination of the SnO_2 catalytic layer.

Application of SnO_2 Oxidation Layer and Nano-Al_2O_3 Gas-diffusion Barrier

In our previous paper (9), we reported that the sensing characteristics of a NiO-SE which exhibited a selective response towards toluene rather than C_3H_6, H_2, CO and NO_2 was unfortunately affected by ethanol, which is a common interfering gas in an indoor atmosphere. We also reported that SnO_2 had selective oxidation characteristics towards ethanol, despite maintaining high toluene concentration. Thus, in this study, an SnO_2 layer was formed on the NiO-SE to improve toluene selectivity by selectively oxidizing ethanol before it reaches to the TPB. Figure 3 represents the response transients towards different gases for the YSZ-based sensor utilizing an SnO_2/NiO(+ 20 wt.% Al_2O_3)-SE at an operational temperature of 450°C under humidified and carbonized conditions (RH≈32%, 400 ppm CO_2). It is apparent that the SnO_2/NiO(+Al_2O_3)-SE is capable of detecting very low concentration toluene which is lower the value outlined in the Japanese guideline (70 ppb), with low interferences from other gases including 80 ppb ethanol which is higher than the average indoor concentration (19). However, it was also reported that the indoor ethanol concentrations sometimes spikes to several ppm levels, suggesting that higher ethanol concentration needs to be evaluated. Unfortunately, from Fig. 3, the toluene detection was found to be affected by high concentration ethanol, because the sensor gave relatively high sensitivities towards 240 ppb and 480 ppb ethanol.

For the laminated-type sensor utilizing an SnO_2/NiO(+Al_2O_3)-SE as shown in Fig. 3, selective ethanol oxidation is facilitated on the SnO_2 catalytic layer which improved the toluene selectivity drastically. However, in this electrode geometry, direct ethanol penetration from the exposed NiO-SE edges occurs without decomposing on the SnO_2 layer, which can result in the interferences by the unoxidized ethanol, as shown in Fig. 3.

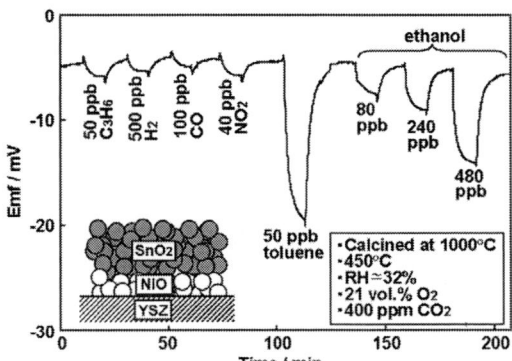

Figure 3. Response transients towards different gases for YSZ-based sensor utilizing an SnO_2/NiO(+ 20 wt.% Al_2O_3)-SE, at an operational temperature of 450°C under humidified and carbonized conditions (RH≈32%, 400 ppm CO_2). Inset graphic: the electrode structure of the fabricated sensor.

VOC Sensing Performance of SnO$_2$/nano-Al$_2$O$_3$/NiO(+Al$_2$O$_3$)-SE

In order to avoid incomplete ethanol oxidation, a nano-Al$_2$O$_3$ layer as a gas-diffusion barrier was edged on the peripheries of the NiO-SE, which helps to encourage further ethanol oxidation by hindering ethanol penetration. As a result, the nano-Al$_2$O$_3$ edging process further decreased ethanol sensitivity, as shown in Fig. 4. The fabricated sensor exhibited high sensitivity towards toluene and *m*-xylene, which are aromatic VOCs, but low sensitivity towards the aldehyde VOCs (formaldehyde). It is also seen from Fig. 4 that the sensor had low interferences from common interfering indoor atmospheric components such as C$_3$H$_6$, H$_2$, CO, NO$_2$, and ethanol, suggesting good potential for aromatic VOCs monitoring in indoor atmospheres. Further detailed experiments will be performed to clarify the mechanism for the selective sensing of aromatic VOCs.

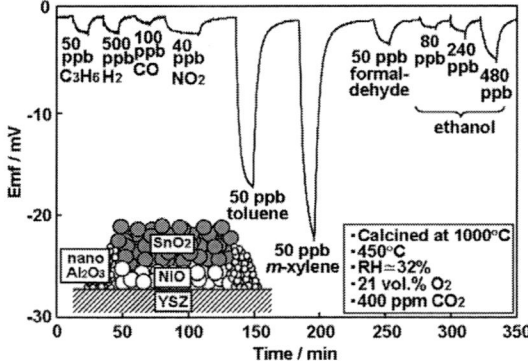

Figure 4. Response transients towards different gases for YSZ-based sensor utilizing an SnO$_2$/nano-Al$_2$O$_3$/NiO(+ 20 wt.% Al$_2$O$_3$)-SE, at an operational temperature of 450°C under humidified and carbonized conditions (RH≈32%, 400 ppm CO$_2$). Inset graphic: the electrode structure of the fabricated sensor.

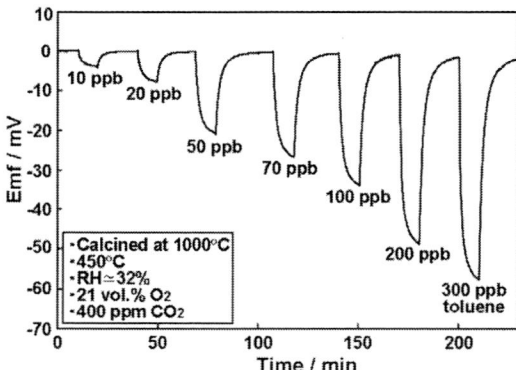

Figure 5. Response transients towards different concentration toluene for YSZ-based sensor utilizing an SnO$_2$/nano-Al$_2$O$_3$/NiO(+ 20 wt.% Al$_2$O$_3$)-SE, at an operational temperature of 450°C under humidified and carbonized conditions (RH≈32%, 400 ppm CO$_2$).

The toluene responses in the concentration range of 10-300 ppb were measured for the fabricated sensor (Fig. 5). As indicated in this figure, the sensor was capable of detecting very low concentrations of toluene, as low as 10 ppb, which is well below the requirements of the Japanese indoor guidelines (70 ppb). Additionally, reversible response and recovery curves were observed in Fig. 5, with stable base emf during the 4 h measurement. The average values for 90% response and recovery time in this toluene concentration range were respectively calculated to be approximately 5 and 7 min, which is thought to be acceptable for the timescale involved in indoor VOC monitoring.

The sensitivity plotted against different toluene concentrations are presented in Fig. 6, which are obtained from the results in Fig. 5. The toluene concentration is plotted on either linear or logarithmic scales. It can be seen that the sensitivity varies linearly both in linear and logarithmic scales, over different concentration ranges; 10 to 70 ppb on a linear scale, 100 to 300 ppb on a logarithmic scale. Garzon *et al.* reported that a sensor utilizing $Pt/Ce_{0.8}Gd_{0.2}O_{1.9}/Au$ showed an analogous behavior to the linear observation presented in this paper. They reported a similar trend of sensitivity being proportional to concentrations of H_2, propylene, CO, propane and methane in linear scales (20). If the diffusion process of sample gases in the SE layer was controlled by the rate of the whole reaction kinetics rather than the electrochemical reaction which occurs at SE/YSZ interface, obtained current values from triple phase boundary (TPB) should be proportional to sample gas concentration ($I \propto C$). In addition, the Butler Volmer equation describes the relationship between overpotential and current ($E \propto i$) at low overpotential, resulting potential are proportional to sample concentrations ($E \propto C$) which explains the sensitivity dependence at low concentration range in this paper. In contrast, we reported a different behavior; that the sensitivity was proportional to the logarithm of CO, H_2 and NO_2 concentrations (21-23), which is a common behavior for the mixed-potential type sensor, that prevails at the higher concentration range (50-300 ppb).

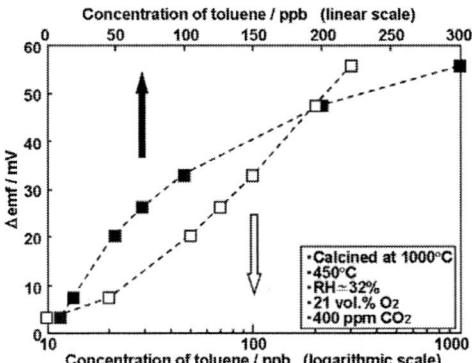

Figure 6. Sensitivity dependence on toluene concentration for YSZ-based sensor utilizing an SnO_2/nano-Al_2O_3/NiO(+ 20 wt.% Al_2O_3)-SE, at an operational temperature of 450°C under humidified and carbonized conditions (RH≈32%, 400 ppm CO_2).

Conclusions

The application of an SnO_2 catalytic layer as well as a nano-Al_2O_3 gas-diffusion barrier to a YSZ-based sensor utilizing an NiO(+Al_2O_3)-SE drastically improved VOCs selectivity by effectively oxidizing high concentration ethanol. The developed sensor was found to have preferable responses towards aromatic VOCs rather than aldehyde VOCs, which potentially showed a possibility to screen VOCs species. In addition, interferences by other common atmospheric gases such as C_3H_6, H_2, CO and NO_2 were negligible. The detection limit of toluene for the obtained sensor was 10 ppb, which is 1/7 of the indoor guideline value that was established by the Japanese government for the prevention of sick building syndrome. Considering the performance of the developed sensor, it has good potential for the monitoring of aromatic VOCs in ppb levels in real indoor environments.

Acknowledgments

This work was partially supported by Kyushu University, G-COE program on "Novel Carbon Resource Sciences", and Grant-in-Aid for Scientific Research (B) (22350095) as well as for JSPS Fellows (22-0353).

References

1. P.K. Sekhar, E.L. Brosha, R. Mukundan, M.A. Nelson, D. Toracco, F.H. Garzon, *Solid State Ionics*, **181**, 947 (2010).
2. S.I. Somov and U. Guth, *Sens. Actuators, B*, **47**, 131 (1998).
3. N. Miura, M. Nakatou and S. Zhuiykov, *Electrochem. Commun.*, **4**, 284 (2002).
4. T. Sato, V.V. Plashnitsa, M. Utiyama and N. Miura, *J. Electrochem. Soc.*, **158**, 175 (2011).
5. T. Sato, V.V. Plashnitsa, M. Utiyama and N. Miura, *Electrochem. Commun.*, **12**, 512 (2010).
6. A. Apter, A. Bracker, M. Hodgson, J. Sidman and W.Y. Leung, *J. Allergy Clin. Immun.*, **94**, 277 (1994).
7. T. Sasahara, H. Kato, A. Saito, M. Nishimura and M. Egashira, *Sens. Actuators, B*, **126**, 536 (2007).
8. K. Kanda, T. Maekawa, *Sens. Actuators, B*, **108**, 97 (2005).
9. T. Sato, M. Breedon and N. Miura, *Sensors*, **12**, 4706 (2012).
10. Y. Fujio, V.V. Plashnitsa, M. Breedon and N. Miura, *Langmuir*, **28**, 1638 (2012).
11. R. Wama, V.V. Plashnitsa, P. Elumalai, T. Kawaguchi, Y. Fujio, M. Utiyama and N. Miura, *J. Electrochem. Soc.*, **156**, J102 (2009).
12. R. Wama, V.V. Plashnitsa, P. Elumalai, M. Utiyama, N. Miura, Solid State Ionics, **181**, 359 (2010).
13. V.V. Plashnitsa, P. Elumalai, Y. Fujio and N. Miura, *Electrochim. Acta*, **54**, 6099 (2009).
14. J. Park, B.Y. Yoon, C.O. Park, W.J. Lee and C.B. Lee, *Sens. Actuators, B*, **135**, 516 (2009).
15. X. Liang, S. Yang, J. Li, H. Zhang, Q. Diao, W. Zhao and Geyu Lu, *Sens. Actuators, B*, **158**, 1 (2011).
16. N. Miura, T. Shiraishi, K. Shimanoe and N. Yamazoe, *Electrochem. Commun.*, **2**, 77 (2000).
17. N. Miura, J. Wang, M. Nakatou, P. Elumalai, S. Zhuiykov and M. Hasei, *Sens. Actuators, B*, **114**, 903 (2006).
18. F. Garzon, I. Raistrick, E. Brosha, R. Houlton and B.W. Chung, *Sens. Actuators, B*, **50**, 125 (1998).

19. E. Gallego, X. Roca, J.F. Perales and X. Guardino, *J. Environ. Sci.*, **21**, 333 (2009).
20. F.H. Garzon, R. Mukundan and E.L. Brosha, *Solid State Ionics*, **136**, 633 (2000).
21. Y. Fujio, V.V. Plashnitsa, M. Breedon and N. Miura, *Langmuir*, **28 (2)**, 1638 (2012).
22. M. Yamaguchi, S.A. Anggraini, Y. Fujio, M. Breedon, V.V. Plashnitsa and N. Miura, *Electrochimica Acta*, in press, doi:10.1016/j.electacta.2012.04.126
23. M. Breedon, S. Zhuiykov and N. Miura, *Materials Letters*, **82 (1)**, 51 (2012).

138

Rapid and Simple Immunoassay Based on Negative Dielectrophoresis with Three-Dimensional Interdigitated Array Electrodes

T. Yasukawa[a, b], H. Shiku[c], T. Matsue[c], and F. Mizutani[a]

[a] Graduate School of Material Science, University of Hyogo, Ako, Hyogo 678-1297, Japan
[b] JST-CREST, Tokyo, 102-0075, Japan
[c] Graduate School of Environmental Studies, Tohoku University, Sendai 980-8579, Japan

Rapid accumulations of particles and living cells with negative dielectrophoresis (n-DEP) have been applied to develop the rapid and simple immunosensing method. Grid formation of electrodes was fabricated by rotating the upper template interdigitated microband array (IDA) electrode by 90° relative to the lower IDA. When AC electric signal was applied to the bands on the upper and lower IDA, island organization was rapidly formed at the intersections with low electric fields. The accumulated particles were fixed through the immunoreactions between the antibody immobilized on the particle surface and analytes in the solution. The presence of the specific antigens allowed the formation of fixed complexes of particles. It is noted that the time required for single sensing is as short as 5 min and separation steps are eliminated in the presented procedure. We demonstrated the rapid and simple immunosensing using the aggregation of particles accumulated with DEP.

Introduction

The wide variety of methods to produce ordered patterns with micro- and nano-subjects has been proposed by a physical and chemical assembly process. Self-assembly monolayer (SAM) can control a well-defined spatial distribution of chemical properties based on an electrostatic interaction, surface tension, wettability and supramolecular assembly. Thus, particles are directed toward specific areas of the substrate that is pre-patterned with SAM through selective attachment or repulsion to form highly ordered particles arrays. The prepatterning processes for the SAM require complex chemical pretreatments of the substrate. Moreover, relatively long duration was required to form the well-ordered array. Therefore, a rapid and simple method with the flexibility to the patterning designs without any chemical pretreatment would be widely applicable to the mass production of micromaterial- and nanomaterial-based systems. We have studied particle manipulations with dielectroporesis (DEP) as a driving force for the rapid fabrication of particle patterns. Line and island patterns with particles and living cells can be rapidly prepared on the untreated substrate (1-3) and membrane (4, 5).

We report a novel method to quickly create island patterns of particles with DEP by using device consisting of three-dimensional construction of upper and lower interdigitated array (IDA) electrodes. When we applied AC voltage to the upper and lower indium tin oxide (ITO) band electrodes of IDA (ITO-IDA) fabricated on the upper and lower substrates, the particles rapidly accumulated at intersections of the electrode grids with relatively low electric fields enclosed with strong electric fields. As a result, the particles formed a pattern following the shape of the pattern of electric field strength.

However, the particle patterns formed by DEP are re-dispersed after the AC voltage is switched off. In our previous works, we used covalent bonding through cross-linking agents to immobilize the patterned particles on the solid supports (1) and encapsulation techniques in the photoreactive hydrogel (2). We applied the patterning with DEP to demonstrate simple immunosensing. We have recently applied dielectrophoretic line formation with particles to develop rapid and simple immunosensing platforms (6-9). When we applied AC voltage to the IDA electrodes, the particles modified with antibody were rapidly accumulated on the opposite substrates modified with antibody to form the line patterns. In the presence of analytes, accumulated particles were irreversibly captured on the substrates due to the formation of the sandwich type of immunocomplexes, while the particles were re-dispersed again after the AC voltage was switched off in the absence of analytes. More recently, we applied this method for rapid detection of cell surface antigen (10). In this work, we used the fixation of particles accumulated at the electrode intersections for rapid and simple immunosensing. The sandwich type immunocomplexes were formed between the accumulated particles modified with antibody in the presence of specific analytes.

Experimental Section

Fabrication of the device for particle accumulation

A dielectrophoretic patterning device was constructed with two ITO-IDA electrodes, which were fabricated by conventional photolithography and chemical etching using an etchant solution (ITO-02, Kanto Chemical, Tokyo, Japan) for 10–15 min under ultrasonication. Each band element was 2.0 mm long, 35 µm wide, and placed at a distance of 35 µm from adjacent bands. A 5-µm thick insulating layers of negative photoresist (SU-8 3005, MicroChem Corp. Newton, MO) were fabricated on the bandarray patterns by 750×750 µm to define the electrode areas exposed to the solution. Figure 1A schematically depicts the DEP patterning device. A 30-µm thick polyester film (Nitto Denko, Osaka, Japan) was used as a spacer. The upper ITO-IDA electrode was mounted on the lower ITO-IDA electrode with the polyester film to be completed the patterning device (Fig. 1A). Grid formation of electrodes was fabricated by rotating the upper template ITO-IDA by 90° relative to the lower ITO-IDA.

Island patterning with particles by n-DEP

Polystyrene latex microparticles (diameter 3 µm, 3.0×10^7 particles mL^{-1}, Polysciences Inc., Warrington, PA) were suspended in pure water for the manipulation of particles by DEP. Manipulation of particles by DEP was performed under an optical and fluorescent microscope (IX70, Olympus, Tokyo, Japan) equipped with a CCD camera (DP72, Olympus). A particle suspension (1.0–1.5 µL) was introduced into the patterning device. The AC voltage was applied between the ITO-IDA band electrodes to form the alternating electric field in the device (function generator 7075, Hioki E.E. Co., Ueda, Japan).

Figure 1B and 1C show the illustration and optical microscopic image of the arrangement of four band elements for particle manipulation. The AC voltage of typically

20 peak-to-peak voltage (V_{pp}) in the n-DEP frequency region (1.0 MHz) was applied to the band element (ii) on the lower ITO-IDA and the band element (b) on the upper ITO-IDA with the same phase, while the band elements (i) and (a) were connected to the ground. Then, the frequency applied to the band (ii) was switched to 500 kHz to remove the particles from the intersections of band (ii) and (b). To form another pattern, the frequency applied to the band (ii) was returned to 1.0 MHz and the frequency applied to the band (i) was switched to 500 kHz.

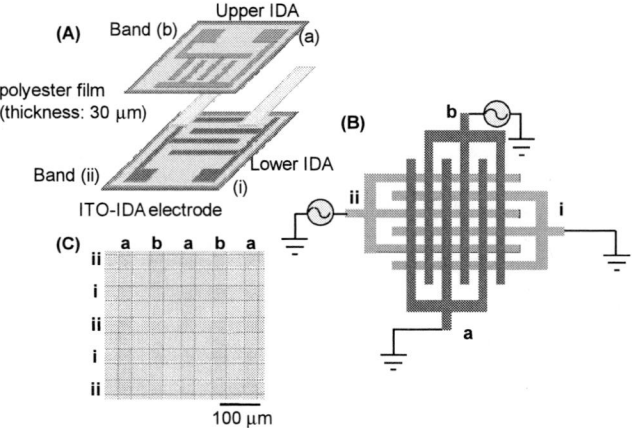

Figure 1. (A) Illustration of the DEP patterning device with upper and lower IDA substrates. (B) Illustration and (C) optical microscopic image of the arrangement of four band elements for particle manipulation in the DEP device.

Modification of particles with the anti-mouse IgG antibody

We modified silica particles (500 nm diameter, 1.5×10^{11} particles mL^{-1}) with goat anti-mouse IgG polyclonal antibody (anti-mouse IgG, Invitrogen, Tokyo, Japan) to develop the simple and rapid immunosensing method based on the particle accumulation. A goat anti-chicken IgG polyclonal antibody (anti-chicken IgG, Bethyl Laboratories Inc., Montgomery, TX) was used as a negative control. The surface of the silica particles were modified with anti-mouse IgG antibody. Silica particles were subsequently incubated into dimethyl sulfoxide (DMSO) containing 5% v/v 3-mercaptopropyltriethoxysilane (Shin-Etsu Chemical Co., Ltd, Tokyo, Japan) for 3 h and 0.5 mM succinimidyl-4-(N-maleimidomethyl) cyclohexane-1-carboxylate (Thermo Scientific) for 2 h to introduce the N-Hydroxysuccinimide ester which can react with primary amine groups on proteins to form stable amide bonds. After washing with DMSO and phosphate buffered saline (PBS), the modified silica particles were treated with 10 mM PBS containing 10 µg mL^{-1} anti-mouse IgG antibody for 12 h at 8 °C. Finally, the particles modified with the antibody were washed with PBS and stored in PBS at 4 °C before use.

Aggregation of particles accumulated at intersections by n-DEP

The different concentrations of mouse IgG which was used as a model target molecule in this work were prepared in the DEP medium containing 2.6 mM PBS and 100 mM sucrose. Then, the antibody-immobilized particles were added into the DEP medium. A few-μL DEP medium of the final suspensions containing the different concentration of analyte, and 1.5×10^{11} particles mL^{-1} silica particles were introduced into the DEP device immediately after adding the analyte to antibody-immobilized particles. We then performed particle accumulation with n-DEP. For the accumulation, an AC voltage of 20 V$_{pp}$, at a frequency of 1.0 MHz, was applied to the band (b) and (ii) for 5 min. The aggregation of particle by n-DEP-induced immunoreactions was investigated based on the dispersion motion of particles.

Results and discussion

Particle patterning with n-DEP

The patterns with particles were created by applying AC voltage in the frequency region for n-DEP to the upper band (b) and lower band (ii). Figure 2A shows a series of optical images showing the formation of first pattern with particles. A suspension of the particles in water was injected into the device to guide the suspended particles to the areas at intersections by the repulsive force arising from n-DEP. Application of an AC voltage (1.0 MHz) to the band (b) and (ii) resulted in the rapid formation of the pattern with particles accumulated at the intersections band (a-i) and (b-ii). Since the AC voltage with an identical frequency and phase was applied to the band (b) and (ii), the strong electric field formed at the intersections (a-ii) and (b-i), causing particles to repel from the areas around these intersections to accumulate them at the intersections (a-ii) and (b-i). It is noted that the time required to form the pattern (Fig. 2A-4) is only 1 s.

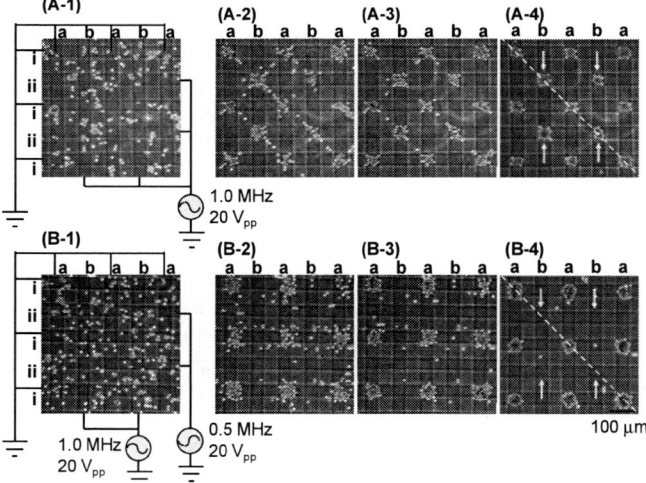

Figure 2. Series of optical images showing the formation of (A) first pattern and (B) second pattern with particles. Images were obtained (A-1) before and (A-2) 0.33, (A-3) 0.67 and (A-4) 1.0 s after the AC voltage was applied, and also obtained (B-1) before and (B-2) 1.0, (B-3) 2.0 and (B-4) 3.0 s after the application.

We also performed another patterning with particles using the same design of patterning device. Figure 2B shows a series of images for the formation of second pattern. For the formation of the second pattern, AC voltages with the frequency of 1.0 MHz and 500 kHz were applied to the band (b) and (ii), respectively. In this case, the electric field formed at the intersections (a-i) is relatively low compared to that at the intersections (a-ii) and (b-1), because both band (a) and (i) are connected to the ground. However, the application of the AC voltage with different frequencies to the band (b) and (ii) gave rise to the formation of the strong electric field at the intersection (b-ii). Thus, no particles were accumulated at the intersections (b-ii) pointed by the four white arrows in Fig. 2B-4. The pattern shown in Fig. 2B-4 was obtained 3 s after the AC voltages were applied to the bands. The average velocity to manipulate particles for the second patterning is slower than that for the first patterning.

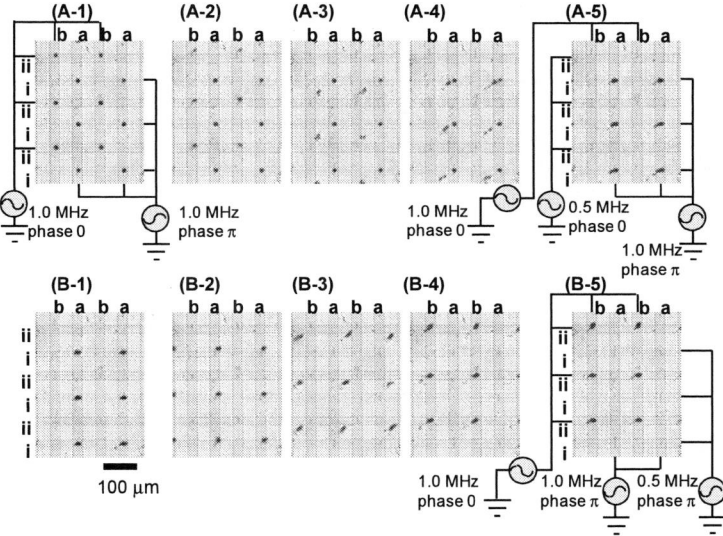

Figure 3. Series of optical images showing (A) the conversion from first pattern to second pattern and (B) the conversion from second pattern to another second pattern. Images were obtained (A-1) before and (A-2) 1.3, (A-3) 2.7 and (A-4) 4.0 s after the frequency for applied AC voltage was switched, and also obtained (B-1) before and (B-2) 1.3, (B-3) 2.7 and (B-4) 4.0 s after switching.

Conversion of particle patterns by switching the applied frequencies

We have demonstrated the conversion of the particle pattern from first pattern to second pattern and the alteration of the accumulated positions in second pattern by switching the applied frequencies. Figure 3A shows a series of optical images showing the conversion of the pattern from first to second. When we applied AC voltage (1.0 MHz, 20 V_{pp}) to the band (b) and (ii) and connected the band (a) and (i) to the ground, the dispersed particles were accumulated to the intersections (a-i) and (b-ii) to allow to the formation of first pattern (Fig. 3A-1). After the particles were formed in first pattern, the frequency for the band (ii) was switched to 500 kHz. As a result, the particles accumulated at the intersection (b-ii) were quickly moved to the intersection (a-i) to form the second pattern (Fig. 3A-5) due to the production of the strong electric field at the intersection (b-ii). The slight fluidic flow from the lower left to the upper right carries the particles accumulated at the intersection (b-ii) to the intersection (a-i) in the upper right. Then, the band (ii) was connected to the ground and the AC voltage with 500 kHz was applied to the band (i) after the production of the second pattern (Fig. 3B-1). The particles accumulated at the intersection (a-i) (Fig. 3B-1) were moved to the intersection (b-ii) to form another second pattern (Fig. 3B-5) in which the particles were accumulated at the different intersections (b-ii). These results indicate that the various particle patterns could be rapidly create in the present patterning device consisted in the three-dimensional interdigitated array electrodes by forming the various electric fields by AC voltage with different frequencies.

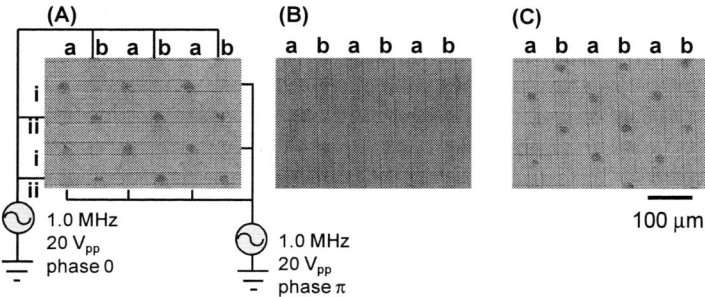

Figure 4. Particles aggregated at the intersections. Silica particles accumulated at the intersections (b-ii) and (a-i) (A) before and 60 s after the AC voltage was turned off (B) in the presence of 1.0 µg/mL mouse IgG and (C) in the absence of mouse IgG.

Application of rapid particle accumulation to the simple immunosensing

We applied the present particle accumulation technique to the rapid and simple immunosensing based on the aggregation behavior via immunoreactions. Silica particles modified with anti-mouse IgG were added to the DEP medium containing 10 ng/mL to 100 µg/mL mouse IgG and immediately injected into the device. The particles were accumulated at the appointed intersections for 5 min. Figure 4 shows the particles aggregated at the intersections after switching off the voltage. In the presence of 100 ng/mL and below mouse IgG, the pattern of accumulated particles broadened gradually and almost disappeared after 60 s (see Fig. 4A and 4B). In contrast, the accumulated particles maintained the clumped form even after switching off the voltage in the presence of 500 ng/mL and above, indicating that the sandwich type of immunocomplexes were formed between the antibodies on each particle and mouse IgG which works as a binder (see Fig. 4C). No particle aggregation was formed by using the particles modified with anti-chicken IgG antibody. The aggregation behavior can be found within 5 min and without any separation and washing-out steps for bounded and unbounded immunoreagents usually required for conventional immunoassay. It should be noted that the time required for immunosensing can be significantly shortened by the n-DEP process. Therefore, the presence of the specific analyte in the sample solution could be simply and rapidly distinguished from the aggregation behavior, although it is difficult to found the concentration of analytes quantitatively in the present procedure.

Conclusion

We investigated the patterning with particles in a DEP device with three-dimensional IDA electrodes as the template. The upper ITO-IDA was mounted on the lower ITO-IDA to fabricate the grid formation by rotating 90° relative. When the AC voltage with the frequency region for the n-DEP was applied to the band (b) and (ii), and the band (a) and (i) were connected to the ground, the particles were rapidly patterned at the intersections (b-ii) and (a-i) to create the pattern. On the other hand, the application of the AC voltages with different frequencies to the band (b) and (ii) caused to form another pattern without the particle accumulation at the intersection (b-ii). We used the particle aggregation via the formation of the immunocomplexes to demonstrate the rapid and simple immunosensing. Particles modified with antibody for mouse IgG quickly accumulated to designated intersections in DEP device. The presence of the specific antigens allowed to the aggregation of particles by the formation of immune-complexes between the antibodies on each particle and analytes in the sample solution. Since the particles redispersed into the bulk solution in the absence of the analytes after the AC voltage was switched off, we can easily found the presence of the target analytes from the aggregation behavior. The time required for single sensing is as short as 5 min and significantly rapid compared to the conventional assay format using the microtiter plate. The separation and washing-out steps usually required for conventional immunoassay are eliminated in the presented procedure. Moreover, removal of the modification steps for the device may allow to the repeated use of the device. The distinctive concentration of mouse IgG in this work is sub-μM level. If we assume the silica particle is a single molecule, the initial concentration of particle is corresponded to approximately 0.2 nM. Therefore, the least concentration required for the particle aggregation (approximately 3 nM) corresponded to one order of magnitude as high as that of the particles prepared. The distinctive concentration level may be improved with the decrease of the initial concentration of particles. We will investigate the relationship between the initial concentration of particles and the detectable concentration of analytes in the next investigation.

Acknowledgments

This work was partly supported by Shimazu Science Foundation.

References

1. M. Suzuki, T. Yasukawa, Y. Mase, D. Oyamatsu, H. Shiku and T. Matsue, *Langmuir*, **20**, 11005 (2004).
2. M. Suzuki, T. Yasukawa, H. Shiku and T. Matsue, *Langmuir*, **23**, 4088 (2007).
3. M. Suzuki, T. Yasukawa, H. Shiku and T. Matsue, *Biosens. Bioelectron.*, **24**, 1043 (2008).
4. H. J. Lee, T. Yasukawa, M. Suzuki, Y. Taki, A. Tanaka, M. Kameyama, H. Shiku and T. Matsue, *Sens. Actuators, B*, **131**, 424 (2008).

5. H. J. Lee, T. Yasukawa, M. Suzuki, S. H. Lee, T. Yao, Y. Taki, A. Tanaka, M. Kameyama, H. Shiku and T. Matsue, *Sens. Actuators, B*, **136**, 320 (2009).
6. H. J. Lee, T. Yasukawa, H. Shiku and T. Matsue, *Biosens. Bioelectron.*, **24**, 1000 (2008).
7. H. J. Lee, S. H. Lee, T. Yasukawa, J. Ramón-Azcón, F. Mizutani, K. Ino, H. Shiku and T. Matsue, *Talanta*, **81**, 657 (2010).
8. J. Ramón-Azcón, T. Yasukawa, H. J. Lee, T. Matsue, F. Sánchez-Baeza, M. –P. Marco and F. Mizutani, *Biosens. Bioelectron.*, **25**, 1928 (2010).
9. J. Ramón-Azcón, T. Yasukawa and F. Mizutani, *Anal. Chem.*, **83**, 1053 (2011).
10. H. Hatanaka, T. Yasukawa and F. Mizutani, *Anal. Chem.*, **83**, 7207 (2011).

Bismuth-film electrodes for Sn^{2+} sensing: the roles of grain size, preferred orientation ratio, and surface roughness

Chien-Hung Lien, Chi-Chang Hu[*,1], Yi-Da Tsai, David Shan-Hill Wang,

Laboratory of Electrochemistry & Advanced Materials
Department of Chemical Engineering
National Tsing Hua University
Hsin-Chu, 30013 TAIWAN

*Corresponding Author: Chi-Chang Hu, Professor
Department of Chemical Engineering
National Tsing Hua University
Hsin-Chu 30013, TAIWAN
Phone & Fax: +886-3-5736027
E-mail: cchu@che.nthu.edu.tw
1: Electrochemical Society Active Member

Abstract

Based on the electrodeposited method and the experimental strategy, the preferred orientation ratio of Bi deposits electroplated under a direct-current (dc) mode has been precisely controlled and predicted. According to the simple but reliable model for changing the preferred orientation ratio, the deposit plated at 28 °C and pH = 4.25 showed the highest f value. Bi deposits with various crystallite sizes, surface roughness, and preferred orientation ratio can be easily prepared and compared by electrodeposited method in this work. Consequently, the sensing ability of Sn^{2+} has been found to strongly depend on the preferred orientation ratio of Bi deposits, but is weakly affected by the crystallite size according to the current response of various bismuth deposits.

Introduction

Bismuth-film electrodes (BFEs) have been used as an alternative to replace mercury-film electrodes (MFEs) for the determination of heavy metal ions since 2000s.[1] The most significant advantage of BFEs is the environmentally friendly property since the toxicity of bismuth and its salts is negligible.[2] However, the unique analytical ability of BFEs in detecting heavy metal ions, roughly comparable to that of MFEs, has been attributed to the formation of "fused alloys" with heavy metals for bismuth. This property, analogous to the amalgams that mercury forms,[3] usually leads to use a stripping solution containing Bi(III)[4] or Bi oxide nanoparticles on substrates[5] for sensing heavy metal ions. The former method is generally complicated because of introducing the detecting solutions into the stripping solution. The second method usually involves the dispersion issue of Bi oxide nanoparticles onto the substrate for promoting the sensing ability. Accordingly, understanding the relationship between the microstructures and the heavy metal ions sensing ability of metallic Bi films is important for the future applications of BFEs.

There are several factors, such as complex agents (e.g., catechol and dimethylglyoxime),[6,7] surface roughness,[8] morphology,[9] crystallite size,[10] etc., were proposed to influence the heavy-metal-ion sensing ability of BFEs but a systematic comparison of these factors is a lack of this field. Fortunately, in our previous work,[11] the key factors for controlling the microstructures of Bi deposits have been found. Accordingly, this work systematically compares the influences of the crystallite size, preferred orientation ratio, and roughness factor of Bi deposits in the tin ion (II) sensing ability, which is important to the future applications of BFEs.

There are several factors affect the bismuth electroplating solutions such as composition,[12,13] pH,[14] complex agents,[15] and additives[16,17] as well as the deposition parameters e.g., current density[11] and deposition mode,[18-20] etc. Although the influences of preparation variables on the preferred orientation ratio and the qualities (e.g., uniformity, good adhesion, etc.) of bismuth deposits are complicated, we can only vary the pH and temperature of the plating bath to control the preferred orientation ratio (denoted as f), Bi(110)/Bi(012) facets of BFEs.

Experimental

Bismuth deposits were electroplated from a solution containing 0.06 M Bi(NO$_3$)$_3$·5H$_2$O (Hayashi, EP), 0.3 M citric acid (CA, Shimakyu, EP), 4000 ppm gelatin (from porcine skin, Type A ~ 300 Bloom (G2500-100G)), 0.3 M ethylenediamine-tetraacetic acid (EDTA, Riedel-deHaen, GA), and variable concentrations of polyethylene glycol (PEG, MW400, Shimakyu, EP) onto copper (99.5%, 1.094×10^{-2} cm^2, 1.18×10^{-1} cm in diameter) circular plates which have been deposited with a nickel film (ca. 2 µm). The pretreatment procedures of Cu/Ni substrates completely followed our previous work.[10-11] The Cu/Ni substrates, rinsed with deionized water, were placed in a 50 ml jacket cell and faced with a dimensionally stable anode (DSA) to electroplate Bi deposits. The dimensions of the stirring bar are 1.5 cm in length and 6mm in diameter. The dimensions of 50 ml jacket cell are 4.5 cm in diameter and 4 cm in height. The pH of plating baths was adjusted with concentrated HCl or NH$_4$OH. After deposition, these electrodes were repeatedly rinsed with deionized water and finally dried in a vacuum oven at room temperature.

The electrochemical analyses were performed with an electrochemical analyzer system, CHI 663a (CH Instruments, USA). All electrochemical analyses were carried out in a three-compartment cell. An Ag/AgCl electrode (Argenthal, 3 M KCl, 0.207 V versus a standard hydrogen electrode (SHE) at 25°C) was utilized as the reference electrode and a platinum wire served as the counter electrode. The solutions for analyses were degassed with purified nitrogen gas for 20 min before the electrochemical measurements and this nitrogen flow was passed over the solution during the measurements. Solution temperature was maintained at 25°C with an accuracy of 0.1°C by means of a water thermostat (Haake DC3 andK20).

Morphologies of all deposits were examined by a field emission scanning electron microscope (FE-SEM, Hitachi S-4700). X-ray diffraction patterns were obtained from an X-ray diffractometer (CuK$_\alpha$, Ultima IV, Rigaku). All solutions were prepared with deionized water produced by a reagent water system (Milli-Q SP, Japan) at 18 MΩ cm

and all reagents not specified were Merck, GR.

Results and Discussion

Figure 1 shows the microstructure difference of four typical BFEs prepared by

direct-current electroplating through controlling five preparation variables (pH, current

density of electroplating, PEG concentration, plating temperature, and stirring rate) in

this work. The main difference between deposits A and B is the crystallite size of Bi,

which can estimated from the full width at the half-maximum (FWHM) through the

XRD patterns. In this work, the crystallite size of the Bi deposits is estimated at 27.3°

(facet Bi(012)) which is the strongest peak on all XRD patterns (see Fig. 2). The

preferred orientation ratio, f, is defined as follow:

$$f = \frac{Bi(110)}{Bi(012)} = \frac{I_{39.7°}}{I_{27.3°}}$$
(1)

where I indicates the intensity of a X-ray diffraction peak. The average surface

roughness of Bi deposits was determined by means of an atomic force microscope

(AFM) at 3 random places (5×5 μm) of the Bi surface. Clearly, the preferred orientation

ratio and roughness factor of deposits A and B are almost the same; consequently, the

effect of grain size on the sensing ability of Sn^{2+} can be compared. Similarly, the

influences of the preferred orientation ratio and roughness factor can be demonstrated by comparing the Sn^{2+}-sensing ability of deposits B and C as well as deposits C and D, respectively. The above results indicate Bi deposits with different grain sizes, surface roughness, and preferred orientation ratio can be easily prepared by varying the electroplating variables.

Figure 3 shows the typical surface morphologies of deposits A-D. From a comparison between the SEM images in Fig. 3 and the f values in Fig. 1, there are certain relationships between the preferred orientation ratio and surface morphology of Bi deposits. For deposits A and B, the flower-like microstructure is visible for these deposits with high f values. However, a prickly rod-like shape is obtained for deposits C and with lower f values. Accordingly, the preferred orientation ratio may be considered as an important index in representing the intrinsic microstructure of Bi deposits.

In this work, the square-wave anodic stripping voltammetry (SWASV) is employed to ascertain which microstructure (i.e., crystallite size, preferred orientation ratio, or surface roughness) significantly affects the Sn^{2+} sensing ability of BFEs. Here, a solution containing 1.9 ppm Sn^{2+} is employed as a standard testing solution for comparing the Sn^{2+} sensing ability of all BFEs. The SWASV current responses of all BFEs are shown in Table 1. From a comparison between Table I and Fig. 1, the order of BFEs with respect to decreasing the peak current density is deposit B (59.7 μA cm^{-2}) > deposit A (22.7 μA cm^{-2}) > deposit C $\approx$ deposit D (~ 0 μA cm^{-2}). Accordingly, the Sn^{2+}

sensing ability of BFEs strongly depends on the preferred orientation ratio of Bi

deposits, but is weakly affected by its crystallite size.

The above results show that the key factor determining the detecting ability of tin

ion is the preferred orientation ratio. Fortunately, we use the experimental strategies to

screen out the key preparation factors affecting the preferred orientation ratio of Bi

deposits in our previous work[11] and a regression model can be used to predict the f value

of Bi deposits:

$$f = 0.321 - 0.040x_A + 0.032\,x_B - 0.114\,x_D - 0.026x_E + 0.056\,x_Ax_B$$

$$+ 0.063\,x_Ax_D + 0.043\,x_Ax_E + 0.030\,x_Bx_D + 0.066\,x_Ax_B\,x_D \qquad (2)$$

where x_A, x_B, x_D, x_E are the coded variables for factors A: pH, B: current density of

electroplating (mA cm^{-2}), D: temperature (^{o}C), and E: stirring rate (rpm). Although

equation (2) is a function of pH, current density, temperature, and stirring rate, factors A,

the f value of deposits will obviously increase with decreasing temperature since the

coefficient of this factor is negative (-0.114). Moreover, the influence of this factor is

the highest among all factors and interactions because the absolute value of its

coefficient is the largest.

Based on equation (2), we can extend the preferred orientation ratio of Bi deposits

by varying factors A and D with the regular step sizes of factors A and D equal to 0.5

(in pH) and 6.25 ^{o}C (in plating temperature), respectively. Such a study is called as the path of the steepest decent/ascent (PSD/PSA) and the corresponding results are shown

in Fig. 4. Clearly, a deposit with the highest f value is obtained under the plating

condition of run 4. Based on the results shown in Fig. 4., the preferred orientation ratio

of Bi(110)/Bi(012) facets for Bi deposits can be simply controlled by varying the pH

value and plating temperature from a simple plating bath basically containing 0.06 M

$Bi(NO_3)_3 \cdot 5H_2O$, 0.3 M citric acid, 4000 ppm gelatin, 0.3 M EDTA, and PEG (20 or 100

mM) under a constant current density (3 mA cm^{-2}).

The morphologies of Bi deposits with the f value varying from 1.32 to 0.01 are

shown in Fig. 5. The insets show the morphologies under a lower magnification. From

all insets in Fig.5, all deposits show uniform morphologies while the deposit with the

lowest f value (Fig. 5d) seems to be obviously rougher in comparison with the other

deposits. In Fig. 5a, irregular/ellipsoid- like particles between 50 and 100 nm are visible

although the particle boundaries are not very clear. With decreasing the f value, the

particles transform into irregular/triangular platelets (see Fig. 5b). In Fig. 5c, piling

triangular platelets up into aggregates is visible and sharp edges of crystallites are

clearly found. The piling of metallic rods is very clear in Fig. 5d, resulting in the

formation of many prickly rods with length longer than 600 nm on this deposit (see

inset). All the above results indicate that controlling the preferred orientation ratio of Bi

deposits leads the simultaneous change in the surface morphology of Bi deposits. Again, the preferred orientation ratio may be considered as an important index in representing the intrinsic microstructure of Bi deposits.

Conclusions

In this work, BFEs with various crystallite sizes, surface roughness, and preferred orientation ratio can be easily prepared by varying the electroplating variables including pH, current density of electroplating, PEG concentration, plating temperature, and stirring rate. From the comparisons among the crystallite size, preferred orientation ratio, and roughness of Bi deposits as well as the SWASV current response of Sn^{2+} sensing, the Sn^{2+} sensing ability of BFEs has been found to strongly depend on the preferred orientation ratio of Bi deposits, but is weakly affected by its grain size. Fortunately, pH and temperature of the plating solution exhibit monotonous influences on the f value in electroplating Bi deposits, which can be employed as the reliable variables for controlling the preferred orientation ratio of Bi deposits. This idea has been confirmed and the preferred orientation ratio of Bi(110)/Bi(012) facets can be gradually changed from 1.32 to 0.01.

Acknowledgments—the financial support of this work, by the National Science Council Taiwan and the Low Carbon Energy Research Center of National Tsing Hua University, is gratefully acknowledged.

References

1. J. Wang, J. Lu, S.B. Ho˘cevar, P.A.M. Farias, *Anal. Chem.*, **72**, 3218 (2000).

2. A. Economou, *Trends in Analytical Chemistry*, **24**, 334 (2005).

3. G.G. Long, L.D. Freedman, G.O. Doak, Bismuth and bismuth alloys, in: Encyclopedia of Chemical Technology, Wiley, New York, USA, 1978, pp. 912–937.

4. J. Wang, J. Lu, Ü. A. Kirgöz, S. B. Hocevar, B. Ogorevc, *Anal. Chim. Acta*, **434**, 29 (2001).

5. R. Pauliukaite, R.Metelka, I. Svancara, A. Krolicka, A. Bobrowski, K. Vytras, E. Norkus, K. Kalcher, *Anal. Bioanal. Chem.*, **374**, 1155 (2002).

6. E. A. Hutton, S. B. Hocevar , L. Mauko, B. Ogorevc, *Anal. Chim. Acta*, ,**580**,244 (2006)

7. S. Legeai ,S. Bois, O. Vittori, *J. Electroanal. Chem.*,**591**,93(2006)

8. C. Kokkino, A. Economoua, I. Raptis,C. E. Efstathiou, T. Speliotis, *Electrochem. Commun.*, 2795 (2007)

9. A. Bobrowski, A. Krolicka, J. Zarebski, *Electroanalysis*,**22**,1421(2010)

10. Y.D. Tsai, C.H. Lien, C.C. Hu, *Electrochimica Acta*, **56**, 7615 (2011)

11. C.H. Lien, C.C. Hu , Y.D. Tsai, and David S.H. Wang, *J. Electrochem. Soc.*, **159**, D260 (2012).

12. Y. Peng, D.H. Qin, R.J. Zhou, H.L. Li, *Mater. Sci. Eng. B*, **77**, 246 (2000).

13. Y.D. Tsai, C.C. Hu, *J. Electrochem. Soc.*, **156**, D58 (2009).

14. C.C. Hu, Y.D. Tsai, C.C. Lin, G.L. Lee, S.W. Chen, T.C. Lee, T.C. Wen, *J. Alloys Compd.*, 472, 121 (2009).

15. Y.D. Tsai, C.C. Hu, C.C. Lin, *Electrochim. Acta*, **53**, 2040 (2007).

16. Y.D. Tsai, C.C. Hu, *J. Electrochem. Soc.*, **156**, D490 (2009).

17. H. Sato, T. Homma, H. Kudo, T. Izumi, T. Osaka, S. Shoji, *J. Electroanal. Chem.*, **584**, 28 (2005).

18. C.G. Jin, G.W. Jiang, W.F. Liu, W.L. Cai, L.Z. Yao, Z. Yao, X.G. Li, *J. Mater. Chem.*, **13**, 1743 (2003).

19. L. Li, Y. Zhang, G. Li, X. Wang, L. Zhang, *Mater. Lett.*, **59**, 1223 (2005).

20. V. Richoux, S. Diliberto, C. Boulanger, J.M. Lecuire, *Electrochim. Acta*, **52**, 3053 (2007).

Captions

Table I The current response for four BFEs measured in 0.1 M citric acid solution containing 26.6 ppm Sn(II).

Figure 1 Four BFEs with various crystallite sizes, preferred orientation ratio, and roughness factor.

Figure 2 The XRD patterns of Bi deposits for sample a, b, c, d.

Figure 3 The FE-SEM photographs of Bi deposits for sample a, b, c, d.

Figure 4 Dependence of the f value of Bi deposits on pH (factor A) and temperature (factor D) of the plating bath in the PSD study.

Figure 5 The FE-SEM photographs of Bi deposits with the f value of (a) 1.32, (b) 0.82, (c) 0.56, and (d) 0.01.

Table I

Deposit	Current response $(\mu A/cm^2)$
A	22.7
B	59.7
C	~ 0
D	~ 0

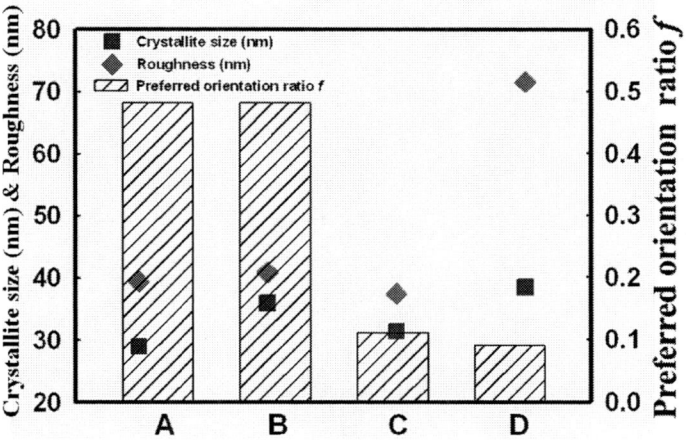

Figure 1

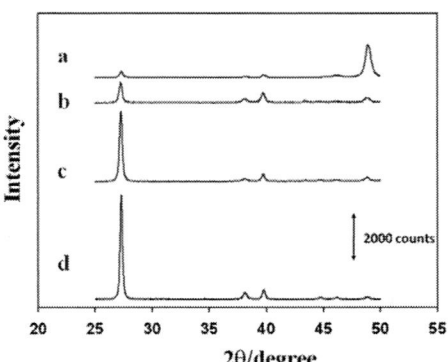

Figure 2

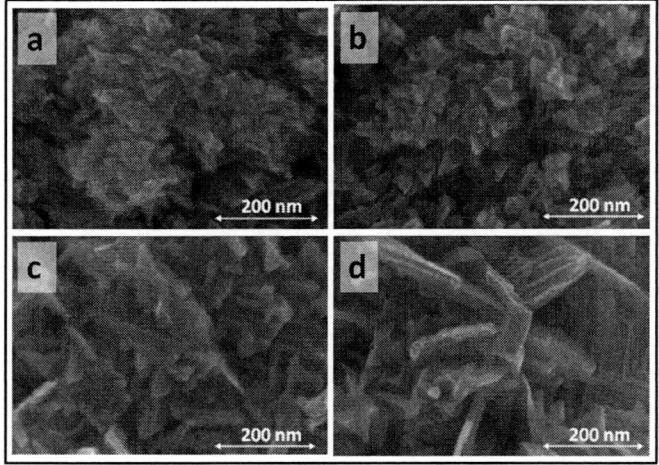

Figure 3

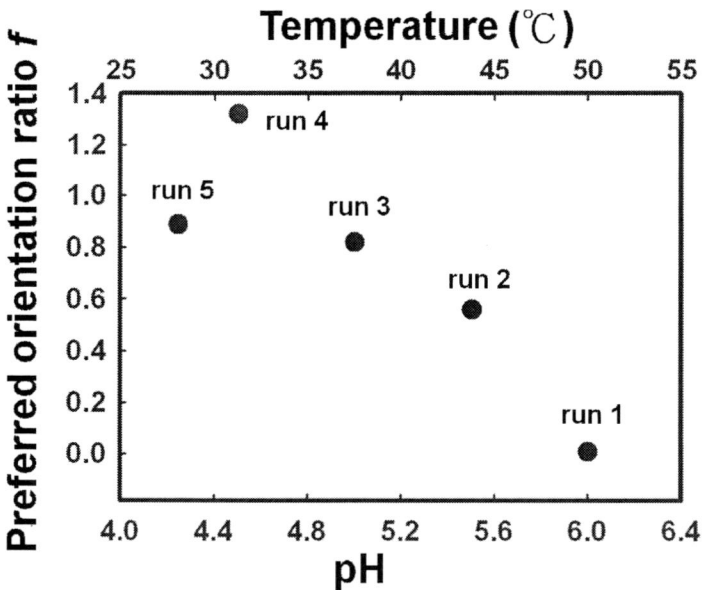

Figure 4

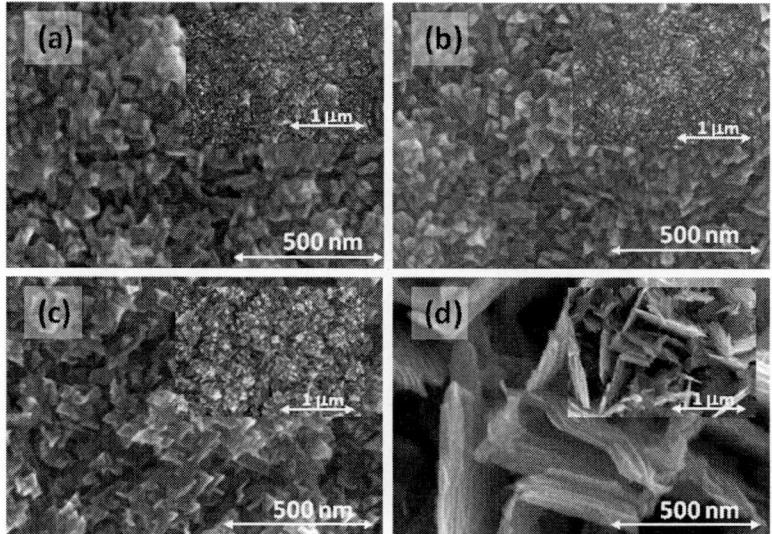

Figure 5

High-Throughput Separation Assay for NO Metabolites in Blood
Using Microfluidic Electrophoresis

S. Wakida[a,*], T. Miyado[a,b], K. Shimazu[c], Y. Shibutani[c],
T. Mizukami[d], K. Nose[d] and A. Shimouchi[d]

[a] Health Research Institute, National Institute of Advanced Industrial Science and
Technology (AIST), Takamatsu, Kagawa 761-0395, Japan
[b] Department of Industrial Chemistry, Kinki Polytechnic College,
Kishiwada, Osaka 596-0103, Japan
[c] Faculty of Engineering, Osaka Institute of Technology, Osaka, Osaka 535-8585, Japan
[d] Department of Cardiac physiology, National Cerebral and Cardiovascular Center,
Suita, Osaka 565-8565, Japan

As an alternative NO assay based on the Griess method, we
investigated a high-throughput separation assay of NO_2^- and NO_3^-
in human blood using an electrophoretic Lab-on-a-Chip with UV
detection directly. To measure the NO metabolites in real human
blood, we optimized the running buffer condition of an on-chip
pre-concentration technique, transient isotachophoresis (tITP) and
also applied voltages for a new dry-etched quartz glass chip in
microfluidic sample-introduction and separation. We achieved
acceptable sensitivity for the rapid NO metabolite separation assay
with 6.5 seconds and 25 seconds for separation and sample
introduction, respectively. As a result for the tITP application, the
limits of detection for NO_2^- and NO_3^- in human pooled serum were
improved to be 2.3 μM from 5.0 μM and 9.1 μM from 14.2 μM,
respectively.

Introduction

Nitric oxide (NO) is a biologically important short-lived reactive nitrogen species and
and has also been identified as an intrinsic mediator (1,2). Close associations between
NO production and cardiovascular diseases as well as lifestyle disease have attracted
much attention (3). Due to its short half-life, the amount of NO is estimated from the
concentrations of NO metabolites, *i.e.* nitrite (NO_2^-) and nitrate (NO_3^-) in clinical
chemistry.

There are many literatures on the determination of NO_2^- and NO_3^- (4). In biological fluids,
the Griess reaction is widely used (5,6), however, this method is relatively time-
consuming because of the complicated chemical reactions. We have developed a high-
throughput separation assay for NO_2^- and NO_3^- in human saliva using an electrophoretic
Lab-on-a-Chip (microchip capillary electrophoresis; MCE) with UV detection (7-10),
however, it is difficult to develop the separation of NO metabolites in human blood (10)
because of the lower sample concentration. In this study, we investigated the application
of an on-chip pre-concentration technique, transient isotachophoresis (tITP) (11), to
increase the sensitivity for the blood NO assay.

Experimental

Instrumentation

A MCE-2010 (Shimadzu, Kyoto, Japan) and a new MCE quartz glass chip (Type Ui) fabricated with a dry etching was used for the increase of optical path length (*c.a.* 1.5-hold) compared with Type U fabricated with a wet etching. The MCE chip was cross-type with a sample injection channel, a separation channel and four reservoirs at the ends of each channel as shown in Fig. 1. Electrophoretic sample injection and separation was performed by applying a voltage to four reservoirs of the chip, using computer-programmed sequencing of the MCE-2010. The separation behavior was observed with direct UV detection at 210 nm for the whole separation channel, using a linear photodiode array detector. A HM-60V pH meter (DKK-TOA, Tokyo, Japan) was also used for the pH adjustment for the running buffer.

Figure 1. A new MCE quartz glass chip (Shimadzu, MCE Type Ui) fabricated with a dry etching for the increase of 1.5-hold longer optical path (30 μm wide × 30 μm deep for all channels).

Chemicals and solutions

Our original running buffer with a similar composition to serum (10) was prepared and adjusted to pH 7.4 with 0.1 M hydrochloric acid. An artificial serum (10) was also prepared and adjusted to pH 7.4. Standard solutions were prepared with an artificial serum sample by adding NO_2^- and NO_3^- solutions (10). All chemicals were of analytical grade and used without further purification. All solutions were prepared in deionized water purified by Milli-Q Jr. (Millipore, MA, USA).

Sample preparation

Plasma and pooled serum (NESCOL-X; Oriental Yeast, Tokyo, Japan) was used as a human sample. Plasma was obtained by removing hemocyte by centrifugation at 3000 rpm for 15 min from whole blood added heparin as an anticoagulant. In the Griess method, plasma was deproteinized by sequential centrifugal ultrafiltration at 10,000×g for 30 min using Microcon Ultracel YM-10 (Millipore).

This study was approved by the Institutional Ethics Committee in AIST.

Results and discussion

Separation of pooled human serum separation without tITP

We have already reported the microfluidic separation under the condition of a novel running buffer similar to the blood serum composition in addition to a novel zwitterionic additive to decrease the adhesion of blood protein (10). As a result, the limits of detection (S/N=3) for NO_2^- and NO_3^- in pooled human serum were 53 μM and 41 μM, respectively. It was necessary to increase the sensitivity by a factor of 10. In order to obtain high sensitivity, under the optimized running buffer condition, we used a new MCE quartz glass chip for the increase of optical path length (*c.a.* 1.5-hold) as shown in Fig. 1. Using the above separation condition, we optimized the applied voltage for microfluidic control. The separation profile of pooled human serum added 50 μM NO_2^- and NO_3^- is shown in Fig. 2. The LOD for NO_2^- and NO_3^- in human pooled serum were considerably improved to be 5.0 μM and 14.2 μM.

Optimization of the serum separation with tITP

To improve the sensitivity, we investigated an on-chip pre-concentration technique, transient isotachophoresis (tITP). Under the strategy of near electrophoretic mobility for analyte with minimum UV absorbance (11), we evaluated several terminal-ion candidates for tITP as shown in Fig. 3. We selected 800 μM MoO_4^- as the proper terminal ion after optimization of concentration of several terminal-ion candidates.

We applied human pooled serum under the optimized condition with tITP. As a result, the limits of detection (S/N=3) for NO_2^- and NO_3^- in pooled human serum were improved to be 2.3 μM from 5.0 μM and 9.1 μM from 14.2 μM, respectively. As shown in Fig. 2, we achieved acceptable sensitivity for the rapid NO metabolite separation assay with 6.5 seconds and 25 seconds for separation and sample introduction, respectively.

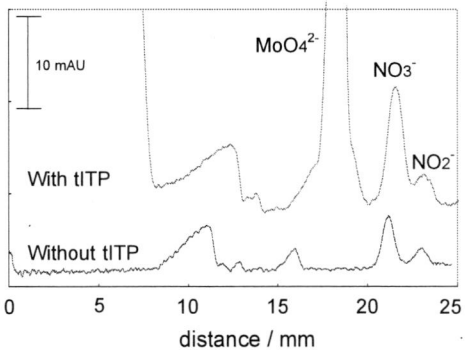

Figure 2. 6.5 second separation profile of pooled human serum added 50 μM NO_2^- and NO_3^- with/without tITP preconcentration technique. The microfluidic conditions were almost the same as in reference (10) without a terminal ion.

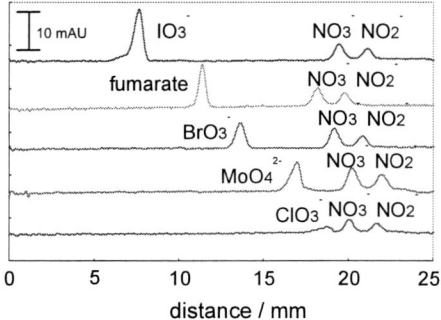

Figure 3. Effect of terminal ion on the MCE separation of 100 μM NO_2^- and NO_3^- in the artificial blood serum. The other conditions were as the same as in Fig. 2.

Challenge for the blood plasma NO assay with tITP

We applied the blood plasma assay under the above-mentioned condition for the pooled human serum. Under the tITP condition, a clear peak of NO_3^- in the human plasma was successfully obtained within 1 minute in total sample analysis, as shown in Fig. 4. Spiked calibration curves of NO_2^- and NO_3^- concentrations against the peak height showed a linear correlation (r>0.98). The relative standard deviations of peak height and peak area were 2.0 % and 5.0 % (n=6), respectively. The correlation coefficients between the MCE method and the conventional NO assay kit based on the Griess method were more than 0.903 (n=17) for the sum of NO_2^- and NO_3^-.

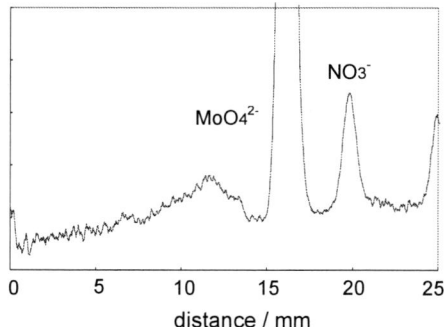

Figure 4. 6.5 second separation profile of human blood plasma with tITP pre-concentration technique. The other conditions were the same as in Fig. 2

Conclusions

We achieved acceptable sensitivity for the rapid NO metabolite separation assay with 6.5 seconds and 25 seconds for separation and sample introduction, respectively. It may be convenient NO assay after the sample pre-treatment from whole blood.

We have been challenging the whole blood separation assay with an on-chip pretreatment of whole blood sample to determine NO_2^- and NO_3^- in fresh blood precisely within only one minute for total analysis including blood sampling. We are now investigating on on-chip separation of plasma and blood red cell from whole blood in the MCE chip to establish the real bed-side monitor of NO metabolites.

References

1. R. M. Palmer, A. G. Ferrige and S. Moncada, Nitric oxide release accounts for the biological activity of endothelium-derived relaxing factor. Nature, 327, 524-526 (1987).
2. S. Moncada, R. M. Palmer and E. A. Higgs, Nitric oxide; physiology, pathophysiology, and pharmacology, Pharmacol. Rev., 43, 109-142 (1991).
3. D. Ozbasar, U. Toros, O. Ozkaya, M. Sezik, H. Uzun, H. Genc and H. Kaya, Raloxifene decreases serum malondialdehyde and nitric oxide levels in postmenopausal women with end-stage renal disease under chronic hemodialysis therapy, Am. J. Obstet. Gynecol. 36, 133-137 (2010).
4. M. J. Moorcroft, J. Davis and R. G. Compton, Detection and determination of nitrate and nitrite: a review, Talanta, 54, 785-803 (2001).
5. M. J. Follett and P. W. Ratcliff, Determination of nitrite and nitrate in meat products, J. Sci. Food Agric., 14, 138-144 (1963).
6. B. T. Alexander, K. L. Cockrell, M. B. Massey, W. A. Bennett and J. P. Granger, Elevations in plasma TNF in pregnant rats decreases renal nNOS and iNOS and results in hypertension, Am. J. Hypertens, 15, 170-175 (2002).
7. T. Miyado, Y. Tanaka, H. Nagai, S. Takeda, K. Saito, K. Fukushi, Y. Yoshida, S. Wakida and E. Niki, Simultaneous determination of nitrate and nitrite in biological fluids by capillary electrophoresis and preliminary study on their determination by microchip capillary electrophoresis, J. Chromatogr. A, 1051, 185-191 (2004).
8. T. Miyado, Y. Tanaka, H. Nagai, S. Takeda, K. Saito, K. Fukushi, Y. Yoshida, S. Wakida and E. Niki, High-throughput microfluidic device for NO assay based on electrophoresis separation, Proc. 3rd Int. IEEE EMBS Special Topic Conf. on Microtechnologies in Medicine and Biology, 66-68 (2005).
9. S. Wakida, T. Miyado, Y. Tanaka, H. Nagai, N. Naruishi, K. Yoshino, K. Matsuoka, K. Yoshida and E. Niki, Development of Lab-on-a-Chip for Rapid Salivary NO Assay, Chemical Sensors, 22, (Supplement B), 94-96 (2006).
10. T. Miyado, S. Wakida, H. Aizawa, Y. Shibutani, T. Kanie, M. Katayama, K. Nose and A. Shimouchi, High-throughput assay of nitric oxide metabolites in human plasma without deproteinization by lab-on-a-chip electrophoresis using a zwitterionic additive, J. Chromatogr. A, 1206, 41-44 (2008).
11. K. Fukushi, N. Ishio, T. Miyado, K. Saito, S. Takeda, S. Wakida and K. Hiiro, Capillary zone electrophoresis with on-line transient isotachophoresis for the determination of nitrite and nitrate in environmental waters: Effect of additional termination ions, Proc. 12th Int. Symp. on Capillary Electroseparation Techniques, 125-130 (2000).

ECS Transactions, 50 (12) 171-178 (2012)
©The Electrochemical Society

Effects of Surface Modification of Noble-metal Electrodes with Au on the H₂-sensing Properties of Diode-type Gas Sensors

T. Hyodo[a], T. Yamashita[b], and Y. Shimizu[a]

[a] Graduate School of Engineering, Nagasaki University,
1-14 Bunkyo-machi, Nagasaki 852-8521, Japan
[b] Graduate School of Science and Technology, Nagasaki University,
1-14 Bunkyo-machi, Nagasaki 852-8521, Japan

Diode-type TiO_2 film gas sensors equipped with noble-metal (Pd or Pt) electrodes (M/TiO_2, M = Pd or Pt) have been fabricated, and the effects of surface modification of the electrodes with Au on their H_2-sensing properties have been investigated at 250°C in both air and N_2 under dry and wet atmospheres. H_2 response of M/TiO_2 sensors in dry N_2 was much larger than that in dry air, but the surface modification of the Pd or Pt electrode with Au (Au/Pd or Au/Pt) improved the H_2 response, especially in dry air. In addition, the sensor with the Au/Pt electrode showed large and O_2-independent H_2 response under wet atmosphere.

Introduction

Hydrogen (H_2) is promising as a clean energy in the next-generation, and thus highly-sensitive and selective detection of H_2 leakage from various electric and electrochemical systems utilizing H_2 gas as an energy source (e.g., fuel cells) and storage systems for H_2 is indispensable for operating them safely and effectively. An exact quantitative analysis of H_2 concentration in breath is also important for monitoring activities of bacteria that inhabit the large intestine (mainly colon) and then diagnosing digestive and absorptive function of carbohydrate. Therefore, numerous efforts are directed to developing different types of H_2 detectors (e.g., semiconductor-type sensors (1), hot wire-type sensors (2), catalytic combustion-type sensors (3), electrochemical sensors (4) and thermoelectric sensors (5)). Our efforts have also been directed to developing diode-type H_2 sensors, by employing anodized TiO_2 thin films and Pd electrodes and we have found that these sensors show large H_2 response and relatively-excellent H_2 selectivity to other inflammable gases in a wide range of H_2 concentration under flowing both air and N_2 atmospheres (6-9). Alloying of Pd with Pt was quite effective in improving the H_2 response and the long-term stability (10-12). However, the H_2 response of these sensors varied markedly with oxygen concentration and humidity. Therefore, we have attempted to modify the surface of Pd-Pt electrodes with various polymers, in order to reduce the cross-sensitivity to other gaseous components (13, 14). In this paper, effects of surface modification of Pd and Pt electrodes with Au on the H_2 sensing properties of diode-type gas sensors have been investigated in both air and N_2 under dry and wet atmospheres.

Experimental

After a Ti plate ($5.0 \times 10.0 \times 0.5$ mm^3) was heat-treated at 600°C for 1 h in air, the half part of the Ti plate was anodized in 0.5 M H$_2$SO$_4$ aqueous solution at 20°C for 30 min at a current density of 50 mA cm^{-2}. A pair of noble-metal (Pd or Pt) electrodes were fabricated on both the TiO$_2$ thin film and the Ti plate by radio-frequency (rf) magnetron sputtering (Shimadzu, HSR-552S). In some cases, a small amount of Au was deposited additionally on the Pd or Pt electrode by the rf magnetron sputtering. Each condition for the rf magnetron sputtering of noble-metals is shown in Table I. Each electrode was connected with Au lead wire by using a Pt paste, and the electrical contact was ensured by firing at 600°C for 1 h in air. The obtained sensors with noble-metal (Pd or Pt) electrodes modified with and without Au were denoted as Au/M/TiO$_2$ and M/TiO$_2$ (M = Pd or Pt), respectively. Morphology of the TiO$_2$ thin film was observed by scanning electron microscopy (SEM; JEOL, JSM-7500F). The thickness of each electrode is listed in Table I. The schematic drawing of the diode-type gas sensor obtained is shown Fig. 1. The change in the chemical state of Au, Pd and Pt of the electrode surface was characterized by X-ray photoelectron spectroscopy using Al K$_\alpha$ radiation (XPS, Kratos, ACIS-TLATRA DLD), and the binding energy was calibrated using the C 1s level (284.5 eV) from usual contamination.

A dc voltage of 100 mV was applied to the sensor under forward bias condition (M(+)-TiO$_2$-Ti(-)) and the sensing properties to 8000 ppm H$_2$ balanced with air or N$_2$ under dry or wet (absolute humidity (AH): 6.80, 9.40 or 12.8 g m^{-3}) atmospheres were measured at 250°C after pre-heat treatment under the same gaseous condition at 600°C for 1 h. The sensor current at 10 min after the exposure to H$_2$ balanced with air or N$_2$ was regarded as H$_2$ response, since a base current in air or N$_2$ was negligibly small in comparison with that

Table I. Sputtering conditions and thickness of electrodes of all sensors.

Sensor	Power to target / W (Time / s)			Thickness of electrode / nm
	Pd	Pt	Au	
Au/Pd/TiO$_2$	300 (300)	-	40 (10)	111
Au/Pt/TiO$_2$	-	300 (360)	40 (10)	84.2
Pd/TiO$_2$	300 (1080)		-	287
Pt/TiO$_2$	-	300 (480)	-	77.7

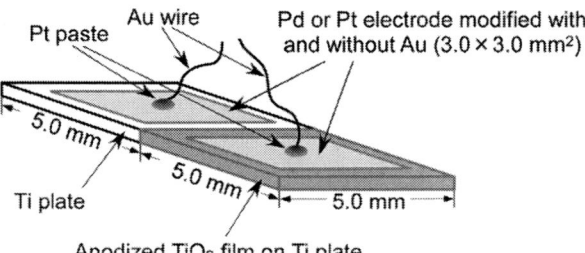

Figure 1. Schematic drawing of a diode-type gas sensor.

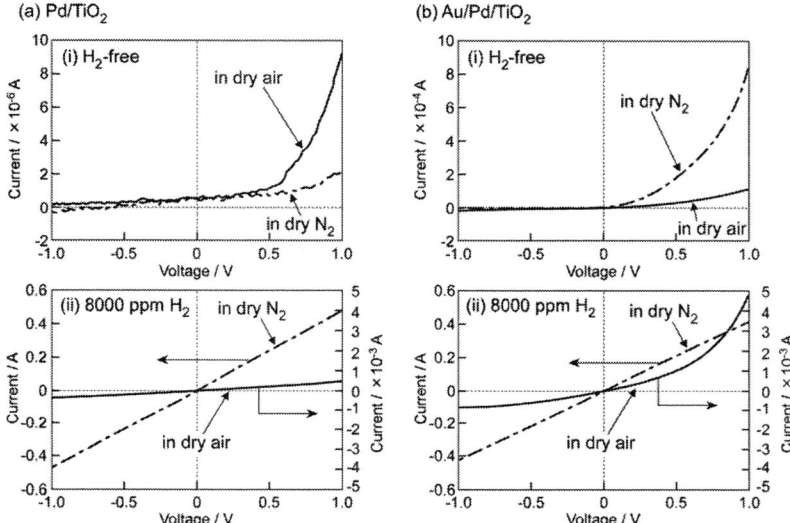

Figure 2. *I-V* characteristics of Pd/TiO$_2$ and Au/Pd/TiO$_2$ sensors in 0 and 8000 ppm H$_2$ balanced with dry air and dry N$_2$ at 250°C.

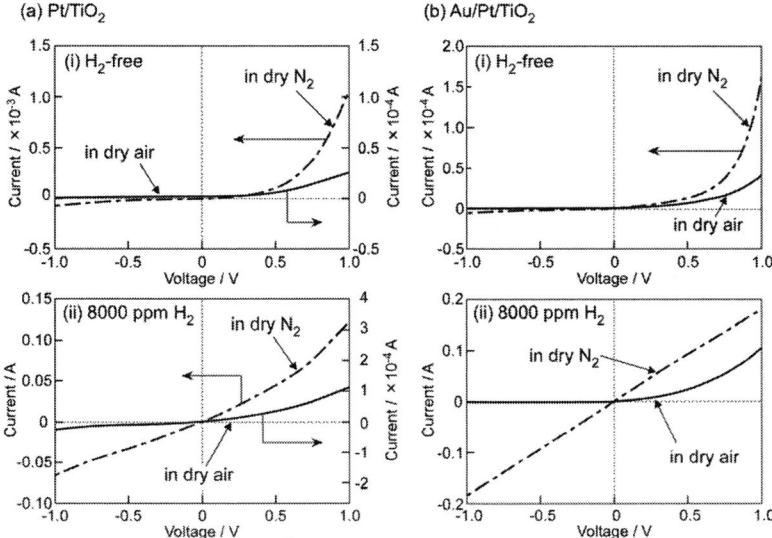

Figure 3. *I-V* characteristics of Pt/TiO$_2$ and Au/Pt/TiO$_2$ sensors in 0 and 8000 ppm H$_2$ balanced with dry air and dry N$_2$ at 250°C.

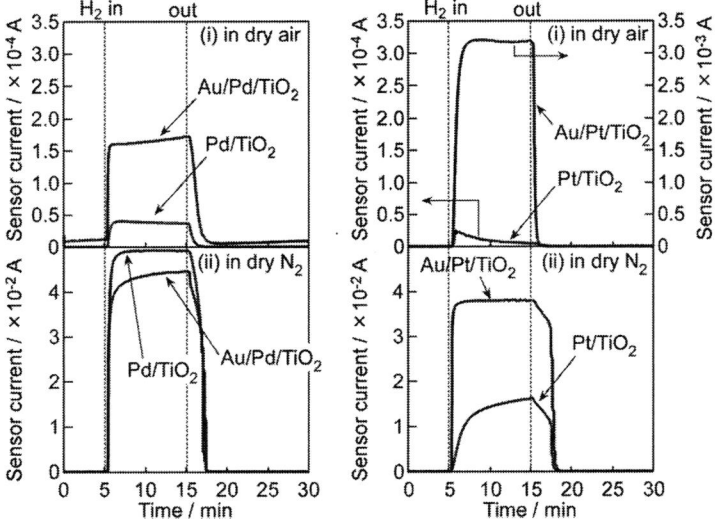

Figure 4. Response transients of all sensors to 8000 ppm H_2 at 250°C in dry air and dry N_2.

in H_2 balanced with air or N_2. Current (I)-voltage (V) characteristics of the sensors were measured in a range of $-1.0 \sim +1.0$ V.

Results and discussion

Figures 2 and 3 show I-V characteristics of all M/TiO$_2$ and Au/M/TiO$_2$ sensors under dry air and dry N_2 atmospheres. All sensors apparently showed nonlinear I-V curves, which is typical for a diode-type sensor, under H_2-free atmospheres. This behavior indicated that the Schottky barrier was formed at the interface between the electrode and the TiO$_2$ film. In addition, the magnitude of current under H_2-free atmospheres was extremely small, even when a forward bias of +1.0 V was applied to the sensors. The modification of the Pd electrode with Au enhanced the magnitude of current at the forward bias under H_2-free atmospheres. The introduction of H_2 into both dry air and dry N_2 drastically enhanced the magnitude of current, and all sensors showed ohmic I-V curves especially in 8000 ppm H_2 balanced with dry N_2, because dissociative adsorption of H_2 molecules on the electrode and subsequent dissolution of H species into the electrode reduced work function of the electrode and thus the height of Schottky barrier between the electrode and the TiO$_2$ film. In addition, the magnitude of current of the Pd/TiO$_2$ sensor was larger than that of the Pt/TiO$_2$ sensor in 8000 H_2 balanced with both dry air and dry N_2, because the amount of hydrogen solubility into Pd is generally larger than that into Pt. The modification of the Pt electrode with Au was also effective in enhancing the magnitude of current at the forward bias especially in 8000 H_2 balanced with dry air, while maintaining the nonlinear I-V characteristics (compare Fig. 3(a)(ii) and Fig. 3(b)(ii)). Then, the

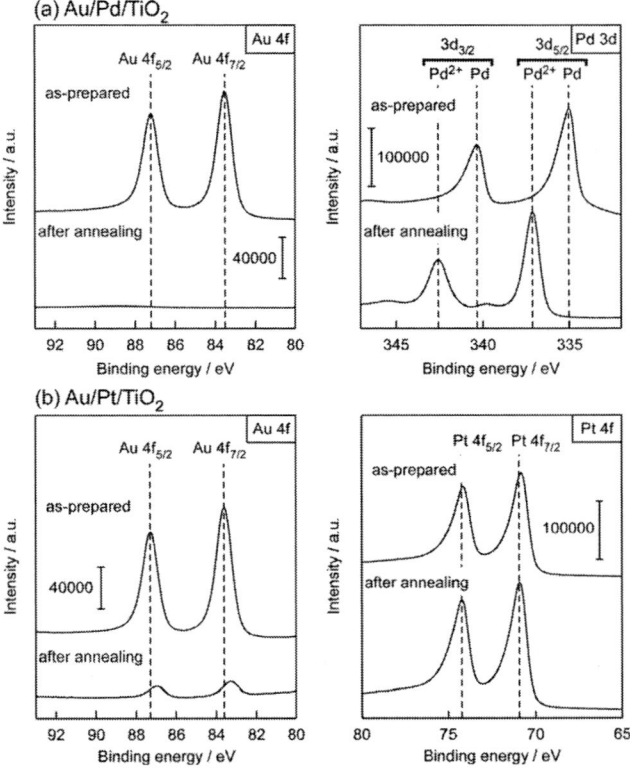

Figure 5. XPS spectra of Au, Pd and Pt on the surface of (a) Au/Pd/TiO$_2$ and (b) Au/Pt/TiO$_2$ sensors ((i) as-prepared and (ii) annealed at 600°C for 1 h in air).

magnitude of current of the Au/Pt/TiO$_2$ sensor became much larger than that of the Au/Pd/TiO$_2$ sensor at the forward bias in 8000 H$_2$ balanced with dry air (compare Fig. 3(b)(ii) and Fig. 2(b)(ii)).

Figure 4 shows response transients of all sensors to 8000 ppm H$_2$ in air at 250°C in both dry air and dry N$_2$. All sensors showed much larger H$_2$ response in dry N$_2$ than that in dry air, as expected from their I-V characteristics (Figs. 2 and 3). In case of M/TiO$_2$ sensors, H$_2$ response of the Pd/TiO$_2$ sensor was relatively larger than that of the Pt/TiO$_2$ sensor in both dry air and dry N$_2$. The surface modification of the Pd or Pt electrode with Au improved the H$_2$ response of M/TiO$_2$ sensors. Namely, the H$_2$ response of the Au/Pt/TiO$_2$ sensor in dry air became much larger than that of the Pt/TiO$_2$ sensor by two orders of magnitude, and this enhanced response value was also larger than that of the Au/Pd/TiO$_2$ sensor. On the other hand, the Au-modification effects were not so significant in dry N$_2$.

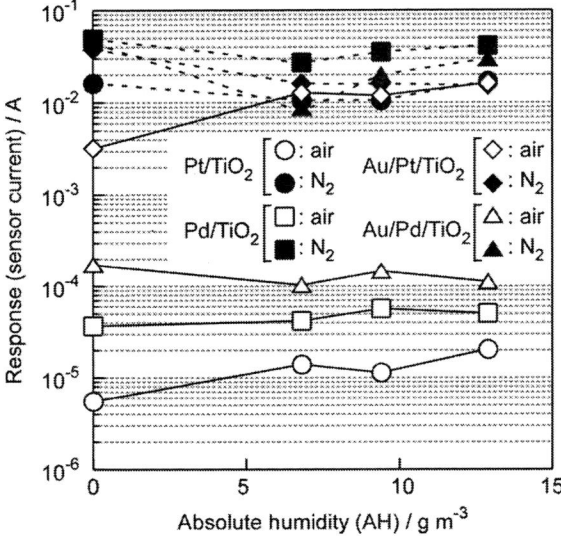

Figure 6. Variations in response of all sensors to 8000 ppm H_2 in both air and N_2 at 250°C with absolute humidity.

XPS spectra of Au, Pd and Pt on the surface of as-prepared and annealed (at 600°C for 1 h in air) $Au/M/TiO_2$ sensors have been investigated in order to clarify the effects of the surface modification of the Pd or Pt electrode with Au on the H_2-sensing properties. From the XPS spectra shown in Fig. 5, existence of Au and Pd metals (15) were confirmed on the electrode surface of the as-prepared $Au/Pd/TiO_2$ sensor, but the Au metal disappeared and the Pd metal was oxidized to PdO after the annealing. On the other hand, the small spectrum of the Au metal was confirmed on the electrode surface of the $Au/Pt/TiO_2$ sensor even after the annealing. In addition, it was also confirmed that the Pt metal (15) was not oxidized even after the annealing. Taking into account of large H_2 oxidation activity of Pd and Pt in comparison with Au and the surface composition of the electrode, the reason for the larger H_2 response of $Au/M/TiO_2$ sensors than that of M/TiO_2 sensors in dry air is considered to be larger amounts of dissolved H species in the Pd or Pt metal owing to lower H_2 oxidation activity of the Au metal at the outer surface of the electrodes. Furthermore, the less effectiveness of the Au modification over the Pd electrode on H_2 response, in comparison with the case of Pt electrode, is considered to be explained by the following scenario: H_2 molecules are likely consumed at the outer surface of the electrode of $Au/Pd/TiO_2$ sensor by the reduction of PdO to Pd and then high catalytic activity of Pd itself due to disappearance of Au and therefore the amount of dissolved H species in the electrode tends to reduce.

Responses of all sensors to 8000 ppm H_2 in wet air and wet N_2 were also investigated at 250°C in this study and variations in response of all sensors 8000 ppm H_2 in both air and N_2 with AH are plotted in Fig. 6. Under dry atmosphere (AH = 0), it was found that all sensors showed much large H_2 response in N_2 in comparison with that in air. In addition, response in N_2 was in almost the same and high level for all sensors, whereas that in air was in low level and varied significantly with the kinds of sensors, i.e. electrode materials. The reason for the similar response level in N_2 for all sensors can be ascribed to little interference from oxygen. Namely, the amounts of oxygen adsorbates and/or oxide species (e.g., PdO) are extremely small on the electrode surface in dry N_2 and thus dissociative adsorption of H_2 molecules and successive dissolution of H species into Pd or Pt easily proceeds, irrespective of the composition of electrode surface. The H_2 response of M/TiO_2 and $Au/Pt/TiO_2$ sensors in air gradually increased and that of $Au/Pd/TiO_2$ sensor in air slightly decreased with an increase in AH, while that of all sensors in N_2 was hardly dependent on AH. It is worth noting that only the $Au/Pt/TiO_2$ sensor showed large and comparable response both in air and N_2 under wet atmosphere, i.e. large and O_2-independent H_2 response. This may arise from limited oxidation of H_2 on the Au electrode covered with adsorbed water even in air and then large amounts of dissolved H species into the Pt electrode which are comparable to that in N_2.

Conclusions

Effects of surface modification of noble-metal (M) electrodes with Au on the H_2 sensing properties of M/TiO_2 sensors have been investigated at 250°C in air and N_2 under dry and wet atmospheres. M/TiO_2 sensors showed much larger H_2 response in dry N_2 than that in dry air, and H_2 response of the Pd/TiO_2 sensor was relatively larger than that of the Pt/TiO_2 sensor in both dry air and dry N_2. The surface modification of the Pd or Pt electrode with Au improved the H_2 response of M/TiO_2 sensors. Especially, the H_2 response of the $Au/Pt/TiO_2$ sensor in dry air was much larger than that of the Pt/TiO_2 sensor. In addition, the $Au/Pt/TiO_2$ sensor showed large and O_2-independent H_2 response.

Acknowledgments

The present work was partly supported by Grant-in-Aid for Scientific Research (B) (No. 23360289) from Japan Society for the Promotion of Science.

References

1. N. Yamazoe, *Sens. Actuators B*, **5**, 7 (1991).
2. A. Katsuki and K. Fukui, *Sens. Actuators B*, **52**, 30 (1998).
3. E.-B. Lee, I.-S. Hwang, J.-H. Cha, H.-J. Lee, W.-B. Lee, J. J. Pak, J.-H. Lee, and B.-K. Ju, Sen, *Sens. Actuators B*, **153**, 392 (2011).
4. G. Korotcenkov, S. D. Han, J. R. Stetter, *Chemical Reviews*, **109**, 1402 (2009).
5. M. Nishibori, W. Shin, N. Izu, T. Itoh, and I. Matsubara, *Sens. Actuators B*, **137**, 524 (2009).
6. Y. Shimizu, N. Kuwano, T. Hyodo, and M. Egashira, *Sens. Actuators B*, **83**, 195 (2002).
7. T. Iwanaga, T. Hyodo, Y. Shimizu, and M. Egashira, *Sens. Actuators B*, **93**, 519 (2003).
8. T. Hyodo, T. Iwanaga, Y. Shimizu, and M. Egashira, *ITE Lett.*, **4**, 594 (2003).

9. H. Miyazaki, T. Hyodo, Y. Shimizu, and M. Egashira, *Sens. Actuators B*, **108**, 467 (2005).
10. Y. Shimizu, K. Sakamoto, M. Nakaoka, T. Hyodo, and M. Egashira, *Adv. Mater. Res.*, **47-50**, 1510 (2008).
11. M. Nakaoka, T. Hyodo, Y. Shimizu, and M. Egashira, *ECS Trans.*, **16**(11), 317 (2008).
12. T. Hyodo, M. Nakaoka, Y. Shimizu, and M. Egashira, *Sens. Lett.*, **9**, 641 (2011).
13. T. Hyodo, M. Nakaoka, Y. Shimizu, and M. Egashira, *IOP Conference Series: Materials Science and Engineering*, **18**, 212006 (2011).
14. G. Yamamoto, T. Yamashita, K. Matsuo, T. Hyodo, and Y. Shimizu, *Proc. of the IMCS 2012 - The 14th International Meeting on Chemical Sensors, May 20-23, Nuremberg, Germany* (2012) pp. 193-196.
15. C. D. Wagner, W. M. Riggs, L. E. Davis, J. F. Moulder, and G. E. Muilenberg, *Handbook of X-ray Photoelectron Spectroscopy*, pp. 110-112 and pp. 152-155, Perkin-Elmer Corp., Minnesota (1978).

Potentiometric YSZ-based Sensors Using Zn-Ta-O-based Sensing Electrode for Selective H$_2$ Detection

S. A. Anggraini[a], M. Breedon[b,c], and N. Miura[c]

[a] Interdisciplinary Graduate School of Engineering Sciences, Kyushu University, Kasuga, Fukuoka 816-8580, Japan
[b] Japan Society for the Promotion of Science, Chiyoda-ku, Tokyo 102-8472, Japan
[c] Art, Science and Technology Center for Cooperative Research (KASTEC), Kyushu University, Kasuga, Fukuoka 816-8580, Japan

An yttria-stabilized-zirconia (YSZ)-based gas sensor utilizing a sensing electrode (SE) comprised of a mixture of co-sintered ZnO and Ta$_2$O$_5$ powders was developed and reported here. The addition of 30 wt.% Ta$_2$O$_5$ into ZnO after sintering at 1000°C, was found to increase the H$_2$ sensitivity when compared with other SEs with different mixture compositions. Unfortunately, the response toward H$_2$ was accompanied by high responses toward CO and hydrocarbons (HCs). The sensitivity as well as the selectivity improvement investigation was extended by elevating the sintering temperature from 1000° to 1300°C. The sensor utilizing ZnO(+ 30 wt.% Ta$_2$O$_5$)-SE that was sintered at 1300°C was found to be capable of generating high response toward 100 ppm H$_2$, with only minor responses toward other gases examined in this study at an operating temperature of 600°C under humid conditions.

Introduction

Among the existing alternative energy choices, the utilization of hydrogen (H$_2$) as an energy source brings many advantages, such as abundant sources of H$_2$ and low green house gas emissions (1-2). Hence, H$_2$ is often championed as a sustainable energy source for future needs. The utilization of H$_2$ as an alternative energy is expected to avert adverse effects on the environment, and reduce dependency on imported fuel, particularly for countries with limited fossil fuel resources. One of the most realistic uses of H$_2$ is in fuel cells and localized power generators (3). Other than as a fuel, H$_2$ has also been used in many manufacturing processes such as those for plastics, fertilizer, metallurgical treatments and many other (4).

Under atmospheric pressure and at room temperature, it is well-known that the lower explosive limit of H$_2$ is about 4 vol.% and the upper explosive limit occurs around 74.5 vol.% (5). H$_2$ can easily leak out from a containment vessel, because of its small molecular size. In addition to that, its low ignition energy (0.017 mJ) with a diffusion coefficient (0.61 cm^2/s) is four times larger than that of gasoline (0.16 cm^2/s), which makes the safe utilization of H$_2$ challenging, due to the potential risks (6). Therefore, to ensure safe utilization of H$_2$, a high-performance H$_2$ sensor for accurate and reliable H$_2$ monitoring, is of importance.

Metal oxides have been playing an important role for several decades as gas sensing materials for many target gases, while, an yttria-stabilized zirconia (YSZ) solid electrolyte is one of the most widely used oxygen ion conductors (7). Its sensing performance has been studied extensively; particularly when YSZ is used in conjunction together with a metal-oxide based SE (8). However, there have been few reports on a sensitive and selective H_2 response using YSZ-based sensors when operated as potentiometric sensors (9).

As an SE material for potentiometric YSZ-based sensors, ZnO was reported to give high response toward H_2 which was also accompanied by responses toward other gases such as hydrocarbons (HCs) and NO_2 (10). Meanwhile, Ta_2O_5 was reported to possess selective catalytic activity, when mixed with another metal oxide (11). When the mixture of Ta_2O_5 and another metal oxide was used as an SE material for a potentiometric sensor, the selectivity was improved (12). Furthermore, the potentiometric sensors utilizing Ta_2O_5-based materials have been reported to exhibit high sensitivity toward H_2 (13). Thus, in this study, Ta_2O_5 was added into ZnO and the sensing performance of the developed YSZ-based sensor using a Zn-Ta-O-based SE, particularly toward H_2, was examined. The effect of sintering temperature on its sensitivity and selectivity was investigated as well.

Experimental

Sensor fabrication

The sensor was fabricated using a hemi-spherically terminated YSZ-tube (8 mol.% Y_2O_3-doped ZrO_2, Nikkato, Japan). Recently, our group has proposed the utilization of an intermediate layer of YSZ, in order to improve mechanical stability at the interface between SE and YSZ solid electrolyte (14). In light of this, an intermediate YSZ layer between the SE and the YSZ solid electrolyte was also fabricated here. In this process, YSZ powder (Tosoh Corp., Japan) was mixed with an organic binder (α-terpineol). The resulting paste was painted onto the surface of YSZ tube.

The SE material was fabricated by mixing commercial ZnO and Ta_2O_5 powder (Konjundo Chemical Lab., Japan) in a mortar. α-terpineol was also added into the mixed powder, making a uniform paste. This paste was applied onto the surface of intermediate YSZ layer, forming a 3 mm-wide band. The reference electrode (RE) was made by applying platinum (Pt) paste (Tanaka, Kikinzoku, Japan) on the inner surface of the end of the YSZ tube. Subsequently, the assembled sensor was sintered at 1000°C for 2 h in air to form the SE and the RE. For sensing characteristic optimization, the sensor was also sintered at 1100, 1200 and 1300°C for 2 h in air.

Sensing characteristic evaluation

The sensor was assembled in a quartz testing cell residing in an electric furnace which was connected to several mass-flow controllers. The RE was always exposed to the ambient air, while the SE was exposed to the sample gas or base gas. The gas sensing characteristics of the fabricated sensors were evaluated at an operational temperature of 600°C under humid conditions (1.35 vol.% or 5 vol.% H_2O). The sample gas was diluted with humidified base gas with N_2 and O_2 (5 vol.% or 21 vol.% O_2). The concentration of sample gases (NO, NO_2, CO, CH_4, C_3H_8, C_3H_6, H_2) was fixed at 400 ppm, with the exception of NH_3 concentration (100 ppm). The concentration of H_2 was varied from 20 to 400 ppm for H_2 dependence measurements. The potential difference between SE and RE, was measured using a digital electrometer (R8240, Advantest, Japan), and recorded as the sensing signal.

Characterization of sensing electrode

The morphology change of the SE was observed by using a field emission scanning electron microscope (FE-SEM, JSM-6340F, JEOL, Japan). The crystal structure of the SE was investigated by using an X-ray diffractometer (XRD, RINT2100VLR/PC, Rigaku, Japan; Cu Kα radiation, λ=1.5406 Å).

Results and Discussion

The effect of Ta_2O_5 additions on H_2 sensitivity.

Initially, the sensing performance of each parent metal oxide was examined. Figure 1 shows the preliminary-test results of the sensor using (a) ZnO-SE; and (b) Ta_2O_5-SE. The results confirmed the ability of the sensor using ZnO-SE to generate moderate responses toward HC (Δemf : about -75 mV) at 600°C. However, the sensor using Ta_2O_5-SE generated minor responses to all gases (Δemf : less than ± 20 mV), indicating that this material may possess the ability to function as an oxidation catalyst. The addition of a catalyst into ZnO is expected to improve the sensing performance of the sensor using ZnO-based-SE. Thus, Ta_2O_5 was introduced into ZnO and the sensing characteristics of the resulting mixtures were examined.

The effect of Ta_2O_5 additions on the H_2 sensitivity is presented in Fig. 2. The addition of 30 wt.% Ta_2O_5 into ZnO-SE was observed to give the highest response to H_2, among other ZnO : Ta_2O_5 mixture ratios and even to the original ZnO- and Ta_2O_5-SE.

The crystal phase and compound stoichiometry composition caused by addition of Ta_2O_5 to ZnO was studied via XRD analysis. After the sintering at 1000°C, the addition of 30 wt.% Ta_2O_5 was found to form a mixture of ZnO (JCPDS no.:36-1451) and $Ta_2Zn_3O_8$ (JCPDS no.: 20-1237) (Fig. 3 (b)). This particular mixed oxide composition and morphology are believed to enhance the H_2 sensitivity. However, the observed XRD patterns revealed that 70 wt.% additions of Ta_2O_5 into ZnO was found to be in good agreement with the peaks identified for $Ta_2Zn_3O_8$, $ZnTa_2O_6$ (JCPDS no.: 39-1484) and Ta_2O_5 (JCPDS no. 25-0922) (Fig. 3 (c)). This electrode composition with rather large and smooth surface particles (Fig. 3(c)) is believed to responsible for lowering the catalytic activity to electrochemical reactions or increasing the catalytic activity to gas-phase reaction across the SE layer, which ultimately resulted in lowering H_2 sensitivity.

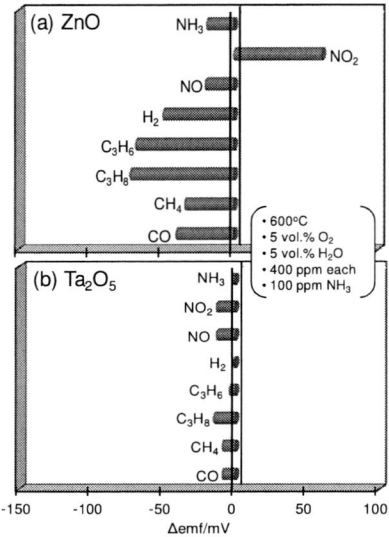

Figure 1. The cross sensitivity of the sensor using (a) ZnO-SE and (b) Ta$_2$O$_5$-SE; operating at 600°C under humid conditions.

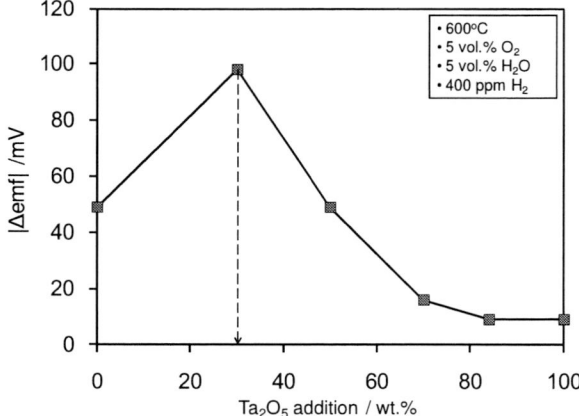

Figure 2. The effect of Ta$_2$O$_5$ additions on H$_2$ sensitivity for the sensor using ZnO-SE, at 600°C under humid conditions.

Since the sensor using ZnO(+ 30 wt.% Ta_2O_5)-SE was found to generate the highest response toward H_2, this SE was chosen for further investigations. However, it should be noted that the high response toward H_2 that was generated by the sensor using ZnO(+ 30 wt.% Ta_2O_5)-SE, was also accompanied by high response toward CO and C_3H_6. In order to optimize the sensing performance in terms of sensitivity and selectivity, the sensor using ZnO(+ 30 wt.% Ta_2O_5)-SE was sintered at different temperatures.

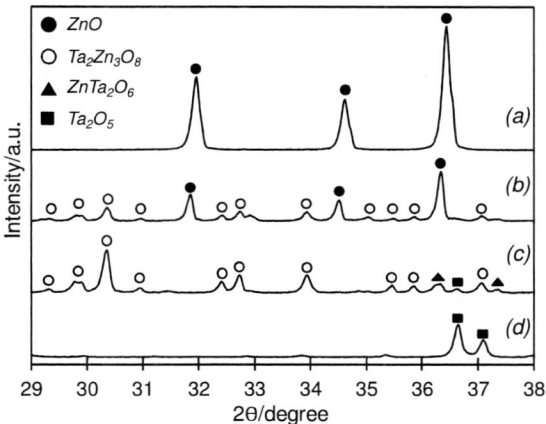

Figure 3. The XRD patterns of (a) ZnO, (b) ZnO(+ 30 wt.% Ta_2O_5), (c) ZnO(+ 70 wt.% Ta_2O_5) and (d) Ta_2O_5 after sintering at 1000°C for 2 h in air.

The effect of sintering temperature on gas sensitivity.

The effect of sintering temperature on the sensing characteristics of the sensor with ZnO(+ 30 wt.% Ta_2O_5)-SE is presented in Fig. 4. The sensor sintered at 1000°C generated a relatively sensitive response to H_2, followed by some response towards CO, C_3H_6 and C_3H_8 at an operating temperature of 600°C (Fig. 4(a)). Similar trends of gas sensitivity were observed for the sensor that was sintered at 1100°C (Fig. 4 (b)). However, the sensitivity itself had decreased, when the sintering temperature was raised by 100°C. The sensor using the same material that was sintered at 1200°C exhibited a sensitive and relatively selective response to NO_2 (Δemf : about 145 mV), while maintaining response to other gases below -50 mV (Fig. 4(c)). The cross-sensitivity characteristic of the sensor sintered at 1300°C is presented in Fig. 4 (d). It can be seen that this sensor is capable of generating selective and sensitive response toward 400 ppm H_2 (Δemf : about -580 mV) with only minor responses toward other examined gases (Δemf : less than 20 mV) under the same operating conditions.

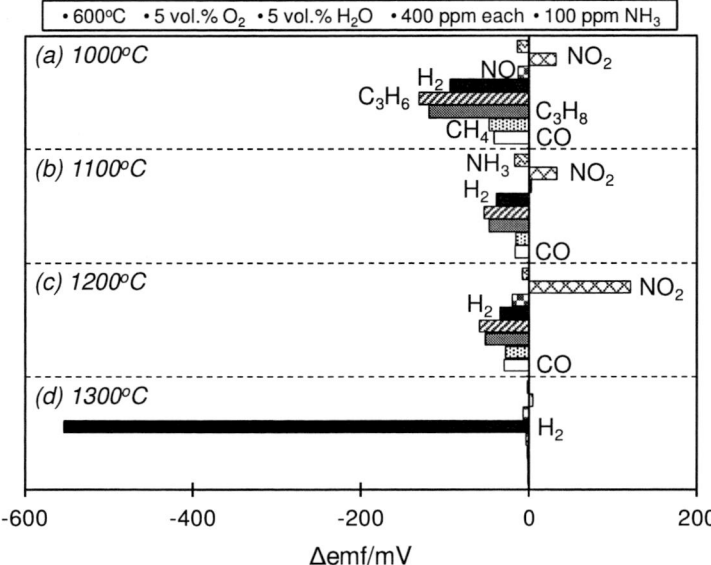

Figure 4. The cross sensitivity of the sensor using ZnO(+ 30 wt.% Ta₂O₅)-SE after sintering at (a) 1000°C, (b) 1100°C, (c) 1200°C and (d) 1300°C, operating at 600°C under humid conditions.

The difference in the sensing behavior was further investigated by observing the morphology change via several SEM observations. Figure 5 shows the morphology changes of the sensing electrode surface after sintering at elevated temperatures. As can be seen here, an increase in the sintering temperature was found to decrease the porosity and enlarge the particle size. When the sensing electrode was sintered at 1300°C (Fig. 5(d)), the resulting morphology was observed to be denser. Such a denser and more tortuous morphology with low porosity is believed to lessen the proportion of gas, particularly for gases which have larger molecular sizes than H_2, capable of reaching and reacting at the interface of SE and YSZ, where the electrochemical reactions that generate sensor emf occur. This results in a decreased O_2 cathodic current, and a slightly increased H_2 anodic current, hence shifting the mixed-potential interception point to increasingly lower potential, increasing H_2 sensitivity, when the sensor was sintered at higher temperature.

The cross sensitivity of the sensor using ZnO(+ 30 wt.% Ta₂O₅)-SE sintered at 1300°C was then examined under humid conditions (1. 35 vol.% H_2O) with 21 vol.% O_2 and the result is shown in Fig. 6. It is clear that the sensor is still capable of generating sensitive response toward H_2 (Δemf = -120 mV) with negligible response towards other examined gases (Δemf < ± 10 mV). Figure 7 shows the dependency of sensitivity on H_2 concentration in the range of 20-100 ppm H_2. The response toward H_2 was found to be linear with H_2 concentration on a logarithmic scale. This linearity indicates that the sensing mechanism of the developed sensor is based on the mixed potential model (8, 10, 14). The sensor was also found to capable of detecting H_2 as low as 20 ppm. The sensing mechanism of the developed sensor is being investigated. However, given the excellent sensing performance toward H_2, this sensor has the potential to be utilized for actual selective H_2 monitoring.

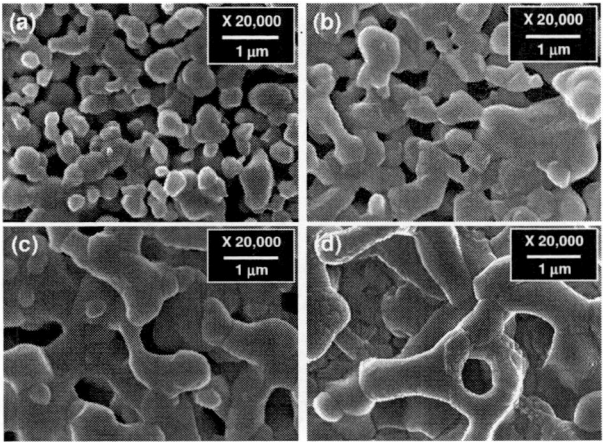

Figure 5. The morphology changes of ZnO(+ 30 wt.% Ta_2O_5)-SE material after sintering at (a) 1000°C, (b) 1100°C, (c) 1200°C and (d)1300°C.

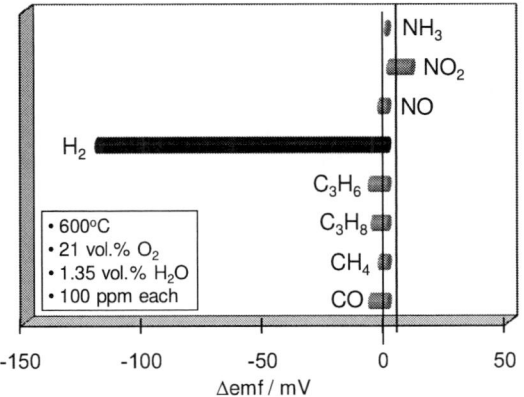

Figure 6. The cross sensitivity of the sensor using ZnO(+ 30 wt.% Ta_2O_5)-SE after sintering at 1300°C, measured under humid conditions.

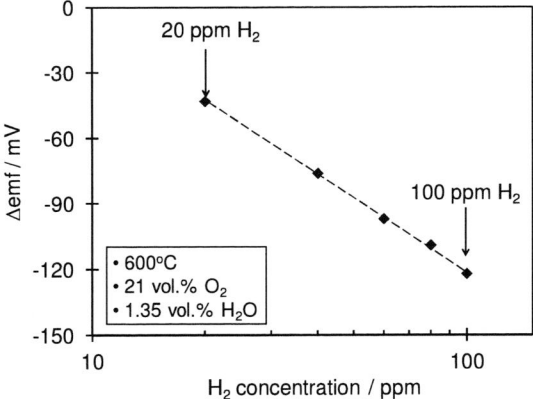

Figure 7. The sensitivity dependency on H_2 concentration of the sensor using ZnO(+ 30 wt.% Ta_2O_5)-SE sintered at 1300°C.

Conclusions

A systematic investigation of the effect of Ta_2O_5 addition into ZnO-SE on the gas sensing performance was performed. The addition of 30 wt% of Ta_2O_5 into ZnO has been proven to be effective in improving the sensing properties of the parental oxide-SE. The addition of 30 wt.% Ta_2O_5 was found to form a mixture of ZnO and $Ta_2Zn_3O_8$. This particular mixture, with its corresponding morphology is believed to be responsible for generating the highest response toward H_2, among the examined oxide mixtures. Sintering of the sensor using ZnO(+ 30 wt.% Ta_2O_5)-SE at 1300°C was found to affect the sensitivity and selectivity toward H_2. The SEM observation revealed that the dense and tortuous morphology that formed after sintering at 1300°C is believed to decrease the proportion of analyte capable of reaching the interface of SE and YSZ. The sensitivity was found to be linear with the H_2 concentration on a logarithmic scale, indicating that the sensing mechanism was based on mixed potential model.

Acknowledgments

This work was partially supported by Kyushu University-Global-COE program "Novel Carbon Resource Sciences" and Grant-in-Aid for Scientific Research (B) (22350095) and for JSPS Fellows (22-0353).

References

1. S. Chiu, J. Tsai, K. Liang, T. Huang, K. Liu, T. Tsai, K. Hsu, and W. Lour, *Sens. Actuators B*, **141**, 532 (2009).
2. S. Verhelst, and T. Wallner, *Progress in Energy and Combustion Science*, **35**, 490 (2009).
3. B. Johnston, M. C. Mayo, and A. Khare, *Technovation*, **25**, 569 (2008).

4. X. Lu, S. Wu, L. Wang, and Z. Su, *Sens. Actuators B*, **107**, 812 (2005).
5. S. Okazaki, H. Nakagawa, S. Asakura, Y. Tomiuchi, N. Tsuji, H. Murayama, and M. Washiya, *Sens. Actuators B*, **93**, 142 (2003).
6. C. Ji, and S. Wang, *Int. J Hydrogen Energy*, **34**, 7823 (2009).
7. P. H. Rogers, G. Sirinakis and M. A. Carpenter, *J. Phys. Chem. C*, **112**, 6749 (2008).
8. V. V. Plashnitsa, T. Ueda, P. Elumalai, and N. Miura, Sens. Actuators B, **130**, 231 (2008).
9. L. P. Martin, A. Q. Pham, and R. S. Glass, *Solid State Ionics*, **175**, 527 (2004).
10. N. Miura, T. Raisen, G. Lu, and N. Yamazoe, *Sens. Actuators B*, **47**, 84 (1998).
11. T. Ushikubo, *Catalysis Today*, **57**, 331 (2000).
12. L. Chevallier, E. Traversa, and E. Di Bartolomeo, *J. Electrochem. Soc.*, **157** (11), J386 (2010).
13. J. Zosel, G. Schiffel, F. Gerlach, K. Ahlborn, U. Sasum, V. Vashook, and U. Guth, *Solid State Ionics*, **177**, 2301 (2006).
14. Y. Fujio, V. V. Plashnitsa, P. Elumalai, and N. Miura, *Talanta*, **85**, 575 (2011).

Solid Electrolyte Type Ammonia Gas Sensor with High Water Durability

S. Tamura, T. Nagai, and N. Imanaka

Division of Applied Chemistry, Graduate School of Engineering, Osaka University,
2-1 Yamadaoka, Suita, Osaka 565-0871, Japan

Highly water durable ammonia gas sensor was fabricated by the combination of Al^{3+} ion conducting $(Al_{0.2}Zr_{0.8})_{20/19}Nb(PO_4)_3$ solid electrolyte and NH_4^+-β-gallate $(NH_4^+$-$Ga_{11}O_{17})$ auxiliary sensing electrode, and its NH_3 gas sensing performance was investigated in humid atmospheres. The present sensor showed an advanced sensing performance of continuous, quantitative and reproducible response which obeys the theoretical Nernst relationship even in a highly humidified condition containing 4.2 vol% H_2O at 230 °C.

Introduction

Ammonia (NH_3) gas is one of the useful gas species in industrial field. However, the toxicity of NH_3 gas is so high that serious accidents happen by the ammonia gas aspiration and, therefore, the NH_3 gas sensor showing an exact gas detection with a rapid response is greatly required. Today, NH_3 gas concentration is measured by using analytical apparatuses based on ion or gas chromatography, but these are not suitable for on-site NH_3 gas sensing tool because their sizes are too large to settle at every emitting site and some pretreatment of sample gas is always required. Therefore, it is necessary to develop a smart ammonia gas sensor realizing on-site monitoring at every emission site.

Among the various types of comact gas sensors proposed, the solid electrolyte type sensor is considered to be a promising candidate for the practical gas sensing device from the view points of high selectivity, quantitative gas detection, compact size, and easy treatment. Up to now, NH_3 gas sensors based on NH_4^+ ion conducting solid electrolyte were proposed (1-3), however, these sensors were not a practical one due to poor sintering nature, low thermal stability, and low electrical conductivity of the NH_4^+ ion conducting solid. On the other hand, there are also some reports on the NH_3 gas sensor combining another cation conducting solid electrolyte and the ammonium salt as the auxiliary sensing electrode. While this type sensor showed the theoretical NH_3 gas detection, the sensing performance deteriorated in a humid atmosphere due to low water durability of applied ammonium salts.

To develop a practically applicable NH_3 gas sensor, it is required to use the water durable sensor components. Recently, we have proposed the solid electrolyte type NH_3 gas sensors (4, 5) applying the Al^{3+} ion conducting $(Al_{0.2}Zr_{0.8})_{20/19}Nb(PO_4)_3$ solid electrolyte (6) with rare-earth ammonium sulfate of $R_2(SO_4)_3 \cdot (NH_4)_2SO_4$ (R: rare-earths) (7) or lanthanum oxysulfate based $La_2O_2SO_4$-$NH_4H_2PO_4$ solid solution as the auxiliary sensing electrode. Although both sensors showed high sensing performances to NH_3 gas obeying Nernst relationship at 230–300 °C or 170–200 °C, respectively, the theoretical NH_3 gas sensing performance was not realized under highly humidified condition containing H_2O over 0.6 vol% (sutrated vapor at 0 °C) because the water durability of these materials are never sufficient for such a severe condition.

In this study, we focused on the NH_4^+-β-gallate (NH_4^+-$Ga_{11}O_{17}$) solid (8) as a candidate for the auxiliary sensing electrode material because the NH_4^+ ions in NH_4^+-$Ga_{11}O_{17}$ are tightly held in the β-gallate type layered structure, suggesting that the NH_4^+ ions in the β-gallate are chemically and thermally stable compared to the common ammonium salts. Here, we fabricated an NH_3 gas sensor based on the $(Al_{0.2}Zr_{0.8})_{20/19}Nb(PO_4)_3$ solid electrolyte combined with NH_4^+-$Ga_{11}O_{17}$ solid as the auxiliary sensing electrode [9], and its NH_3 gas sensing performance was investigated under high humid atmospheres.

Experimental

$(Al_{0.2}Zr_{0.8})_{20/19}Nb(PO_4)_3$ was prepared by mixing the starting materials of $Al(OH)_3$, ZrO_2, Nb_2O_5, and $(NH_4)_2H(PO_4)_3$ in a molar ratio of 8:32:19:114. The mixed powder was heated at 1000 °C for 12 h, 1200 °C for 12 h, and then, 1300 °C for 12 h in air. The $(Al_{0.2}Zr_{0.8})_{20/19}Nb(PO_4)_3$ obtained was pelletized and sintered at 1300 °C for 12 h in air. NH_4^+-$Ga_{11}O_{17}$ was obtained by the ion-exchange method using (K^+, Rb^+)-$Ga_{11}O_{17}$ solid. (K^+, Rb^+)-$Ga_{11}O_{17}$ was synthesized by mixing K_2CO_3, Rb_2CO_3, and Ga_2O_3 (molar ratio is 1:1:10) and calcined at 1320 °C for 2 h in air. The sample obtained were then pulverized, pelletized, and sintered at 1320 °C for 6 h in air. NH_4^+-$Ga_{11}O_{17}$ was obtained by the ion-exchange of (K^+, Rb^+)-$Ga_{11}O_{17}$ in molten NH_4NO_3 at 180 °C for 25 days. The obtained NH_4^+-$Ga_{11}O_{17}$ was washed three times with ultrapure water until K^+, Rb^+, and NH_4NO_3 were rinsed off.

The NH_4^+-$Ga_{11}O_{17}$ sample obtained was identified by X-ray powder diffraction (XRD) using Cu-Kα radiation in the 2θ range from 10 to 50° with step width of 0.02° (Multiflex, Rigaku). Thermal analysis (DTA-60H, Shimadzu) was carried out in a synthetic air atmosphere. The heating rate was 5°C·min^{-1}. The electrical conductivity of NH_4^+-$Ga_{11}O_{17}$ and (K^+, Rb^+)-$Ga_{11}O_{17}$ was measured by the complex impedance method in the frequency region from 5 to 13M Hz (Precision LCR meter 8284A, Hewlett Packard) and in the temperature range from 50 to 250 °C.

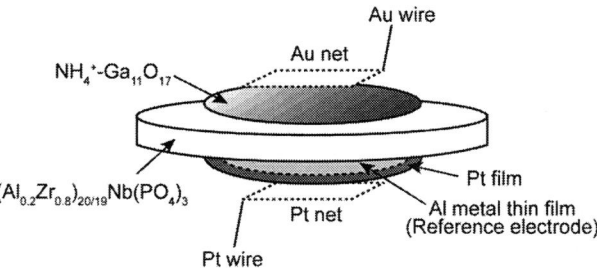

Figure 1. Schematic illustration of the NH_3 gas sensor element by the combination of Al^{3+} ion conducting solid electrolyte with the NH_4^+-$Ga_{11}O_{17}$ auxiliary sensing electrode. The whole sensor cell was set in the electrical furnace.

Figure 1 illustrates a schematic illustration of the present NH_3 gas sensor. Aluminum metal thin film was prepared on one side of Al^{3+} ion conducting solid electrolyte pellet [6] as a reference electrode, and the Al metal was covered with Pt sputtered film to prevent oxidation of Al. The NH_4^+-$Ga_{11}O_{17}$ auxiliary sensing electrode was set on the opposite side of Al^{3+} ion conductor. NH_3 gas sensing properties were investigated in the atmosphere where NH_3 gas concentration was regulated by mixing 1% NH_3 diluted with N_2 gas and humidified air. The humidified air (0.6–4.2 vol% H_2O) was obtained by passing dry air through H_2O at 0–30 °C. The total gas flow rate was kept constant at 100 ml·min^{-1}. The oxygen gas pressure (PO_2) was fixed at 2.1×10^4 Pa. The sensor output EMF was monitored with an electrometer (Advantest, R8240).

Results and Discussion

The NH_4^+-$Ga_{11}O_{17}$ solid obtained was identified to hold the β-gallate type structure from the XRD patterns, and the diffraction peaks of NH_4^+-$Ga_{11}O_{17}$ shifted to lower angles compared to those of (K^+, Rb^+)-$Ga_{11}O_{17}$ as shown in Fig. 2. Furthermore, the NH_4^+-$Ga_{11}O_{17}$ solid exhibited c.a. 1.7wt% weight loss from 270 to 400 °C, whereas no weight loss was observed for the (K^+, Rb^+)-$Ga_{11}O_{17}$ solid. Since the estimated amount of NH_4^+ ion in the NH_4^+-$Ga_{11}O_{17}$ solid is 1.8 wt% if K^+ and Rb^+ ions in the (K^+, Rb^+)-$Ga_{11}O_{17}$ were fully replaced by NH_4^+ ion, the observed weight loss found to be by the release of NH_4^+ ions from the solid. These results strongly indicate that the K^+ and Rb^+ ions in (K^+, Rb^+)-$Ga_{11}O_{17}$ were successfully ion-exchanged with NH_4^+ ions, and NH_4^+ ions present in the resultant NH_4^+-$Ga_{11}O_{17}$ solid.

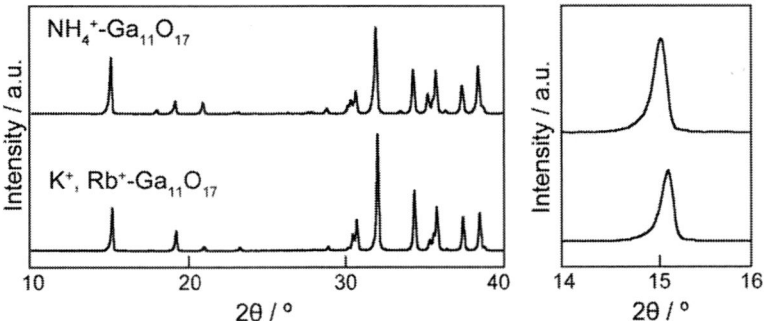

Figure 2. XRD patterns of the (K^+, Rb^+)-$Ga_{11}O_{17}$ and NH_4^+-$Ga_{11}O_{17}$.

To determine the temperature at which the NH_3 vaporizes from NH_4^+-$Ga_{11}O_{17}$ solid completely, electrical conductivity of the NH_4^+-$Ga_{11}O_{17}$ solid was measured because it is essential to contain the NH_4^+ ions in the auxiliary sensing electrode for detecting the NH_3 gas concentration. Figure 3 depicts the electrical conductivity for the (K^+, Rb^+)-$Ga_{11}O_{17}$ and NH_4^+-$Ga_{11}O_{17}$ solids. It is confirmed that the NH_4^+-$Ga_{11}O_{17}$ solid showed one order of magnitude low conductivity compared to that of (K^+, Rb^+)-$Ga_{11}O_{17}$. Although the conductivity of NH_4^+-$Ga_{11}O_{17}$ increased linearly with temperature up to 230 °C, the conductivity at 250 °C was slightly decreased compared to the value estimated from the conductivities between 100 and 230 °C. Furthermore, the conductivities in cooling process from 250 °C were lower than those in heating process. These results clearly indicate that the NH_4^+ ions start to remove from the NH_4^+-$Ga_{11}O_{17}$ solid at ca. 250 °C, resulting in the reduction of conductivity in cooling process. From the results of TG and conductivity measurements, we decided the operating temperature of the sensor applying NH_4^+-$Ga_{11}O_{17}$ auxiliary sensing electrode is set at 230 °C, which is 20 °C lower than the limiting temperature at which NH_4^+ ions remain in the solid.

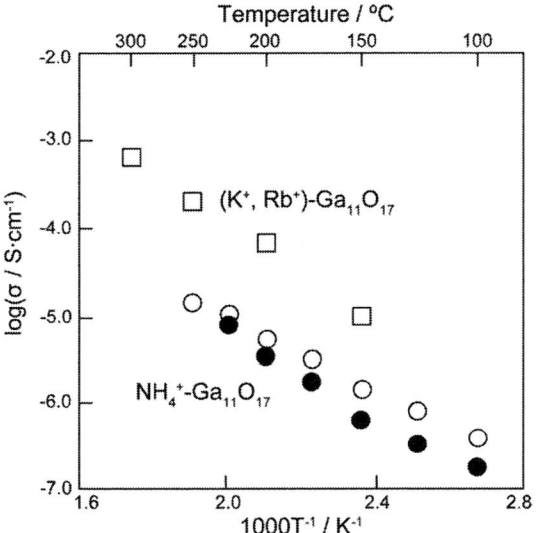

Figure 3. Electrical conductivity of the (K^+, Rb^+)-$Ga_{11}O_{17}$ ($\square$) and NH_4^+-$Ga_{11}O_{17}$ ($\circ$: in heating process, $\bullet$: in cooling process).

For the present sensor, the following plausible reactions are considered to occur at the auxiliary sensing electrode, the interface between the auxiliary sensing electrode and the Al^{3+} ion conductor, and at the Al metal reference electrode.
At auxiliary sensing electrode:

$$NH_4^+ \text{ (in } NH_4^+\text{-}Ga_{11}O_{17}) \leftrightarrow NH_3 + H^+ + e^- \qquad [1]$$

At interface between the auxiliary sensing electrode and the $(Al_{0.2}Zr_{0.8})_{20/19}Nb(PO_4)_3$ electrolyte:

$$H^+ + 19/12(Al_{0.2}Zr_{0.8})_{20/19}Nb(PO_4)_3 \leftrightarrow 19/12(H_{0.6}Zr_{0.8})_{20/19}Nb(PO_4)_3 + 1/3Al^{3+} \quad [2]$$

At reference electrode (Al metal thin film):

$$1/3Al^{3+} + e^- \leftrightarrow 1/3Al \qquad [3]$$

From the Eqs. [1] to [3], total chemical reaction can be expressed as follows.

$$NH_4^+ \text{ (in } NH_4^+\text{-}Ga_{11}O_{17}) +19/12(Al_{0.2}Zr_{0.8})_{20/19}Nb(PO_4)_3 \leftrightarrow$$

$$1/3Al + 19/12(H_{0.6}Zr_{0.8})_{20/19}Nb(PO_4)_3 + NH_3 \qquad [4]$$

The following Nernst equation can be obtained from the Eq. [4].

$$E = E_0 - (RT/nF) \cdot \ln\{(a_{Al})^{1/3} \cdot (a_{(H_{0.6}Zr_{0.8})_{20/19}Nb(PO_4)_3})^{19/12} \cdot (P_{NH_3}) \cdot$$
$$(a_{NH_4^+-Ga_{11}O_{17}})^{-1} \cdot (a_{(Al_{0.2}Zr_{0.8})_{20/19}Nb(PO_4)_3})^{-19/12}\} \qquad [5]$$

where, a and P terms are the activity of the solid materials and the pressure of the gas, respectively, and R, F, and n are the gas constant, Faraday's constant, and the number of electrons participating in the reaction (here, n = 1.00). Since the activities of the solids are constant at fixed temperature, the Nernst equation (Eq. [5]) can be simplified as follows.

$$E = C - (RT/nF) \ln (P_{NH_3}) \quad (C: constant) \qquad [6]$$

Figure 4 shows a representative sensor response curve observed for variation of the NH_3 gas concentration from 200 to 500 ppm and vice versa at 230 °C in a humidified atmosphere containing 4.2 vol% H_2O (saturated vapor at 30 °C). The response time defined as the time to attain a 90% total response was within 8 min (including the gas replacement time in the test tube). The sensor EMF outputs for increasing and decreasing the NH_3 gas concentrations were the same, which means that the sensor with NH_4^+-$Ga_{11}O_{17}$ as the auxiliary sensing electrode exhibits a continuous and reproducible response for the change of the NH_3 gas concentration even under highly humid conditions of 4.2 vol% H_2O.

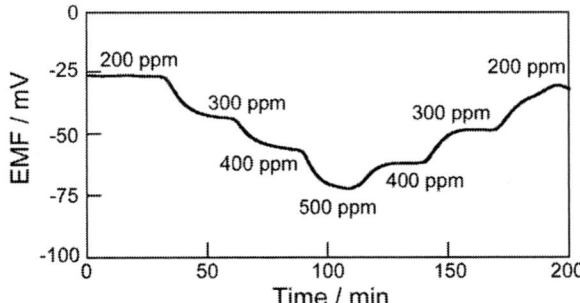

Figure 4. Representative sensor response curve for the sensor with a NH_4^+-$Ga_{11}O_{17}$ auxiliary sensing electrode at 230 °C. The NH_3 gas concentration was varied from 200 to 500 ppm and vice versa under humidified atmosphere (4.2 vol% H_2O).

Figure 5 depicts the relationship between the sensor EMF output and the logarithm of the NH_3 gas concentration at 230 °C for the sensor with the NH_4^+-$Ga_{11}O_{17}$ auxiliary sensing electrode under several humidified atmospheres (0.6, 2.3, and 4.2 vol% H_2O). The theoretical slope (n = 1.00) estimated from Eq. [6] is also presented as a solid line. The sensor EMF outputs decrease linearly with increasing the NH_3 concentration and a 1:1 linear relationship is observed between the EMF and the logarithm of the NH_3 gas concentration. The n values calculated from the slope were 1.03 (0.6 vol% H_2O), 0.98

(2.3 vol% H_2O), and 1.01 (4.2 vol% H_2O), which coincide well with the theoretical value (n = 1.00), and the sensor EMF outputs at each NH_3 concentration are almost the same, regardless of the water vapor content. This result provides strong evidence that the present sensor with the NH_4^+-$Ga_{11}O_{17}$ auxiliary sensing electrode exhibits a theoretical response for NH_3 gas without any interference of water vapor.

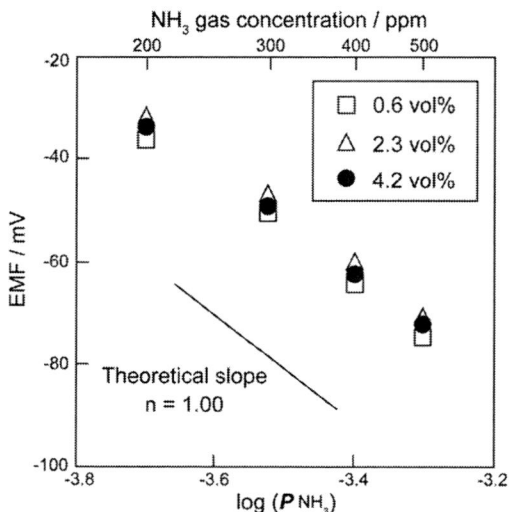

Figure 5. Representative sensor response curve for the sensor with the NH_4^+-$Ga_{11}O_{17}$ auxiliary sensing electrode at 230 °C. The NH_3 gas concentration was varied from 200 to 500 ppm and vice versa under humidified atmosphere (4.2 vol% H_2O).

References

1. N. Miura and W.L. Worrell, *Chem. Lett.*, **16**, 319 (1987).
2. N. Imanaka, S. Tamura, and G. Adachi, *Electrochem. Solid-State Lett.*, **1**, 282 (1998).
3. N. Imanaka, S. Yoshikawa, T. Yamamoto, and G. Adachi, *Electrochem. Solid-State Lett.*, **2**, 352 (1999).
4. T. Nagai, S. Tamura, and N. Imanaka, *Sens. Lett.*, **9**, 552 (2011).
5. T. Nagai, S. Tamura, and N. Imanaka, *Sens. Actuators B*, **147**, 735 (2010).
6. N. Imanaka, Y. Hasegawa, M. Yamaguchi, M. Itaya, S. Tamura, and G. Adachi, *Chem. Mater.*, **14**, 4481 (2002).
7. L.D. Iskhakove, I.E. Sukhova, O.P. Chernova, I.V. Shakhno, and V.E. Plyushchev, *Russ. J. Inorg. Chem.*, **20**, 193 (1975).
8. H. Ikawa, T. Tsurumi, K. Urabe, and S. Udagawa, *Solid State Ionics*, **20**, 1 (1986).
9. T. Nagai, S. Tamura, and N. Imanaka, *Electrochemistry*, **79**, 450 (2011).

Direct Comparison of Anti-interference Property for Bimetallic PtAu, PtIr, and PtRu Nanoparticle Catalysts in Amperometric Detection for H_2O_2 based Biosensors

M. Janyasupab[*a], Y. Zhang[b], C.-W. Liu[c], and C.-C. Liu[a]

[a] Department of Chemical Engineering, Case Western Reserve University, Cleveland, Ohio 44106 USA
[b] Department of Chemistry, Shanghai University, Shanghai 200444, China
[c] Institute of Materials Science and Engineering, National Central University, Jhong-Li 320, Taiwan

[*]Correspondence should be addressed to Metini Janyasupab, metini@case.edu

Platinum based bimetallic nano electrocatalysts such as PtAu, PtIr and PtRu appear to be very unique toward H_2O_2 oxidation due to their composition, morphology, shape, surface active site and other properties. Enzymatic reactions often produce H_2O_2 as a by-product. Therefore, an electrocatalyst reacts favorably with H_2O_2 will enhance the detection of the analyte in biosensing. This study describes the synthesis and characterization of these bimetallic electrocatalysts and their electrochemical oxidation with H_2O_2 and their reactions with essential interference species, such as ascorbic acid, uric acid and others. A minute interference by PtIr and PtRu nano electrocatalysts are found minimizing the effect contributed by ascorbic acid, uric acid, dopamine and acetaminophen. The results of this study will aid in the design of H_2O_2 sensing based biosensors using the platinum based bimetallic electrocatalysts enhancing their performance.

Introduction

Amperometric electrochemical biosensors are attractive for enhancing high sensitivity, fast response and eliminating a requirement of expensive equipment or elaborative laboring. One approach of these biosensors is to detect the H_2O_2 generated which is often a by-product of an enzymatic reaction with the analyte. The produced H_2O_2 can be oxidized electrochemically resulting in an oxidation current, which in turn can then be used to quantify the analyte. For practical applications, the measurements should not require any mediator or other chemicals and have minimum interference.

Platinum (Pt) is a well-known and traditional heterogeneous catalyst using in many sensing devices. [1-4] The reactivity and selectivity of heterogeneous catalysts usually depend on their surface active sites, and therefore the synthesis of nanoparticles with controlled shapes and sizes is critical for tuning the surface active sites and important for improving their applications [5-7] Well-dispersed Pt nanoparticles with different morphologies, such as polyhedral, truncated cubic and cubic nanocrystals have been controllably synthesized and applied to biosensors [8-10]. However, an addition of a second metal has shown further enhancement as a catalyst improvement than that for the single Pt catalyst for H_2O_2 oxidation detection at a lower potential. [11, 12] Therefore, bi-metallic platinum based nano-catalysts represent a new class of catalysts which can be used in accomplishing the objectives described leading toward a new potential generation of biosensors.

One of the major concerns in the biosensor development is to overcome interference problems. Several studies have been shown that many electroactive species exist in human body can be oxidized on the surface working electrode of biosensors. [13-16] Recent develepment of bimetallic biocatalysts shows a certain degree of anti-interference. PtIr sensors were reported for glucose detection and found no interference toward uric acid and acetaminophen. [12, 17, 18] Bo and coworkers reported that PtPd nanocomposite exhibited a good capability of anti-interference to ascorbic acid, uric acid, and dopamine. [19] Chen *et al.* also proposed MnO_2 on multiwall carbon nanotube to eliminate those interferents and chloride anions effects. [20] However, few studies was reported the anti-interference property toward the same synthesis technique of those bimetals.

In this study, we have prepared and characterize platinum based bi-metallic nano-partcle catalysts, PtM (M= Au, Ir, and Ru). We have carried out the electrochemical detection of H_2O_2 in phosphate buffer solution (PBS), and potential interference study by other species. Several interferents that commonly found in human body such as ascorbic acid, uric acid, dopamine, acetaminophen, acetylsalicylic acid, L-cystein, and chloride anions are tested in amperometry for each metal. The experimental results will be useful to design the proper bi-metallic catalyst for the construction of the proper amperometric biosensor.

Experimental

Nanoparticle Preparation

Platinum based bi-metallic nano-particles could be prepared by various methods. However, for comparable purpose, we would only describe one identical method in preparing these bi-metallic nano-particles. In this study, single Pt and bimetallic PtAu, PtIr, and PtRu were synthesized using a highly dielectric solvent, ethylene glycol, (EG, Fisher, AR) assisted with microwave irradiation. In a typical run, 0.75 ml of an aqueous solution of 0.05 M $K_2PtCl_6 \cdot 6H_2O$ (Aldrich, ACS Reagent) was mixed with 25 ml of EG, and 0.75 ml of 0.4 M KOH in a 300 mL beaker. The beaker was then placed in the center of a household microwave oven (Panasonic, 1250 W) and heated for 36 s. For the preparation of PtRu and PtIr bimetallic nanoparticles, the starting mixture contained 25 ml EG, 0.75 ml 0.05 M $K_2PtCl_6 \cdot 6H_2O$ aqueous solution, 0.75 ml 0.05 M $RuCl_3$ (Aldrich, ACS Reagent) or 0.05 M $IrCl_3$ (Aldrich, ACS Reagent) aqueous solution respectively and 0.75 ml 0.04 M KOH solution. In the preparation of PtAu nanoparticles, the concentration of Pt and Au precursors were reduced to 0.025 M, the other reagents were maintained the same. The resulting pure Pt and bimetallic solutions (pure Pt: 11.60 mL, PtRu: 10.97 mL, PtIr: 5.96 mL, PtAu: 11.76 mL) were mixed with 16 mg of carbon powder (Vulcan XC 72R, Cabot, USA) by ultrasound for 2 hours, respectively. Finally, these Pt-based bimetallic nanoparticles were mixed with the carbon powder, producing a printable ink to be evaluated and used for electrochemical characterization. The nanoparticle metal and carbon mixtures were washed with acetone and ethanol in sequence three or four times. After centrifuging and drying at 60 °C under vacuum for 12 h, the mixed nano-particle electrocatalysts were obtained, and ready to be used.

Material Characterizations

The morphology, size, and elemental composition of the bimetallic and the single platinum nano-particles were characterized by transmission electron microscopy (TEM, Techai F30, using an accelerating voltage of 300 kV), and energy-dispersive X-ray spectroscopy (EDX). The crystal phase of as-synthesized products was obtained by powder X-ray diffraction (XRD) analysis (Cu Kα, $\lambda = 1.54056$ Å, Scintag X-1, USA), and the XRD data were collected at a scanning rate of 0.02° s^{-1} for 2θ in a range from 10° to 80°.

Electrochemical and Interference Studies

Modified GCE preparation. A three-electrode system of glassy carbon electrode (GCE, 0.5 cm. in diameter, AFE2M050GC, Pine Instrument), Pt mesh counter electrode (1 cm. x 1 cm.), and Ag/AgCl reference electrode was employed in this study. Prior to each experiment, GCE was clean with acetone and ethanol in sequent, and then polished with 50 μm alumina powder. For a typical run, 8 mg of each electrocatalyst was dispersed in ethanol (200 μL) and Nafion solution (100μL, 5% wt.) in an ultrasonic bath for 30 minutes. Eight microliters of the mixture (the ink) was deposited onto the GCE and dried in ambient air for 3 minutes. The working electrode was operated at the rotational speed of 900 rpm. The electrochemical performances were investigated with various concentrations of H_2O_2 in a phosphate buffer solution (0.1 M PBS) of pH 7.4. Amperometric measurements (Electrochemical Workstation CHI 660B) were performed determining the electrochemical response of H_2O_2.

Interference study. Species of electroactive interferents commonly found in human body were prepared. First, each catalyst was quantified the biosensing performance toward H_2O_2 without any interference at +0.25 V versus Ag/AgCl. Second, H_2O_2 at 0.5, 1, and 1.5 mM was added to PBS for amperometric studies at +0.3V versus Ag/AgCl. Then, the experimens were repeated with the mixture of 5 mg/L ascorbic acid (AA) and H_2O_2 at 0.5, 1, and 1.5 mM in comparison. Finally, all other essential interference species, such as uric acid (UA), L-cystein (L-cys), dopamine (DP), acetaminophen (AP), acetyl salicylic acid (SA), potassium chloride (KCl), and sodium choride (NaCl) were injected in various concentrations, commonly found in physiological range. The interference results of these species were chosen as examples and presented. Each study was repeated for three separate trials.

Results and Discussion

Figure 1 shows the TEM images of single Pt (a) and bimetallic PtRu (b), PtAu (c) and PtIr (d) nanoparticle. The morphology and size of different metals can be obtained by controlling the concentration of metal precursor and time of irradiation. In Figure 1a, the single Pt with diameter ranging from 3 to 5 nm was resulted by microwave irradiation for 36 s. Similarly, bimetallic PtRu and PtIr nanoparticles with a diameter of about 2-3 nm were formed upon microwave irradiation for 28 and 20 s respectively. Decreasing the concentration of Pt and Au precursor to 0.025 M yielded nanoparticles with a diameter of about 2-4 nm. All bimetallic nanoparticles also exhibited internetworking between spacing and neighboring clusters, which were distinguish to the single Pt particles. Furthermore, elemental compositions of each metal were characterized by EDX analysis (Figure S1, supplemental). Each sample was well mapping and revealed the co-existence of Pt, Au, Ir, and Ru, accordingly. In addition, XRD pattern of each metal was obtained to confirm the bimetallic formation (Figure S2). Both PtIr and PtRu revealed a slightly right shifted toward larger degree 2θ of Pt(111) peak, indicating the decreased d-spacings in the Pt fcc lattice due to the incorporation of the smaller metal atoms. On the other hand, the Pt(111) peak of PtAu shifted to smaller degree 2θ, suggesting that the incorporation of larger Au atoms into the Pt system.

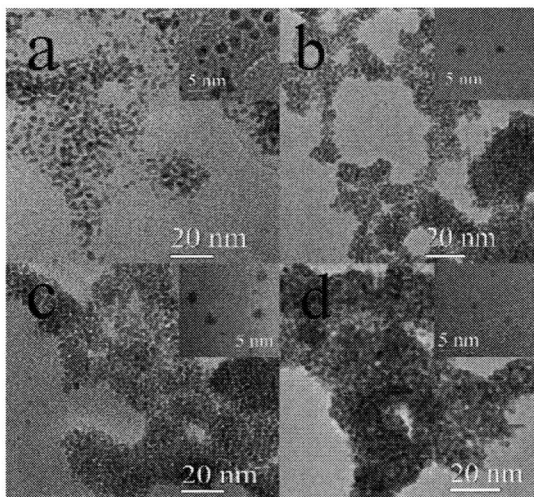

Figure 1. The TEM images of pure Pt (a) and bimetallic PtRu (b), PtAu (c) and PtIr (d) nanoparticles

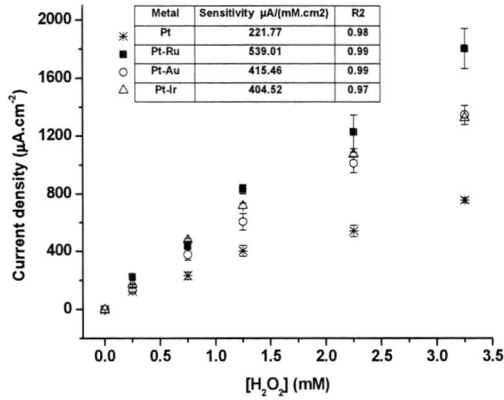

Figure 2.

Figure 2. Current-concentration of H_2O_2 calibration plots at + 0.25 V vs. Ag/AgCl

Figure 2 shows the current-concentration plot of bimetallic compounds from amperometric titration to detect H_2O_2 over the concentration range of 0.25 mM to 3.25 mM. In comparison, the sensitivities of PtRu, PtAu, and PtIr are estimated to be 539.01, 415.46, and 404.52 $\mu A.mM^{-1}.cm^{-2}$, respectively, higher than that of single Pt nanoparticles (221.77 $\mu A.mM^{-1}.cm^{-2}$). In six separate trials, PtIr and PtRu showed remarkable sensitivity, and also high reproducibility with RSD of 0.5% and 2.0% towards the response of 0.25 mM H_2O_2, respectively. The detection limit of PtRu, PtAu and PtIr were estimated to be 2.0, 1.7 and 0.8 μM (SNR=3), which was lower than 3.3μM on the single Pt nanoparticles. The R^2 values of all catalysts indicated a good linear relationship between oxidation current and H_2O_2 concentration. Among the bimetals, PtRu outperformed the other two bimetallic compounds with the highest sensitivity with a R^2 value = 0.99, providing a desirable characteristics to detect pure H_2O_2 at this low applied potential. The low applied potential was desirable, for it would minimize oxidation of any other species in the biological system. Consequently, this would minimize potential interference.

Investigation of interference toward H_2O_2 detection was also carried out for biosensing in this study. It was commonly known that ascorbic acid (AA) often affected in electrochemical reaction resulting in interference in the H_2O_2 measurement. In order to investigate the effect of AA, the applied potential was increased to be +0.3 V where oxidation of most electroactive species occurred. In Figure 3, the direct comparison between pure H_2O_2, and mixture of H_2O_2 with high concentration of AA (5mg/L) for each bimetallic electrocatalyst was presented. The oxidation current at 0.5, 1.0, and 1.5 mM H_2O_2 of the pure Pt shows the interference effect of AA by 30% increase of oxidation current. However, a minute interference of AA was found in PtIr and PtRu with excellent sensitivity of 1359.7 $\mu A.mM^{-1}.cm^{-2}$ ($R^2 = 0.99$), and 1236.4 $\mu A.mM^{-1}.cm^{-2}$ ($R^2 = 0.99$), respectively.

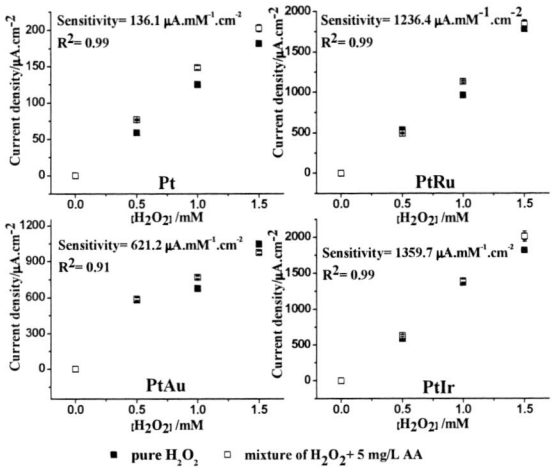

■ pure H₂O₂   □ mixture of H₂O₂+ 5 mg/L AA

Figure 3. Effect of 5 mg/L AA to H_2O_2 detection at +0.3 V vs Ag/AgCl

Further study was carried out to investigate the bimetals toward other interferents. Figure 4 showed amperometric responses of each catalyst to a series addition of interferents at commonly presented in physiological system [13]. The interferents were selected at the high concentration of 10 mg/L ascorbic acid, 0.50 mM uric acid, 0.01 mM dopamine, 0.20 mM acetaminophen, 0.20 mM acetylsalicylic acid, and 0.015 mM L-cystein. Also, 0.35 M KCl, and 0.35 M NaCl were used to test Cl⁻ anion response. As shown in the Figure, after AA injection a significantly increasing current of Pt catalyst could be observed. On the other hand, PtRu and PtIr exhibited negligible increase of current.

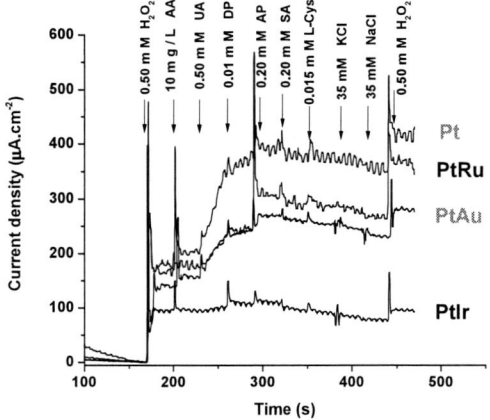

Figure 4. Amperometric response at +0.30 V of 0.5 mM H_2O_2 and other interferents: 10 mg/L ascorbic acid, 0.50 mM uric acid, 0.01 mM dopamine, 0.20 mM acetaminophen, 0.20 mM acetylsalicylic acid, 0.015 mM L-cystein, 0.35 M KCl, and 0.35 M NaCl

Furthermore, some interferents such as UA and DP resulted in a significant increase of oxidation current in the pure Pt. However, only a minute increase of current was found in PtAu, PtRu, and PtIr. Especially, PtIr could maintain a minimum increase of AA, UA, DP, and AP response. In addition to those interferents, SA, L-Cys, and Cl$^-$ ions showed more pronounced effects toward all bimetals in comparison to the pure Pt. After all of eight interferents species addition, 0.50 mM H$_2$O$_2$ was added again. Only 15-22% increase of current was found among all bimetallic nanocatalysts.

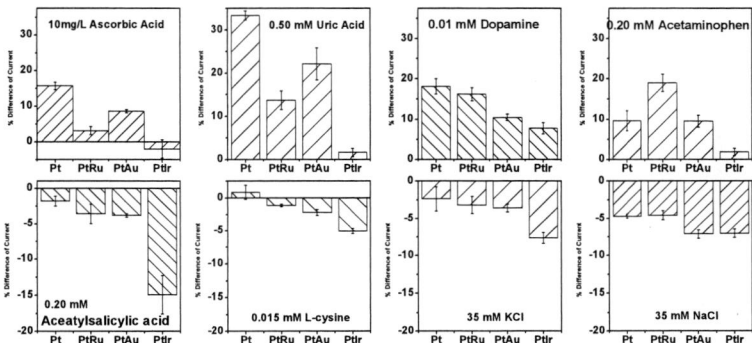

Figure. 5 Percentage difference of current after interference addition for each catalyst (n=3)

Figure 5 summarized the effect of each interferent based on the amperometric response, repeated in three separate trials. It was found that 0.50 mM UA contributed the most effect by increasing 35 % current in the pure Pt. Some other interferents such as AA, DP, and AP could also be oxidized in the pure Pt, but only a minute effect was pronounced in the bimetals, especially PtIr. However, SA could affect the performance of PtIr as shown by 15% decrease in current. Further study of long-term stability was also carried out during five consecutive days, quantifying 1.5 mM H$_2$O$_2$ responses on carbon printed screen sensor prototypes. The single Pt and PtAu catalysts showed 84.6%, and 77.4% of catalytic activity, respectively. On the contrary, PtIr and PtRu could maintain 96.7% and 98.6% of the activity after five days. It is understood and recognized that the long term stability of the nano bimetallic catalysts is important for the practical utilization of the bimetallic catalysts in H$_2$O$_2$ detection, and we believe that improvement of its bimetallic composition would certainly provide more insightful aid in further development for H$_2$O$_2$ base biosensors. Overall, both PtIr and PtRu by far exhibited an improvement of anti-interference property and could be promising candidate for future biocatalysts toward H$_2$O$_2$ detection.

Supplemental Figures

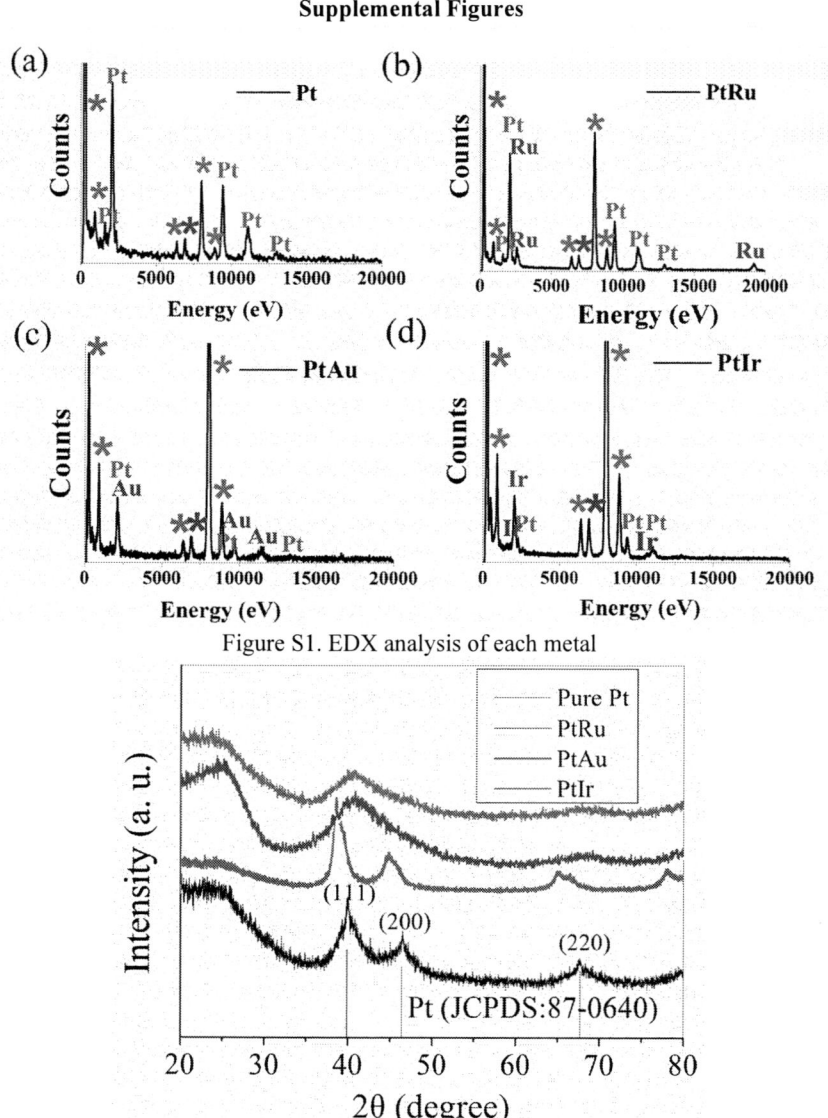

Figure S1. EDX analysis of each metal

Figure S2. XRD pattern of bimetallic catalyst: (top to bottom) PtIr, PtRu, PtAu, and Pt

Conclusions

Platinum based bimetallic electrocatalysts, namely, PtAu, PtIr and PtRu were synthesized and characterized in this study. Dielectric solvent, ethylene glycol assisted with microwave irradiation was used in the synthesis process. TEM, EDX and XRD analysis were employed in the characterization of these materials. The detection of H_2O_2 and the reactions with essential interference species, such as ascorbic acid, uric acid, and dopamine of each bimetallic electrocatalyst were carried out and assessed. Among these catalysts, PtIr and PtRu showed an improvement of anti-interference property, capable of developing a new class of biocatalyst in H_2O_2 detection.

Acknowledgments

This study is supported by Royal Thai Government Scholar Fellowship, China Scholarship Council Postgraduate Scholarship Program, Taiwan National Central University Fellowship, Delta Environmental and Educational Foundation, and the NSF Grant no. 1000768 and MURI-2011-Nanofabrication, DOD-Air Force office of Scientific Research are gratefully acknowledged.

References

1. Bak, T., J. Nowotny, M. Rekas, and C.C. Sorrell, *International Journal of Hydrogen Energy*, **19**, 27 (1), 2002.
2. Wang, A., X. Ye, P. He, and Y. Fang, *Electroanalysis*, **1603**, 19 (15), 2007.
3. Ebbing, A., O. Hellwig, L. Agudo, G. Eggeler, and O. Petracic, *Physical Review B*, **012405**, 84 (1), 2011.
4. Mao, L., R. Yuan, Y. Chai, Y. Zhuo, and W. Jiang, *Analyst*, **1450**, 136 (7), 2011.
5. Song, H., F. Kim, S. Connor, G.A. Somorjai, and P. Yang, *The Journal of Physical Chemistry B*, **188**, 109 (1), 2004.
6. Gontard, L.C., L.-Y. Chang, C.J.D. Hetherington, A.I. Kirkland, D. Ozkaya, and R.E. Dunin-Borkowski, *Angewandte Chemie International Edition*, **3683**, 46 (20), 2007.
7. Wang, C., H. Daimon, T. Onodera, T. Koda, and S. Sun, *Angewandte Chemie*, **3644**, 120 (19), 2008.
8. Ye, J.-S., A. Ottova, H.T. Tien, and F.-S. Sheu, *Bioelectrochemistry*, **65**, 59 (1–2), 2003.
9. Chu, Z., Y. Zhang, X. Dong, W. Jin, N. Xu, and B. Tieke, *Journal of Materials Chemistry*, **7815**, 20 (36), 2010.
10. Wang, S., L. Lu, and X. Lin, *Electroanalysis*, **1734**, 16 (20), 2004.
11. Janyasupab, M., Y. Zhang, P.-Y. Lin, B. Bartling, J. Xu, and C.-C. Liu, *Journal of Nanotechnology*, 2011 2011.
12. Chen, K.J., K. Chandrasekara Pillai, J. Rick, C.J. Pan, S.H. Wang, C.C. Liu, and B.J. Hwang, *Biosensors and Bioelectronics*, **120**, 33 (1), 2012.
13. Toghill, K.E. and R.G. Compton, *International Journal of Electrochemical Science*, **1246**, 5 (9), 2010.
14. Bai, Y., W. Yang, Y. Sun, and C. Sun, *Sensors and Actuators B: Chemical*, **471**, 134 (2), 2008.
15. Ali, S.R., R.R. Parajuli, Y. Ma, Y. Balogun, and H. He, *The Journal of Physical Chemistry B*, **12275**, 111 (42), 2007.
16. Su, Y., R. Hu, W. Huang, and K. Hu, *Microchimica Acta*, **19**, 173 (1), 2011.
17. Holt-Hindle, P., S. Nigro, M. Asmussen, and A. Chen, *Electrochemistry Communications*, **1438**, 10 (10), 2008.

18. Qiang, L., S. Vaddiraju, D. Patel, and F. Papadimitrakopoulos, *Biosensors and Bioelectronics*, **3755**, 26 (9), 2011.
19. Bo, X., J. Bai, L. Yang, and L. Guo, *Sensors and Actuators B: Chemical*, **662**, 157 (2), 2011.
20. Chen, J., W.-D. Zhang, and J.-S. Ye, *Electrochemistry Communications*, **1268**, 10 (9), 2008.

A high-throughput assay for evaluation of embryoid bodies using local redox cycling-based electrochemical chip device

K. Ino[a], T. Nishijo[a], Y. Kanno[a], H. Shiku[a], T. Matsue[a,b]

[a]Graduate School of Environmental Studies, Tohoku University, Japan
[b]Advanced Institute of Materials Research, Tohoku University, Japan

We have previously proposed an excellent method to incorporate many electrochemical sensors into a single chip device using two sets of microelectrode array to induce local redox cycling, and we have designated the novel methodology as local redox cycling-based electrochemical (LR-EC) system. In this study, we developed a LRC-EC chip device consisting of 256 electrochemical sensors. At the electrochemical sensors, ring-type interdigitated array (IDA) electrodes were placed to induce local redox cycling. The LRC-EC chip device was applied to evaluate three-dimensional (3D) culture cells, such as embryoid bodies (EBs).

1. Introduction

We have previously developed a local redox cycling-based electrochemical (LRC-EC) chip device for high-throughput electrochemical detection [1-4]. In the LRC-EC chip device, row and column electrodes are arranged orthogonally and these electrodes are connected to comb-type interdigitated array (IDA) electrodes [1-3] or ring-ring electrodes [4] to form n^2 crossing points with only $2n$ bonding pads for external connection. By applying proper potential to these electrodes, local redox cycling can be induced at the desired electrodes, and the comb-type IDA electrodes or the ring-ring electrodes can be used as individual electrochemical sensors. Therefore, many electrochemical sensors can be incorporated into a single chip by using the system. In this study, we fabricated a LRC-EC chip device consisting of ring-type IDA electrodes and the LRC-EC chip device was applied to evaluate three-dimensional (3D) culture cells.

In this study, we used embryonic stem (ES) cells. ES cells can differentiate into any body tissues, by forming 3D tissue organs, such as embryoid bodies (EBs) for developing the ES cells into cardiomyocytes. Since the degree for differentiation can be evaluated through their activity of alkaline phosphatase (ALP) on the EBs, the differentiation level of the EBs was evaluated via their ALP activity using the LRC-EC chip device.

2. Materials and Methods

The general outline is shown in Figure 1. The LRC-EC chip device consisted of 16 row and 16 column electrodes to form 256 (16×16) electrochemical sensors by using 32 (16+16) connector pads. At individual electrochemical sensors, ring-type IDA electrodes [5] were placed. EBs were prepared according to our previous report [1, 6, 7]. Briefly, mouse ES cells were suspended in medium without mouse leukemia inhibitory factor (mLIF). The droplets (20 µl) of the ES cell suspension were hung from the dish cover and were incubated for 2-10 days to fabricate cell aggregates, called EBs. The EBs were introduced into the LRC-EC chip device and the electrochemical detection was performed after trapping the EBs into the microwells. The scheme for ALP detection using the LRC-EC device is described in our previous paper [1]. Briefly, *p*-aminopheny phosphate (PAPP) was used for substrate. PAPP was catalytically hydrolyzed by ALP on the EBs and *p*-aminophenol (PAP) was yielded. The generator electrode (+0.30 V vs. Ag/AgCl) was used for oxidizing PAP and the oxidation product, *p*-quinone imine (PQI), was then reduced back to PAP at the collector electrode (-0.30 V vs. Ag/AgCl). The scanning process is shown in our previous paper [1].

The device fabrication process is described in our previous paper [1]. The device consisted of 4 layers. Briefly, the row and column electrodes were fabricated on glass substrates with a conventional photolithography method [1-4, 8]. An approximately 100 nm layer of Ti and approximately 100 nm layer of Pt were fabricated on the substrate by sputtering to create a Ti/Pt multilayer. SU-8 2002 was used to insulate excess electrodes. For completing electrodes, Ti/Pt electrodes were fabricated on the SU-8 layer by sputtering. Finally, SU-8 3050 layer was fabricated to create microwells.

The electrochemical performance of the individual ring-type IDA electrodes was characterized by amperometry in ferroceme methanol (FcCH$_2$OH) solution using a potentiostat with an Ag/AgCl saturated KCl electrode as the reference electrode. The scheme for evaluating the chip performance is described in our previous paper [1]. Briefly, the potential of one column electrode was stepped from 0.00 to 0.50 V to oxidize FcCH$_2$OH while the potential of all other electrodes was kept at 0.00 V to reduce FcCH$_2$OH$^+$ back to FcCH$_2$OH. The electrochemical signals from one row electrode were acquired to evaluate the chip performance.

The detection scheme for evaluation of differenciation level of ES cells is shown in Figure 2. The scheme of the scanning process and the electrochemical detection for ALP is described in our previous paper [1]. After electrochemical detection, the EBs were collected and reseeded onto gelatin-coated dishes. After a further 3-day culture, the differentiated ES cells were observed to check whether the differentiated ES cells beat spontaneously or not.

3. Results and discussion

Figure 3 showed that the LRC-EC chip device had 256 sensors. Images of the ring-type IDA electrodes showed that the width of each electrode finger was approximately 5 μm and the gap between the fingers was approximately 5 μm, and the number of the pairs of electrode fingers was 9.

The electrochemical signals at the individual sensors were proportional to the concentration of FcCH$_2$OH (Figure 4). Thus, the present LRC-EC chip device can be used for quantitative determination of redox compounds. Since the electrochemical signals are amplified by inducing redox cycling, the LRC-EC chip device can be applied as sensitive assays.

In this study, we prepared long-term and short-term cultured EBs (Figure 2). The initial cell concentrations were controlled to prepare the EBs of the same size (Figure 5). Even if these EB sizes were same, the ALP activity on the short-term cultured EBs was higher than that on long-term cultured EBs, indicating that the long-term cultured EBs differentiated. Since PAP is oxidized to PQI by a two-electron reaction while FcCH$_2$OH is oxidized to FcCH$_2$OH$^+$ by a one-electron reaction, the PAP concentration from the 2-day cultured EBs near the electrode was roughly estimated at 200 μM by using the carburation curve in Figure 4. The value was similar to that in our previous paper [1]. Since the long-term cultured EBs developed into cardiomyocytes (Figure 6), the differentiated EBs shrank and expanded (beat) spontaneously (Table 1). These results showed that the differentiated EBs had low ALP activity, which are reasonable results compared to previous studies. Thus, the LRC-EC chip device can be applied for cell analyses, such as detection of differentiation level of EBs.

In conclusion, the LRC-EC chip device was applied for evaluating EBs. We investigated the relationship between their ALP activity and their differentiation into cardiomyocytes, and clearly showed the good relationship. Since the LRC-EC chip device contained 256 electrochemical sensors and comprehensive electrochemical detection can be performed, we believe that the device can provide high-throughput electrochemical assays on cell analysis.

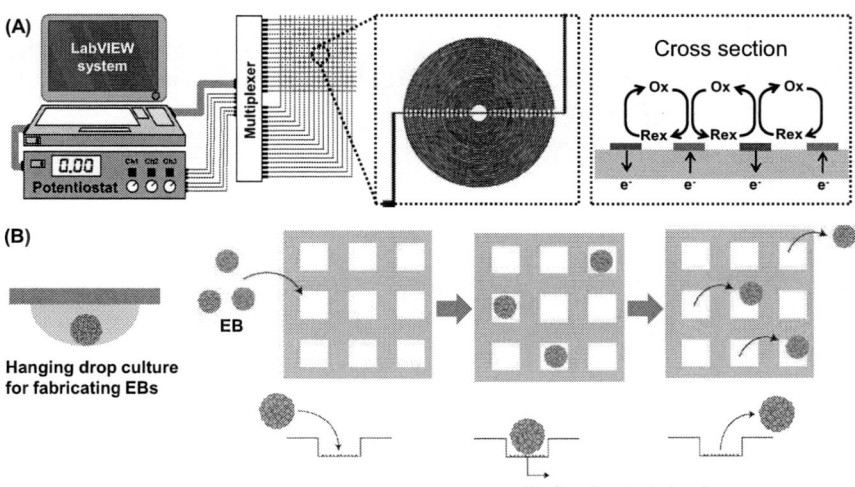

Figure 1 (A) General outline of the LRC-EC chip device. The potentiostat is connected to the row and column electrodes through the multiplexer and PC. Local redox cycling is induced only at the desired crossing point. (B) Detection scheme. EBs are fabricated by using a hanging drop culture method. The EBs are introduced and electrochemically detected, and collected.

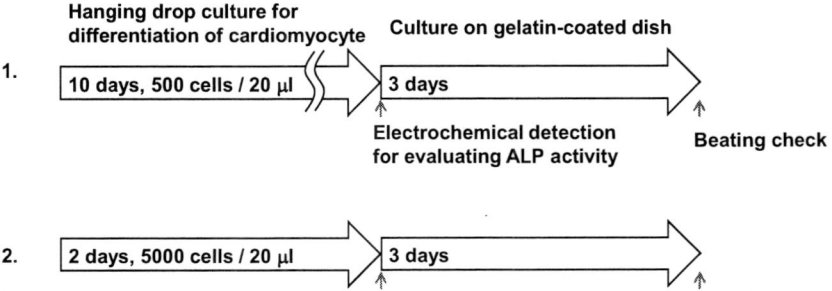

Figure 2 Scheme for preparing EB samples, ALP detection and check of differentiation level.

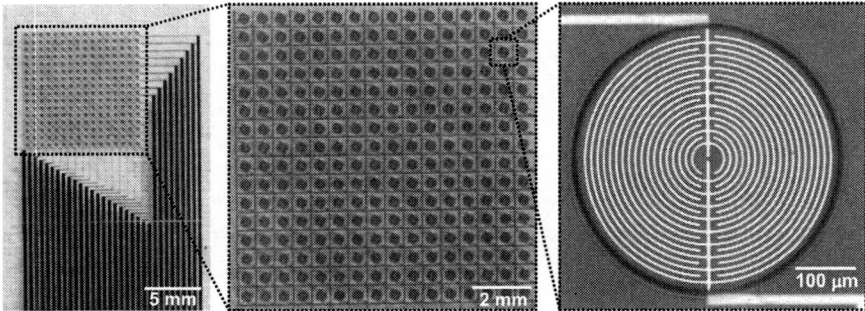

Figure 3 Device images. (A) The device had 256 sensor points with only 32 connector pads. At the crossing point, the ring-type IDA electrodes were placed. Individual ring-type IDA electrode consisted of 9 pairs of finger (width, 5 μm: gap, 5 μm). The center-to-center between the electrochemical sensors was 650 μm.

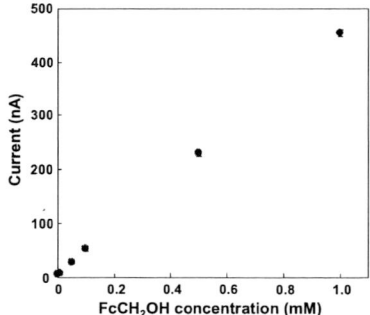

Figure 4 Dependence of the electrochemical signals from single electrochemical sensors on the FcCH$_2$OH. The detection scheme is described in our previous paper [1].

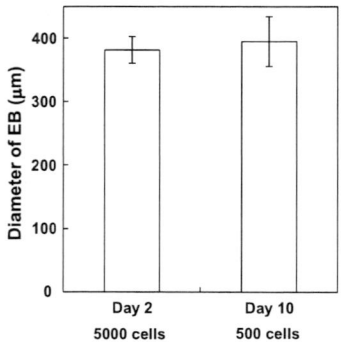

Figure 5 Size of long-term and short-term cultured EBs.

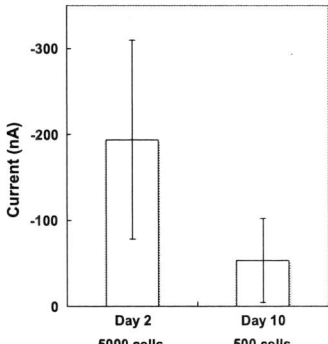

Figure 6 ALP activity on long-term and short-term cultured EBs.

Table 1. Diameter of EB, ALP activity and percentage of beating EB

	Day 2 5000 cells	Day 10 500 cells
Diameter of EB (μm)	381±21	394±39
ALP activity (nA)	-193±117	-53±49
Beating EB (%)	0	60

Acknowledgments

This work was supported in part by a Grants-in-Aid for Scientific Research (A) (No. 22245011) and for Young Scientists (B) (No. 23760745) from the Japan Society for the Promotion of Science (JSPS). This study was also supported by the Asahi Glass Foundation and the Mitsui Sumitomo Insurance Welfare Foundation. This work was partly supported by the Cabinet Office, Government of Japan, through its "Funding Program for Next Generation World-Leading Researchers".

References

1. K. Ino, T. Nishijo, T. Arai, Y. Kanno, Y. Takahashi, H. Shiku, T. Matsue. *Angew. Chem. Int. Ed.*, **51**, 6648-52 (2012).
2. K. Ino, W. Saito, M. Koide, T. Umemura, H. Shiku, T. Matsue, *Lab Chip*, **11**, 385-388 (2011).
3. K. Ino, Y. Kanno, Taku. Nishijo, T. Goto, T. Arai, Y. Takahashi, H. Shiku, T. Matsue. Electrochemical detection for dynamic analyses of a redox component in droplets using a local redox cycling-based electrochemical (LRC-EC) chip device. *Chem. Commun.*, 2012, in press.
4. M. Takeda, H. Shiku, K. Ino, T. Matsue, *Analyst*, **136**, 4991-6 (2011).
5. X. Zhu, C.H. Ahm. *IEEE Transactions on Nanobioscience*, **4**, 164-9 (2005).

6. K. Ino, A. Ito, Y. Wu, N. Saito, E. Hibino, O. Takai, H. Honda. *J. Biosci. Bioeng.*, **104**, 420-3 (2007).
7. H. Kurosawa. *J. Biosci. Bioeng.*, **103**, 389-98 (2007).
8. K. Ino, Y. Kitagawa, T. Watanabe, H. Shiku, M. Koide, T. Itayama, T. Yasukawa, Matsue T. *Electrophoresis*, **30**, 3406-12 (2009).

Printed Amperometric Gas Sensors

M. T. Carter, J. R. Stetter, M. W. Findlay and V. Patel

KWJ Engineering, Inc., 8430 Central Avenue, Ste. 2C, Newark, CA 94560 USA

A new type of amperometric gas sensor, based on screen printing
and lamination strategies, has been fabricated and tested. These
sensors use thin film electrodes, printed on plastic substrates to
form a small, low power (μW) amperometric sensor package,
which uses a very small volume of electrolyte. These sensors have
been demonstrated to have improved performance compared to
amperometric gas sensors of conventional design, when
conventional electrolytes are used. The sensors have also been
demonstrated to have promising performance, compared to
conventional technology, using room temperature ionic liquid
electrolytes (RTILs). The printed sensors have been demonstrated
successfully for monitoring of carbon monoxide (CO), ammonia
(NH_3).

Introduction

Amperometric gas sensors are a well-known and versatile type of gas sensor with many
desirable characteristics, including good sensitivity, selectivity and detection limit for
electrochemically oxidizable or reducible toxic gases, and relatively low power
requirements (1). This paper covers fundamentals and selected applications of a new
type of amperometric gas sensor developed at KWJ Engineering (2). This sensor is
fabricated using screen printing and printed electronics manufacturing approaches to
produce a high performance gas detection device, with very low production cost and a
number of advantageous characteristics, including microwatt power requirements, a tiny
form factor compatible with cell phone technology, and competitive, and in some case
superior, performance compared to current, commercially available amperometric gas
sensors. The screen-printed electrochemical sensor provides a new avenue to high-
performance, low-cost monitoring of electroactive toxic gases. The platform provides
high sensitivity, high selectivity, low detection limit, long-life and durability in a variety
of applications. Its performance surpasses what is possible with other typical low-cost
sensors, such as those incorporating heated metal oxide semiconductor (HMOS)
transducers. There is typically a significant performance gap between small, low cost,
mass producible gas sensors with low selectivity and low sensitivity, and high cost
sensors or analytical instrumentation, which provide high performance at a premium
price. The higher cost analytical methods also usually require significant power and are
more difficult to miniaturize. Our vision for the printed devices described here is to
bridge the cost-performance gap for a wide spectrum of gas monitoring applications,
making high performance gas sensing more widely available than has been possible
previously.

The KWJ printed gas sensor uses the same scalable, low cost printing approach used
in the electronics industry to achieve large area flexible electronics at low cost. Further,

by virtue of the unique aspects of its design and fabrication approach, the technology provides the ability to use novel electrolytes, including room temperature ionic liquids, to tailor the response characteristics to the application.

This report will discuss selected recent results for gas sensing using the new, printed platform with conventional (aqueous acid) and developmental (room temperature ionic liquid, RTIL) electrolytes, for the gases carbon monoxide (CO) and ammonia (NH$_3$). Sensor performance is improved for conventional amperometric CO gas sensing by the printed format. We also show that RTILs, which have grown in popularity for sensor applications in recent years, can be used effectively to produce amperometric NH$_3$ sensors with excellent performance characteristics (3-10).

Experimental

Sensor Fabrication

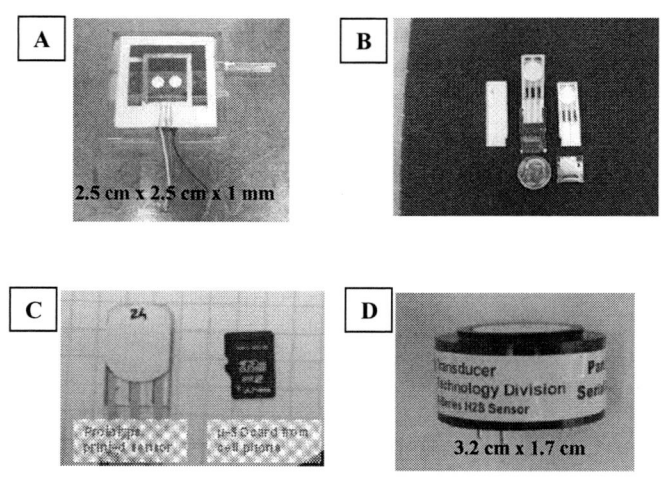

Figure 1. (A-C) Examples of recent printed amperometric gas sensor prototypes fabricated at KWJ as described in the Experimental section, compared to (D) a conventional, canister type format (KWJ R Series housing). The prototype shown in (A) was used primarily for NH$_3$ measurements and contained RTIL electrolyte. Sensors as in (B) and (C) were used primarily for CO measurements and contained acidic aqueous electrolytes. Fabrication details are described in the text.

Some details of sensor fabrication and structure are proprietary information of KWJ Engineering and are part of a patent in process (2). The general fabrication process has been described elsewhere (2). Examples of the resulting sensors are illustrated in Figure 1. The description provided here refers specifically to the 1 in^2 device shown in the upper left photograph in Figure 1, but is generally the same for all other printed formats shown in Figure 1. A commercial CO module, KWJ R Series CO, is shown for comparison in the lower right photograph of Figure 1.

Examples of several different designed printed sensors are shown in Figure 1, compared with a conventional amperometric gas sensor package (KWJ R Series). Smaller devices are made by analogous approaches to those used to print larger area devices. All the printed sensors of Figure 1 were fabricated at KWJ Engineering by the general procedure described here. Briefly, the gas inlet side of the device contains an array of 16 holes, each 0.2" in diameter, in a 20 mil polycarbonate piece. This is laminated with a Mupor membrane upon which the electrode catalyst layer has been printed. The compositions of the screen printing inks are proprietary, but contain standard electrocatalysts materials for CO sensors, i.e., Pt, and can be made from a variety of Electrocatalytic metals (1). Electrode contacts (runners) are defined on a second polycarbonate piece which is laminated on top of the electrode layer, followed by the porous silicate wick material. The sensor housing in this case is laminated to the electrodes so as to encapsulate it and form a thin electrolyte chamber. The electrolyte is held in place by capillary forces. The volume of RTIL or aqueous acid electrolyte in the filled sensor was ca. 500 µL. After filling, the fill holes are sealed. The entire device is 1 in^2 and about 1mm thick. Devices of any arbitrary size can be fabricated using this process. Aqueous acidic electrolytes, as described below, were used for CO measurements with sensors of type (B) and (C) of Figure 1. RTIL electrolytes were used with sensors of type (A) of Figure 1 for NH$_3$ measurements.

Electrochemical Cells and Electrolytes

All gas sensors, either printed or standard commercial packages, were three electrode cells comprising a printed working electrode for gas detection and Pt auxiliary and quasi-reference (QRE) electrodes. In addition to the developmental, thin printed sensor format, we also performed some characterization surveys, particularly for down-selection of electrolytes and working electrodes, using a conventional KWJ-manufactured R Series gas sensor housing, which is shown in Figure 1. This approach is simply for convenience: the R Series housing and standard sensors made with it have been extensively characterized and so provide a well-understood comparative system for testing components of the developmental sensor. The sensors were typically tested at constant voltage bias vs. QRE sufficient to drive CO or NH$_3$ oxidation under diffusion limited conditions.

CO sensors were prepared using acidic electrolytes, which are commonly employed in such devices, for example, 4 M H$_2$SO$_4$. NH$_3$ sensors were constructed with room temperature ionic liquid (RTIL) electrolytes and their performance was compared to commercially available amperometric ammonia sensors from other manufacturers.

RTILs were investigated with the aim of improving NH$_3$ sensing properties compared to commercial aqueous acid or nonaqueous or gel electrolytes. The RTILs investigated here included 1-ethyl-3-methylimidazolium bis(trifluoromethylsulfonyl)imide, 1-butyl-3-methylimidazolium bis(trifluoromethylsulfonyl)imide and 1-hexyl-3-methylimidazolium bis(trifluoromethylsulfonyl)imide, hereafter denoted EMIMTf$_2$N, BMIMTF$_2$N and HMIMTf$_2$N, respectively. These were purchased from EMD Chemicals (Gibbstown, NJ) and were used as received. While the viscosity of the RTILs is substantially greater than that of aqueous solutions, their surface tensions and wetting properties on the sensor structures are such that they wick easily into the printed device, filling the electrolyte bucket volume and fully wetting the silicate wick to provide electrolytic contact with the electrode catalyst layers.

All gases (CO, NH$_3$, air) were obtained either in pure form or as mixtures with air and then further diluted dynamically to obtain the final, desired concentrations.

Sensor Test System

The system for automated testing and evaluation of gas sensors is shown in Figure 2. The system includes Lab View™-based control of mixing, dilution, electrochemical parameters and data acquisition. The environmental test chamber can accommodate up to 32 gas sensors, controlled by multiple 8 channel potentiostat boards, which were fabricated in-house. Our testing equipment allows us to easily evaluate all specifications for new sensors by repeatedly exposing each sensor to varying concentrations of target gases in the presence of a selected background, with logging all results for analysis of response time, linearity, range, resolution, repeatability and noise at a selected voltage bias and relevant range of temperatures and humidities (RH). A commercial RTD (resistance temperature detector), RH (relative humidity) sensor and pressure sensor (P) are resident in the gas test chamber. The chamber also has an internal fan for gas mixing.

All sensor response times cited below are t$_{90}$ values, defined as the time required for the sensor signal to increase from 10% to 90% of full scale.

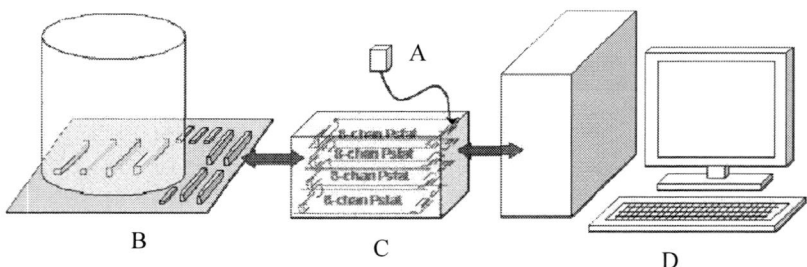

Figure 2. Gas test system for parallel testing of replicate amperometric gas sensors. (A) Power; (B) Test fixture for 3 electrode sensors (4 cards, 8 sensors/card) with T, RH and P sensors; (C) 32 channel potentiostat rack/card cage; (D) PC with DAQ and I/O cards.

Results and Discussion

Printed Sensors for Carbon Monoxide

Figure 3 shows some typical response characteristics for three CO printed sensors from one preparation batch and a comparison to the same amperometric measurement using a conventional R-Series sensor. In all cases, the amperometric measurements were performed at 0V vs. Pt QRE and the working electrode was high surface area Pt prepared by screen printing a custom colloidal Pt ink.

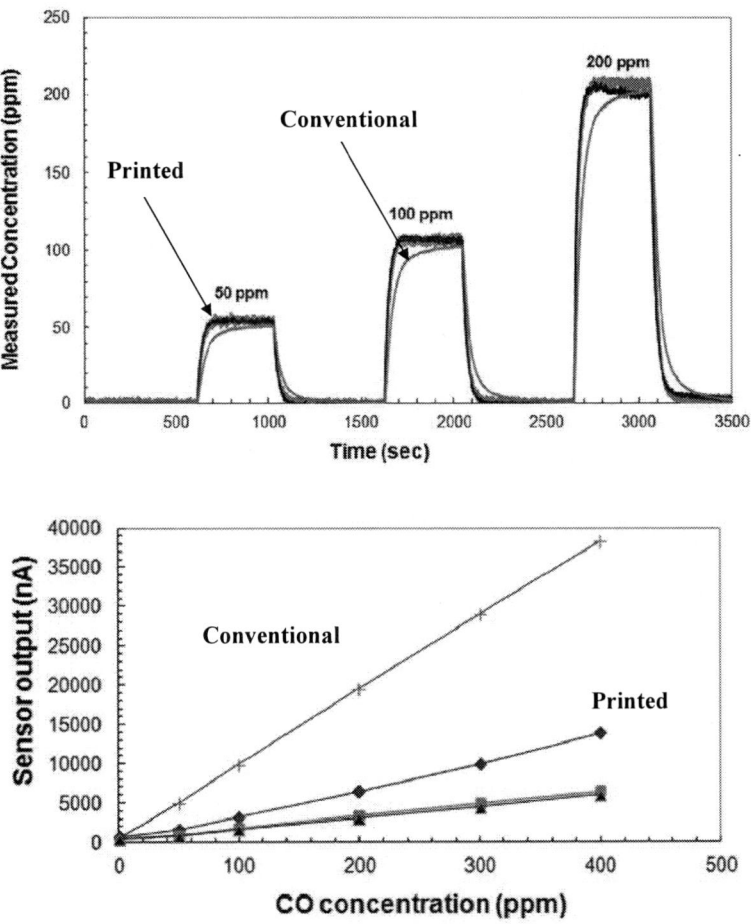

Figure 3. Raw sensor responses for three replicate printed CO sensors with comparison to a conventional CO sensor configuration (R Series). Note that the measured signal in upper plot is rendered as measured CO concentration (i.e., a practical, calibrated instrumental output in normal operation), whereas the lower plot shows sensor output signal in raw current units. The lower plot shows calibration curves for the R-Series conventional sensor (+) and three replicate printed sensors (♦, ■, ▲).

The linear ranges of the printed and conventional sensors were comparable, 0-500 ppm CO in dry air. The printed sensor response time was 70 s at 500 ppm CO, compared to 90 s for the commercial sensor. As can be seen in Figure 3, both the approach to steady state response and return to baseline were more rapid for the printed format compared to the conventional format. This feature is a distinct improvement of the new, printed technology over existing approaches. The lower limit of detection (LLD), estimated as 3 times the standard deviation of baseline noise, was 2 ppm. This is comparable to the LLD for a conventional sensor design. The sensitivity, as measured by the slope of the calibration curves in Figure 3, was somewhat greater for the conventional format (95.2 nA/ppm CO for the R Series sensor compared to 14.2,15.6 and 33.1 nA/ppm CO for the printed sensors) and current magnitudes were larger for the R Series format at all CO concentrations. The latter observation is caused by the significantly larger total electrode area of the conventional format.

Batch to batch repeatability of the printed sensors at 100 ppm CO was 1-3 % (relative standard deviation), compared to 1.2% for a single batch of R series devices. Zero drift (baseline) was comparable for both types of sensors over a 350 s test (± 2%). Span drift at constant exposure to 100 ppm CO for 4000 s was also comparable between the developmental and conventional sensor platforms (± 3%).

Printed Sensors for Ammonia

Down Selection of Electrodes and Ionic Liquids. The knowledge base for development of NH_3 sensors based on RTIL electrolytes is limited compared with the long development history of amperometric CO sensors (1,5,11-13). Therefore, we first examined several candidate RTILs and printed electrode materials in order to determine the best candidates for further studies. We examined 10 combinations of three RTILs ($EMIMTf_2N$, $BMIMTF_2N$ and $HMIMTf_2N$) and 5 printed electrode materials (sputtered, low surface area platinum film (LSA Pt), sputtered, low surface area gold film (LSA Au), fuel cell grade, high surface area Pt black (HSA Pt), carbon nanotube (CNT) coated sputtered gold (CNTAu), and glassy carbon spheres doped with CNTs (GC-CNT). Sensitivity, measured in nA/ppm, was determined for 200 ppm NH_3 in air at R Series sensors. The electrode/RTIL combinations HSA Pt/$BMIMTf_2N$ ($E_{applied}$ = 480 mV vs. Pt), CNT Au/$BMIMTf_2N$ ($E_{applied}$ = 880 mV vs. Pt) and GC-CNT/ $EMIMTf_2N$ ($E_{applied}$ = 880 mV vs. Pt) provided the best sensitivities at the indicated biases. The raw biases can be converted to V vs. SHE by adding 1.02V. While the maximum sensitivity for HSA Pt/ $BMIMTf_2N$ occurred at about 1.5 V vs. SHE or about +480 mV bias, we found that the sensors could be operated perfectly adequately at 0 V bias in the printed format, which is the typical bias used for commercial ammonia amperometric sensors (14,15). All subsequent printed sensor work was done using the HSA Pt electrode and $BMIMTf_2N$ electrolyte operated at 0V bias.

Printed NH_3 Sensor Performance. Printed amperometric gas sensors were fabricated in the 1 in x 1in x 1mm format with array of 16 holes (0.2 in dia) in the gas inlet. Assembled sensors were filled with RTIL electrolyte (ca. 500 µL) by capillary action in just the same manner as the sensors would be filled with conventional aqueous electrolytes. These sensors function in an entirely analogous manner to gas sensors with conventional electrolytes.

The HSA Pt/BMIMTf$_2$N printed sensor was tested for a number of typical performance characteristics. The dynamic range of the sensor exceeded 0-1000 ppm (Figure 4), with the highest sensitivity and best linearity in the region 0 – 200 ppm NH$_3$. The detection limit was estimated to be 2.0 ppm, based on the highly linear 0-50 ppm NH$_3$ region and the noise of the measurement (3δ). This was also the practical resolution of the Phase I sensor. The prototype printed sensor was demonstrated to be sufficiently sensitive and noise free to differentiate NH$_3$ concentrations at environmentally relevant levels (16).

The response times of the printed sensor (t$_{90}$, time to reach 90% of full scale from 10% of full scale; t$_{10}$, time for reverse of the t$_{90}$ process) were a function of NH$_3$ concentration, as expected. The t$_{90}$ varied from 71 s at 1000 ppm to 178 s at 50 ppm NH$_3$ (bias = 0V). These were much faster than R-Series conventional sensor tests, but slower than the published manufacturer specification for a typical benchmark commercial sensor (ca. 30 s, Sensoric) (14,15).

Printed sensor precision was ± 5.2% at 1000 ppm and 18.1% at 100 ppm NH$_3$ (3δ of the mean as a percentage). Baseline (zero) drift was observed in some sensors, amounting to about a 12% signal decrease over 1 h of continuous testing with repeated 100 ppm NH$_3$ exposure and air purge cycles.

Printed sensors were tested in 0% to 95% relative humidity (RH) air. Increasing humidity resulted in increased sensitivity to NH$_3$. Cross sensitivities ranged from 0.44 ppm NH$_3$/%RH 0 to 2.8 ppm NH$_3$/% RH over several tests at various NH$_3$ concentrations and RH values. Sensor calibration for RH will probably be needed to fully account for humidity changes for practical devices. When dried to constant weight at 5% RH, the printed sensors take up significant moisture on exposure to 95% RH air. The weight increase is larger than would be expected based on the solubility of water in BMIMTf$_2$N and related media (17), suggesting that water adsorption on the high surface area porous glass wick component of the device may contribute.

The printed sensor with RTIL electrolyte was compared directly to a commercial NH$_3$ sensor (Sensoric 3E100SE) (15). The commercial sensor's average sensitivity was 136 nA/ppm at 10 ppm NH$_3$ and that of the printed sensor was 6.3 nA/ppm at the same NH$_3$ concentration. The signal magnitude for the printed sensor was about 5% of that of the commercial sensors (a relative electrode area effect), however the response times were comparable (180 s for the printed sensor and 190 s for the Sensoric at 50 ppm NH$_3$). The printed RTIL sensor response shape was also very similar to the commercial sensor (Figure 5). The sensitivity of the printed sensor per unit estimated nominal electrode surface area was about 2% that of the commercial sensor. Therefore, we conclude that the printed sensor already has competitive response performance compared to commercial sensors, but optimization of the triple phase boundary (i.e., ensuring the absence of electrode flooding) will be required to improve the overall signal-to-noise and sensitivity of the RTIL filled devices.

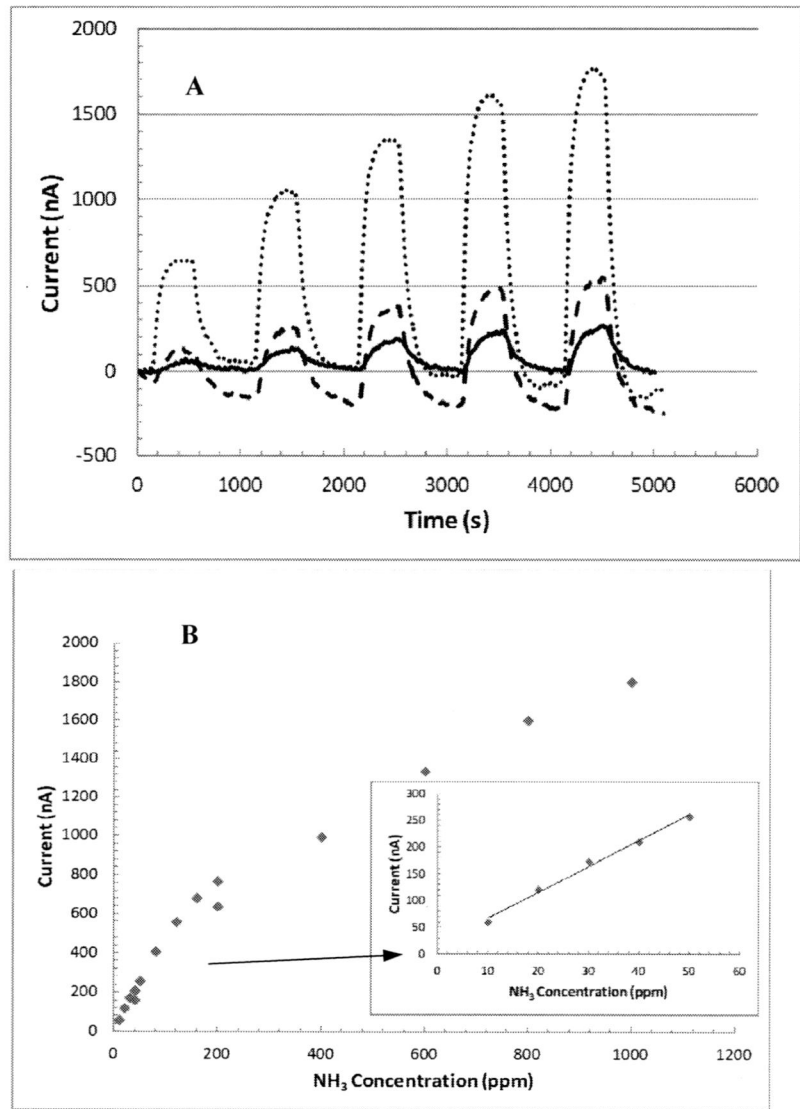

Figure 4. (A) Raw NH$_3$ responses for printed sensors (high surface area Pt) with RTIL electrolyte (BMIMTf$_2$N) : 0 – 50 ppm NH$_3$ (solid line; specific gas exposures, left to right, 10, 20, 30, 40, 50 ppm NH$_3$), 40-200 ppm NH$_3$ (long dashed line; specific gas exposures, left to right, 40, 80, 120, 160, 200 ppm NH$_3$), 200-1000 ppm NH$_3$ (dotted line; specific gas exposures, left to right, 200, 400, 600, 800, 1000 ppm NH$_3$); (B) Dynamic range and low concentration linearity for the printed sensor.

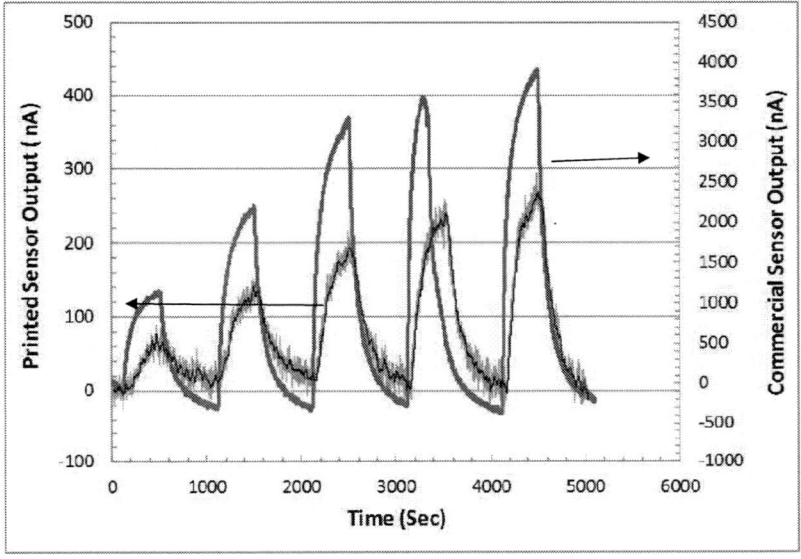

Figure 5. Direct comparison of the Pt/BMIMTf$_2$N printed sensor to a commercial ammonia sensor (Sensoric 3E100 SE) at 0 to 50 ppm NH$_3$ in air.

Conclusions

Printed amperometric gas sensors with practical, commercializable designs were fabricated and characterized for typical performance measures. We found that the printed CO sensor, using acidic aqueous electrolytes, was very competitive with existing state of the art commercial technology. Further, the printed format was demonstrated to have superior response time features compared to the conventional format.

Printed ammonia sensors were demonstrated using ionic liquid electrolytes in a manufacturable format and their performance was compared directly to commercial ammonia sensors. These sensors had performance characteristics that were in many respects competitive with existing conventional commercial technology, but further work on gas sensor design will be needed to fully explore the possibilities for an optimized device that competes favorably with commercial technology in all respects.

Acknowledgments

This work is supported by the National Science Foundation, Award No. 1058563 the National Institutes of Health, Grant Nos. 1R43ES019385-01 and 1R43ES02176-01 and Department Energy Grant No. DE-SC0007530.

References

1. J. R. Stetter and J. Li, *Chem. Rev.*, **108**, 352 (2008).
2. J. R. Stetter, E. R. Stetter, D. Ebeling, M. W. Findlay and V. Patel, "Printed Gas Sensor", U.S. Patent Application, US20120125772 A1, May (2012).
3. X.-J. Huang, L. Aldous, A. M. O'Mahony, F. J. del Campo and R. G. Compton, *Anal. Chem.*, **82**, 5238 (2010).
4. M. Nádherná, F. Opekar and J. Reitera, *Electrochim. Acta*, **56**, 5650 (2011).
5. K. Murugappan, J. Lee and D. S. Silvester, *Electrochem. Commun.*, **13**, 1435 (2011).
6. M. C. Buzzeo, C. Hardacre and R. G. Compton, *Anal. Chem.*, **76**, 4583 (2004).
7. M. Koel, *Crit. Rev. Anal. Chem.*, **35**, 177 (2005).
8. S. Pandey, *Anal. Chim. Acta*, **556**, 38 (2006).
9. D. Wei and A. Ivaska, *Anal. Chim. Acta*, **607**, 126 (2008).
10. N. V. Shvedene, D. V. Chernyshov, and I. V. Pletnev, *Russian J. Gen. Chem.*, **78**, *2507 (2008)*.
11. X. B. Ji, C. E. Banks, D. S. Silvester, L. Aldous, C. Hardacre and R. G. Compton, *Electroanalysis*, **19**, 2194 (2007).
12. M. Buzzeo, D. Giovanelli, N. S. Lawrence, C. Hardacre, K. R. Seddon and R. G. Compton, *Electroanalysis*, **16**, 888 (2004).
13. X. B. Ji, D. S. Silvester, L. Aldous, C. Hardacre and R. G. Compton, *J. Phys. Chem. C*, **111**, 9562 (2007).
14. Sensoric Ammonia 3E100 Product Data Sheet, http://www.citytech.com/PDF-Datasheets/nh33e100.pdf.
15. Sensoric Ammonia 3E100SE Product Data Sheet, http://www.citytech.com/PDF-Datasheets/nh33e100se.pdf.
16. CDC-NIOSH Pocket Guide to Chemical Hazards, http://www.cdc.gov/niosh/npg/npgd0028.html.
17. C. Hilgers and P. Wasserscheid in *Ionic Liquids in Synthesis*, P. Wasserscheid and T. Welton (Eds.), Wiley-VCH, Weinheim, Chap. 2.2, p. 27 and references therein (2003).

Interaction of Water Vapor with SnO$_2$ Sensor Materials: a Comparison of DRIFTS and Resistance Measurements

R.G. Pavelko[a], K. Grossmann[b], N. Barsan[b], K. Shimanoe[a]

[a] Kyushu University, Department of Energy and Material Sciences, Fukuoka, Japan
[b] University of Tübingen, Institute of Physical Chemistry, Tübingen, Germany

> Operando DRIFT spectroscopy together with resistance measurements was used to study sensitivity (here response change upon changing analyte concentration) of undoped SnO$_2$ materials to water vapors. The materials synthesized using different protocols were found to manifest very different sensitivity to water, as well as to H$_2$ and CO in dry and humid air. While for one of the materials (SnO$_2$ Ac) low sensitivity to water combines with similarity between evolution of two bands in OH region and resistance, another material (SnO$_2$ Cl), with high sensitivity to water vapors, manifests similarity with the resistance only for the broad band between 3590 and 3040 cm^{-1}. Together with high sensitivity to water vapors the latter material shows higher sensitivity to CO in dry and humid air, while SnO$_2$ Ac showed higher sensitivity to H$_2$.

Introduction

Studies of water effect in the field of chemoresistive gas sensors date back to the first commercial Taguchi sensors integrated into gas-leakage detectors. At that time, in Figaro laboratories water was found to be one of the reasons for "hypersensitivity", leading to false alarms, long-term drift of sensor response and serious deterioration of sensitivity during storage under humid conditions (1, 2). First researchers, studying water effects on commercial and home-made SnO$_2$ based sensors, revealed that responses toward water, being dependent on operation temperature and additives, were caused by water chemisorption and subsequent formation of surface hydroxyls (3-8). The strongest water effects were observed at operation temperatures below 400 °C. Doping with noble metals helped to shift the water responses in low-temperature region (increasing the response though), thereby remarkably decreasing response to water in commercial gas sensors working at operation temperature 450-300 °C (1, 9, 10).

However, nowadays almost all new electric devices, aimed at mobile applications or employment of sustainable energy sources, face a requirement of low-power consumption. In the case of chemoresistive gas sensors, this implies not only use of micro-hotplates to reduce sensor power consumption, but also a considerable decrease of sensor operation temperature. This means that water effects should be understood much better in all aspects of gas sensing phenomenon to find a solution for selective and reliable low-power devices.

In situ and *operando* IR spectroscopy has been proved to be a powerful technique to study surface chemistry of gas sensing materials in general and SnO$_2$ based materials in particular (11-14). All spectroscopic studies performed on SnO$_2$ either in humid or dry air report changes in the region of OH stretching vibrations, suggesting that even traces of water in the gas phase play important role in sensing phenomenon.

In the case of pure SnO_2, increase of CO concentrations both in humid and dry air causes an absorbance loss at *ca.* 3700 cm^{-1} and between 3650 and 3000 cm^{-1} (13). Similar situation was observed for hydrogen interacting with undoped SnO_2 (15). While the band at 3700 cm^{-1} represents typically a narrow and intense peak, the absorbance loss at higher frequencies is very broad and unspecific (sometimes with several narrow peaks), suggesting either hydrogen bonding between OH groups, heterogeneous nature of OH bonds, or absorbance by free carriers (12, 16, 17). The relationship between spectroscopic changes and resistance can give us a clue (in a direct or indirect way) about surface species involved in the charge transfer.

This study reports *operando* investigation of SnO_2 based materials in the presence of water vapors, performed with the help of DRIFT spectroscopy and resistance measurements. Evolution of DRIFT spectra and sheet resistance of pure SnO_2 were studied and compared as a function of water vapor concentration in air at 300 °C. Together with sensitivity of the materials to H_2 and CO in dry and humid air the results contribute to better understanding of how SnO_2 surface changes in the presence of water vapors and how these changes are related to the sensor response.

Experimental Part

Material synthesis

Two types of blank SnO_2 materials were used: the one was synthesized from tin(IV) hydroxide acetate and the other one – from tin(IV) chloride. In the case of the former precursor, the solution of NH_3H_2O (12.5%, puriss., Aldrich) was added dropwise to the precursor solution in acetic acid under external cooling and stirring, until pH of the solution was 6.3. The transparent solution was heated up to 70 °C to cause hydrolysis of the precursor. The milky white colloid was separated by centrifugation at 6000 rpm and washed three times with deionized water (Trace SELECT™ Ultra, Aldrich) at 80 °C. The washed colloid was dried at 90 and at 200 °C for 8 hours at each temperature and then the powder was annealed at 400 °C for 24 hours. More details can be found in (18). The synthesized material will be referred to as SnO_2 Ac.

Another tin oxide material was synthesized also via wet chemistry, starting from an aqueous solution of $SnCl_4$. The hydrolysis of the latter was realized upon drop-wise addition of ammonium hydroxide. The precipitate was subsequently dried for 12 hours (80°C). The final calcination step was performed at 1000°C for 8 hours, more details can be found in (19). This synthesized material will be referred to as SnO_2 Cl.

Sensor fabrication

For DRIFTS and resistance measurements the synthesized materials were deposited on the front side of the planar Al_2O_3 substrate, provided with Pt electrodes for the read out of the electrical resistance. For this purpose the materials were mixed with 1-2-propanediol, used as organic vehicle, to prepare a homogenous paste, which was screen-printed onto the front side. Pt heaters on the backside of the substrates allow one to reach and control the desired temperature of the sensors. After the screen printing the sensors were kept at room temperature for 12 hours and then underwent a final thermal treatment in a belt oven (400-600 °C).

Another substrate, provided only with gold electrodes for the read out of the electrical resistance, was used to study responses of the materials to H_2 and CO in dry and humid air. The ink for the screen printing was prepared using also 1-2-propanediol. The sensors were dried and annealed similarly to the procedure described above.

DRFTS and resistance measurements

DRIFTS, resistance and humidity measurements were performed simultaneously within one experiment on a sensor operated at 300 °C. DRIFT spectra were recorded with an evacuable Bruker Vertex 80 V. Before the measurements, the sensor was stabilized in a flow of dry air (less than 7 ppm H_2O) for 12 hours at the operation temperature 300 °C. Water vapor concentration was changed from 90 to 1700 ppm in air with a dwell of 2 h. A series of 80 consecutive spectra were recorded with a spectral resolution of 4 cm^{-1}, 1024 scans per spectrum (ca. 12 min) and with 3 min delay time. The total gas flow in the system was set to 200 ml/min. Data acquisition for resistance as well as for dew point (Vaisala) was performed every 5 s.

CO and H_2 tests

For CO and H_2 gas tests, four sensors were placed in the quartz chamber located in the resistive oven in such a manner that the whole quartz chamber together with the sensors was heated. Prior to the measurements, the sensors were stabilized in dry air at 300 °C for 24 h. Following concentrations were used for both gases: 55, 100, 150, and 200 ppm in air. The measurements were performed in dry (less than 100 ppm H_2O), and humid air with total gas flow rate of 100 ml/min. Water concentration in humid air was set to 900 ppm and 1700 ppm (3.5 and 6.3 % RH @ 25 °C).

Results and Discussion

DRFTS and resistance measurements

Figure 1 shows DRIFT spectra and resistance change of the materials in question. Two materials manifest differently directed tendency for the spectra evolution upon increasing humidity rate. If we compare the spectra with the ones reported for the same materials interacting with D_2O (15), we should conclude that the broad bands observed between 3600 and 2700 cm^{-1} are shifted about 50-100 cm^{-1} towards higher frequencies in respect with the bands recorded in 700 ppm D_2O and assigned as consumed surface OH groups. Also the shapes of the bands reported here differ from those observed in D_2O. This probably suggests that the reported here features should not be considered as the absolute and their positions as well as the directions are determined by the background correction (built-in concave rubberband correction in OPUS 5.5 with 5 iterations and 25 baseline points), performed prior to the absorbance calculation. Nevertheless, the spectral *changes* registered upon humidity rise seem to be intrinsic for the materials and should not be dependent on the same background correction.

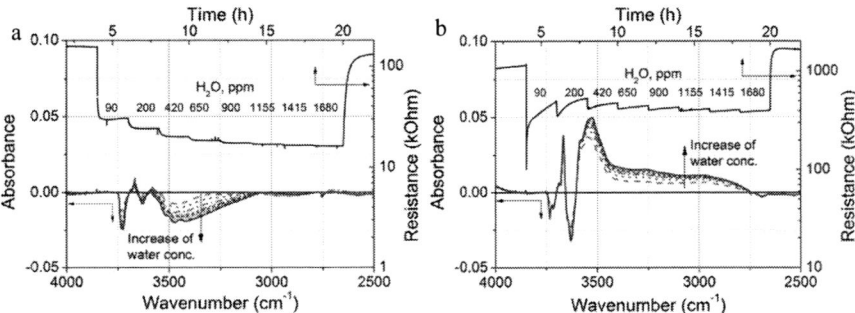

Figure 1. DRIFT spectra and resistance change upon step-wise increase of water vapor concentration in air for SnO_2 Cl (a) and SnO_2 Ac (b).

Apart from the broad bands, both materials manifest a negative doublet centered at 3730 cm^{-1}. Position of this band as well as its doublet nature was reported not only for these materials in H-D experiments [15], but also for SnO_2 synthesized differently [12] and Pd-doped SnO_2 [13]. Accordingly, the band position seems to reflect fairly well the dynamics of the surface species, assigned as isolated or terminal OH groups.

To compare evolution of DRIFT spectra with the resistance changes we calculated integrals of the two main features observed. The broad band was taken between 3600 and 2730 cm^{-1} for SnO_2 Ac, and between 3590 and 3040 cm^{-1} for SnO_2 Cl. Integrals for the band at 3730 cm^{-1} was taken between 3760 and 3705 cm^{-1} for SnO_2 Ac, and between 3760 and 3690 cm^{-1} for the SnO_2 Cl. The results, plotted against water concentration, are shown in Figure 2 and compared with the sensor response, defined as R_0/R_g (R_0 – is the resistance in a reference gas, R_g is the resistance in a target gas). The experimental values in Figure 2 were fitted with the Langmuir-Freundlich isotherm, which reduces to Freundlich isotherm at low pressures, and to Langmuir isotherm in the case of homogeneous surfaces [20]. The equation is shown in Figure 2 (a). Exponent n together with Langmuir adsorption constant b reflects sensitivity of the material, which is defined here as a change of the response as a function of change of target gas concentration. For the sake of the brevity, we will use mainly exponent n to estimate sensor sensitivity and to compare it with the evolution of the DRIFT spectra.

Note that exponent n shows how much sensor signal changes upon increase in the target gas concentration by 1 ppm. However, since sensor signal (or band integral) is considered here as dimensionless parameters, exponent n becomes also dimensionless.

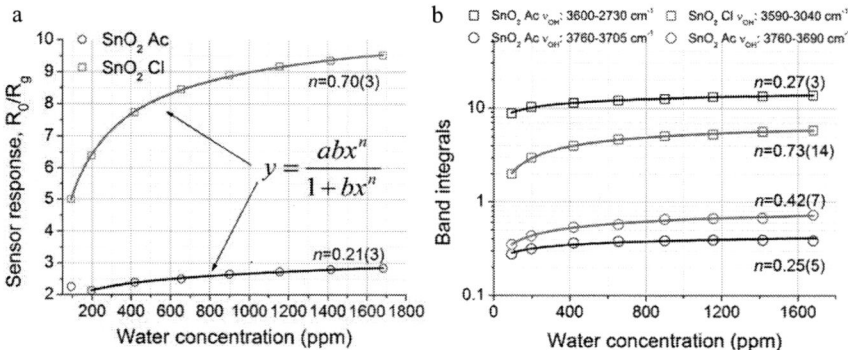

Figure 2. Sensor response (a) and integrals of the bands (b) plotted as a function of water vapor concentration for materials in question.

Figure 2 (a) gives a comparison of sensor responses to water for two materials. Exponent n was found around 2 times higher for SnO_2 Cl than for SnO_2 Ac, indicating that the former manifests higher sensitivity to this range of water vapor concentration regardless the fact that crystallite size for SnO_2 Cl is much higher in comparison with SnO_2 Ac (70 against 5 nm). Interesting to note, that b amounts to 0.35 and 0.15 for Cl and Ac respectively, suggesting higher binding energy of water molecules on SnO_2 Cl material.

Values of n obtained from integration of the broad band were found to be similar to those calculated from resistance measurements for both materials. The same we can say about evolution of the narrow band for SnO_2 Ac. This band in the case of SnO_2 Cl was found to evolve under humid conditions with exponent equal to 0.42 which differs greatly from the resistance measurements. Hence, for SnO_2 Ac both spectral features evolve similarly to resistance, while for SnO_2 Cl the similarity was found only for the high-frequency broad band.

Similarity between spectroscopic changes in the OH region and resistance change can be interpreted as IR absorbance induced by free carries (9, 17, 21). Few studies have been undertaken to separate possible contribution of free carriers, which can be done using *e.g.* isotopic exchange on SnO_2 surface with minimal variation of the resistance (15, 22). As it was shown in (15), bands appeared upon H_2O-D_2O exchange (both 210 ppm) on the same SnO_2 materials are much more intense compared with the ones emerged after simple exposition to 250 ppm H_2, 240 ppm CO or water. This suggests that IR absorbance (again calculated after the background correction) caused by H-D exchange on the surface is much more intense than possible absorbance increase due to free carriers.

However, even if we assume that electronic effects together with other sources of the unspecific absorbance are removed by background correction, the assignment of the bands is not straightforward still, since background correction apparently introduces similar but dramatic changes in the spectra.

CO and H_2 tests

Two materials manifest different sensitivity to water vapors, which would be interesting to compare with sensitivity to reducing gases and how it changes in the presence of water vapor. Figure 4 compares responses to CO and H_2 in dry and humid air. Apart from dramatic decrease of the response, water remarkably increases response and recovery rates, making both sensors faster.

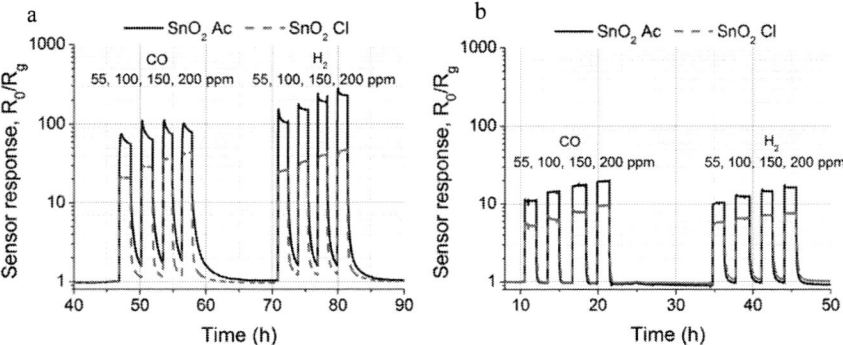

Figure 4. Sensor responses to CO and H_2 in dry air (a) and humid air with 1700 ppm H_2O (b).

Regardless very high responses, SnO_2 Ac manifests rather low sensitivity to 50-200 ppm CO in dry air with n equal to 0.27, which is similar to the case of water vapors, reported above. One should expect that for this material maximum of the sensitivity to CO in dry air is shifted well below 55 ppm. On the other hand, sensitivity to hydrogen for SnO_2 Ac is ca. 2 times as much as the previous value, suggesting that material is sensitive to rather wide range of H_2 concentrations. In the case of SnO_2 Cl, its sensitivity to CO is higher than to hydrogen (the same concentration, see Figure 5), but both are lower than that, found for water vapors (compare Figure 2a).

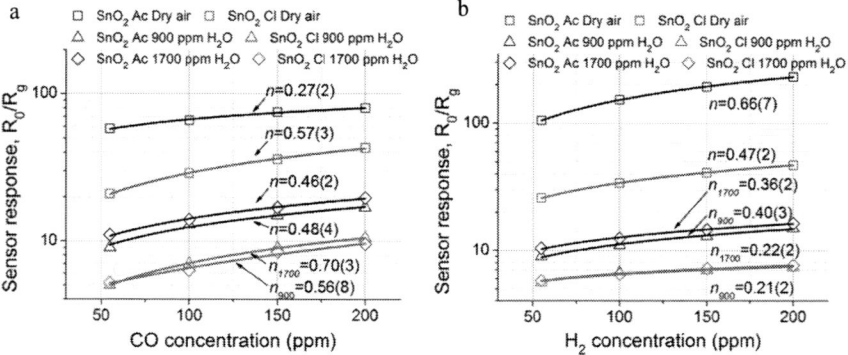

Figure 5. Sensor responses plotted as a function of target gas concentration: for CO (a) and for H_2 (b).

Exponent values are summarized in Figure 6. If water concentration rises, n for both materials increases in case of CO and decreases in case of H_2. In spite of the fact that responses to CO and H_2 in humid air drop significantly compared to dry air, sensitivity of the materials seems to behave differently and depends on type of the gas. Sensitivity increase for 10-1000 ppm CO in humid air has been already reported in the literature for undoped SnO_2 (23-25). However, together with the increase of sensitivity an increase of sensor response to CO is also reported. The difference can be related to the fact that our sensors were stabilized in dry and not in the ambient air. Very high responses to CO in dry air support this assumption (Figure 4a).

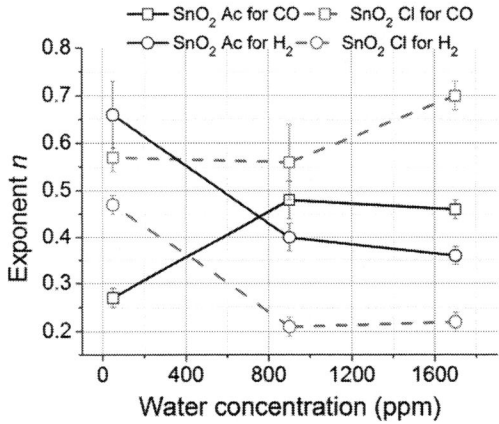

Figure 6. Exponent n (sensitivity) calculated from sensor responses to 50-200 ppm CO and H_2 plotted as a function of water concentration in air.

The highest sensitivity to 55-200 ppm CO was found in the case of SnO_2 Cl, while the highest increase of the sensitivity is observed for SnO_2 Ac. Note that rise of the sensitivity occurs at lower water concentration for SnO_2 Ac, which is in line with the assumption of higher hydroxylation degree of this material (26). In the case of 55-200 ppm H_2, the materials show similar drop at the same water concentration, but the highest sensitivity is observed for SnO_2 Ac.

Conclusions

Two undoped SnO_2 materials synthesized using different protocols were found to manifest very different sensitivity to water vapors, hydrogen and CO. In the case of SnO_2 Ac both spectral features observed between 3600 and 2700 cm^{-1} and at 3730 cm^{-1} evolve upon increasing humidity similarly to resistance change, with exponent n equal to 0.2. The low n value suggests that sensor responses weakly depend on water concentration. For SnO_2 Cl only the broad band between 3590 and 3040 cm^{-1} was found to evolve similarly to resistance with n equal to 0.7, while the doublet at 3730 cm^{-1} exhibited lower dependency on water vapor – 0.4. Both values indicate rather high dependency of sensor response on humidity.

Together with higher sensitivity to 90-1700 ppm H_2O, SnO_2 Cl shows higher sensitivity to 55-200 ppm CO in dry and humid air. Regardless the drop of sensor responses to CO in humid air, sensitivity to this gas was found to be increased in the presence of water vapors. In contrast, sensitivity to 55-200 ppm H_2 for both materials decreases under humid conditions.

Acknowledgments

R.G.P. is grateful to the JSPS Postdoctoral Fellowship For Foreign Researchers.

References

1. K. Ihokura and J. Watson, *The Stannic Oxide Gas Sensor Principles and Applications*, CRC Press Inc., Florida (1994).
2. Y. Nakamura, in *Chemical Sensor Technology*, T. Seiyama, Editor, Vol. 2., Kodansha Ltd and Elsevier Science Publishers B.V., Tokyo, Amsterdam (1989),
3. N. Yamazoe, J. Fuchigami, M. Kishikawa and T. Seiyama, *Surface Science,* **86** (0), 335-344 (1979).
4. J. F. McAleer, P. T. Moseley, J. O. W. Norris, D. E. Williams, P. Taylor and B. C. Tofield, *Materials Chemistry and Physics*, **17** (6), 577-583 (1987).
5. J. F. McAleer, P. T. Moseley, J. O. W. Norris and D. E. Williams, *J. Chem. Soc., Faraday Trans. 1,* **83** (4), 1323-1346 (1987).
6. P. G. Harrison and M. J. Willett, *J. Chem. Soc., Faraday Trans. 1*, **85** (8), 1921-1932 (1989).
7. K. Takahata, in *Chemical Sensor Technology*, T. Seiyama, Editor, Vol. 1, Kodansha Ltd and Elsevier Science Publishers B.V., Tokyo, Amsterdam (1988).
8. G. Heiland, *Sens. and Act. B*, **2** (0), 343-361 (1981).
9. G. Ghiotti, A. Chiorino, G. Martinelli and M. C. Carotta, *Sens. and Act. B*, **25** (1-3), 520-524 (1995).
10. G. Martinelli, M. C. Carotta, L. Passari and L. Tracchi, *Sens. and Act. B*, **26** (1-3), 53-55 (1995).
11. D. A. Popescu, J.-M. Herrmann, A. Ensuquea and F. Bozon-Verduraz, *Phys. Chem. Chem. Phys.*, **3**, 2522-2530 (2001).
12. N. Sergent, P. Gélin, L. Périer-Camby, H. Praliaud and G. Thomas, *Phys. Chem. Chem. Phys.*, **4**, 4802–4808 (2002).
13. S. Harbeck, A. Szatvanyi, N. Barsan, U. Weimar and V. Hoffmann, *Thin Solid Films*, **436** (1), 76-83 (2003).
14. R. G. Pavelko, H. Daly, C. Hardacre, A. A. Vasiliev and E. Llobet, *Phys. Chem. Chem. Phys.*, **12** (11), 2639-2647 (2010).
15. K. Grossmann, R. G. Pavelko, N. Barsan and U. Weimar, *Sens. and Act. B*, **166–167** (0), 787-793 (2012).
16. E. W. Thornton and P. G. Harrison, J. Chem. Soc., *Faraday Trans. 1*, **71**, 461 - 472 (1975).
17. A. Chiorino, G. Ghiotti, F. Prinetto, M. C. Carotta, G. Martinelli and M. Merli, *Sens. and Act. B*, **44** (1–3), 474-482 (1997).
18. R. G. Pavelko, A. A. Vasiliev, E. Llobet, V. G. Sevastyanov and N. T. Kuznetsov, *Sens. and Act. B*, In Press, Corrected Proof (2011).
19. A. Diéguez, A. Romano-Rodríguez, J. R. Morante, J. Kappler, N. Bârsan and W. Göpel, *Sens. and Act. B*, **60** (2-3), 125-137 (1999).
20. I. Quiñones and G. Guiochon, *Journal of Chromatography A*, **796** (1), 15-40 (1998).
21. P. Emelie, J. Phillips, B. Buller and U. Venkateswaran, *Journal of Electronic Materials*, **35** (4), 525-529 (2006).

22. C. A. Wolden, T. M. Barnes, J. B. Baxter and E. S. Aydil, *Journal of Applied Physics*, **97** (4), 043522 (2005).
23. G. Huyberechts, P. Szecówka, J. Roggen and B. W. Licznerski, *Sens. and Act. B,* **45** (2), 123-130 (1997).
24. S. Capone, P. Siciliano, F. Quaranta, R. Rella, M. Epifani and L. Vasanelli, *Sens. and Act. B,* **77** (1-2), 503-511 (2001).
25. T. Sahm, W. Rong, N. Bârsan, L. Mädler, S. K. Friedlander and U. Weimar, *Journal of Materials Research,* **22** (04), 850-857 (2007).
26. R. G. Pavelko, A. A. Vasiliev, E. Llobet, X. Vilanova, N. Barrabes, F. Medina and V. G. Sevastyanov, *Sens. and Act. B,* **137** (2), 637-643 (2009).

Development of Micro Hydrogen Gas Sensor Utilizing Polymerized Gel with Ionic Liquid as a Solvent

T.Yamauchi[a, c], T. Matsui[b], T. Nishiyama[b], K. Tsunashima[d], N. Tsubokawa[a, c] and S. Harada[a]

[a]Graduate School of Science and Technology, Niigata University
[b]Faculty of Engineering, Niigata University
[c]Center for Transdisciplinary Research, Niigata University,
8050, Ikarashi 2-nocho, Nishi-ku, Niigata 950-2181, Japan
[d]Wakayama National College of Technology, 77 Nadachounoshima, gobou-shi,
Wakayama 644-0023, Japan

> In this study, the gel beads contained ionic liquid were applied to electromotive force (EMF) hydrogen sensor which has a high hydrogen gas detecting ability. Ionic gel beads were prepared by suspension polymerization and their diameters were 100-2000 μm. The sensor utilizing ionic liquid gel has flexibility and heat endurance. The weight and shape of gel beads were maintained not only in high temperature condition but also under vacuuming condition. The hydrogen micro sensor was able to detect 2 % hydrogen concentration. It has high grade gas selectivity and can detect hydrogen gas in the range of explosion limit. It is considered that the sensor can be used as superior hydrogen sensor.

Introduction

Recently, environmental issues, which are caused by utilization of huge amounts of fossil fuels, are a global problem. Hydrogen gas has attracted much attention as an energy source for next generation. However, sensing system for leaking hydrogen is required because of the flammability of hydrogen. The electromotive force (EMF) hydrogen sensor is quite a new type of sensors which has rapid and safe sensibility.

The EMF hydrogen sensor that detects hydrogen by noting changes in the electromotive force (1,2). While conventional hydrogen sensors have operated with some risks of explosion, and they have a big error range, EMF hydrogen sensor has rapid sensing speed, and a higher accuracy rate for detection of hydrogen. To avoid global warming, CO_2 reduction has been required. At the same time, novel technology which emits no carbon dioxide has become increasingly recognized as a measure to prevent warming, thus hydrogen is prospective energy and desired to use safely, but unfortunately hydrogen has an explosive gas and its explosion range is 4-75 % under the room temperature. EMF hydrogen sensor is simply composed of a reference electrode, a sensing electrode and electrolyte, however, if uses a fragile crystalline solid electrolyte thus its application is limited. It is thought that soft and flexible material curved surface sensor is efficient for vehicle vibration of a hydrogen machine such as hydrogen car. In order to improve flexibility of the sensor, we focused on ionic liquid as solvent in gel beads. Ionic liquids have been actively researched for application to electro deposition, batteries and fuel cells, and solvents (3). Ionic liquids are liquid salt at room temperature, in general, thermally and chemically stable and have negligible vapor pressure, high ionic conductivity and possibilities to change molecular design. In recent years, ionic liquids have been studied intensively as new electrolytes, reaction solvent including enzyme reaction and organic synthesis, lubricant agent, batteries, actuators, and many other things for various chemical, electrochemical, biological, physical applications on the basis of their unique properties. It is also known that use of phosphonium-based ionic liquids provide chemical and thermal stabilities in various applications.

Polymer gels are soft and flexible material that consist of three dimensional structure of polymer chain and solvent, thus gels containing ionic liquids would be used to new material including electrolytes and capacitors (4,5). As mentioned above, it is expected that the gel containing ionic liquid is applied to a new type of solid electrolyte (6,7).

Therefore, we developed electromotive force (EMF) hydrogen sensor as an application of the gels. This sensor is the quite new type of sensor which detects hydrogen gas through changes of electromotive force[4]. It can detect only hydrogen gas and the sensing speed is high, moreover it can be easily miniaturized. In this study, the gel beads were prepared by suspension polymerization and investigated its ability for electrolyte of micro hydrogen gas sensor.

Experimental

Preparation of ionic liquid gel beads by suspension polymerization

A Polymer gel beads containing ionic liquid were prepared by suspension polymerization. 1 mol/l of vinyl monomer such as methyl methacrylate (MMA), styrene (St) , and N-isopropylacrylamide (NIPAM), 0.1 mol/l of ethylene glycol dimethacrylate as cross linker and 0.03 mol/l of azobisisobutyronitrile as initiator were dissolved in 1 ml of ionic liquid Tri-n-butyloctylphosphonium tetrafluoroborate (P4448-BF4).

The mixture solution was dropped into continuous phase composed of PVA and distilled water. The polymerization was carried out at 70 $^{\circ}$C and suspension for 4 h. For the evaluation of thermal properties, the weight of gel beads was investigated under vacuum condition for 24 h. For evaluation of thermal endurance, the thermal decomposition temperatures of the gel beads were measured by thermo gravimetric analysis (TGA) device.

Development and evaluation of EMF hydrogen sensor

PMMA gel beads were prepared by suspension polymerization. The sensor was prepared by sandwiching these gel beads between platinum electrode and copper electrode. The electrodes were connected to digital multi-meter. The changes on EMF were evaluated while various gases including oxygen, carbon dioxide, argon, nitrogen and hydrogen were dispersed into the atmosphere in room temperature. The EMF for each gas was measured by calculation of electric potential difference between platinum and copper electrodes. Hydrogen/air mixture explodes in the range between 4-75 %. Hydrogen was mixed with air in ratios of 2, 4 and 6 %. The sensor was sprayed with each mixture and the changes of EMF were evaluated.

Results and Discussion

Preparation of ionic liquid gel beads by suspension polymerization

PMMA gel beads were prepared by suspension polymerization and their diameters are 100-2000 μm (Figure 1). These diameters can be controlled by changing stirring speed. The increased with stirring speed at 100, 200, 300 rpm, the size of gel beads were reached 2060, 497, 186 μm, respectively. The diameters of gel beads have the exponent function with stirring speed (Figure 2). Even though the polymers were changed from Poly methyl methacrylate (PMMA) to poly *N*-isopropylacrylamid (PNIPAM) or poly styrene (PS), their diameters of gel beads were the exponent functions with stirring speed.

(a)　　　　　　　　　　(b)　　　　　　　　　　(c)

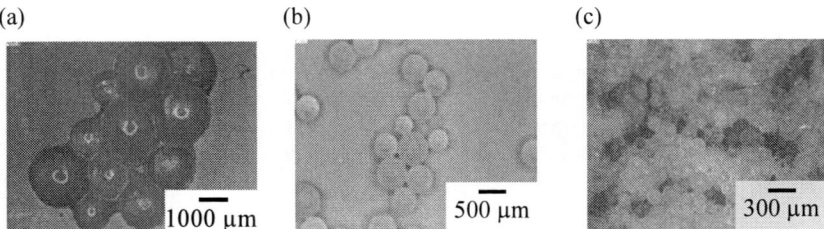

Figure 1 Photographs of gel beads containing ionic liquid. Stirring speed (a)100rpm, (b) 200rpm, (c) 300 rpm

During thermal properties evaluation, the weight and shape of gel containing ionic liquid was kept even under vacuum condition. On the other hand, the weight of the gel containing methanol decreased drastically and most of the gel solution evaporated under vacuum condition. The weight alteration modulus of gel containing distilled water is 8.84 % after vacuum drying. This result shows about 90 % of solvent of gel was evaporated in vacuum drying. On the other hand, the weight alteration modulus of gel containing ionic liquid is 100 %. The difference of weight alteration moduli between gel containing distilled water and gel containing ionic liquid is come from the differences of boiling points between distilled water and ionic liquid. The boiling point of methanol is 64.51 °C and that of ionic liquid is over 300 °C.

The thermal endurance experiment shows the weight and shape of gel beads were maintained even in temperature above 200 °C.

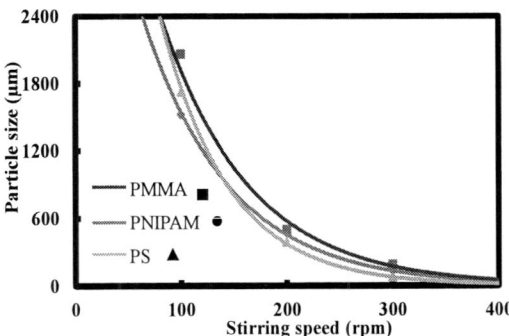

Figure 2 The relationship between particle size and stirring speed.

Young's moduli of PMMA, PS and PNIPAM are respectively 24.1, 21.2 and 20.6 KPa. This result shows PMMA gel has more flexible than PS and PNIPAM gels. PMMA gel is superior to not only flexibility but also formability.

The thermal decomposition temperatures of 5 % weight loss of PS gel, PMMA gel and PNIPAM gel were respectively 369, 323 and 306 °C. This result shows the gels containing ionic liquid can be used high temperature condition that is at even over 200 °C.

Development and evaluation of EMF hydrogen sensor

The EMF hydrogen sensors operate according to the following mechanism: the sensor commonly has a fixed electromotive force that is diminished as hydrogen molecules near the surface of platinum, where hydrogen atoms are separated into, proton and electron by platinum as catalysis in the presence of electrolyte. After the recombination of the hydrogen ion and the dispersion of the gas away from the electrode, electromotive force is increased and recovered (2).

Figure 3 shows the response of hydrogen sensor toward hydrogen. The value of EMF was decreased rapidly after hydrogen gas was dispersed around sensor in room temperature. The sensor detected the hydrogen quickly. On the other hand, EMF hardly changed in presence of the other gases such as oxygen, carbon dioxide, argon, and nitrogen. This indicates the EMF hydrogen sensor detected only hydrogen gas. It was appeared that this sensor has hydrogen selectivity and worked as novel micro device.

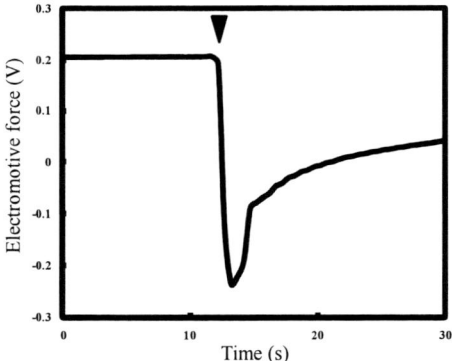

Figure 3 Response of hydrogen sensor utilizing gel beads.

Figure 4 shows the response of hydrogen sensor to different hydrogen ratios. The EMF decreased rapidly when the sensor was sprayed with the mixtures. This result shows the hydrogen sensor utilizing gel beads was able to detect up to 2 % hydrogen concentration which is less than explosion limit.The amounts of change were respectively 7.9, 14.8, 15.9 mV when the hydrogen was dispersed around the sensor in concentration 2 %, 4 % and 6 % respectively.

As can be seen, the sensor responded to low concentration hydrogen. Which meets one of the requirements for effective hydrogen sensors.

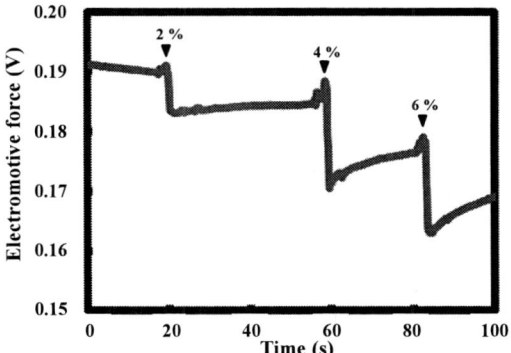

Figure 4 Response of hydrogen sensor towards low concentration of hydrogen.

Conclusion

The gel beads were prepared by suspension polymerization and their diameters were 100-2000 μm. As a result of thermal properties evaluation, the weight and shape of gels were kept even in temperature over 200 °C condition. The weight and shape of gel beads were maintained not only in high temperature condition but also under vacuuming condition.

To evaluate gas reactivity, various gases were injected. The EMF sensor responded toward only hydrogen and don't responded toward other gases. The hydrogen sensor was able to detect 2 % hydrogen concentration which is less than explosion limit. These results show that this sensor can be used as the hydrogen micro sensor.

References

(1) N. Matsuyama, S. Okazaki, H. Nakagawa, H. Sone, K. Fukuda, Response kinetics of a fiber-optic gas sensor using Pt/WO3 thin film to hydrogen, *Thin Solid Films* 517 (2009)

(2) H. Akakic, S. Harada, M. Kubota, A search for quantum phenomena of hydrogen in palladium at low temperatures; specific heat and torsional oscillator experiments, *J.Phys. Chem. Solid.* 66 (2005)

(3) K Tsunashima, Novel Quaternary Phosphonium Ionic Liquids and Their Electrochemical Applications, 2nd International Congress on Ionic Liquids (2007)

(4) T. Ueki, M. Watanabe, Lower Critical Solution Temperature Behavior of Linear Polymers in Ionic Liquids and the Corresponding Volume Phase Transition of Polymer Gels, Langmuir, 23(2007)

(5) P. Snedden, A. I. Cooper, K. Scott, and N. Winterton, Cross-Linked Polymer-Ionic Liquid Composite Materials, Macromolecules, 36 (2003)

(6) H. Tokuda, K. Hayamizu, K. Ishii, M. A. B. H. Susan, and M. Watanabe, Physicochemical Properties and Structures of Room Temperature Ionic Liquids, *J. Phys. Chem. B*, 108 (2004)

(7) Z. Tang, L. Qi, G. Gao, Dynamic mechanical properties of gel polymer electrolytes containing ionic liquid, *Solid State Ionics* 179 (2008)

The Influences of Quenching Times on the Property of Thermal Oxidized Iridium pH Sensor

F. -F. Huang[a], Y. Jin[a], L. Wen[a], D. -B. Mu[b], and M. -M. Cui[a]

[a] National Center for Materials Service Safety, University of Science and Technology Beijing, Beijing, 100083, CHINA
[b] School of Chemical Engineering and Environment, Beijing Institute of Technology, Beijing 100081, CHINA

The iridium/iridium oxide pH sensors were fabricated by thermal oxidation, during which one, two and three times of quenching were performed in deionized water, respectively. The characteristics and properties of the sensors developed with different quenching times were examined extensively, which include surface morphology, section structure, porosity distribution, as well as the linear range of E-pH response, pH response rate, and the long time stability, etc. It is found that the sensors quenched for three times appear to show better performance. Electrochemical impedance spectra (EIS) of the iridium oxide electrodes were investigated in pH buffer to analyze the pH response mechanism of the electrode.

Introduction

pH is a so important parameter for daily life and industries that the fast and precise detection of pH value in the environment is of great importance. The most widely used and generally satisfactory pH sensor is the glass electrode. However, they have several disadvantages due to the intrinsic nature of the glass membrane. For example, it shows high input impedance, it is easily broken and inadaptable to HF solutions, and errors may arise in an alkaline electrolyte. Furthermore, it is difficult to miniaturize, which withhold their application in micro-environment and on-line detection in vivo. Therefore, many efforts have been made to develop other kinds of pH sensors, as well as clarify the pH response mechanism of them (1-6). In recent years, metal/metal oxide pH electrodes have been paid increasing attention. Comparing with the glass electrodes, the metal/metal oxide electrodes are much easier to fabricate, miniaturize and maintain, show low-cost, fast response to pH changes, high strength and rigidness, and can be applied in more severe environments such as high temperature, high pressure systems and HF solutions.

Review of IrO_x Electrodes Fabrication Methods

The earliest metal/metal oxide electrodes were antimony oxide or Sb-Pb oxide electrodes. Although being commercialized, it was finally replaced by other metal/metal oxide electrodes due to its undesirable E-pH linear relationship. Then, PtO_2, IrO_2, RuO_2, OsO_2, Ta_2O_5, RuO_2, TiO_2, SnO_2, etc. were investigated in the literature. Fog et al. made a synthetic evaluation of the above metal/metal oxide pH electrodes (1, 2). By comparing the sensitivity, Nernst response range, the ion selectivity, the interference of oxidation and reduction, and the potential drift of the above metal/metal oxide electrodes, the IrO_x electrode was considered to be the most promising one for pH detection.

The performance of the IrO_x pH electrode is determined by its micro-structure, composition and the fabrication method of the electrode. Up to now, the main methods for fabricating the IrO_x electrode include the electrochemical cyclic voltammetry (CV), electrodeposition, radio frequency (RF) magnetron sputtering deposition, the high temperature carbonate oxidation and thermal oxidation.

L. D. Burke et al. produced so called anodic iridium oxide film (AIROF) by CV with the potential scan range of 0~1.5V (vs SHE) in $1MH_2SO_4$ (2, 4). It was indicated that IrO_x film can only grow within a certain potential range. Differing from the cases of Pt, Ru, or Ni, the thickness of the oxide film on iridium changes with the cyclic time and scanning rate (3). The IrO_x film developed by this approach is amorphous, highly hydrated, and in a gel state. The sensitivity of AIROF is 60-80mV/pH, which is super Nernst response. According to the Nernst formula,

$$E = E_0 - 2303RTpH/F \qquad [1]$$

$$E = E_0 - 59.16pH \qquad [2]$$

in which, the standard electrode potential, E_0, is 681mV(SCE). At 25°C, and the Nernst response slope is -59.16mV/pH. It was believed that more than one H^+ participated in the reaction leading to the E-pH response sensitivity larger than 59mV/pH. Kinoshita et al. fabricated the IrO_x electrodes by CV which presented E-pH linear response in the pH range of 2.5~8.5. The IrO_x electrode produced by Hitchman et al. by similar approach broaden its E-pH linear range to 2~12 (1, 5). The application of AIROF electrodes was restrained due to its unsatisfied reproducibility.

RF magnetron sputtering deposition is another widely used method to fabricate IrO_x electrode (7). The adhesion of the IrO_x film to the substrate developed by this method is good. The response sensitivity has good reproducibility, and its value is close to 59mV/pH. However, the open circuit potential of the electrode and the interference of the redox ions rely greatly on the sputtering parameters. In addition, the comprehensive costs of this process are relatively high.

Other approaches, such as electrodeposition, high temperature carbonate oxidation, $IrCl_3$ thermal decomposition were also applied to fabricate IrO_x electrode (8-13). Among which, Lin prepared the IrO_x electrode by electrodeposition (8). The response sensitivity is 70mV/pH. The disadvantages of the pH electrode developed by this method are that the adhesion of the IrO_x to the substrate is poor, and the plating solution is difficult to prepare and maintain. Sheng Yao et al. and Chen Xu et al. fabricated the IrO_x electrodes by high temperature carbonate oxidation (9-11). The sensitivity of the electrodes was close to Nernst response. Although there is quite large deviation in potential value among electrodes, the reproducibility of their pH electrodes is acceptable. It was indicated that the condition of heat treatment had influences on the morphology, uniformity of the electrodes, which had further impacts on the electrode stability and the response rate (11).

Some researchers fabricated the IrO_x electrodes by thermal oxidation (4, 5, 14). The oxide film developed by this approach is dry film which is similar to the IrO_x made by RF magnetron sputtering deposition. The IrO_x electrodes show wide E-pH linear range, good stability and high response sensitivity which is close to Nernst response. The drawbacks of thermal oxidation include the relative poor adhesion of the film to the substrate and the potential drift of the electrode caused by hydration is large etc.

To sum up, the method and craft of fabricating the iridium oxide electrode are still in development, and the pH response behavior which is related to the form of the oxide film and other factors needs to be further studied. In this article, thermal oxidation was adopted to fabricate the IrO_x electrode. Besides the improvement made in the electrode fabrication, the influences of quenching times were intensively investigated by examining the morphology, structure, composition of the electrodes, as well as their performance in pH sensing.

Experimental

After being cleaned, iridium (Ir) wires (0.5mm in diameter, $>$99.9%, purchased from Cuibelin Non-Ferrous Metal Technology Development Center) of 10mm were soaked in 5M NaOH solution for more than 24 hours. Then, the oxidation of the Ir wires were performed at 800℃ for 30min in air, and immediately followed by the quenching process in deionized water. The above heating and quenching process were repeated to fabricate the electrodes that were quenched for two or three times. Subsequently, the wires were soaked in deionized water for more than 2 days for hydration. As the result, a layer of dark-blue film was formed on the surface of the Ir wires. After wiring and sealing procedures, 5 mm of Ir wire was exposed for pH response. The pH electrodes were maintained in deionized water when not used.

The top image and the cross-section of the Ir/IrO_x electrodes were observed by FEI Quanta250. The composition of the iridium oxide layer was analyzed by EDS.

Prior to the pH measurements, PHS-3CT digital pH meter with glass pH electrode was calibrated in the standard solutions. 0.01M H_3PO_4, 0.01M H_3BO_3, 0.01M CH_3COOH and 0.1M KCl were used as the pH buffer solution. $0.1mol·L^{-1}$ NaOH and $0.1mol·L^{-1}$ HCl solutions were used for pH adjustment, respectively. Solutions with pH equal to 1~13 were prepared, respectively. The electrochemical tests were performed by CHI650D with three-electrode system. In the three-electrode system, saturated calomel electrode (SCE) was used as reference electrode, and the platinum plate was the auxiliary electrode. All the tests were performed under the room temperature of 22 ± 3℃. For electrochemical Impedance Spectra (EIS) tests, the scan frequency was in the range of 10^5Hz~0.01Hz, and ZSimpWin was used as the software to carry out data fitting. All the potentials in this article are versus SCE.

Results and Discussion

IrO$_x$ Electrode Characterization

Pictures in Fig. 1 are the surface morphology of the iridium oxide electrodes after being quenched for one, two or three times and after hydration. Top images illustrate that IrO$_x$ are in particles on the electrode surface, with the particle size ranging from a few hundreds of nanometers to several micros. The uniformity of the surface film for all the three quenching conditions are all satisfied. The film is different from the iridium oxide films that were fabricated by other methods such as the electrochemical CV and the thermal decomposition method. The oxide films made from those approaches are usually loose and full of cracks (15). Comparing with the other cases, the IrO$_x$ on three-time-quenched electrode show 3D grow pattern, more layers of particles could be recognized on the picture (Fig.1 c).

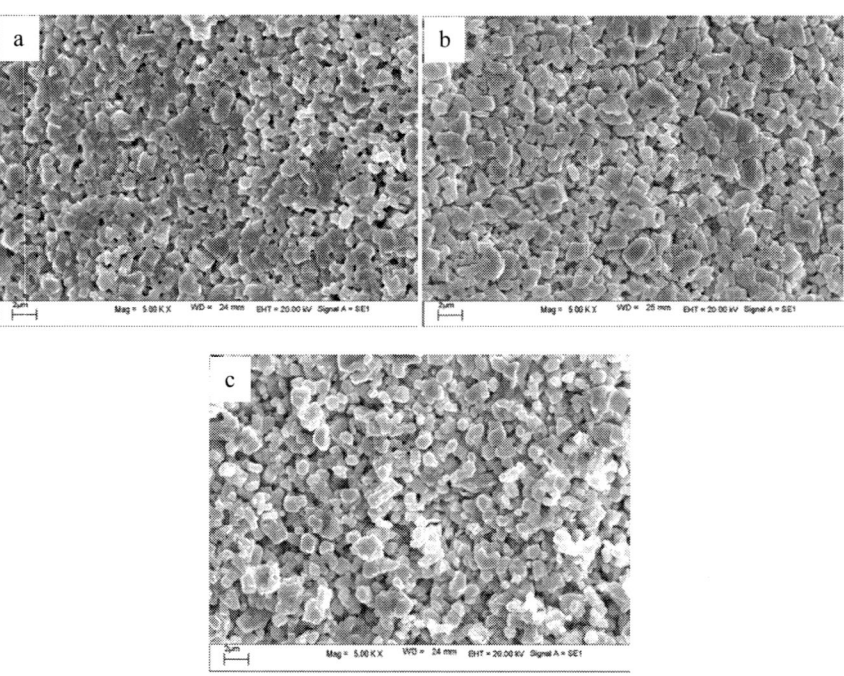

Figure 1. Surface morphology of the iridium oxide electrodes after being quenched for one (a), two (b) or three times (c), respectively.

From Fig. 1, it is seen that the element compositions of the iridium oxide include Ir, O and sodium (Na). Among them, the element O came from the oxygen of the air, whereas the Na came from NaOH solution. The EDS results of the electrodes quenched for different times were almost the same.

In order to study the structure and composition distributions of the cross section of the film along the depth direction, the morphology and the intensive EDS analysis of the electrodes quenched for different times were examined.

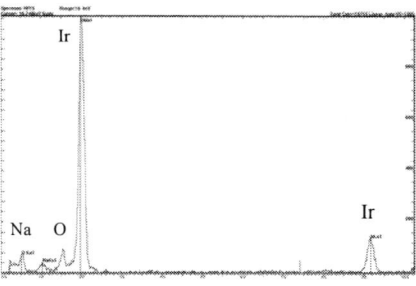

Figure 2. The EDS analysis of the electrode.

Figure 3 shows that the oxidation film thickness and the section morphology are all varying due to different quenched times. The thickness and the morphology of the oxide film depend on the condition of the thermal treatment, such as the temperature, the duration time, quenching times and the atmosphere of the thermal treatment, etc. (11).

Based on the results of our experiments, the film consists of two parts (Fig. 3c), i.e. the inner homogeneous and dense layer, and the outer hydrated and porous layer. The oxygen content was on the rise from the inner part to the outer part. With the increase of the quenching times, the interface of the inner layer and the outer layer become obvious, and the thickness of the outer layer increases while the inner layer does not increase with the quenching times. The out layer of the electrodes that quenched for three times is porous, thus has more pores and will provide more convenient channels for hydration and the pH response reaction which will be discussed later. When the electrodes were soaked in the deionized water, the hydration either dissolves the outer layer leading to pore or leads to phase transformation. The mechanism needs further study.

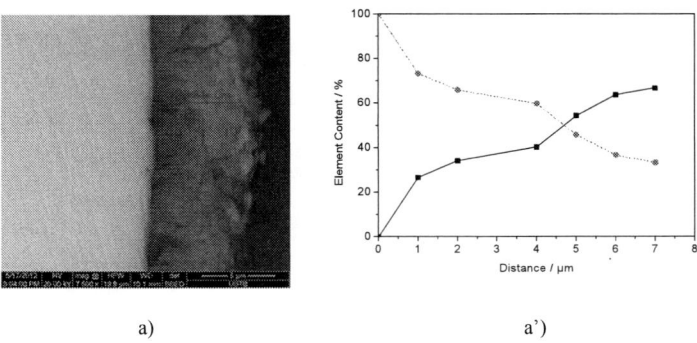

a)   a')

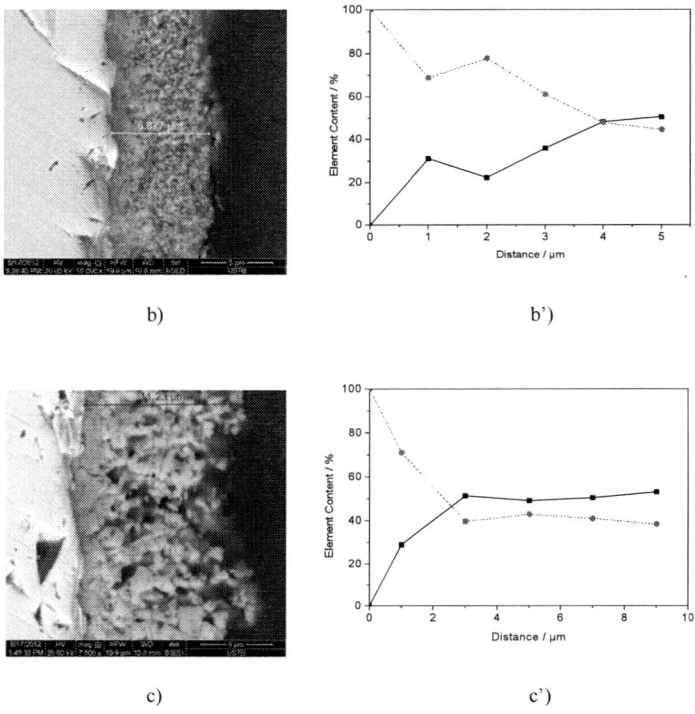

Figure 3. The cross section images and the corresponding element distribution analyses of the electrodes quenched for one (a), two (b) or three times (c), respectively. a´, b´ and c´ show the element contents of the IrO_x film changing with the distance away from the Ir substrate, in which the solid line and square represent the content distribution of oxygen, and the dash and circle represent the content of iridium.

IrO_x Electrode Property Examination

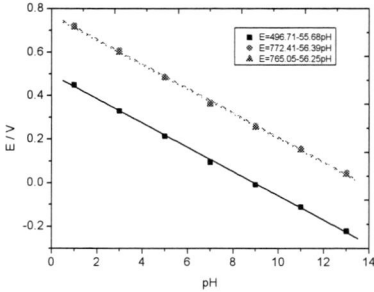

Figure 4. E-pH relationship of the electrodes quenched for different times. Quenching times: one time (■ —), two times (●---), three times(▲…).

E-pH Response. Figure 4 illustrates the typical pH response of the IrO_x electrodes produced with once, twice and three times of quenched, respectively. It is seen that all the IrO_x electrodes fabricated by the thermal oxidation approach show good linear E-pH relationship in the entire examined pH range of 1 ~ 13. The pH sensitivity is around -55~-57 mV/pH, which is close to the Nernst response.

Response Rate. In order to evaluate the response rate of the electrodes, the potential-time curve in the different pH buffer solutions were tested. Taking the results of pH=5 and pH=11 as two examples, figure 5a and figure 5b show the response curves of the electrodes prepared with different quenching times.

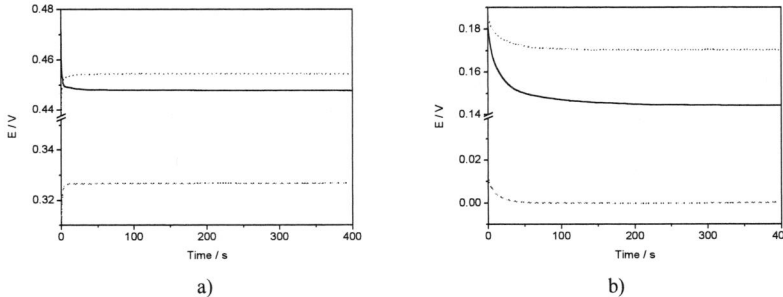

Figure 5. Potential response of electrodes quenched for different times in buffer solutions. a. pH=5, b. pH=11. Quenching times are one time (—), two times (---), and three times (...), respectively.

According to the rules of International Union of Pure and Applied Chemistry (IUPAC), the electrode can only be considered as achieving steady state when the change rate of the electrode potential reduces to less than 1 mV/min.

It is clear that the response rate of the electrode increases with the quenching time, i.e., the time needed for the electrode potential to become stable decreases with the quenching times of the electrode. It takes less than 10 sec in pH=5 solution and less than 40 sec in pH=11 solution to reach stable stage for the electrode quenched three times. This response rate is faster than many other pH electrodes, which show more than 1min for stabilization (3, 11).

Figure 6 shows the response time distribution error bar map of the IrO_x electrodes quenched for different times. It is apparent that the electrodes fabricated by thermal oxidation could respond in different pH buffer solutions, with almost all the electrodes reach steady state in less than 120 sec. Among them, the behavior of electrodes quenched for two times and three times were better than that of the one-time-quenched electrode. And the three-time-quenched ones seem the best. Figure 6 also indicates that the IrO_x electrodes response faster in acid environment than in alkaline environment. Taking the electrodes quenched for three times as the examples, the response time of them in acid solutions is less than 10 sec while the response time in alkaline solutions is around 30~60 sec.

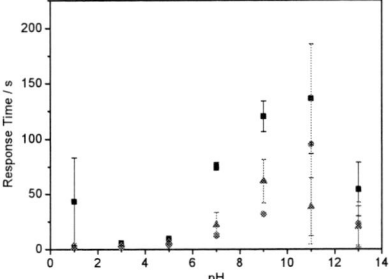

Figure 6. The response time distribution error bar map of electrodes quenched for different times. Quenching times: one time (■), two times (✳), three times (▲).

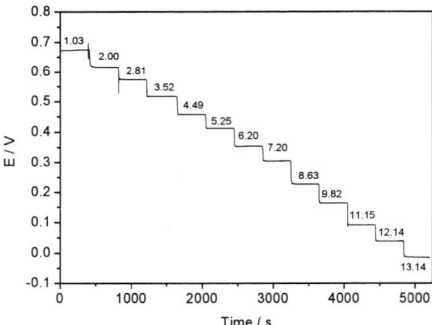

Figure 7. Potential response of an IrO$_x$ electrode (quenched for three times) to solution pH changes.

In addition, the potential response of three-time-quenched electrode was recorded when the pH buffer solution was adjusted step by step from pH=1.03 to pH=13.14. The result was presented in Fig. 7. The solution used for pH adjustment is NaOH. From the figure, the IrO$_x$ electrode can continuously trace the fast pH changes in all examined pH range.

Long Term Stability Test. The long-term stability of the electrodes is also an important parameter of the electrode. Figure 8 illustrates the long-term stability test result of the pH sensitivity of the electrodes quenched for different times. The sensitivity of the electrodes was found to rise slightly with the prolongation of time. It is concluded that for all the IrO$_x$ electrodes fabricated by thermal oxidation, the stability of pH sensitivity was good. Among which, the three-time quenched ones had the best performance, with pH sensitivity 56 ± 1mV/pH in three months.

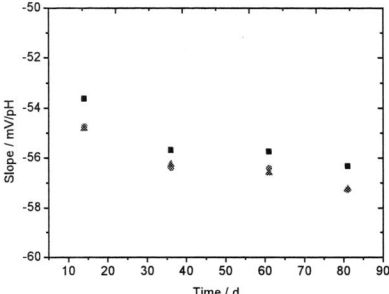

Figure 8. The long time stability of the electrodes quenched for different times. Quenching times: one time (■), two times (*), three times(▲).

Destructive Test. The destructive test was done on the three-time-quenched electrode by scraping off a piece of the oxide film from the substrate. Then, the potential responses of it were examined in pH=1~13 buffer solutions. The result still indicated a linear E-pH relationship. This destructive test implies a good durability in practical engineering applications.

EIS Tests. In order to evaluate the electrode response behavior in the solutions with various pH values, EIS tests were measured in pH=4 and pH=11 buffer solutions, respectively. The EIS results of IrO_x electrodes immersed in pH=4 buffer solution are shown in Fig. 9. It can be found that there are two time constants in the EIS plots, the one appearing at high frequency corresponding to the electrochemical response of IrO_x film in pH=4 buffer solution, while the other appearing at low frequency corresponding to the hydrated reaction between the oxide/solution interface. Meanwhile, it can be noticed that each Nyquist plot in Fig. 9b has an arc at the high frequency side with a change to a diffusion tail at the low frequency. The occurrence of diffusion impedance at low frequency indicates that the electrochemical behavior of the electrode was controlled by mass transport process, during which the reaction rate is determined by the rate of H^+ transferring to the film.

Based on the discussion above, equivalent circuit (EC) and physical model of IrO_x electrodes immersed in pH=4 buffer are presented in Fig. 10. The EIS experimental results can be simulated by using the equivalent circuit shown in Fig. 10a, where R_s is the solution resistance, the pair of R_1 and Y_1 corresponds to electrochemical response of the IrO_x film, the pair of R_{ct} and Y_0 corresponds to the process of hydrated reaction, and W is the diffusion impedance.

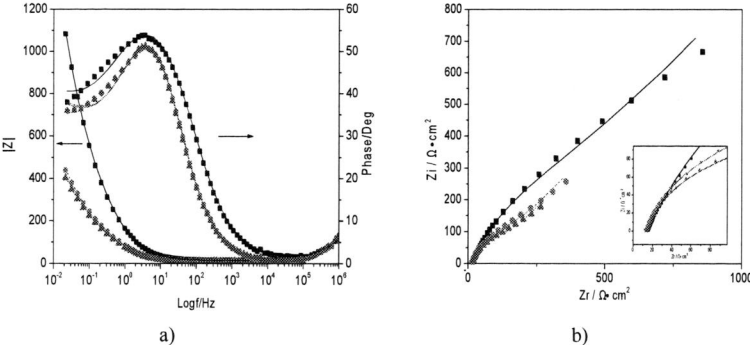

a)   b)

Figure 9. Bode plot (a) and Nyquist plot (b) of IrO$_x$ electrode in pH=4 buffer solution. The points are the experimental impedance values, while the lines are the simulation results. Quenching times: one time (■), two times (✳), three times (▲).

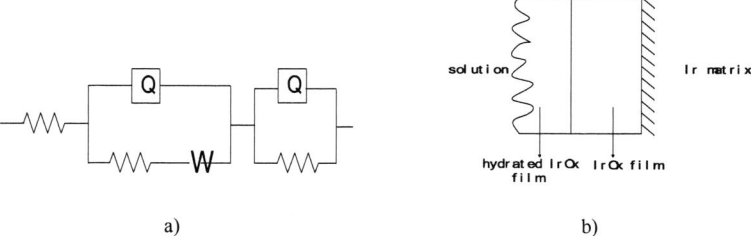

a)   b)

Figure 10. Equivalent circuit model (a) and interface modulation (b).

TABLE I. Parameters Obtained from the Experiment Data Fitting to the Equivalent Circuit Model in pH=4 and pH=11 Buffer Solutions, respectively.

Solution	Quenching times	R_s/Ω	Y_0	n_0	R_{ct}/Ω	W	Y_1	n_1	R_1/Ω
pH=4	1	3.76E-2	1.392E-3	0.7674	558.0	0.00269	1.065E-6	0.6648	15.59
	2	1.00E-5	2.232E-3	0.8488	162.1	0.00734	1.785E-7	0.8087	12.86
	3	9.15E-7	3.091-3	0.8751	126.4	0.01103	1.370E-7	0.8244	15.41
pH=11	1	5.31E-3	2.073E-3	0.7419	711.4	0.00216	1.552E-6	0.6633	10.66
	2	1.99	2.388E-3	0.8010	310.2	0.00519	2.108E-7	0.8451	8.06
	3	8.25E-4	2.891E-3	0.8501	228.7	0.00927	3.583E-7	0.7726	9.60

Note: The units of Y and W are $\Omega^{-1}\cdot cm^{-2}\cdot s^{-n}$ and $\Omega^{-1}\cdot cm^{-2}\cdot s^{-0.5}$, respectively.

The experimental results can be well fitted by using EC in Fig. 10a, and the fitted results are shown in Table I. It can be found that the film resistance and reaction resistance of the electrodes made by the thermal oxidation are relatively small, indicating that the electrode reaction is easy to happen. The resistance of oxide film R_1 is almost ten orders of magnitude lower compared with the traditional ones (11, 16). Particularly, the reaction resistance is in inverse proportion to the thickness of the outer layer, the thicker of the outer layer, the smaller the reaction resistance is (16, 17). As discussed above, the thickness of the outer layer increases with the increasing quenching times, resulting in the refinement of oxide particles and increase of membrane pores. Therefore, the real area of electrode contacting with the solution increase with the increasing quenching times, resulting in the increment of the active sites, thus the electrode reaction resistance decrease and the rate of the electrode potential response increases.

pH Response Mechanism Discussion

The iridium in iridium oxidation presents in various valences, such as +1, +2, +3, +4 and +6 (18). The transformation among different valences of IrO_x leads to the ability for pH response.

$$Ir_2O_3+6H^++6e^- =2Ir+3H_2O \qquad [3]$$

$$IrO_2+4H^++4e^-= Ir+2H_2O \qquad [4]$$

$$2IrO_2+2H^++2e^- =Ir_2O_3+H_2O \qquad [5]$$

The research by Chen et al. indicated that the iridium in IrO_x is mainly IrO_2 and Ir_2O_3 (19, 20). Thus, the pH response reaction is dominantly the reaction [5]. Wang et al. used XPS to study the composition of the oxide and found that the surface of the electrode consists of H_2O, OH^- and O_2^{2-}(16). The hydration of iridium oxide is revealed by reactions [6] and [7]. After being hydrated, the pH response reactions are proposed as reactions [8].

$$IrO_2+4H_2O= IrO_2 \cdot 4H_2O \qquad [6]$$

$$IrO_2 \cdot 4H_2O= [IrO_2(OH)_2 \cdot 2H_2O]^{2-}+2H^+ \qquad [7]$$

$$2\{[IrO_2(OH)_2 \cdot 2H_2O]^{2-} \cdot 2H_f^+\}+2e^-+2H_s^+ =[Ir_2O_3(OH)_3 \cdot 3H_2O]^{3-} \cdot 3H_f^++3H_2O \qquad [8]$$

Where H_f^+ is the H^+ inside the hydrated film, H_s^+ is the free H^+ in the solution.

Conclusion

IrO_x electrodes were fabricated by thermal oxidation approach. The surface top images show that the films consist of micro and sub-micro particles, cross-section morphology observations illustrate the presence of dense inner layer and porous outer layer. The three time quenched electrode show thicker outer layer with larger pores.

The properties of the IrO_x electrodes fabricated by thermal oxidation and quenched for different times were evaluated intensively, including the linear range of E-pH response, the response rate, long term stability, destructive test and EIS test. The electrodes with three quenching times appear to have the best integrated performance, which might attribute to the effective hydration and more active sites in the porous outer layer.

Acknowledgements

It is grateful that the present research has been performed under the great support of two projects (ID NCET-10-022 and ID FRF-TP-10-005B) sponsored by the Ministry of Education, P. R. China, and a project (ID YYXM-1412-0001) from the National Development and Reform Commission, P. R. China. The authors also thank Beijing Municipal Science and Technology Commission for the financial support on attending the conference under Project ID 2009B16.

References

1. A. Fog and R. F. Buck, *Sens. Actuators*, **137**, 5 (1984).
2. L. D. Burke and D. P. Whelan, *J. Electroanal. Chem.*, **121**, 162 (1984).
3. D. C. Chen, J. C. Zheng and Z. Y. Fu, *Rare Met. Mater. Eng.*, **831**, 33(8) (2004).
4. L. D. Burke, J. K. Muleahy and D. P. Whelan, *J. Electroanal. Chem.*, **117**, 163 (1984).
5. K. Kinoshita and M. Madou, *J Electrochem Soc.*, **1089**, 131(5) (1984).
6. M. L. Hitchman and R. Subramaniam, *Talanta*, **137**, 39(2) (1992).
7. K. G. Krerder and M. J. Tarlor, *Sens. Actuators B*, **167**, 28(3) (1995).
8. C. J. Lin, J. L. Luo and H. Y. Sun, *Electrochem.*, **373**, 2(4) (1996).
9. S. Yao and M. Wang, *J. Electrochem. Soc.*, H29, 148(4) (2001).
10. M. Wang and S. Yao, *Sen. Actuators*, **313**, B81 (2002).
11. X. Chen, C. W. Du and X. G. Li, *J. Univer. Sci. & Tech. Beijing*, **200**, 33(2) (2011).
12. B. Bestaoui, E. Prouzet and P. Deniard, *Thin Solid Films*, **35**, 235 (1993).
13. G. M. da Silva, S. G. Lemos, L. A. Pocrifka, P. D. Marreto, A. V. Rosario and E. C. Pereira, *Analytica Chimica Acta*, **36**, 616 (2008).
14. B. Z. Du, X. Y. Li, L Xue and D. X. Zhang, *J. Anal. Sci.*, **567**, 23(5) (2007).
15. P. G. Pickup and V. I. Birss, *J. Electrochem. Soc.*, **126**, 135 (1988).
16. D. C. Chen, J. C. Zheng and C. Y. Fu, *Mater. Mech. Eng.*, **23**, 30(1) (2006).
17. C. N. Cao and J. Q. Zhang, *Electrochemical impedance spectroscopy introduction*, p. 151-166, Science Press, Beijing (2002).
18. S. L. Xu, H. Ma and Z. Y. Liu, *Inorganic Chemistry: 9thVolume*, Science Press, Beijing (1996).
19. M. Wang and S. Yao, *Electroanal.*, **1606**, 15(20) (2003).
20. D. C. Chen, C. Y. Fu and J. C. Zheng, *Rare Met. Mater. Eng.*, **637**, 36(4) (2007).

Surface-Enhanced Raman Scattering on Ordered Metal Nanodot Array
Obtained Using Anodic Porous Alumina

T. Kondo[a], K. Nishio[a,b], and H. Masuda[a,b]

[a] Kanagawa Academy of Science and Technology, Sagamihara, Kanagawa 252-0131,
Japan
[b] Department of Applied Chemistry, Tokyo Metropolitan University, Hachioji, Tokyo
192-0397, Japan

> The fabrication of an ordered array of Au nanodots using anodic
> porous alumina as an evaporation mask, and its application to a
> substrate for the measurement of surface-enhanced Raman
> scattering (SERS) were studied. One of the advantageous points of
> using anodic porous alumina as a template to fabricate
> nanostructures is that the size, shape and arrangement of the
> obtained nanostructures can be controlled by changing the
> geometrical structures of the porous alumina. Au nanodot arrays
> were obtained by thermal evaporation method. The SERS signals
> of pyridine molecules adsorbed on the nanodots were detected.
> The intensity of the SERS signals was strongly dependent on the
> arrangement of the nanodots. The enhancement factor of the
> intensity of the incident light on Au nanodots was analyzed by
> numerical calculations based on finite-difference time-domain
> (FDTD) method. The obtained SERS substrates are expected to be
> used for Raman spectra measurement with high sensitivity.

Introduction

Porous materials are useful as a template for the preparation of arrays of nanostructures
(1,2). The geometrical structures of anodic porous alumina can be controlled easily by
changing the anodizing conditions. The fabrication of ordered arrays of metal
nanoparticles has attracted increasing attention owing to their capability of enhancing the
electric field of incident light based on localized surface plasmon resonance (LSPR).
Various applications of metal nanoparticle arrays, such as sensing devices and surface-
enhanced Raman scattering (SERS), have been proposed (3-5). Precise control of the
size, shape and arrangement of the nanoparticles is important because the properties of
LSPR are strongly dependent on the geometrical structure of the nanoparticle arrays (6-8).
There have been numerous reports on the preparation of ordered metal nanostructures for
optimizing LSPR properties. However, a simple process for precisely controlling the
structure of metal particles has not been established. In this paper, the fabrication of a
geometrically controlled nanodot array using anodic porous alumina and its application to
a substrate for SERS measurements are discussed. Furthermore, the enhancement factor
of the light intensity around nanodots was analyzed by finite-difference time-domain
(FDTD) method.

Experimental

Figure 1 shows the fabrication scheme of Au nanodoto array. Anodic porous alumina was used as an evaporation mask (1,9-11). The anodic porous alumina was formed by anodizing Al in acidic solution. Porous alumina masks with through holes were obtained by the selective dissolution of Al in saturated I_2 methanol solution followed by the dissolution of the bottom part of the porous alumina films in phosphoric acid solution. The hole sizes of the alumina mask were adjusted by postetching treatment in phosphoric acid solution. Au was deposited onto a substrate (Si or glass) using a vacuum evaporator through the nanoholes of the alumina mask, which was set on the substrate. The thickness of the alumina mask was controlled by changing the duration of anodization. After the vacuum deposition, the Au nanodot array on the substrate was obtained by removing the alumina mask mechanically. The Raman scattering spectra of pyridine molecules were measured using a Raman microscope equipped with a He-Ne laser (wavelength: 633 nm) as a light source. The substrate with the Au dot array was dipped in pyridine solution (HPLC grade, > 99.9%) and dried in air before the measurement. The light intensity was calculated by FDTD method using commercial software (Crystalwave; PhotonDesign). The enhancement factor was obtained by comparing the light intensity around Au nanodot with the intensity of the incident light.

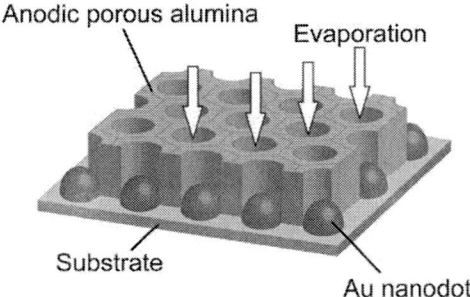

Figure 1. Fabrication scheme of metal nanodot array using anodic porous alumina as an evaporation mask.

Results and Discussion

Figure 2 shows SEM images of a typical Au nanodot array formed on a Si substrate. It was observed that the uniform-sized nanodots were ordered in a triangular lattice with regular intervals. The diameter and the interval between Au dots were 70 and 100 nm, respectively. These values corresponded to the geometrical structure of the porous alumina mask used for evaporation. The height of the nanodots was controlled by changing the nominal thickness of the evaporated Au using a thickness monitor during the vacuum evaporation. The height of the Au dots was 60 nm. The conical shape of the Au dots results from the shadowing effect of the mask during the vacuum deposition of Au.

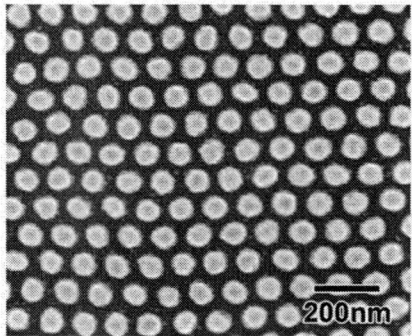

Figure 2. Front SEM images of Au nanodot array. The diameter and height of the nanodots were 70 and 60 nm, respectively. The interval of between the nanodots was 100 nm.

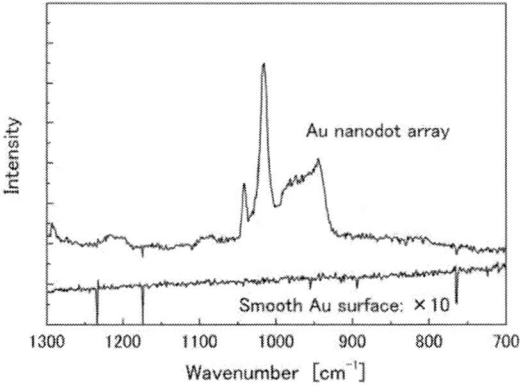

Figure 3. SERS spectra of pyridine molecules adsorbed on Au nanodot arrays.

Figure 3 shows the typical SERS spectra of pyridine molecules measured using the Au dot array formed on the Si substrate. Compared with the spectrum obtained from a smooth Au surface, the intensity of Raman scattering was confirmed to be enhanced on the Au dot arrays, that is, two characteristic peaks from the adsorbed pyridine molecules were observed at 1014 and 1040 cm^{-1}. The broad signal observed around 950 cm^{-1} originated in the Si substrate.

It is considered that the intensity of the incident light was enhanced efficiently on the Au nanodot array. The enhancement factor of the light intensity on the nanodot array was calculated by FDTD method. The diameter and height of the dots were 70 and 60 nm, respectively. The substrate was a glass. The wavelength of the incident light used

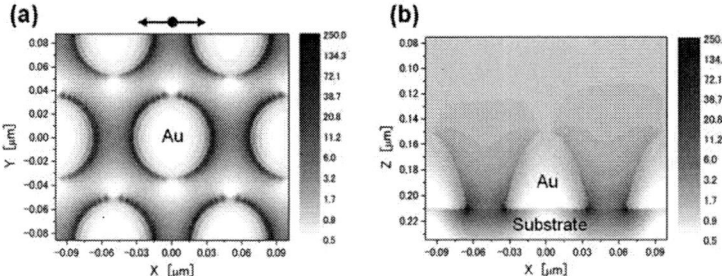

Figure 4. Enhancement factor of light intensity on Au nanodot array. (a) Front and (b) cross-sectional images. The wavelength of the incident light was 633 nm. The incident light was a plane wave and polarized along the x-axis. The light was propagated in the +z-direction.

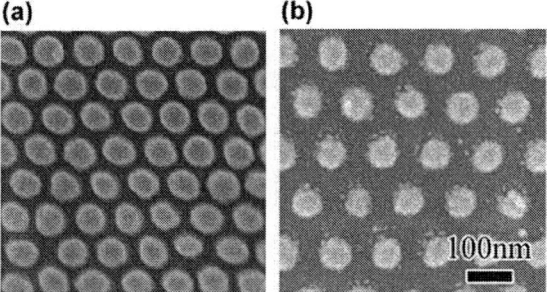

Figure 5. Au nanodot arrays with nanogap sizes of (a) 10 nm and (b) 45 nm. The diameter and height of the dots were 80 and 30 nm, respectively.

for the calculations was set to 633 nm, which is the wavelength used for the Raman spectra measurements. The light propagated in the +z-direction and was polarized along the x-axis. Figure 4a shows a front view of the enhancement factor at a distance of 1 nm above the glass substrate. The light intensity was enhanced near the surface of the dots. Figure 4b shows the cross-sectional view of the Au nanodot array. It was observed that the light intensity was strongly enhanced around the bottom edge of the nanodots.

The nanogaps formed between nanoparticles are effective for the enhancement of the electric field of the incident light. Au nanodot arrays with size-controlled nanogaps were fabricated using the anodic porous alumina masks. The geometrical parameters of the porous alumina, i.e., the diameter and the interval between nanoholes, were precisely controlled by changing the anodizing conditions. Figure 5 shows Au nanodot arrays with nanogap sizes of (a) 10 nm and (b) 45 nm. The diameter and height of the dots were 80 and 30 nm, respectively.

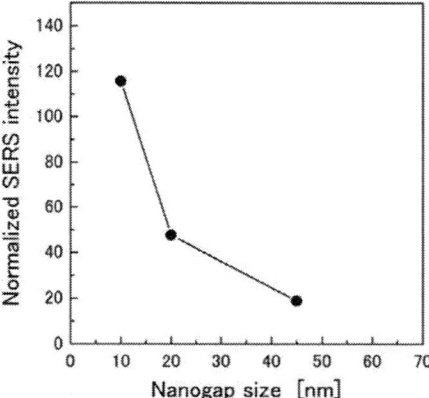

Figure 6. Dependence of SERS intensity on the size of nanogap between Au nanodots. The SERS intensities were normalized by the surface area of the Au nanodot array.

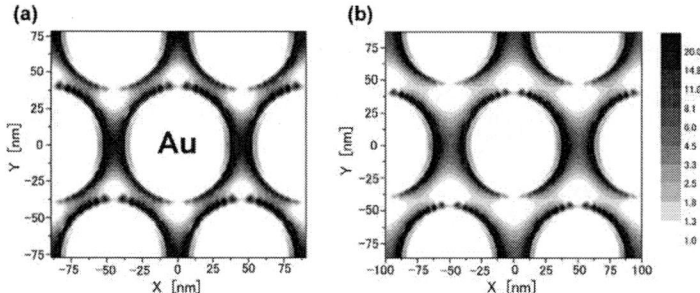

Figure 7. Enhancement factor of the light intensity around Au nanodots with nanogap sizes of (a) 10 nm and (b) 20 nm. The diameter and height of dots are 80 nm an 30 nm, respectively.

The Raman spectra of pyridine molecules were measured on Au nanodot arrays with different nanogaps. Figure 6 shows the dependence of the SERS intensity at 1014 cm^{-1} on the nanogap size. The intensity was normalized by the surface area of the Au nanodot array. The intensity of the SERS signals when the nanogap size was 10 nm was 2.4 and 6 times higher than that when the nanogap sizes were 20 and 45 nm, respectively. The intensity of the SERS signal was increased with decreasing nanogap sizes.

Figure 7 shows the enhancement factor of the light intensity around the Au nanodots with nanogap seizes of (a) 10 nm and (b) 20 nm. The light intensity when the nanogap size was 10 nm was enhanced 1.5 times higher than that when the nanogap size was 20 nm. The dependence of the enhancement factor on the nanogaps sizes was in agreement with the results of SERS measurements.

Conclusions

The SERS intensity from ordered Au nanodot arrays prepared using an anodic porous alumina mask was examined. The intensity was strongly dependent on the arrangement of the dots. The process described in the present study allows the preparation of nanodot arrays of other metals on various types of substrates. The present process enables the easier preparation of SERS substrates compared with the process employing the electron beam lithographic technique. In addition, the geometrical structure of the dots can be tuned to optimize the enhancement of SERS intensity. The obtained SERS substrates are expected to be used for Raman spectra measurement with high sensitivity.

Acknowledgments

This work was partially supported by Grant-in-Aid for Scientific Research No.19049013 on Priority Area "Strong Photons-Molecules Coupling Fields (470)" from the Ministry of Education, Culture, Sports, Science and Technology of Japan.

References

1. H. Masuda and M. Satoh, *Jpn. J. Appl. Phys.*, **35**, L126 (1996).
2. H. Masuda, K. Yada, and A. Osaka, *Jpn. J. Appl. Phys.*, **37**, L1340 (1998).
3. K. Ueno, S. Juodkazis, M. Mino, V. Mizeikis, and H. Misawa, *J. Phys. Chem. C*, **111**, 4180 (2007).
4. B. Nikoobakht, J. Wang, and M. A. El-Sayed, *Chem. Phys. Lett.*, **366**, 17 (2002).
5. Y. Sawai, B. Takimoto, H. Nabika, K. Ajito, and K. Murakoshi, *J. Am. Chem. Soc.*, **129**, 1658 (2007).
6. C. L. Nehl, H. Liao, and J. H. Hafner, *Nano Lett.*, **6**, 683 (2006).
7. F. Wang and Y. R. Shen, *Phys. Rev. Lett.*, **97**, 206806 (2006).
8. L. J. Sherry, R. Jin, C. A. Mirkin, G. C. Schatz, and R. P. Van Duyne, *Nano Lett.*, **6**, 2060 (2006).
9. F. Matsumoto, M. Ishikawa, K. Nishio, and H. Masuda, *Chem. Lett.*, **34**, 508 (2005).
10. T. Kondo, F. Matsumoto, K. Nishio, and H. Masuda, *Chem. Lett.*, **37**, 466 (2008).
11. T. Kondo, H. Miyazaki, K. Nishio, and H. Masuda, *J. Photochem. Photobiol. A*, **221**, 199 (2011).

Enzyme-encapsulated Quantum Dot Hydrogels in the Development of Biosensors: A Multifunctional Platform for Both Bio-catalysis and Fluorescent Probing

J. Yuan, N. Gaponik, and A. Eychmüller

Physical Chemistry/Electrochemistry, TU Dresden, Bergstrasse 66b, 01062 Dresden, Germany

> We have recently fabricated enzyme encapsulated CdTe quantum dot (QDs) hydrogels using the sol-gel method, a versatile way in the QD gel formation. The porous three dimensional QD hydrogel turned out to be an adequate encapsulation medium for enzymes and furthermore, the optoelectronic properties of the individual QDs were still retained in the QD hydrogels. The as-prepared enzyme-encapsulated QD hydrogel incorporated both a bio-catalysis unit and a fluorescence signaling unit, and was taken as a multi-functional platform in the development of optical biosensors.

Introduction

Quantum dots (QDs) have gained great interest in both fundamental research and in technical applications due to their unique size-dependent physical and optical properties. The assembly of QDs into functional architectures and further applications of these nanomaterials in photovoltaics and in the field of sensing are currently amongst the priorities of QD research (1, 2). Gels and aerogels manufactured from a variety of metal nanoparticles available in colloidal solutions have recently proven to provide an opportunity to marry the nanoscale world with that of materials of macro dimensions which can be easily manipulated and processed, whilst maintaining most of the nanoscale properties (3). The materials carry an enormous potential for applications. This is largely related to their extremely low density and high porosity providing access to the capacious inner surface of the interconnected nanoobjects they consist of (3-6). Particularly, these gels are expected to be ideal materials for the development of biosensors, as the meso- and macro-sized pores in the gel are appropriate for encapsulation of biomolecules (7).

Results and Discussion

We fabricated enzyme encapsulated mercaptosuccinic acid (MSA) capped CdTe QDs hydrogels using the sol-gel method, a versatile way in the QD gel formation. MSA capped CdTe QDs were synthesized according to a previous report (8), and the obtained CdTe QDs were precipitated by ethanol to partially remove capping ligands, and the resultant precipitate was then redissolved in 50 mM phosphate buffer solution. Figure 1 shows the UV-vis absorption and fluorescence (FL) spectra of the MSA capped CdTe QDs. The obtained QDs exhibited a broad absorption spectrum (< 600 nm) with a characteristic peak at 554 nm and a narrow emission peak at 589 nm with the excitation wavelength at 450 nm.

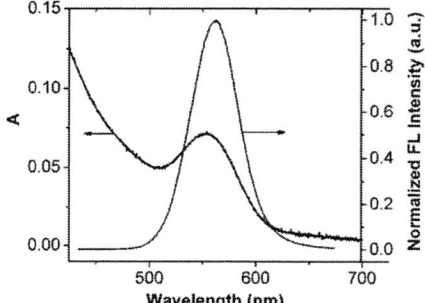

Figure 1. UV-vis absorption and fluorescence spectra of the synthesized MSA capped CdTe QDs.

The resulting CdTe QD sol in phosphate buffer was left at room temperature for certain periods of time to form a CdTe QD gel. The color of the QD sol turned from orange to red after one day. The sol became turbid after storage for several days and then started to coagulate in a gel. A red shift of about 7 nm was observed after gelation in the 50 mM phosphate buffer with pH of 7.1. Similar to the previous report on the CdTe QD gel formed in water (9), the obtained QDs hydrogel can undergo sol-gel switching. The facile gelation of the QDs and the sol-gel switching made it a suitable platform in the immobilization of enzymes.

To keep the native structure and activity of the enzyme is a challenge when trying to immobilize the enzyme on a substrate (10). The mesopores inside the gel network not only provide a small enough space to entrap enzymes but also allow for easy diffusion of substrates and products. Herein, we try to use the CdTe QD gel networks to encapsulate horseradish peroxidase (HRP) to obtain enzyme encapsulated QD hydrogels that can be utilized as bi-functional platforms with both fluorescence probing and bio-recognition as well as bio-catalysis functions in one material. The enzyme solution was added into the CdTe QD sol obtained from sonicating the QD gel, and the mixture was stored at 4 °C. During the gelation process, we suppose that enzyme can be encapsulated in the mesopores of the CdTe QD gel network. Herein we used UV-vis absorption spectra to validate the encapsulation of enzyme in the QD gels by scanning the absorption of benzoquinone at 400 nm produced from the HRP catalyzed oxidation of catechol by hydrogen peroxide. The enzymatic reaction is expressed as equation 1:

$$\text{catechol} + H_2O_2 \rightarrow o\text{-benzoquinone} + H_2O \qquad [1]$$

In the experiments, 0.1 mL 2.4 mg/mL HRP was added into 1.0 mL QD sol obtained by sonication of the QD gel. After the gelation of the QDs, the gel was washed more than 6 times in order to remove the free enzyme molecules. Firstly, we compared the absorption of 1.0 mM catechol after incubation with free HRP, gel and the supernatant, respectively. From the UV-vis absorption spectra shown in Figure 2, it is found that the absorption of 1.0 mM catechol at 400 nm after incubation with 2.0 μL 2.4 mg/mL free HRP solution (4.8 μg HRP, curve a) and 20 μL supernatant (is calculated to contain about 4.4 μg HRP, curve b) for 2 min were nearly the same. Curve c in Figure 2 indicates that only a few HRP molecules were present in the supernatant after the thorough washing. The QD gel did not cause any interference on the enzymatic reaction characterization by the UV-vis absorption method, because the QD gels were on the bottom of the quartz cuvette, and the supernatant above itself did not present any UV-vis absorption.

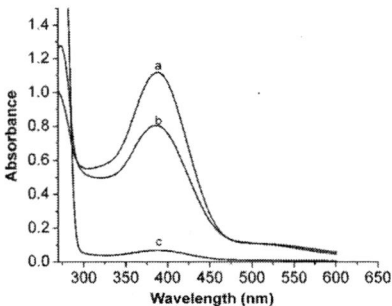

Figure 2. UV-vis absorption spectra of 1 mM catechol and 1 mM H_2O_2 after incubation for 2 min with 2.0 μL 2.4 mg/ml of free enzyme (a), with 20 μL gel (b), and with 20 μL supernatant after washing 6 times (c).

The above phenomenon indicates that after washing the gels thoroughly, few enzyme molecules existed in the supernatant, while the enzyme in the QDs will not easily be washed away and the QD gel network provides a good medium for the enzyme immobilization. Moreover, it has to be pointed out that the activity of the enzyme that was immobilized in the QDs gel is also comparable to that of free enzyme. The enzyme-encapsulated CdTe QD hydrogel enzyme hybrid material serves as both signal transforming and recording unit in the biosensor development. To validate the sensing ability of the enzyme-encapsulated QD hydrogel, catechol was taken as an example analyte in this work. However, the oxidant used in the HRP enzymatic reaction system, H_2O_2, has a quenching effect by itself on the QDs fluorescence. Thus, tyrosinase (TRS) that catalyzes the oxidation of catechol by dissolved oxygen was taken as another example enzyme for the biosensor development. According to the HRP encapsulation, 0.1 mL of 2.0 mg/mL TRS was added into 1.0 mL QD sol to form a TRS-encapsulated QDs hydrogel. TRS in the QDs gel will catalyze the oxidation of catechol by dissolved oxygen into o-benzoquinone, and the QD gel itself served as a quinone probe as its fluorescence can effectively be quenched by quinone. Figure 3A shows the quenching effects of catechol with different concentrations on the fluorescence of the CdTe QD hydrogel. Catechol with the concentration of 0.2 mmol/L can quench the fluorescence of the TRS-encapsulated QD hydrogel to about 50% after 9 min. Control experiments showed that dopamine had no obvious effect on the QD hydrogel in the absence of TRS. Figure 3B shows the relationship between the quenching effect (I_0/I, I_0 and I refer the fluorescence intensity of QDs in the absence and presence of analyte, respectively) and the concentration of catechol. The extent of quenching can be used for the quantification of the analyte.

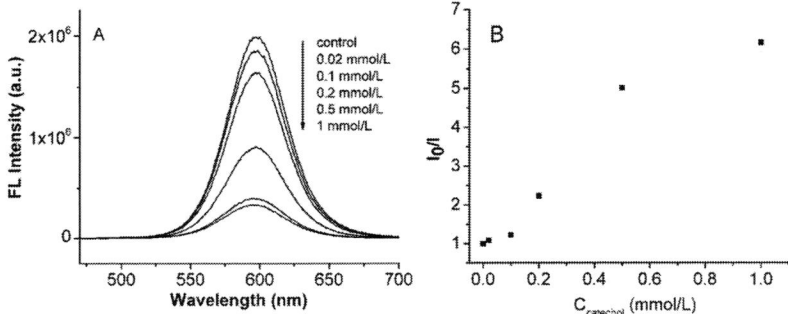

Figure 3. (A) Fluorescence spectra representing the quenching effect of catechol with different concentrations on TRS-encapsulated CdTe QD hydrogel. Incubation time, 9 min; 50 mM phosphate buffer solution (pH 6.8). (B) Relationship between I_0/I and the concentration of catechol.

In conclusion, the porous three dimensional QD hydrogel proved to be an adequate encapsulation medium for enzymes. Both enzyme activity and the optoelectronic properties of the QDs were well retained in the QD hydrogels. Sensing of catechol was successfully achieved using the as-prepared enzyme-encapsulated QD hydrogel. Biomolecule encapsulated QD hydrogels as multi-functional platforms have great potential in the development of optical biosensors.

Acknowledgments

J. Yuan appreciates the support from the Alexander von Humboldt Foundation.

References

1. C. L. Choi and A. P. Alivisatos, *Ann. Rev. Phys. Chem.*, **61**, 369 (2010).
2. P. V. Kamat, K. Tvrdy, D. R. Baker and J. G. Radich, *Chem. Rev.*, **110**, 6664 (2010).
3. I. U. Arachchige and S. L. Brock, *Acc. Chem. Res.,* **40**, 801 (2007).
4. N. C. Bigall, A. K. Herrmann, M. Vogel, M. Rose, P. Simon, W. Carrillo-Cabrera, D. Dorfs, S. Kaskel, N. Gaponik and A. Eychmüller, *Angew. Chem. Int. Ed.*, **48**, 9731 (2009).
5. V. Lesnyak, A. Wolf, A. Dubavik, L. Borchardt, S. V. Voitekhovich, N. Gaponik, S. Kaskel and A. Eychmüller, *J. Am. Chem. Soc.*, **133**, 13413 (2011).
6. W. Liu, A.-K. Herrmann, D. Geiger, L. Borchardt, F. Simon, S. Kaskel, N. Gaponik and A. Eychmüller, *Angew. Chem. Int. Ed.,* **51**, 5743 (2012).
7. A. Lukowiak and W. Strek, *J. Sol-Gel Sci. Technol.,* **50**, 201 (2009).
8. H. F. Bao, E. K. Wang and S. J. Dong, *Small,* **2**, 476 (2006).
9. S.-H. Jeong, J. W. Lee, D. Ge, K. Sun, T. Nakashima, S. I. Yoo, A. Agarwal, Y. Li and N. A. Kotov, *J. Mater. Chem.*, **21**, 11639 (2011).
10. J. Yuan, N. Gaponik and A. Eychmüller, *Anal. Chem.,* **84**, 5047 (2012).

Adaptive Chemical Sampling Device Inspired by Crayfish

R. Takemura, K. Takahashi, T. Makishita, and H. Ishida

Graduate School of Bio-Applications and Systems Engineering,
Tokyo University of Agriculture and Technology
2–24–16 Nakacho, Koganei, Tokyo 184–8588, Japan

Crayfish are known to generate water jets by waving their special appendages with a fan-like shape. The generated jets entrain surrounding water, and thus inflow converging toward their olfactory organs is induced. Therefore, the jets are considered to help crayfish collect water samples to search for the smell of a prey. Although crayfish are also known to be able to change the direction of the jet discharge, its implication in olfactory search has not been well understood. To investigate this issue, we have developed a chemical sampling device equipped with a jet discharger and electrochemical sensors. Results of computational fluid dynamics analyses and chemical detection experiments are presented to show that the angular range of water-sample collection can be adjusted by changing the jet discharge directions. The results suggest that the olfactory search efficiency can be improved by narrowing down the angular sample collection range as the search progress.

Introduction

Many species of animals move fluid across or through their olfactory organs when they smell something. For example, aquatic animals collect water samples from their surroundings to their olfactory organs to detect chemical substances dissolved in water (1)–(4). Here, we call this active behavior to acquire chemical signals as chemical sampling. Since the chemical signals contain rich information, e.g., the nearby existence of foods or mates, their detection is essential for the survival of aquatic animals. However, even if an animal has a highly sensitive nose, such chemical signals are difficult to obtain without actively performing chemical sampling. Olfactory organs respond only to odorant molecules that have actually reached their reactive surfaces. Such contact is not always achieved spontaneously. In aquatic environments, the molecular diffusion process is extremely slow: the diffusion length in one hour is calculated to be only five millimeters (3). Chemical substances can be hardly detected even if the nose is brought to a few centimeters from the chemical source.

Although the ways of chemical sampling are more or less different among spices, the chemical sampling behavior is observed in most animal species. The capabilities of collecting chemical samples might have been gained out of necessity and refined through the evolutionary adaptation. Artificial chemical sensing systems can also benefit by having chemical sampling functions. For example, the chemical detection capability can be enhanced using an actively generated flow to collect chemical substances. The flow can be generated, for example, by sucking the surrounding fluid, which is probably the most popular way of chemical sampling in animal behavior. When an animal inhales fluid, the generated flow brings water samples from the surroundings to the animal. In contrast, crayfish exhibit a unique way of chemical sampling. Instead of sucking water, a crayfish generates jets of water. Since the jets entrain surrounding water, inflows converging toward the jets are induced. From the fluids engineering point of view, chemical sampling using jets have several advantages over simply sucking water. However, the details of chemical sampling by crayfish have not yet been fully revealed. The aim of our research is to gain better understandings on the chemical sampling behavior of crayfish and to develop an effective and efficient chemical sampling device by mimicking it. Here we describe a device that can generate water jets as crayfish do. The device is equipped with an array of electrochemical sensors to detect the collected chemical samples. Results of our investigation on the effects of jets are described after a brief introduction of the chemical sampling behavior of crayfish.

Chemical Sampling Behavior of Crayfish

A chemical substance is dispersed in a fluid medium by molecular diffusion and convection. As described in the previous section, the diffusion rate of chemical species in water is extremely small. Therefore, under stagnant flow conditions, the released chemical mostly stays in the vicinity of its source. Almost no chemical is detected unless the sensors are brought within the proximity of the source (4).

Most species of crayfish live at the bottom of lakes or ponds. The flow conditions in their habitats are stagnant. Nevertheless, they can track down foods from distant places by using their olfaction. The key for their success is the fan organs located below the frontal sensory organs. Exopodites of crayfish's maxillipeds have a fan-like shape (3). By waving the maxillipeds, crayfish generate jets, as shown in Figure 1. The jets drag the surrounding water by the effect of the viscosity of water, and therefore, inflow towards the jets is induced. Using this phenomenon known as jet entrainment, crayfish collect odor-laden water samples from the surroundings to their olfactory organs.

Compared with simply sucking water, the jet entrainment has several advantages in collecting water samples. First, the induced inflow velocity is larger for the entrainment except in the vicinity. This is because the velocity decay of the flow induced by suction is

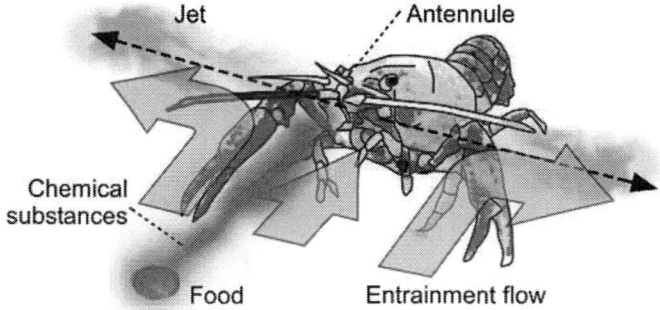

Figure 1. Crayfish collect odor-laden water samples from the surroundings to their olfactory organs using the jet entrainment. In this illustration, two jets are generated horizontally to both sides. As a result, inflow is induced from the front of a crayfish.

much faster than that induced by the jet entrainment (5). Second, the pattern of the flow field induced by the jet entrainment can be entirely altered by changing the directions of the jets. In contrast, the flow pattern generated by suction is almost isotropic. Therefore, a change in the direction of a suction port hardly affects the flow pattern. In fact, crayfish have been observed to generate several different flow patterns by changing the jet discharge directions. Although the details are not clear, there is a possibility that crayfish adaptively change the flow pattern to increase the chemical sampling efficiency.

Chemical Sampling Device

The proposed device is shown in Figure 2. The device was designed to be the same size as the carapace of a crayfish in order to match the Reynolds number. The device consists of a jet discharger and chemical sensors. The jet discharger is connected to a water pump. Water is sucked by the pump from an inlet port (ϕ 2.8 mm) on the frontal face of the device. Several outlet nozzles (ϕ 2.8 mm) are placed on the sides of the cylindrical jet discharger. The orientation of each nozzle is set differently. Pumped water is discharged to generate jets from two of these nozzles at a rate of 150 ml/min for each nozzle. The total mass flow rate discharged from the pump can be adjusted by pulse width modulation of the supply voltage for the pump. By changing the connections between the pump and outlet nozzles, the jet discharge directions can be switched manually. In future, this switching will be done automatically by using motorized nozzles.

The primary olfactory organ of crayfish is a pair of antennules, each having an inner and an outer branch. When a crayfish tries to detect odorants drifting near the floor, the antennules are brought down in front of the maxillipeds. To mimic this configuration, four carbon-rod electrodes are aligned laterally in front of the jet discharger. They are

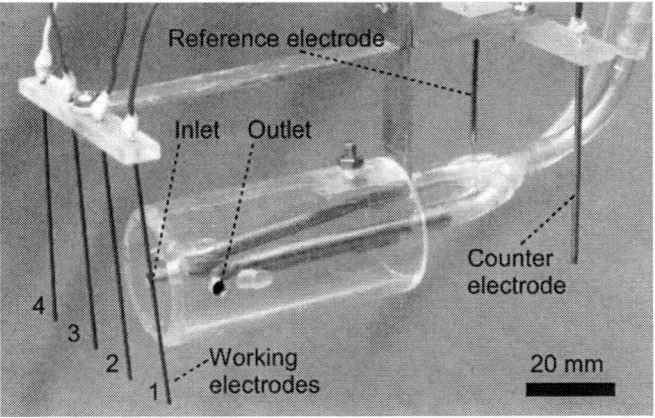

Figure 2. Chemical sampling device developed in this work. Water is sucked by a pump from an inlet port on the frontal face of the device, and is discharged from the sides to generate jets. By changing the tube connections, the directions of the jets can be altered. Four carbon electrodes are aligned laterally for amperometric electrochemical detection of chemical substances.

connected to a custom-made potentiostat circuit and serve as working electrodes. Silver wire and a stainless-steel tube are used as a reference electrode and a counter electrode, respectively. In the chemical detection experiments described later, ascorbic acid was used as a detection target. The potential of the four carbon working electrodes were kept at 0.7 V against the reference electrode for amperometric electrochemical detection of ascorbic acid (6).

Computational Fluid Dynamics Analyses

Firstly, the computational fluid dynamics (CFD) analyses were conducted to investigate the flow fields generated by the jets discharged in several different directions. Turbulent flow fields were numerically simulated using a commercial CFD package, ANSYS Fluent 12.0. In the three-dimensional geometric models used in the simulations, the chemical sampling device was assumed to be set in 100-mm deep water at 10 mm high from the floor. The flow fields generated by the jets discharged horizontally to both sides and 45° upward were simulated. These jet discharge directions were chosen from the results of the ink visualizations of the flow around live crayfish presented in (4).

An unstructured grid was used to discretize the computational domain in the model.

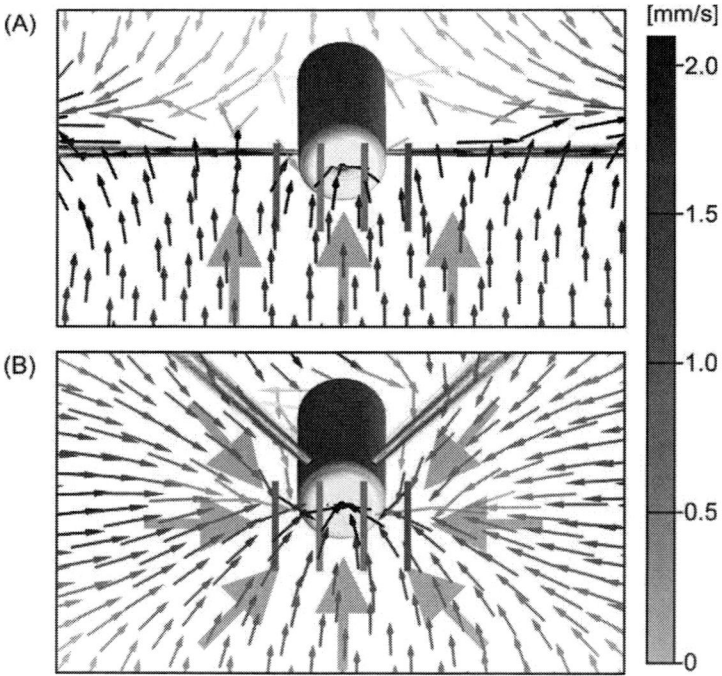

Figure 3. Results of CFD analyses of jets discharged (A) horizontally to both sides and (B) to 45° upward. Small arrows show the flow directions on a horizontal plane passing through the center of the device. Flow velocities are shown in gray scale.

Finer cells were placed in the areas close to the outlet nozzles because of the high velocity gradients in these areas. Large eddy simulation (LES) with a dynamic Smagorinsky model was performed using a SIMPLE solver of the second order of accuracy. No-slip boundary conditions were applied at the bottom of the water tank and the walls of the chemical sampling device. At the inlet port, a uniform velocity distribution was assumed. The velocity distribution at each outlet nozzle was assumed to be parabolic since the flow could be considered to be fully developed in the tube connected to the outer nozzle. At the water surface, free-slip condition was applied. For the other boundaries, open boundary conditions were used. A flow field obtained by a steady-state calculation without using a turbulence model was used as the initial condition for the transient calculation. Time integration was calculated with a second-order implicit method until the effect of the initial condition had sufficiently disappeared. For validation of the analyses, the numerical calculation results were compared with the analytical solution of the cross-sectional velocity profile in turbulent jets (7). Since a good match was achieved, it was confirmed that the flow fields obtained from our CFD analyses were at least qualitatively correct.

Figure 3 shows perspective views of the flow fields obtained by the CFD analyses. The flow field induced by the jets discharged horizontally to both sides is shows in Figure 3(A), and that induced by the jets discharged to 45° upward is shown in Figure 3(B). It should be noted that only the flow directions on a horizontal plane passing through the center of the device are shown using small arrows. When the jets were discharged to the sides, as shown in Figure 3(A), water samples were drawn to the vicinity of the sensor electrodes only from a limited angular range (i.e., from the front of the chemical sampling device). When the jets were discharged to 45° upward, as shown in Figure 3(B), water samples were collected to the sensors from a wide angular range Even the water samples on the backside of the device were drawn to the sensors.

Chemical Detection Experiments

After the characterization of the flow fields induced by jets through the CFD analyses, the effects of the jet discharge directions on the sensor responses were experimentally evaluated. Chemical detection trials were conducted in a small water tank (800 mm × 600 mm) filled with 0.1 mol/l salt water up to a depth of 100 mm. The chemical sampling device was hung from a bridge placed over the tank. The height of the device was set to 10 mm from the bottom of the tank. The chemical source was set at 60 mm from the inlet of the chemical sampling device at an angle θ. 0.01 mol/l ascorbic acid solution colored with a fluorescent dye (10 mg/l of rhodamine 6G) was used as a detection target, and was released at a rate of 1 ml/min from a stainless-steel tube. The outlet of the tube was fixed at one of the seven locations indicated in Figure 4 by the black circles. It was shown in the previous paper that rhodamine 6G has no adverse effect on the detection of the ascorbic acid (6).

In each chemical detection trial, the release of the chemical substance, the discharge of the jets, and the measurement of the sensor responses were all started simultaneously after waiting for the background flow to cease. The reception rate of the chemical substance is defined as the rate of above-threshold chemical detection within a 200-second time period. The threshold value of each sensor response in each trial was obtained by multiplying the mean sensor response value in the first 5-second time period by five. The background response of the sensor was 1.0 μA at most. Three trials were conducted for each source location, and the average reception rate for the three trials was calculated.

Figures 5(A) and (B) show the chemical reception rates for the jets discharged horizontally to both sides and to 45° upward, respectively. In Figure 5(B), high reception rates were attained for the chemical source in the angular range from −135° to 135°. In Figure 5(A), the angular rage of high chemical reception rates was from −45° to 45°. These results agree well with the results of the CFD analyses. Furthermore, it can be seen in Figure 5 that the sensors located closer to the source showed higher reception rates. If the source was set on the left side, for example at 30° to the device, the sensors located on the left (sensors 1 and 2) showed higher reception rates than the sensors on the other side. The results indicate that the chemical source direction can be estimated from the sensor responses.

Discussions

To investigate the effects of changing the jet discharge directions in the chemical sampling behavior of crayfish, the CFD analyses and the chemical detection experiments were conducted. The flow fields induced by the jets were simulated in the CFD analyses. In the experiments, collection of chemical samples at each sensor location was evaluated using the electrochemical sensors. The results showed two major features. First, the angular range of chemical sample collection can be effectively adjusted by changing the directions of the jets. Second, the sensors closer to the chemical source show higher chemical reception rates.

In the flow field induced by the jets discharged to 45° upward, water samples are drawn to the sensors from a wide angular range. This flow pattern maximizes the probability of detecting a chemical substance. Since the chemical source location is unknown at the beginning of the search, a high detection probability is preferable at the initial stage of the search. However, the direction of the chemical source can be estimated

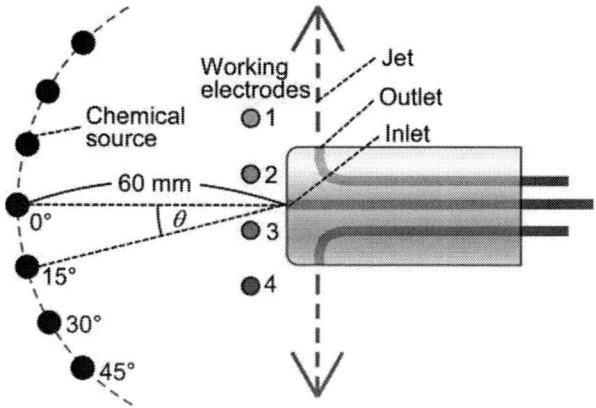

Figure 4. Arrangement of the chemical sampling device and a chemical source in the experiments. Black circles indicate the positions of the chemical source tested in the experiments.

only roughly in this flow field. Water samples are drawn to each sensor from a an angular range of approximately 60°.

Once a chemical substance is detected, the searcher should turn toward the estimated direction of the chemical source. Discharging the jets horizontally to the sides enables focusing of chemical sampling on the area of interest. In the flow field induced by the

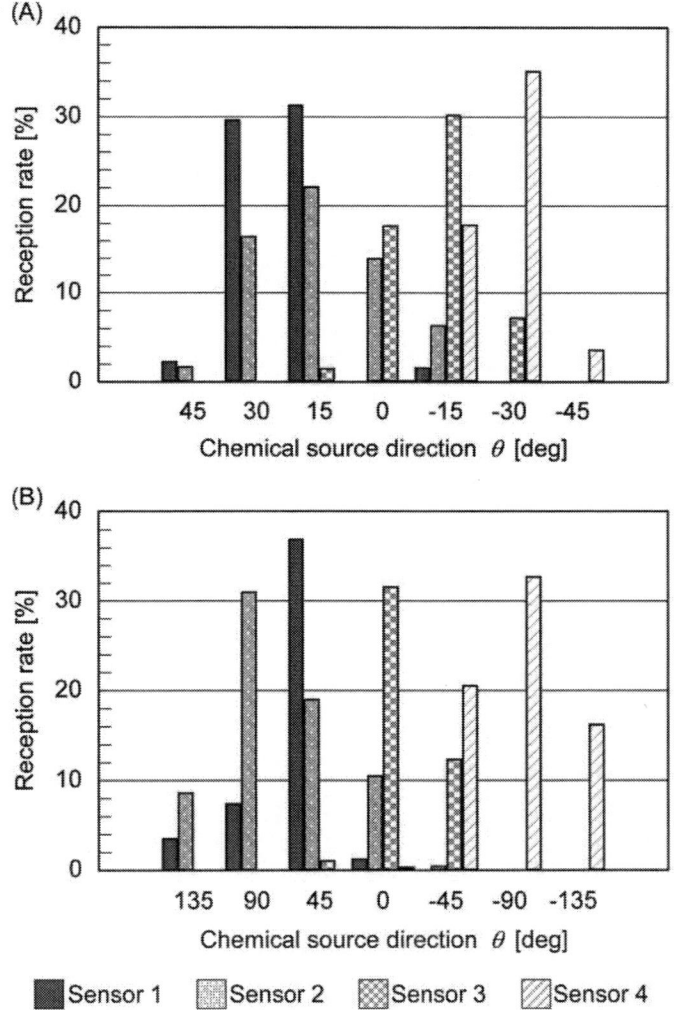

Figure 5. Reception rate of ascorbic acid at each sensor in the experiments for the jets discharged (A) horizontally to both sides and (B) to 45° upward. The chemical source direction, θ, indicates a counterclockwise angle from the front of the device to the chemical source location. The chemical reception rate was defined as the rate of above-threshold chemical detection in a 200-second time period.

horizontal jets, water samples are drawn to the sensors from a narrow area in the front of the device. By comparing the responses of the four sensors, the chemical source direction can be estimated with higher angular resolution. The efficiency of the olfactory search could be improved by adaptively changing the jet discharge directions. It is not clear at this moment if crayfish change the jet discharge directions in such a way. Detailed behavioral observation is awaited to prove our hypothesis.

Recently, autonomous underwater vehicles (AUVs) equipped with chemical sensors are being developed aiming to accomplish autonomous search for chemical sources, e.g., environmental pollutants, hydrothermal vents, and unexploded ordnance (8). The proposed chemical sampling device could be applied to such AUVs. Their chemical sensing capabilities could be improved by actively generating water currents to collect water samples to the sensors.

Conclusion

The chemical sampling device inspired by crayfish was developed. The device discharge water jets in a similar way to crayfish. The CFD analyses were conducted to investigate the effect of changing the jet discharge directions on the flow field induced by the jet entrainment. The experiments were conducted to evaluate the effects of the jet discharge on the chemical detection. The jets discharged to 45° upward help collect water from the wide angular range. The induced flow field maximizes the probability of getting chemical information from the surroundings. In the flow field induced by the jets horizontally discharged to both side, water samples are drawn to the sensors only from the frontal area. This flow pattern is useful in estimating the chemical source direction with high resolution. These results suggest that the olfactory search efficiency can be enhanced by adaptively changing the jet discharge directions.

References

1. J. P. L. Cox, *J. R. Soc. Interface*, **5**, 575 (2008).
2. G. Settles, *J. Fluids Eng.*, **127**, 189 (2005).
3. T. Breithaupt, *Biol. Bull.*, **200**, 150 (2001).
4. P. Denissenko, S. Lukaschuk, and T. Breithaupt, *J. Exp. Biol.*, **210**, 4083 (2007).
5. G. R. Hunt and D. B. Ingham, *Ann. Occup. Hyg.*, **40**, 171 (1996).
6. T. Kikas, H. Ishida, and J. Janata, *Anal. Chem.*, **74**, 3605 (2002).
7. N. Rajaratnam, *Turbulent Jets*, p. 40, Elservier Scientific Publishing, Amsterdam, Netherlands (1976).
8. W. Li, J. A. Farrell, S. Pang, and R. M. Arrieta, *IEEE Trans. Robot.*, **22**, 292 (2006).

CO Sensing Properties of Electrochemical Gas Sensors Using an Anion-conducting Polymer as an Electrolyte

T. Goto[a], T. Hyodo[b], K. Kaneyasu[c], H. Yanagi[d], and Y. Shimizu[b]

[a] Faculty of Engineering, Nagasaki University
[b] Graduate School of Engineering, Nagasaki University
1-14 Bunkyo-machi, Nagasaki 852-8521, Japan
[c] Figaro Engineering Inc., 1-5-11 Senbanishi, Minoo, Osaka 562-8505, Japan
[d] Tokuyama Corp., 40 Wadai, Tsukuba, Ibaraki 300-4247, Japan

An anion-conducting polymer (ACP) with large OH⁻ conductivity
is very promising as an electrolyte of electrochemical gas sensors
working at RT. Therefore, CO sensing properties of an
electrochemical gas sensor using ACP as an electrolyte and carbon
black (CB) powders loaded with n wt% noble metals (EC(nM/CB-
ACP), n: 0.1~10, M: Pt, Pd or Au) as an electrode material were
investigated in this study. Among the sensors tested,
EC(10Pd/CB-ACP) and EC(10Pt/CB-ACP) showed comparatively
large response and fast response and recovery speeds to CO in wet
air (relative humidity: 57%) at 30°C. The magnitude of response
of the sensors tends to decrease slightly with a decrease in RH.

Introduction

CO is a colorless and scentless gas although it exerts very harmful influences on
human health. Therefore, development of CO sensors with high selectivity and
sensitivity detectable at sufficiently-low concentration not to exert harmful influences is
highly requested. And response speed is also important for realizing quick detection of
CO produced by the incomplete combustion of fuels. Recently, various electrochemical
gas sensors operable at room temperature (RT) have been developed by employing liquid
(1), ceramics (2) or polymers (3, 4) as an electrolyte. Among them, Nafion, one of
proton-conducting polymers, has been presently applied as an electrolyte of commercial
electrochemical CO sensors. On the other hand, anion-conducting polymers (ACP) with
large OH⁻ conductivity have also been developed by some companies. Especially,
polymer electrolyte fuel cells (PEFC) fabricated with ACP as an electrolyte are very
promising because they showed high current density as well as high power density with
relatively-low overpotential, and their performance was quite comparable to those of
PEFC employing proton-conducting polymers in the preliminary research stage (6-11).
In our previous study, ACP equipped with Pt-loaded CB electrodes was found to show
CO response at RT (12). In the present study, our efforts have been directed to
evaluating the effects of the kinds of the noble metals loaded on CB electrodes and
relative humidity in the measurement atmosphere on CO response properties in a wide
range of CO concentration.

Experimental

Preparation of noble metal-loaded carbon black powders

Carbon black (CB; Sigma Aldrich Co. LLC., A7455) powders loaded with n wt% noble metal (nM/CB, n: 0.1, 1 or 10, M: Pt, Pd or Au) were prepared by an impregnation method. An appropriate amount of CB powders was added into aqueous solution of $PtCl_4$, $Pd(NO_3)_2 \cdot 2H_2O$ or $HAuCl_4 \cdot 4H_2O$, and then they were ultrasonicated for 60 min. The suspensions obtained were dried on a hot plate. After the dried powders were ground in an agate mortar, they were heat-treated under H_2 stream at 350°C for 1 h.

Sensor fabrication

A schematic drawing of a sensor element is shown in Fig. 1. The nM/CB powder prepared was mixed with ACP solution (AS-4, Tokuyama Corp.) at a weight ratio of nM/CB:ACP = 1:1. The nM/CB-ACP paste obtained was applied on the surface of both sides of an ACP membrane (A201, Tokuyama Corp.) as sensing and counter electrodes by screen printing, and then it was dried at ca. 70°C for 2 h. The electrochemical gas sensor obtained was denoted as EC(nM/CB-ACP). For comparative purpose, an electrochemical gas sensor with CB sensing and counter electrodes without noble metal loading (EC(CB-ACP)) were also fabricated.

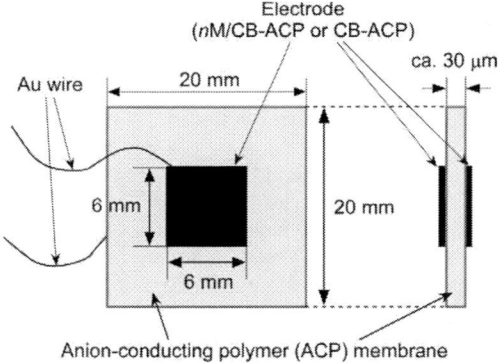

Figure 1. Schematic drawing of a sensor element.

Gas-sensing measurement

The sensors elements were sandwiched with two tubes to set up a gas measurement system which allow us to flow different gases over sensing- and counter-electrodes. Gas-sensing properties of all sensors were evaluated by measuring the electromotive force (EMF) at 30°C by flowing 100~3000 ppm CO balanced with air at the sensing-electrode side, while compressed air was flown at the counter-electrode side. Relative humidity (RH) at both the electrode sides was controlled to be 0~100% at 30°C. The magnitude of sensor response is defined as a change in EMF (ΔEMF) induced by a sample gas.

Results and discussion

Figure 2 shows response transients of EC(nPt/CB-ACP) and EC(CB-ACP) to 500 ppm CO in air at 30°C (57%RH). EC(CB-ACP) showed almost no response to CO, but the loading of Pt to CB enhanced the CO response markedly. The magnitude of CO response of EC(nPt/CB-ACP) increased with an increase in the loading amount of Pt, n. In addition, the increase in n drastically improved the response and recovery speeds, too. Therefore, a loading amount of 10 wt% was adapted in the successive measurements. Response transients of EC(10M/CB-ACP) to 500 ppm CO in air at 30°C (57%RH) were investigated and these results are shown in Fig. 3. EC(10Pt/CB-ACP) and EC(10Pd/CB-ACP) showed large response to 500 ppm CO (EC(10Pt/CB-ACP): -79 mV, EC(10Pd/CB-ACP): -85 mV), and the response and recovery speeds of EC(10Pd/CB-ACP) (90% response time (res): 3.1 min, 90% recovery time (rec): 2.1 min) were much faster than those of EC(10Pt/CB-ACP) (res: 4.5 min, rec: 28.2 min). On the other hand, EC(10Au/CB-ACP) showed almost no CO response, as was observed for EC(CB-ACP).

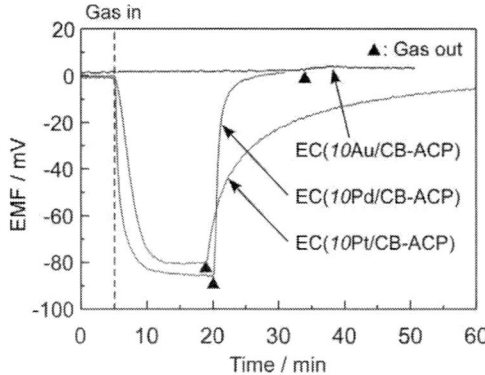

Figure 2. Response transients of EC(nPt/CB-ACP) and EC(CB-ACP) to 500 ppm CO in air at 30°C (57%RH).

Figure 3. Response transients of EC(10M/CB-ACP) to 500 ppm CO in air at 30°C (57%RH).

Figure 4 shows response transients and CO concentration dependence of response of EC(*10*Pd/CB-ACP) and EC(*10*Pt/CB-ACP) in air at 30°C (57%RH). Both EC(*10*Pd/CB-ACP) and EC(*10*Pt/CB-ACP) showed negative response to CO, and the absolute magnitude of CO response, |ΔEMF|, increased with an increase in CO concentration. In addition, ΔEMF of EC(*10*Pd/CB-ACP) and EC(*10*Pt/CB-ACP) negatively changed in proportion to the logarithmic CO concentration and the slope, i.e. CO sensitivity, of EC(*10*Pd/CB-ACP) and EC(*10*Pt/CB-ACP) were ca. -19.3 and ca. -24.8 mV/decade, respectively. However, EC(*10*Pt/CB-ACP) showed overshooting response behavior to high concentration of CO and very fast response and recovery speeds, while the recovery speed of both sensors was enormously slow in a comparatively low CO concentration range (100~300 ppm).

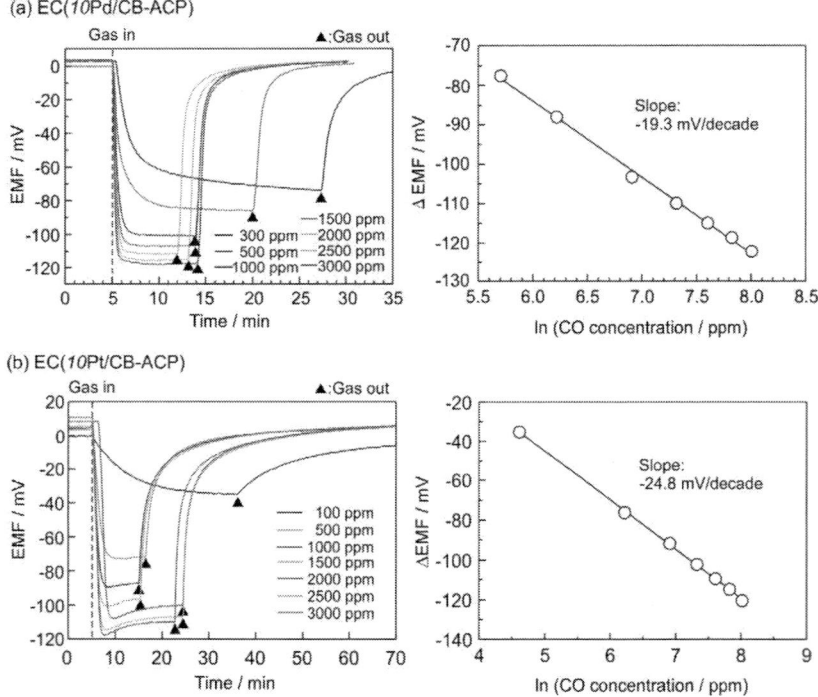

Figure 4. Response transients to CO and CO concentration dependence of response of EC(*10*Pd/CB-ACP) and EC(*10*Pt/CB-ACP) in air at 30°C (CO concentration: 100~3000 ppm, 57%RH).

Figure 5 shows response transients of EC(*10*Pd/CB-ACP) and RC(*10*Pt/CB-ACP) to 500 ppm CO in air at 30°C under various relative humidity (RH). Response of both sensors to CO decreased with a decrease in RH and ΔEMF could not be measured under dry atmosphere (≈ 0%RH), probably because the decrease in OH⁻ conductivity which was caused by the decrease in the amount of moisture in the ACP electrolyte.

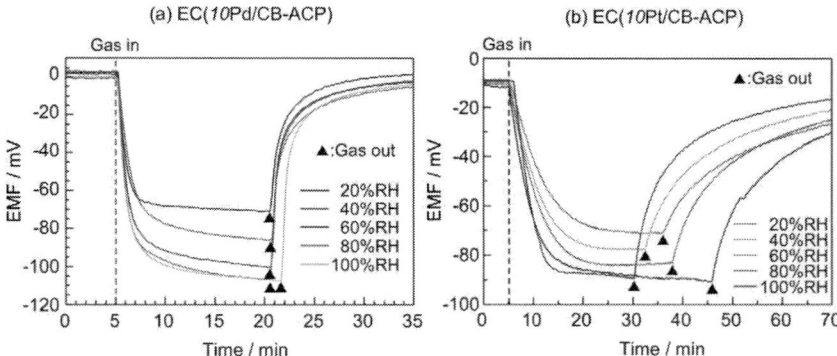

Figure 5. Response transients of EC(10Pd/CB-ACP) and EC(10Pt/CB-ACP) to 500 ppm CO in air at 30°C (20~100%RH).

Conclusions

CO sensing properties of EC(nM/CB-ACP) were investigated in this study. EC(10Pd/CB-ACP) and EC(10Pt/CB-ACP) showed much large response to 100~3000 ppm CO at RT (57%RH) in comparison with those of EC($n10$Au/CB-ACP) and EC(CB-ACP). The response speed of EC(10Pd/CB-ACP) and EC(10Pt/CB-ACP) was fast in the CO concentration range of 500~3000 ppm, while that was extremely slow in the range of less than 300 ppm CO. In addition, ΔEMF of EC(10Pd/CB-ACP) and EC(10Pt/CB-ACP) negatively changed in proportion to the logarithmic CO concentration and CO sensitivities were ca. -19.3 mV/decade and ca. -24.8 mV/decade, respectively. However, responses of both sensors decreased with a decrease in RH, because of the decrease in OH⁻ conductivity which was caused by the decrease in the amount of moisture in the ACP.

References

1. T. Ishiji, *Anal. Chem.*, **65**, (1993) 2736.
2. N. Miura, K. Kanamura, Y. Shimizu, and N. Yamazoe, *Solid State Ionics*, **40-41**, (1990) 452.
3. N. Miura, H. Yamazoe, and T. Seiyama, *Denki Kagaku (presently Electrochemistry)*, **50**, (1982) 858.
4. N. Miura, H. Kato, N. Yamazoe, and T. Seiyama, *Proc. of Intern. Meeting on Chem. Sens., Fukuoka, Japan* (1983) pp. 233-238.
5. H. Imaya, K. Okamura, N. Nakano, K. Nagashima, Y. Suzuki, and K. Takahashi, *Electrochemistry*, **79** (2011) 140.
6. K. Asazawa, K. Yamada, H. Tanaka, A. Oka, M. Taniguchi, and T. Kobayashi, *Angew. Chem. Int. Ed.*, **46** (2007) 8024.
7. H. Yanagi and K. Fukuda, *ECS Transactions*, **16** (2008) 257.
8. K. Fukuda, H. Inoue, S. Watanabe, and H. Yanagi, *ECS Transactions*, **19** (2009) 23.

9. M. Unlu, J. Zhou, and P. A. Kohl, *J. Electrochem. Soc.*, **12** (2009) B27.
10. K. Matsumoto, T. Fujigaya, H. Yanagi, and N. Nakashima, *Adv. Funct. Mater.*, **21** (2011) 1089.
11. M. Manoukian, A. B. LaConti, L. A. Tempelman, and J. Forchione, *U.S. Patent* 2011/0005928 (2011).
12. T. Hyodo, C. Ishibashi, K. Matsuo, K. Kaneyasu, H. Yanagi and Y. Shimizu, *Electrochimica Acta*, in press.

NO$_2$ Sensing Properties of Porous In$_2$O$_3$-based Powders Preapred by Utilizing
Ultrasonic-spray Pyrolysis Employing PMMA Microsphere Templates
: Effects of the Size of the PMMA Microspheres on their Gas-sensing Properties

E. Fujii, T. Hyodo, K. Matsuo, and Y. Shimizu

Graduate School of Engineering, Nagasaki University
1-14 Bunkyo-machi, Nagasaki 852-8521, Japan

PMMA microspheres were synthesized in distilled water by
ultrasonic-assisted emulsion polymerization. The average particle
size of PMMA microspheres was dependent markedly on the kind
of surfactant used. Porous (pr-) In$_2$O$_3$ powders were prepared by
ultrasonic-spray pyrolysis of In(NO$_3$)$_3$ aqueous solution containing
the PMMA microspheres synthesized. The NO$_2$ response of pr-
In$_2$O$_3$ was much larger than that of conventional In$_2$O$_3$ powder
prepared by the similar technique employing PMMA-free In(NO$_3$)$_3$
aqueous solution. The introduction of controlled macroporous
structure into the powder of the sensor material was found to be
effective for improving NO$_2$ response properties.

Introduction

As nitrogen oxide (NO$_2$) is one of the most harmful gases to the human body and also a
cause of air pollution, highly sensitive and selective NO$_2$ sensors at low cost, small size
and good reliability are indispensable for the detection to low concentration of NO$_2$ in the
atmosphere. Therefore, numerous efforts have been directed to optimizing the
microstructural morphology of various gas-sensing materials with different sizes of well-
developed pores to improve their gas-sensing properties, because optimization of the size
and the amount of pores in the gas sensor materials are effective in controlling gas
reactivity (1-6). We also have prepared mesoporous and macroporous oxides by utilizing
the assembly of surfactant (several nanometers in size) and commercial
polymethylmethacrylate (PMMA) microspheres ($\geq$150 nm in size) as a template,
respectively (7-13). However, we have not yet established a preparation method to
prepare gas-sensing materials having well-developed middle-sized pores with a diameter
of several nm~150 nm. Such technique is absolutely essential for optimizing the
microstructural morphology of various gas-sensing materials. In this study, we have
attempted to synthesize smaller PMMA microspheres with controlled particle size
(several nm~150 nm in diameter) by microwave-assisted emulsion polymerization
employing different surfactants. In addition, porous (pr-) In$_2$O$_3$ powders were prepared
by ultrasonic-spray pyrolysis employing the synthesized PMMA microspheres, and their
NO$_2$-sensing properties have been investigated.

Experimental

Synthesis of PMMA microsphere dispersion by ultrasonic-assisted emulsion polymerization

PMMA microspheres were synthesized in deionized water (100 cm^3) by ultrasonic (19.5 kHz)-assisted emulsion polymerization (50 min) employing methylmethacrylate monomer (8 g) as a polymer source, ammonium persulfate (0.3 g) as an initiator and a surfactant (0.1 g) and subsequent stirring (400 rpm) at 60°C for 6 h. The surfactant used was sodium lauryl sulfate (SLS, $CH_3(CH_2)_{11}OSO_3Na$), Triton X-100 (Triton, $(C_2H_4O)_{10}C_{14}H_{22}O$) or P123 ($EO_{20}PO_{70}EO_{20}$, EO: polyethylene oxide, PO: polypropylene oxide). The PMMA microspheres in the resultant dispersions were denoted as PMMA(M) (M: surfactant employed for the synthesis of PMMA microspheres (SLS, Triton or P123)).

Preparation of porous In$_2$O$_3$-based powders by utilizing ultrasonic-spray pyrolysis employing PMMA microspheres

The PMMA microsphere dispersion (37.5 cm^3) was mixed with 0.05 mol dm^{-3} In(NO$_3$)$_3$ aqueous solution (62.5 cm^3) to make a precursor solution. After the precursor solution was set in a plastic container, the mist of the precursor solution was generated by ultrasonic irradiation (2.4 MHz) and then it was fed to an electric furnace at 1100°C under air flowing (1.5 dm^3 min^{-1}) by using a feeding system as shown in Fig. 1 and then was directly heat-treated in the furnace. The pr-In$_2$O$_3$ powder obtained was denoted as pr-In$_2$O$_3$(M). Conventional In$_2$O$_3$ powder (c-In$_2$O$_3$) was also prepared by the similar technique employing PMMA-free precursor solution.

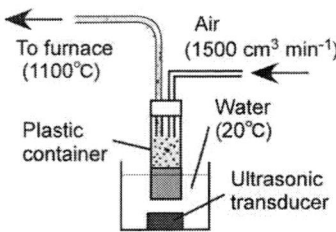

Figure 1. Schematic drawing of a feeding system of mist of precursor solution atomized by ultrasonication (2.4 MHz).

Characterization of PMMA microspheres and In$_2$O$_3$ powders

The particle size distribution of the PMMA microspheres obtained was measured by dynamic light scattering (DLS, Malvern Instrument Ltd., HPPS). The microstructure of PMMA microspheres and In$_2$O$_3$ powders was observed with scanning electron microscope (SEM; JEOL Ltd., JSM-7500F). The specific surface area (SSA) and pore size distribution of pr-In$_2$O$_3$(M) and c-In$_2$O$_3$ powders were measured by the Brunauer-Emmett-Teller (BET) and Barret-Johner-Halenda (BJH) methods, respectively, using a N$_2$ adsorption isotherm (Micromeritics Instrument Corp., Tristar3000).

Gas response measurement

Thick film sensors were fabricated by screen printing employing the paste of each In_2O_3 powder prepared with organic lacquer (Goo Chemical Co., Ltd., OS-4530) on an alumina substrate equipped with a pair of interdigitated Pt electrodes and subsequent heat-treatment at 550°C for 5 h. Gas response properities of these sensors were measured to 10 ppm NO_2 in air at 150~500°C. The magnitude of NO_2 response was defined as the ratio (R_g/R_a) of sensor resistance after 10 min exposure to NO_2 balanced with air (R_g) to that in air (R_a).

Results and discussions

Figures 2 and 3 show particle size distributions and SEM photographs of all PMMA microspheres synthesized by the microwave-assisted emulsion polymerization, respectively. The average particle size of PMMA microspheres was dependent markedly on the kind of the surfactant used. The particle size distribution of PMMA(SLS) was narrow, and the mean particle size of PMMA(SLS) was ca. 55.5 nm. As shown in SEM photographs, the particles of PMMA(SLS) are relatively uniform and the size was slightly smaller than that estimated by the DLS (Fig. 2). On the other hand, the particle sizes of PMMA(Triton) and PMMA(P123) were relatively large (ca. 106 nm: PMMA(Triton) and ca. 159 nm: PMMA(P123)) and their particle size distributions were broader than that of PMMA(SLS). Their SEM photographs showed that particle sizes of PMMA(Triton) and PMMA(P123) were uniform and larger than that of PMMA(SLS).

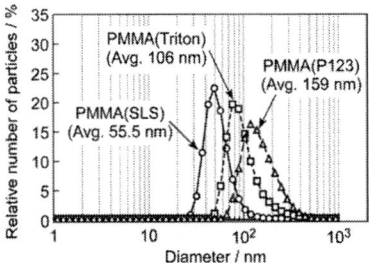

Figure 2.   Particle size distributions of PMMA(M) microspheres synthesized by microwave-assisted emulsion polymerization, together with their average diameter in parentheses.

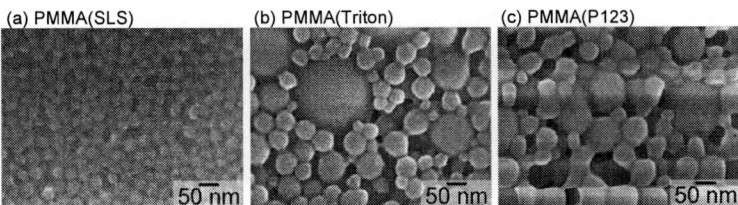

Figure 3.   SEM photographs of PMMA(M) microspheres synthesized by microwave-assisted emulsion polymerization.

Figures 4 and 5 show pore size distributions together with specific surface area (SSA) and SEM photographs of pr-In_2O_3(M) and c-In_2O_3 powders, respectively. The c-In_2O_3 powder showed small SSA (3.3 m^2 g^{-1}) and pore volume, and its morphology was almost spherical (100~600 nm in diameter) and relatively dense. On the other hand, SSA and pore volume of the pr-In_2O_3(M) powders were larger than those of the c-In_2O_3 powder, and the SEM photographs revealed that well-developed pores were formed in the pr-In_2O_3(M) powders. However, the morphology of pores and the pore size distributions were dependent on the kind of PMMA microspheres used. Namely, the pr-In_2O_3(SLS) powder had medium pores with a centered diameter of ca. 30 nm, which was probably originated from the morphology of PMMA(SLS), as shown in Fig. 4, and small pores with a diameter less than ca. 10 nm hardly existed in the pr-In_2O_3(SLS) powder. On the other hand, pr-In_2O_3(Triton) and pr-In_2O_3(P123) powders had well-developed small pores (less than 10 nm in diameter) as well as medium pores (ca. 50 nm in centered diameter) which were confirmed from Fig. 4, along with large pores (more than 100 nm in diameter) which were confirmed from Fig. 5 (c) and (d).

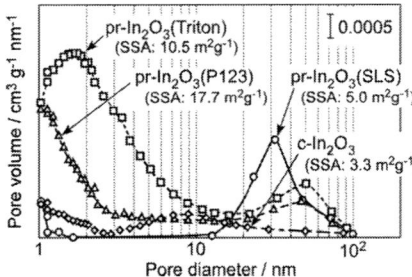

Figure 4. Pore size distribution of pr-In_2O_3(M) and c-In_2O_3 powders.

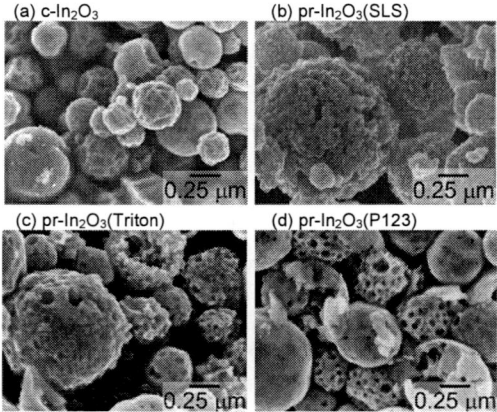

Figure 5. SEM photographs of pr-In_2O_3(M) and c-In_2O_3 powders.

Figure 6 shows response transients of all sensors to 10 ppm NO_2 in air at 400°C. The magnitude of NO_2 responses of all pr-In_2O_3(M) sensors was much larger than that of a c-In_2O_3 sensor, while the sensor resistance of all the pr-In_2O_3(M) sensors in air was much smaller than that of the c-In_2O_3 sensor. Among all the pr-In_2O_3(M) sensors, the pr-In_2O_3(SLS) sensor showed the largest response to 10 ppm NO_2 (R_g/R_a = 36.0) and the fastest response and recovery speeds.

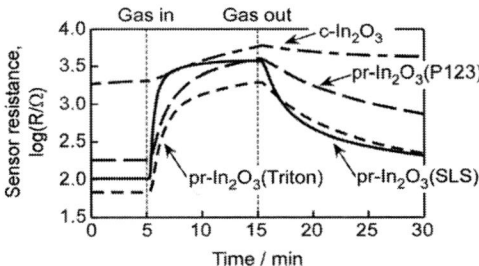

Figure 6.   Response transients of all sensors to 10 ppm NO_2 in air at 400°C.

Figure 7 show operating temperature dependence of the magnitude of response (a) and 90% response time (b) of pr-In_2O_3(M) and c-In_2O_3 sensors to 10 ppm NO_2 in air. The magnitude of NO_2 responses of all the pr-In_2O_3(M) sensors was much larger than that of c-In_2O_3 sensor in the temperature range of 250~500°C. All the pr-In_2O_3(M) sensors showed the largest response at 300°C, and the response at lower temperatures ($\leq$ 200°C) was relatively small. On the other hand, all sensors showed extremely slow response speeds (over 8 min) at lower temperatures ($\leq$ 300°C). However, the response speed of pr-In_2O_3(SLS) sensors was faster than other sensors at higher temperatures ($\geq$ 300°C).

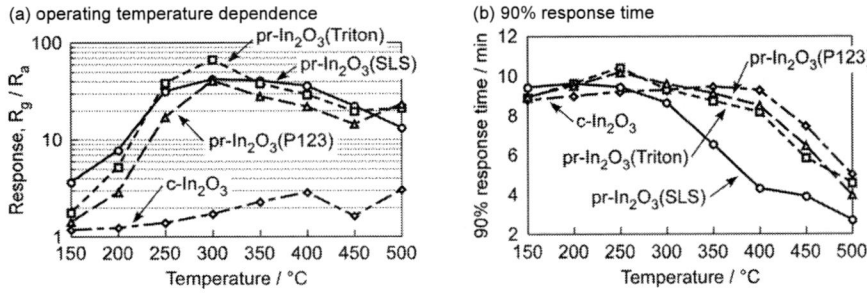

Figure 7.   Operating temperature dependence of the magnitude of response (a) and 90% response time (b) of pr-In_2O_3(M) and c-In_2O_3 sensors to 10 ppm NO_2 in air.

Conclusions

PMMA microspheres with relatively-controlled particle diameter (PMMA(M)) were easily synthesized by microwave-assisted emulsion polymerization employing a typical surfactant (M: SLS, Triton or P123). The PMMA(SLS) was the most uniform among them. pr-In$_2$O$_3$ powders were prepared by ultrasonic-spray pyrolysis employing the synthesized PMMA microspheres. The morphology of pores and the pore size distributions were dependent on the kind of PMMA microspheres used. The magnitude of NO$_2$ responses of all the pr-In$_2$O$_3$(M) sensors was larger than that of c-In$_2$O$_3$ sensor. The introduction of controlled macroporous structure into the powder of the sensor material was found to be effective for improving NO$_2$ response properties.

References

1. M. Hu, J. Zeng, W. Wang, H. Chen, and Y. Qin, *Appl. Surf. Sci.*, **258**, 1062 (2011).
2. I. Hotovy, V. Rehacek, P. Siciliano, S. Capone, and L. Spiess, *Thin Solid Films*, **418**, 9 (2002).
3. T. Zhang, L. Liua, Q. Qi, S. Li, and G. Lu, *Sens. Actuators B*, **139**, 287 (2009).
4. M. Ali, Ch.Y.Wang, C.-C. RoHlig, V. Cimalla, Th. Stauden, and O. Ambacher, *Sens. Actuators B*, **127**, 467 (2008).
5. G. X. Wang, J. S. Park, M. S. Park, and X. L. Gou, *Sens. Actuators B*, **131**, 313 (2008).
6. S. B. Patil, P. P. Patil, and M. A. More, *Sens. Actuators B*, **125**, 126 (2007).
7. T. Hyodo, N. Nishida, Y. Shimizu, and M. Egashira, *Sens. Actuators B*, **83**, 209 (2002).
8. T. Tsumura, T. Hyodo, and Y. Shimizu, *Sensor Letters*, **9**, 646 (2011).
9. T. Hyodo, K. Sasahara, Y. Shimizu, and M. Egashira, *Sens. Actuators B*, **106**, 580 (2005).
10. K. Hieda, T. Hyodo, Y. Shimizu, and M. Egashira, *Sens. Actuators B*, **133**, 144 (2008).
11. M. Hashimoto, H. Inoue, T. Hyodo, Y. Shimizu, and M. Egashira, *Sensor Letters*, **6**, 887 (2008).
12. A. A. Firooz, T. Hyodo, A. R. Mahjoub, A. A. Khodadadi, and Y. Shimizu, *Sens. Actuators B*, **147**, 554 (2010).
13. T. Hyodo, H. Inoue, H. Motomura, K. Matsuo, T. Hashishin, J. Tamaki, Y. Shimizu, and M. Egashira, *Sens. Actuators B*, **150**, 265 (2010).

Redox-Active Alkali Insertion Materials as
Inner Contact Layer in All-Solid-State Ion-Selective Electrodes

Shinichi Komaba,* Chihiro Suzuki, Naoaki Yabuuchi,
Tatsuya Akatsuka, Shintaro Kanazawa, and Taku Hasegawa

Department of Applied Chemistry, Tokyo University of Science, Tokyo 162-8601, Japan
*correspondence to komaba@rs.kagu.tus.ac.jp

Lithium insertion materials, such as $LiFePO_4$, have been widely studied as the electrode materials for rechargeable batteries. Alkali ions can be reversibly inserted/extracted into/from their crystal lattices associated with electrochemical redox. In this study, we apply lithium, sodium, and potassium insertion materials as inner contacting layers for the all-solid-state ion-selective electrodes which are covered with the outer ion-selective poly(vinyl chloride) membrane containing ionophore and plasticizer. The introduction of alkali insertion materials layer between Pt substrate and ion selective membrane is highly efficient for the stabilization of membrane potential and its prompt response when alkaline ion activity in analyte changes compared to those of the electrode without alkali insertion materials. Electrochemical measurements reveal that alkali insertion materials as the inner contact layers stabilize the membrane potential because of drastic reduction of interface resistance at the electrode, and thus improve ion-sensing performance.

Introduction

Ion-selective electrodes (ISEs), e.g. pH-selective glass electrode, are widely used in clinical and environmental analysis (1). In general, the ion-selective electrodes consist of ion-selective membrane, internal solution, and an internal reference electrode. By using ISEs, we can determine concentrations of targeted ions by measuring membrane potential, that depends on difference in the activity of targeted ions between internal solution and analyte (2). Although the measurement of the membrane potential requires two reference electrodes, the internal reference electrode and solution equipped with ISE hinder miniaturization of entire ISEs, and thus disadvantageous for many important applications (3). To miniaturize the ISEs, a coated wire electrode; so-called CWE, which is a metallic wire electrode directly covered with the ion-selective membrane, has been developed. CWEs generally show, however, insufficient potential stability and reproducibility, which possibly originate from the absence of charge transfer process at the membrane/electrode interface and the omission of an inner reference electrode (4).

To simultaneously satisfy potential stability, reproducibility, and miniaturization of ISEs, electroactive π-conjugated polymers have been utilized for all-solid-state ion-selective electrodes. According to our knowledge, Cadogan et al. first reported the application of polypyrrole (PPy) film doped with tetrafluoroborate anion as an inner contact layer of Na^+-ISE (5). Furthermore, Ivaska's research group utilized polyaniline (6) and poly(3,4-ethylenedioxythiophene) (PEDOT) (7) as internal mediating layers for all-solid-state ISEs, which effectively reduce the impedance of ISEs and stabilize the electrode potential. One of us successfully adapted PPy-poly(4-styrenesulfonate) composite film, which shows potassium ion exchange ability, as solid reference electrode for K^+-ISE (8). It was found that the K^+ exchangeable and conducting PPy composite film successfully connects both ionic and electronic transfer between platinum substrate and ion-selective membrane to improve the potentiometric and long-term stability, resulting from stabilization of membrane potential (9, 10).

On the other hand, our research group studies on lithium and sodium insertion inorganic materials Li-ion and Na-ion batteries (11, 12). In 1980s – 1990s, inorganic materials, such as transition metal oxides, as well as π-conjugated polymers were researched for the application to active materials in secondary lithium batteries. Because of higher insertion capacity (mAh cm^{-3}) and superior chemical stability of the inorganic compounds compared to the polymers, transition metal oxide and phosphate have been intensively studied since the commercialization of graphite//LiCoO$_2$ cells known as Li-ion batteries. Among lithium insertion materials, olivine-type Li$_x$FePO$_4$ (13) and spinel-type Li$_x$Mn$_2$O$_4$ (14, 15), which are widely studied as cost friendly positive electrode materials for Li-ion batteries, can reversibly exclude/uptake lithium ions from/into their crystal lattices accompanied by oxidation change of the transition elements during electrochemical redox. Since the electrode potential of the alkali insertion materials are governed by Nernst's law on the basis of redox reaction, the response of redox potential can be applied to determination of alkali ion species as the ion-sensitive electrodes (16-19). However, the electrode potential of the insertion materials is influenced by not only alkali ion but also any redox species dissolved in the analyte, resulting in interference with the detection.

In this study, we investigate fundamental redox properties and potential response of three insertion materials, LiFePO$_4$, Na$_{0.33}$MnO$_2$, and birnessite-type hydrous manganese (di)oxide (K$_x$MnO$_2$), to Li$^+$, Na$^+$, and K$^+$ ions, respectively, in aqueous analytes. Based upon the understanding of redox property of these insertion electrodes, we combine the insertion materials and ionophore-containing PVC membrane to assemble liquid-free (all-solid-state) ion-selective electrodes, in which the Li, Na, and K insertion materials are employed as inner solid contact layers covered with the ion-selective membrane containing ionophores for the alkali ions as schematically drawn in Figure 1.

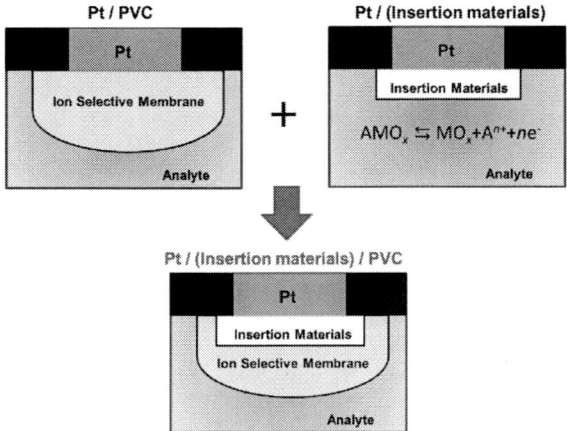

Figure 1. Schematic illustrations of liquid-free (all-solid-state) ion-selective electrodes with alkali insertion materials as inner contact layers.

Experimental

Triphylite-type $LiFePO_4$, $Na_{0.33}MnO_2$, and birnessite-type K_xMnO_2 powders are synthesized and used as each alkali-ion insertion host. $LiFePO_4$ was prepared by hydrothermal process, and $Na_{0.33}MnO_2$ and K_xMnO_2 were prepared by a solid-state reaction of precursors, and used as inner contacting layers. These alkali-ion insertion materials were cast on a Pt disk electrode using slurry consisting of the insertion material, mixed with 10 wt% acetylene black (AB) as conductive additive and 10 wt% poly(vinylidene difluoride) (PVDF) as a binder dispersed in N-methylpyrrolidinone solvent. After casting the slurry, the electrode was dried for 24 h and then immersed for at least 1 day in aqueous solution containing 0.01 mol dm^{-3} of the each alkali-ion chloride salt for conditioning the electrode. After the conditioning, the electrode was dried in a desiccator for at least 1 day. A viscous tetrahydrofuran (THF) solution consisting of ion-selective membrane components, ionophore, o-nitrophenyloctylether (NPOE) as a plasticizer, poly(vinyl chloride) (PVC) as a membrane matrix, and potassium tetrakis(4-chlorophenyl)borate (K-TCPB) as a lipophilic anion, was further cast on the top of thus formed alkali-ion insertion materials film and dried in a desiccator for at least 12 h to form a plasticized PVC membrane. The membrane compositions studied are shown in Table 1. Thus prepared electrodes were conditioned before measurements in aqueous electrolyte solution containing 0.01 mol dm^{-3} of each alkali ion for at least 1 day. Electrochemical measurements were performed by using a three-or two-electrode cell at room temperature. The ion-selective electrode was utilized as a working electrode with a Ag/AgCl electrode as a reference electrode and Pt wire as a counter electrode. The ion selective electrodes were evaluated by potential measurement, cyclic voltammetry, chronopotentiometry, and AC impedance measurement.

Table 1. Compositions of the plasticized PVC membranes for ISEs, given as % (w/w).

	ionophore	NPOE	PVC	K-TCPB
Li^+ selective membrane	1.5	69.3	28.1	0.97
Na^+ selective membrane	0.9	67.1	31.7	0.34
K^+ selective membrane	0.9	65.5	33.3	0.3

Results and Discussion

We selected three alkali insertion materials, $LiFePO_4$ (20), $Na_{0.33}MnO_2$ (21), and K_xMnO_2 (22), as Li^+, Na^+ and K^+ insertion hosts, respectively. Figure 2 shows powder X-ray diffraction patterns and crystal structures of the insertion materials. In the diffraction patterns, almost all diffraction lines are identified as Bragg lines due to the crystalline phases of $LiFePO_4$, $Na_{0.33}MnO_2$, and K_xMnO_2, indicating that the products predominantly consist of the crystalline phase of each desired insertion framework.

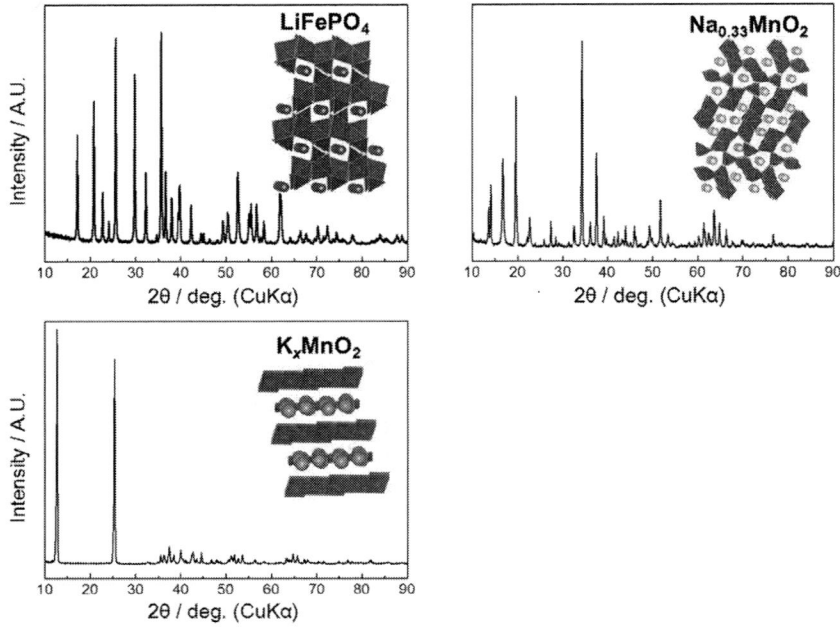

Figure 2. Powder XRD patterns and schematic illustrations of the crystal structure of alkali insertion materials used in this study.

Figure 3 shows the electrochemical behavior of insertion materials electrodes on Pt substrate. Each alkaline ion can be inserted/extracted into/from insertion materials in aqueous electrolyte solution with reversible manner. These cyclic voltammograms indicate that alkaline cations can be inserted/extracted reversibly associated with the electrochemical redox of transition metals. For $LiFePO_4$, one significant redox couple is observed because of electrochemical lithiation and delithiation with the two-phase reactions in Li_2SO_4 aqueous solution (23). Several phase evolutions appear for $Na_{0.33}MnO_2$ because of several redox pairs in the voltammograms which agree well with the previous reports (18). In the case of hydrated oxide of K_xMnO_2, potassium ions and protons are probably inserted and extracted (24).

As the alkali-ion insertion materials are redox active, their redox potentials should change as a function of concentration (precisely, activity) of alkaline ion in electrolyte, and they are governed by Nernst's law. Thus equilibrium potential linearly depends on the logarithm of ion concentration of analyte according to Nernstian equation. Therefore, we examined potential response of the alkali ion insertion materials to the alkali ion-selective electrodes for ion sensing. Figure 4 compares the potential response of the Pt / (insertion materials) electrodes to alkali-ion activity. All electrodes show Nernstian response for each alkaline ion; Li^+, Na^+, and K^+. The slopes obtained are ranged from 40-55 mV decade^{-1}, and they are close to the theoretical value of 59 mV decade^{-1} at 298 K.

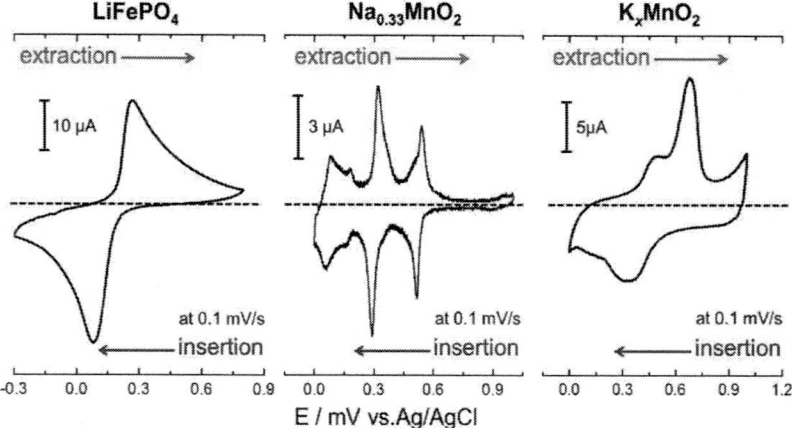

Figure 3. Cyclic voltammograms of Pt / (insertion materials) electrodes in 0.5 mol dm^{-3} A_2SO_4 (A = Li, Na, and K) aqueous solution.

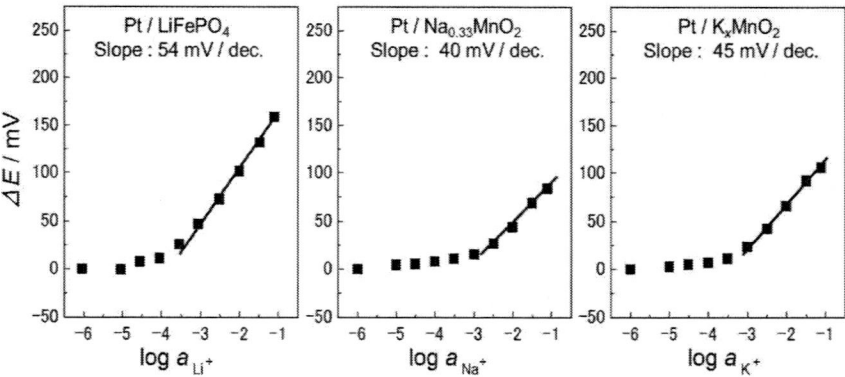

Figure 4. Logarithm of alkali ion activity vs. electrode potential plots of for the Pt / (insertion materials) electrodes.

Furthermore, when the layers of alkali-ion insertion materials are covered with ion selective PVC membrane to fabricate liquid-free ISEs, an ideal Nernstian slope as a function of the activity of each ion is found as shown in Figure 5. In addition, the detection limit in lower ion concentration is significantly improved by covering the insertion materials with ion selective membrane. By intercalating the insertion materials between Pt substrate and PVC membrane, the charge transfer connection with electrochemical redox reaction are supposed to be contribute stabilization of the ionic response of the all-solid-state ISEs.

We investigated long-term potential stability of the Li$^+$-ISEs of Pt / PVC (i.e., CWE) and the Pt / Li$_x$FePO$_4$ / PVC (x = 0, 0.5, and 1 in Li$_x$FePO$_4$ electrodes, in which lithium content was fixed by adjusting charge quantity of electrochemical oxidation for LiFePO$_4$ electrode. The stability was evaluated in 0.01 mol dm^{-3} LiCl aqueous solution and is compared in Figure 6. For the Pt / PVC electrode, large potential drift of ca. $\pm$ 200 mV with overshoot or undershoot was irregularly observed during several days, whereas the potential stability of the Pt / Li_xFePO_4 / PVC electrodes is highly improved with potential drift within $\pm$ 20 mV. In Figure 6, we compared three different oxidation conditions for Li_xFePO_4 (x = 1.0, 0.5, and ~0). For Li_xFePO_4, a two-phase equilibrium reaction takes place during lithium extraction from $LiFePO_4$, and thus a voltage plateau at 3.4 V vs. Li is usually observed on galvanostatic charge/discharge processes (13). This character is suitable as inner contacting layers because inner potential is highly stabilized by the constant redox potential of $Fe^{III}PO_4$ / $LiFe^{II}PO_4$ two-phase couple, and is not easily disturbed for x = 0.5 in Li_xFePO_4 by possible oxidation/reduction, resulting in the superior long term stability (within $\pm$ 5 mV in Fig. 6) to fully lithiated or delithiated iron phosphates. Indeed, the potential stability for x = 0.5 is better than fully reduced (x = 1.0) or oxidized (x = ~ 0) sample as the inner contacting layers. It was confirmed that similar results for the Pt / Na_xMnO_2 / PVC and the Pt / K_xMnO_2 / PVC electrodes were also obtained.

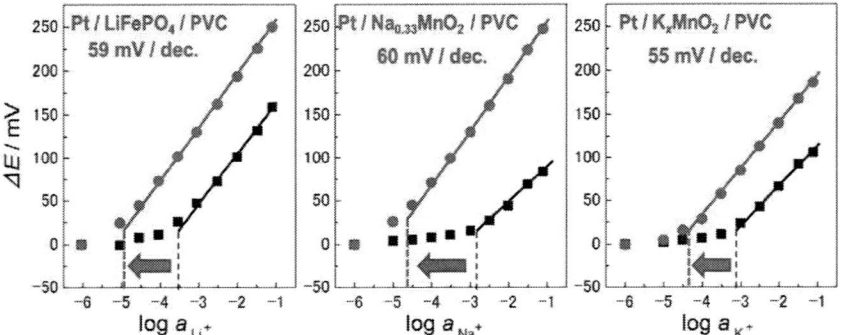

Figure 5. Potentiometric response of the Pt / (insertion materials) / PVC electrodes as a function of logarithm of alkaline ion activity compared with Pt / (insertion materials) electrodes.

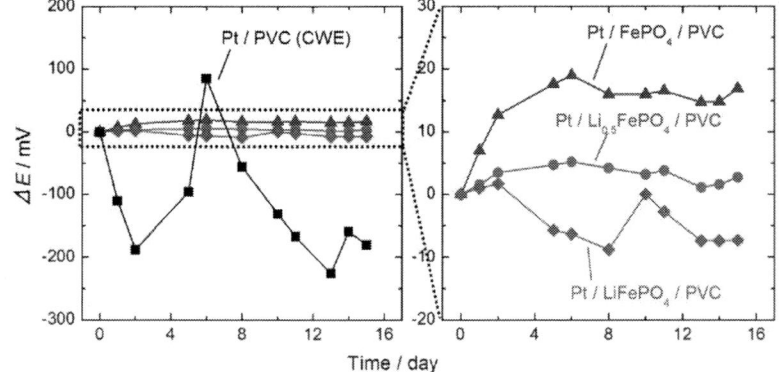

Figure 6. Long-term potential stability of the Li^+-selective electrodes; Pt / PVC and Pt / Li_xFePO_4 / PVC (x = 1.0, 0.5, and ~ 0) in 0.01 mol dm^{-3} LiCl aqueous solution.

In addition, the interface resistance of electrodes is effectively reduced by using the insertion materials as inner layers. Chronopotentiometry was conducted on the Pt / PVC and Pt / LiFePO$_4$ / PVC electrodes for anodic and then cathodic polarization at ±1 nA. In Figure 7, chronopotentiograms show clear difference in the potential response between two electrodes. The potential of the Pt / PVC electrode almost linearly changes as a function of time. In contrast, the Pt / LiFePO$_4$ / PVC electrode shows excellent potential stability, which changes less than ± 0.3 mV in the same experimental condition. Because the potential change is suppressed from 200 to 0.3 mV, the electrode resistance becomes smaller in about three decades by the inner LiFePO$_4$ layer. These results suggest that the lithium insertion materials as the inner contacting layer effectively and drastically reduce the impedance of the all-solid-state electrode. We observed similar reduction of electrode resistance for Na$^+$ and K$^+$-ISEs by employing the corresponding insertion materials. From AC impedance measurement, reduction of electrode impedance was also observed in the whole frequency range by installation of alkali-ion insertion materials as the inner contact layer. The lowered impedance also contributes to the higher potential stability as ion-selective electrode.

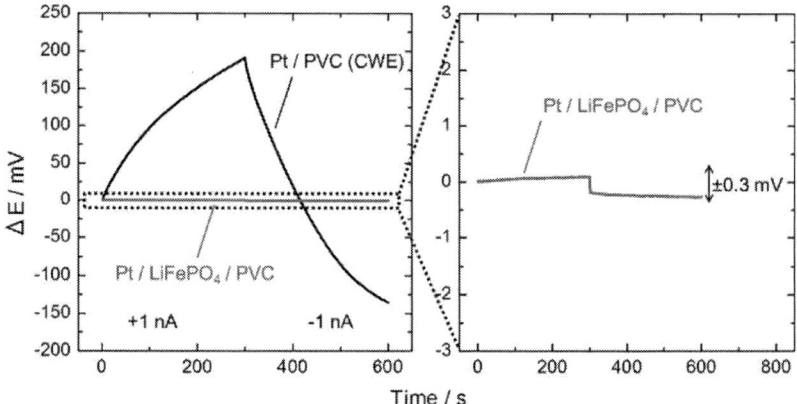

Figure 7. Chronopotentiograms of lithium ion selective electrodes; Pt / PVC (dash line) and Pt / LiFePO$_4$ / PVC (solid line) at reversal constant current of +1.0 and – 1.0 nA for 0 – 300 and 300 – 600 s, respectively, in 0.5 mol dm^{-3} Li$_2$SO$_4$ aqueous solution.

Schematic illustrations of a proposed charge transfer process for the stabilization of the electrode potential using the alkali insertion materials are shown in Figure 8. For the Pt /PVC electrode, the interface at the PVC membrane and the analyte solution is ionically connected due to cation exchange to form ionophore-alkali coordination compound. However, any reversible charge transfer crossing the Pt / (PVC membrane) interface is prohibited, resulting in higher resistance and unstable potential of the electrode. One can notice that alkali insertion materials are an ion/electron mixed conductor. Therefore, the installation of the insertion materials layer makes ionic and electronic communication possible at the interface between the ion-selective membrane and Pt substrate for all-solid-state ion-sensing electrodes. As a result, the electrode impedance is significantly reduced and electrode potential is effectively stabilized with good selectivity to sensing ions.

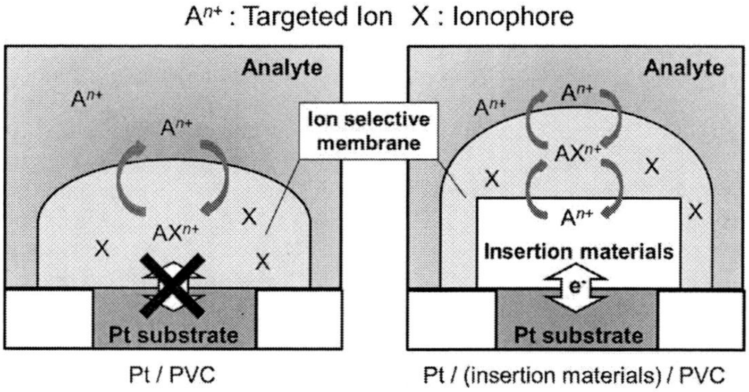

A^{n+} : Targeted Ion   X : Ionophore

Figure 8. Schematic illustrations of interfacial charge transfer in the Pt / PVC (coated wire electrodes; CWE) and the Pt / (insertion materials) / PVC electrodes as the all-solid-state ion-sensing electrode.

Acknowledgments

This study was in part granted by the Japan Society for the Promotion of Science (JSPS) through the "Funding for NEXT Program", initiated by the Council for Science and Technology Policy (CSTP).

References

1. E. Bakker, *Anal Chem*, **76**, 3285 (2004).
2. K. Tohda, S. Yoshiyagawa, M. Kataoka, K. Odashima and Y. Umezawa, *Anal Chem*, **69**, 3360 (1997).
3. G. A. Khripoun, E. A. Volkova, A. V. Liseenkov and K. N. Mikhelson, *Electroanal*, **18**, 1322 (2006).
4. P. Sjoberg-Eerola, J. Bobacka, T. Sokalski, J. Mieczkowski, A. Ivaska and A. Lewenstam, *Electroanal*, **16**, 379 (2004).
5. A. Cadogan, Z. Q. Gao, A. Lewenstam, A. Ivaska and D. Diamond, *Anal Chem*, **64**, 2496 (1992).
6. T. Lindfors, H. Aarnio and A. Ivaska, *Anal Chem*, **79**, 8571 (2007).
7. M. Vazquez, P. Danielsson, J. Bobacka, A. Lewenstam and A. Ivaska, *Sensor Actuat B-Chem*, **97**, 182 (2004).
8. T. Momma, S. Komaba, M. Yamamoto, T. Osaka and S. Yamauchi, *Sensor Actuat B-Chem*, **25**, 724 (1995).
9. T. Momma, M. Yamamoto, S. Komaba and T. Osaka, *J Electroanal Chem*, **407**, 91 (1996).
10. S. Komaba, J. Arakawa, M. Seyama, T. Osaka, I. Satoh and S. Nakamura, *Talanta*, **46**, 1293 (1998).
11. N. Yabuuchi, M. Kajiyama, J. Iwatate, H. Nishikawa, S. Hitomi, R. Okuyama, R. Usui, Y. Yamada and S. Komaba, *Nature Materials*, **11**, 512 (2012).
12. Y. Kawabe, N. Yabuuchi, M. Kajiyama, N. Fukuhara, T. Inamasu, R. Okuyama, I. Nakai and S. Komaba, *Electrochemistry*, **80**, 80 (2012).

13. S. T. Myung, S. Komaba, N. Hirosaki, H. Yashiro and N. Kumagai, *Electrochim Acta*, **49**, 4213 (2004).
14. S. T. Myung, H. T. Chung, S. Komaba, N. Kumagai and H. B. Gu, *J Power Sources*, **90**, 103 (2000).
15. S. T. Myung, S. Komaba and N. Kumagai, *J Electrochem Soc*, **148**, A482 (2001).
16. Y. Tani and Y. Umezawa, *Mikrochimica Acta*, **129**, 81 (1998).
17. N. Le Poul, E. Baudrin, M. Morcrette, S. Gwizdala, C. Masquelier and J. M. Tarascon, *Solid State Ionics*, **159**, 149 (2003).
18. F. Sauvage, E. Baudrin and J. M. Tarascon, *Sensor Actuat B-Chem*, **120**, 638 (2007).
19. F. Sauvage, J. M. Tarascon and E. Baudrin, *Microchimica Acta*, **164**, 363 (2009).
20. T. Nakamura, Y. Miwa, M. Tabuchi and Y. Yamada, *J Electrochem Soc*, **153**, A1108 (2006).
21. F. Sauvage, L. Laffont, J. M. Tarascon and E. Baudrin, *Inorg Chem*, **46**, 3289 (2007).
22. A. C. Gaillot, V. A. Drits, A. Plancon and B. Lanson, *Chemistry of Materials*, **16**, 1890 (2004).
23. J. Y. Luo, W. J. Cui, P. He and Y. Y. Xia, *Nature Chemistry*, **2**, 760 (2010).
24. A. Ogata, S. Komaba, R. Baddour-Hadjean, J. P. Pereira-Ramos and N. Kumagai, *Electrochim Acta*, **53**, 3084 (2008).

**Sensing Characteristics of a Fiber Bragg Grating Hydrogen Gas Sensor
Using Sol-gel Derived Pt/WO₃ Film**

S. Okazaki[a], Y. Maru[b], and T. Mizutani[c]

[a] Yokohama National University, Faculty of Engineering, 79-5 Tokiwadai, Yokohama,
240-8501, Japan
[b] Japan Aero Space Exploration Agency, Institute of Space and Astronautical Science,
3-1-1 Yoshinodai, Sagamihara, 229-8510, Japan
[c] Japan Aero Space Exploration Agency, Tsukuba Space Center, 2-1-1 Sengen, Tsukuba,
305-8505, Japan

A fiber-optic hydrogen gas sensor based on a shift of Bragg
wavelength induced by reaction heat and strain was developed.
Platinum-supported tungsten trioxide (Pt/WO_3) film which was
utilized as hydrogen sensitive material was derived by sol-gel
method. In this study, two types of the sensor structures were
fabricated and evaluated. The sensor device where fiber Bragg
grating (FBG) was fixed on the quartz glass substrate coated with
the Pt/WO_3 film using adhesive tape showed good sensitivity. It
was found that both reduction of WO_3 and oxidation of tungsten
bronze were exothermic reaction. The heat generated by the
oxidation process was considerably larger than that of reduction.
The response behavior of the other type FBG sensing device
directly coated with the Pt/WO_3 film was complicated. It was
suggested that there seemed to be competitive response
mechanisms related to generation of reaction heat and strain in
reduction and oxidation process of sensing film.

Introduction

Because of worldwide environmental concern, hydrogen has been expected as a key
material for renewable and clean energy systems. Global warming gases are not produced
during combustion or fuel cell reaction of hydrogen. However, there are serious
explosion risks associated with hydrogen usage since the minimum ignition energy is
very low and the combustion range is very wide. The development of hydrogen leakage
detector is quite important for achievement of safe operation of hydrogen energy system.
Furthermore, more cost effective and reliable sensor devices which can be applied to
multipoint (quasi-distributed) or distributed leakage monitoring are required in future for
securing the safety of huge infrastructure such as electrolysis, storage, and energy
production plants. Various types of sensor devices have been designed and practically
used. Among them optical methods have the advantages of explosion-proof and
immunity to electromagnetic noises. A fiber-optic hydrogen sensor using FBG is
promising for real-time and multipoint monitoring. FBG has been developed as a kind of
optical passive filter whose refractive index of fiber core is periodically modified. The
changes in this period or effective index of FBG result in shift of Bragg wavelength
(center wavelength of back-reflected light peak). This mechanism is extensively applied

to quasi-distributed strain or temperature sensing. It is deduced that FBG hydrogen sensor could be realized by combination of suitable hydrogen-sensitive material and device structure. In early stage, Pd-coated FBG sensor based on mechanical stress that is induced by absorption of hydrogen was reported (1). Recently, side-polished FBG sensors coated with Pd (2) or Pd-WO$_3$ (3) film were proposed. Furthermore, it was found that highly sensitive H$_2$ monitoring was achieved utilizing temperature change induced by exothermic oxidation reaction of hydrogen on Pt/WO$_3$ catalyst (4, 5). However, detailed sensing behaviors are not fully clarified so far. In this study, sensing characteristics of FBG hydrogen gas sensor in various environments were evaluated.

Experiments

FBG device

A FBG inscribed into 125 μm standard single-mode optical fiber was purchased from SHINKO ELECTRIC WIRE Co., LTD. The Bragg wavelength was 1539.90 nm (halfwidth: 0.22 nm) and the reflectivity was above 90 %. The grating length was about 10 mm and this FBG was placed on the center of portion with 100 mm long where protective polymer coating was stripped.

Sensor Structures

In this study, two types of the sensor devices were fabricated, as shown in Fig. 1. One was indirect type FBG sensor (*Type A*, Fig. 1(a)) where the hydrogen sensitive material (Pt/WO$_3$) was immobilized on quartz glass substrate (5 x 20mm, thickness: 1 mm) and then FBG device was fixed on the reverse side of Pt/WO$_3$ film with fluoroplastic adhesive tape. The other was direct type FBG sensor (*Type B*, Fig. 1(b)). In this type of sensor device, the Pt/WO$_3$ was directly immobilized on the periphery of the FBG device by dip-coating.

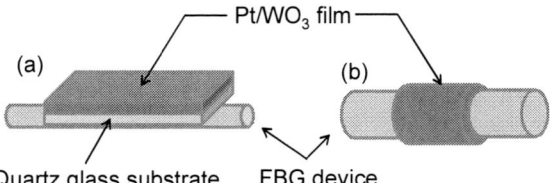

Figure 1. Schematic diagram of the sensor devices (*Type A* (a) and *Type B* (b)).

Hydrogen Sensitive Film

The Pt/WO$_3$ film was prepared by the following sol-gel method. The sodium tungstate aqueous solution (0.5 M) was passed through cation exchange resin (AmberliteTM IR120B) in hydrogen form. A 6.5 ml of the obtained colloidal tungstic acid solution was mixed with 10 ml of the ethanol/water solution 80 % (v/v) containing 0.09 M hexachloroplatinic acid as catalyst precursor and stirred for 60 sec. The FBG device or quartz glass substrate was dipped into this solution and manually pulled at nearly-constant speed (ca. 1 cm/s). The obtained sensor device was dried at room temperature and then annealed at 500 °C for 1 h in air condition.

Characteristics of the FBG device for sensor application

In the case of Type B sensor devices, it is necessary to evaluate the high-temperature endurance of the FBG device in annealing process mentioned above. Fig. 2 shows the stability of Bragg wavelength at 500 °C for 3 h. It was found that severe degradation of the FBG device was not observed. Therefore, the direct immobilization of the Pt/WO_3 film on the periphery of FBG portion would be possible.

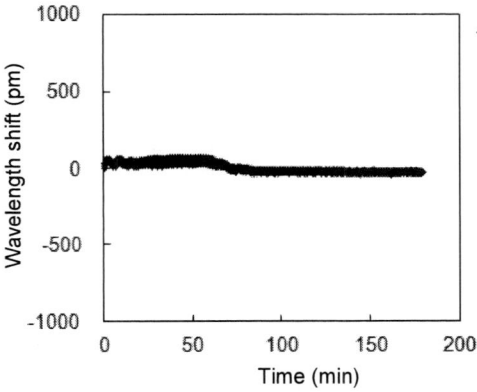

Figure 2. High-temperature endurance of FBG device at 500 °C

Fig. 3 represents temperature dependence of Bragg wavelength in the range from 20 to 500 °C. The relationship between Bragg wavelength shift and temperature was almost linear in this experimental condition. The temperature coefficient was about 12 pm/°C. Using this relation, temperature change of the sensor device by chemical reaction heat could be estimated.

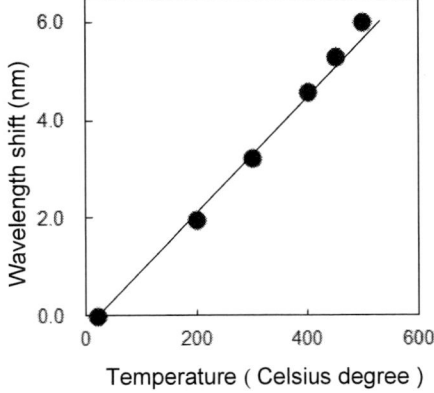

Figure 3. The relationship between Bragg wavelength shift and temperature.

Experimental Procedures for Evaluation of the Sensing Performances

The sensor device was placed in an airtight gas chamber (30 mL). Test gases (pure H_2, H_2/N_2 balance gas and air) were fed to the chamber at constant flow rate (200 L/min). In addition, the response tests in which the sensor devices were exposed to hydrogen gas blowing down from the exit of gas pipe in open-air atmosphere were also conducted. The sensor responses were recorded using FBG Swept Laser Interrogator (MICRON OPTICS, Inc.). All experiments were conducted at room temperature and atmospheric pressure.

Results and discussion

The sensing behavior of Type A sensor is illustrated in Fig. 4. The sharp increase of Bragg wavelength was observed immediately after the exposure to hydrogen gas. This increase would result from temperature elevation around the FBG device. It also indicated that the reduction of WO_3 by hydrogen is exothermic reaction. However, this small response gradually disappeared and the signal returned to its baseline level in spite of the existence of hydrogen. In this region, it is surmised that the WO_3 is completely reduced to tungsten bronze and the generation of reaction heat ceased. The large and transient response appeared after the hydrogen gas was replaced with air. Although this behavior is similar to response to hydrogen, the results indicates that the reaction heat generated by oxidation of tungsten bronze is considerably larger than that of hydrogen reduction.

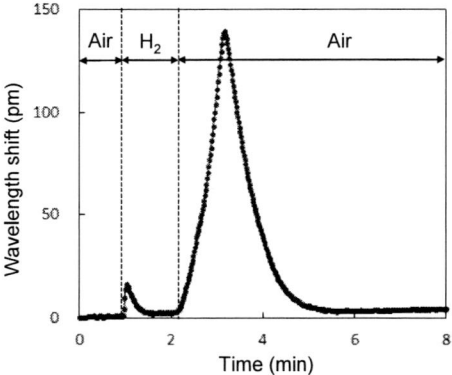

Figure 4. The response characteristics of *Type A* sensor device.

In order to obtain more detailed information about the sensing behavior, sensor response with the exposure to hydrogen gas flowing from tube end in air atmosphere was measured. Fig. 5 shows that the response characteristic in this case was quite different from those obtained in airtight gas chamber. The value of wavelength shift increased with the blowing duration and reached to steady state within 1 min. It is suggested that heat generation in this condition is continuous because of the coexistence of hydrogen and oxygen. Furthermore, the signal normally recovered without transient response behavior after stop blowing of hydrogen gas. The recovery process had almost the same speed as response one. Although the data is not shown, it was found that the sensitivity (wavelength sift) strongly depended on the area of the Pt/WO_3 film. It is under investigation and will be reported in future.

The typical response of Type B sensor to hydrogen (50) – air (50) mixture gas is represented in Fig. 6. The detectable change in Bragg wavelength was observed. However, the response behavior was complicated and there seemed to be multiple response mechanisms. It is likely that the response related to the change in strain accompanied by hydrogen insertion into WO_3 lattice would overlap the response curve. On the other hand, this strain change could not conduct from the Pt/WO_3 film to FBG portion in the case of Type A sensor structure because the strain would be released in the portion of polymer tape. Considering above two factors (temperature change and strain) that influence response characteristics, the initial and rapid rising would result from strain. The following rising was slow and it would be attributed to conduction rate of reaction heat. For recovery process, the spike-like response was reproducibly observed shortly after the replacement of hydrogen gas with air. It is surmised that the anomalous behavior would be caused by the competitive response associated with the disappearance of strain and exothermic reaction heat in oxidation process of tungsten bronze. After this transient process, the sensor device gradually cooled and the output signal recovered.

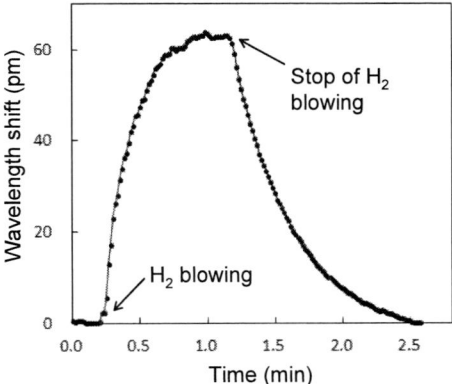

Figure 5. The response of *Type A* sensor device when the device was exposed to blowing hydrogen gas in air atmosphere.

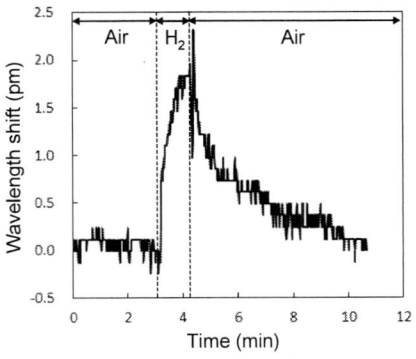

Figure 6. The response of *Type B* sensor device.

Conclusion

A fiber-optic Bragg grating hydrogen sensor device was fabricated and characterized. The Pt/WO_3 film prepared by sol-gel process was utilized as sensing film. In the presence of hydrogen gas, Bragg wavelength shift based on the change in temperature at the grating portion was observed. This sensor has the potential for low-cost and robust quasi-distributed hydrogen leak monitoring although there is need for optimization and improvement of sensor structures.

Acknowledgments

This research was supported by KAKENHI (24560516).

References

1. B. Sutapun, M. Tabib-Azar and A. Kazemi, *Sensors and Actuators*, **B60**, 27 (1999).
2. C-L. Tien, H-W. Chen, W-F. Liu, S-S. Jyu, S-W. Lin and Y-S. Lin, *Thin Solid Films*, **516**, 5360 (2008).
3. J. Dai, M. Yang, Y. Chen, K. Cao, H. Liao, and P. Zhang, *Optics Express*, **19**, 6141(2011).
4. C. Caucheteur, M. Debliquy, D. Lahem and P. Megret, *IEEE Photonics Technology Letters*, **20**, 96 (2008).
5. C. Caucheteur, M. Debliquy, D. Lahem and P. Megret, *Optics Express*, **16**, 16854 (2008).

Zirconia-based Electrochemical Oxygen Sensor for Accurately Determining Water Vapor Concentration

Richard E. Soltis

Chemical Engineering Department, Ford Motor Company, Research & Innovation Center, Dearborn, Michigan 48124, USA

By varying the control voltages of commercially available zirconia-based automotive exhaust gas oxygen sensors, other oxygen-containing molecules such as H_2O and CO_2 can be measured. Typical water concentration sensors that measure relative humidity via swelling of certain polymers have a narrow temperature range of operation (typically < 120°C) and limited accuracy of about ± 0.4% absolute. Initial measurements with modified zirconia oxygen sensors, both in the laboratory and on engine dynamometers, have shown that water vapor concentration in a gas stream can be measured accurately to ± 0.1% over the range from 0-20 vol% H_2O at temperatures up to 800°C.

Introduction

Zirconia-based electrochemical sensors for measuring the oxygen concentration in the automotive exhaust have been used on vehicles for over 30 years. The oxygen sensor is an electrochemical cell based on oxygen-ion conducting zirconia. The sensor consists of a single electrochemical cell which has one electrode exposed to a reference atmosphere, (typically air), and the other electrode exposed to the measurement (exhaust) gas and operates as a simple Nernst cell. The open circuit voltage of this zirconia electrochemical cell changes with variations in the oxygen partial pressures existing adjacent to its two electrodes. The emf generated by this Nernst cell gives a measurement of the concentration of oxygen in the exhaust gas. This measurement can be used by the engine control unit to adjust the air-to-fuel ratio so as to maintain it close to stoichiometry, thereby minimizing the unwanted emissions. All gasoline (and some diesel) vehicles produced today utilize an oxygen sensor to control their emission systems.

Although successfully used for stoichiometric A/F control for many years, Nernst-type oxygen sensors are not useful for applications away from stoichiometry because of their low sensitivity (i.e. weak logarithmic dependence of the emf on oxygen partial pressure). However, double-zirconia-cell sensors have been developed which have much higher sensitivity and thus a wider range of operation and are therefore applicable to A/F measurement and control from very rich to very lean A/F mixtures (1,2). These sensors are based on oxygen pumping, which involves the transfer of oxygen from one side of a zirconia cell to the other by passing an electric current through the device. Since these sensors operate over a wide range, they are referred to as "wide-range oxygen" or "universal exhaust gas oxygen" (UEGO) sensors. Currently, most gasoline vehicles, as well as diesel powertrains which operate lean, employ UEGO sensors as part of the engine A/F control and aftertreatment systems to meet the ever-stricter emission requirements mandated by law.

Figure 1 shows a schematic of the operating principle of a typical UEGO sensor (heater not shown). The UEGO sensor consists of essentially two zirconia cells (Ip cell and Vs or Nernst cell) with porous platinum electrodes on the opposite sides of each cell. The cells are positioned to form an enclosed cavity. The measurement gas(es) enter the gas detection cavity through a porous diffusion layer. A potential is applied across the pump cell (Ip cell) to pump oxygen into or out of the cavity. The second cell, which acts as a Nernst cell, detects the residual oxygen in within the cavity. The Nernst potential generated by this cell is compared to an external reference potential, which is typically 450 mV for a standard oxygen sensor. If the Nernst potential is less than the reference voltage, oxygen is pumped out of the cavity. If the Nernst potential is greater than the reference voltage, oxygen is pumped into the cavity to maintain a constant level of oxygen at few ppm. The pumping current (Ip) is a measure of how much oxygen is pumped out of or into the cavity. The diffusion barrier is designed so that it limits the gas flow into the cavity. Under steady state conditions, the pumping current is directly related to the concentration of oxygen in the measurement gas.

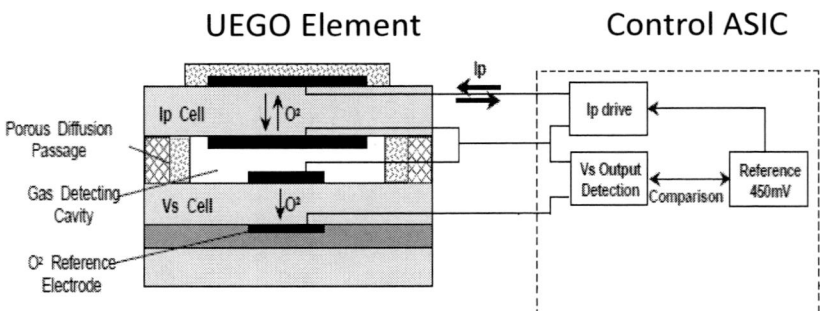

Figure 1. Schematic of dual zirconia cell UEGO sensor.

A double-cell oxygen-pumping device, such as a UEGO sensor, can also be used to measure the concentration of other oxygen containing molecules in the exhaust gas, e.g. H_2O, CO_2, SO_2, and NO_x (3). In order to measure the H_2O concentration for example, a voltage, which is greater than the dissociation voltage for H_2O, is applied to the pump cell. Oxygen generated from the decomposition of water ($H_2O \rightarrow H_2 + \frac{1}{2}O_2$) can be measured via a pumping current through the Ip cell, thereby providing a measure of the H_2O concentration in the gas. In the exhaust gas, this measurement is complicated by the fact that oxygen (and other molecules) is also dissociated. If the concentrations of these other molecules were constant, then the concentration of H_2O could still be determined unambiguously by subtracting the contribution of these other molecules. However, if the concentrations of the different oxygen-containing molecules vary, (as is the case in the exhaust gas under various engine operating conditions) the pumping current will be proportional to the sum of the constituents. This problem can be overcome by measuring the pumping currents at various dissociation potentials. At a lower potential the concentration of oxygen can be measured. A second voltage larger than the dissociation potential of the measurement gas, H_2O e.g., can be applied to measure the concentration of oxygen plus water. The water concentration can be determined by subtracting the pumping current due to oxygen from the pumping current due to the combination. Other gases, CO_2 e.g., can also be measured in a similar fashion. In principle, these measurements can be achieved without any modification to the existing sensor hardware. A simple modification to the electronic control circuitry is all that is required to perform this type of measurement.

Experimental

Several commercially available automotive UEGO sensors were utilized for this study. The actual sensors were unchanged, although the UEGO control circuitry (most vehicles use a dedicated ASIC) was modified in order to adjust the sensing voltage level Vs. These sensors were primarily characterized in the laboratory on a gas flow bench. The flow bench consists of several small stainless steel sensor mounting vessels with internal volume of several milliliters. The sensor mounts were connected in series to a gas mixing system comprised of calibrated mass flow controllers. The gas handling system is capable of mixing up to ten gases to simulate various exhaust gas compositions. Additionally, water can be injected at concentrations up to 16 mole%. The temperature of the gas can be varied up to 750°C, at flow rates up to 50 liters per minute (LPM) through an 8 mm I.D. tube. Most of the measurements for the present study were performed at 200°C gas temperature, utilizing flow rates between 30 and 50 LPM. The entire flow system and data acquisition system is controlled by a PC.

Results

Figure 2 shows the sensitivity of a commercially available UEGO sensor for the dissociation of H_2O as a function of the reference voltage Vs. As previously mentioned, a typical operating value for Vs for oxygen measurement is 450 mV. The plot below was generated from data measured in the lab at various water concentrations in a carrier gas of dry nitrogen. From the plot it is evident that water dissociation is occurring at potentials less than 800 mV. Additionally, the output begins to saturate at a potential approaching 1100 mV, suggesting that full dissociation of the water has been achieved. Figure 3 shows the sensitivity (pumping current) of the same sensor as a function of water concentration for Vs = 1080 mV. Once again the carrier gas is dry nitrogen. The output (pumping current) is linear in water concentration, and the sensitivity is approximately one half that of oxygen, since there are 2 oxygen atoms per oxygen molecule and only one oxygen atom per water molecule ($H_2O \rightarrow H_2 + \frac{1}{2}O_2$).

Figure 4 shows the response of the sensor to varying amounts of water in air. The water concentration was varied from 5 mole% to 3.5 mole%, then down to 0.5 mole% in increments of 0.5 mole%. The reference voltage Vs was switched between 450 mV and 1080 mV at each water level. When the Vs was switched to 450 mV, the sensor output (pumping current) only responded to the oxygen concentration in the mixture. Higher water concentrations corresponded to lower oxygen concentrations due to dilution effects. When the Vs was switched to 1080 mV, the sensor responded to both the water and the oxygen in the gas mixture. It should also be noted that the response time of the sensor to

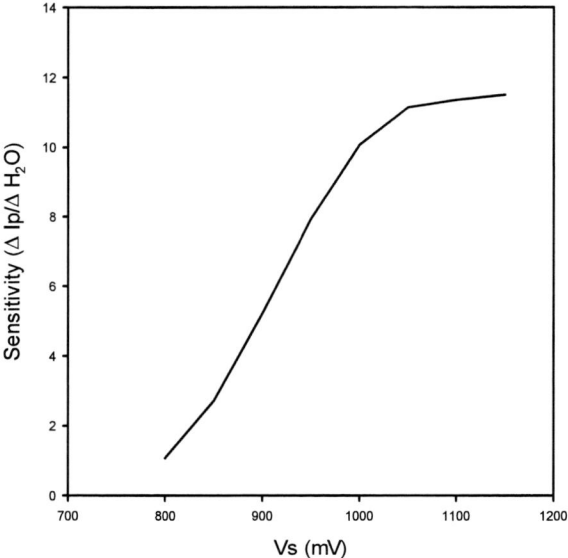

Figure 2. Sensitivity of water dissociation to reference voltage Vs.

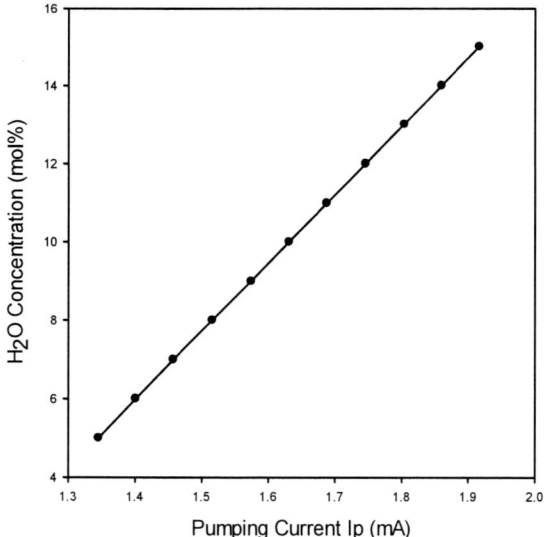

Figure 3. Water calibration curve for production UEGO sensor for dissociation voltage Vs=1080 mV in carrier gas of nitrogen.

a step change in Vs is relatively fast. The data indicate that the sensor has a response time of less than 1 sec. and that this time is currently limited by the control electronics used to switch Vs. Further tests with optimized control electronics are needed to determine the actual response time of the sensor to variations in Vs.

From the plot in Fig. 4 it is evident that the sensor can readily resolve low concentrations of water in air (0.5 mole% water in air) for Vs = 1080 mV. In fact with the currently available ASIC electronic control modules used on-board vehicles, it is possible to accurately measure water concentrations of less than 0.1% in air. Most commercially available humidity sensors that exist on the market at the present time utilize a polymer membrane that swells and expands upon exposure to water. By measuring the dielectric constant of this material, the ambient relative humidity can be measured. Together with a separate measurement of the temperature, the absolute water concentration or dew point can be determined. Due to the fact that two separate measurements are required to determine the dew point, the accuracy of this type of humidity sensor is limited to approximately 0.4 mole%. The level of accuracy/resolution for absolute water concentration measurement with the UEGO sensor utilizing a variable Vs far exceeds the state-of-the-art humidity sensors

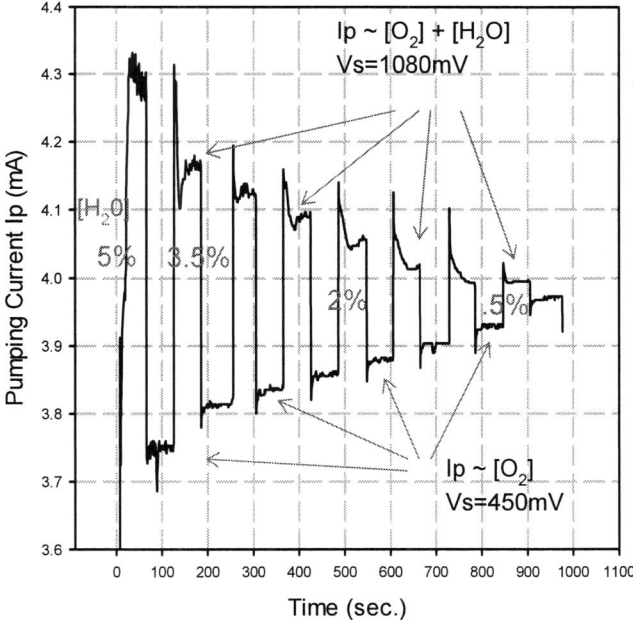

Figure 4. Response of production UEGO sensor to air mixtures with varying amounts of water at Vs = 450 mV and Vs = 1080 mV.

Conclusions

Highly sensitive and accurate water concentration sensors have been developed for automotive applications. These devices are capable of resolving water concentration to ± 0.1 mole% in the presence of high levels of oxygen. An automotive water vapor sensor can be useful in a number of applications. For example, when placed in the exhaust gas of a flex fuel vehicle which can utilize gasoline/ethanol as a fuel, it can be used to determine the fuel composition. The concentration of ethanol in the fuel can be inferred because the concentration of water in the exhaust gas (operated at the stoichiometric A/F ratio) is a unique function of the ethanol content in the ethanol/gasoline fuel mixture. The limitation of this sensing approach is that the fuel composition measurement is made after the engine has been running. This may not be a severe drawback since the fuel concentration does not change instantaneously after refueling due to the fuel stored in the lines and fuel rail. Therefore, this approach might be still useful, possibly in combination with a less accurate capacitive-type fuel sensor which measures the fuel composition directly in the fuel line. A water vapor sensor can also be used to measure the amount of EGR by measuring the water content in the intake-air/EGR mixture. As was the case with the aforementioned ethanol/gasoline fuel sensor, this type of measurement is possible because of the fact that the concentration of water vapor in this mixture is directly proportional to the amount of EGR. Variable humidity in the ambient air is a major noise factor and decreases the accuracy of both of these measurements. A separate measurement of the ambient humidity may also be possible with this sensor under certain conditions to correct for this noise factor.

References

1. W.C. Vassell, E.M. Logothetis, and R.E. Hetrick, SAE paper 841250, *SAE Meeting*, Oct. 1-4, Dearborn, MI, (1984).
2. T. Takeuchi, *Proceed. of 2nd Inter. Meeting on Chemical Sensors*, p. 69, Bordeaux, France, (1986).
3. E.M. Logothetis and R.E. Soltis, U.S Patent 5,288,375 (1994).

Fabrication of Surface Enhanced Raman Scattering (SERS)-Active Substrates by Using Dip-Pen Nanolithography

F. P. Lee[a, b, c], K. Chao[c, d], K. H. Hsieh[c, e], and K. L. Ou[c, d, f]

[a] Department of Otolaryngology, Taipei Medical University Wan-Fang Hospital, Taipei Medical University, Taipei 110, Taiwan
[b] Department of Otolaryngology, School of Medicine, Taipei Medical University, Taipei 110, Taiwan
[c] Research Center for Biomedical Implants and Microsurgery Devices, Taipei Medical University, Taipei 110, Taiwan
[d] Graduate Institute of Biomedical Materials and Tissue Engineering, Taipei Medical University, Taipei 110, Taiwan
[e] Industrial Technology Master Program in Biomedical Devices, Taipei Medical University, Taipei 110, Taiwan
[f] Research Center for Biomedical Devices and Prototyping Production, Taipei Medical University, Taipei 110, Taiwan

In this study, we propose an effective strategy to prepare a stable surface enhanced Raman scattering (SERS)-active substrate by using dip-pen nanolithography (DPN) with top-down wet chemical etching. This method was made by DPN to pattern arrays, 16-mercaplohexadecaoic acid (MHA) on Au/Cr/Si substrates and then ferri/ferrocyanide etching was adopted to remove the exposed gold. It is found that a nanostructure with ϕ 300 nm gold dots and the distance between dot and dot is 400 nm can be obtained on the surface of substrates after etching. These nanostructure gold dots can enhance local electromagnetic fields, inducing enhancement of analytes' Raman signal. This process supplies the effective and stable SERS-active substrate which can be used as biomedical applications, such as protein chip, and microorganisms' discrimination chip.

Introduction

Raman spectroscopy is a powerful analytical tool for biological and chemical sensing (1, 2). However, the Raman scattering is extremely inefficient. The inelastic scattering is just only 1 in 10^7 photons. Until 1977, Jeanmaire(3) research group proposed that the Raman signal could be enhanced in a rough surface of silver. This enhancement is called as Surface-Enhanced Raman Scattering (SERS). The SERS technology has been extensively developed for the past 30 years(4). The intensity of SERS depends on the wavelength and strength of the plasmon resonance and electric field at the surface of the structure(5). The molecules on a rough metal surface or nanostructure surface can enhance Raman signal intensity by 10^6 times or more. Therefore, a large number of methods have been used to fabricate various metallic nanostructures for SERS-active substrates. Such as the nano-sphere lithography, rough precious metals surface by electrochemical plating, and electron beam lithography (6-9). However, rare technologies for producing high-sensing, high stability and high yield ratio have been reported. Besides, the dip-pen nanolithography (DPN) is also an effective and mass-production technology to manufacture the nanostructure on metal surface.

In 1999 years, A. Mirkin (10) research team reported a new lithography strategy to fabricate a functionalized nanostructure on Au surface. The lithography is called dip-pen nanolithography (DPN), and it has been reported in Journal of Science. DPN is a direct-writing lithography in which an AFM probe is use as a tip, chemical molecules as an ink, and the substrate as a paper. The preliminary result indicated that DPN could pattern self-assembled monolayers (SAMs) on gold and semiconductor surface with high resolution (11, 12). DPN generated SAMs (ODT or MHA) pattern can also be used as a resist pattern to creating 3D solid states sub-micrometer structure by wet chemical etching (13-15). In 2009(16), SERS has been an effective tool for the directed placement of molecules on a SERS surface. Detection by SERS provides a highly sensitive, rapid readout system than the DPN method. However, there are still few researches for this application.

Until now, there is no research using a combined technique of DPN and chemical etching to fabricate SERS-active substrate. In the present research, we aim at investigating and developing a lithography technique to produce a SERS-active substrate. The experiment shows that DPN with chemical etching process can supply effectively the stable SERS-active substrates which can be applied in biomedical field, such as the protein chip (17-19) and microorganisms' discrimination chip (20-22).

Experiment procedure

Materials. Silicon wafer (100) was procured from Silicon Valley Microelectronics Inc. 16-mercaplohexadecaoic acid (MHA) (90.0 %) was purchased from Sigma. $K_3Fe(CN)_6$ (98.0 %), $K_4Fe(CN)_6$ (99.0 %), KOH (85.0 %) were obtained from ACROS, and SHOWA. Milli-Q water (> 18 MΩ) has been used for all substrate cleaning steps.

Preparation of Substrates. The substrate for dip-pen nanolithography (DPN) experiment contained three layers (Si/Cr/Au). The lowest layer was silicon wafer cleaned by using piranha solution (H_2SO_4 : H_2O_2 = 3:1) for 10 minutes, and rinsed for several times with Milli-Q water, then dried with N_2. The middle layer was an adhesion layer between Si and Au layers. The adhesion layer was coated with chromium thin film (about 5 nm) by using e-beam evaporation. The top Au layer film (about 50 nm) was coated by using thermal evaporation. After preparation of the Au/Cr/Si substrates, the Dip-Pen Nanolithograph experiment was carried out as soon as possible.

Dip-Pen Nanolithography (DPN) and Wet Etching. The DPN sample was patterned by 16-mercaplohexadecaoic acid (MHA) with a concentration of 5 mM. The MHA coated tip (A-type probe NanoInk, Inc.) was prepared by immersing Si_3N_4 cantilever in alcohol solution with MHA for ~10 seconds. The array pattern was executed using the commercial lithography software (InkCAD. NanoInk Inc.). The parameters of DPN experiment were set point = 0.5 nN, temperature = 23-24 °C, and humidity = 56 %.

The DPN substrates patterned with MHA were immersed in the ferri-/ferrocyanide etching solution for over 30 minutes. The etching solution includes etchants of 1.0 M KOH, 0.01 M $K_3Fe(CN)_6$, 0.001M $K_4Fe(CN)_6$, and 0.1 M $Na_2S_2O_3$, with a volume ratio of 1:1:1:1. Then, the samples were washed with Milli-Q water and dried with nitrogen.

Analysis. All of the samples were analyzed by atomic force microscope (NSCRIPTOR^{TM} 5000 NanoInk Inc.), to observe the nanostructure. Wettability examinations were performed using the sessile drop method using a GBX DGD-DI contact angle goniometer. The liquid deionized water was adopted in the test. Contact angle measurements were measured using at least five drops for the sample on defferent region in order to get good statistical averages. Raman spectra were acquired by the UniRAM-Raman microscope (Protrustech Co., Ltd) equipped with a thermoelectrically-cooled charge-coupled device of 1024×128 pixels and a diode-pumped solid-state laser at a wavelength of 532 nm.

Result and discussion

The surface roughness of Au/Cr/Si substrate is a very important parameter for DPN experiment. The 2D and 3D topography image of Au film are shown in Figure 1. It can be seen in Figure 1 that the Au surface is very smooth with Ra = 1.42 nm. The surface roughness of Au/Cr/Si substrate is suitable for the DPN experiment.

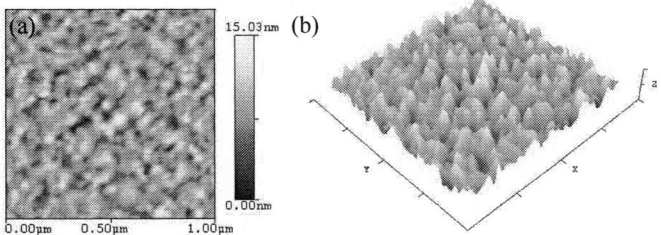

Figure 1: The topography image of Au/Cr/Si substrate, (a) 2D, (b) 3D.

The contact angle is the easiest method to measure the surface energy, and also indicated the surface property of the substrate. In Figure 2, the average contact angle of DPN substrate is 83.9°±2.2°. The surface of Au film is almost hydrophilic. However, the hydrophilic surface affected mechanism of SAMs. So the environment condition, such as temperature and humidity, will affect the diffusion of the MHA molecular not only transfer to the surface of Au film, but also form SAMs on Au surface. Using contact angle, the measurement is efficient to filter the substrate of DPN experiment.

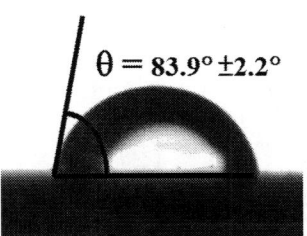

Figure 2: The average contact angle of the Au/Cr/Si substrate.

Figure 3 the LFM image shows that the MHA molucules are imbolized on the surface of Au film by using DPN technology. This is a 4×4 arrays, and the diameter of dots is 300 nm and the distance between dot and dot is 400 nm. The MHA molucules can be used as a mask to resist the ferri/ferrocyanide etching.

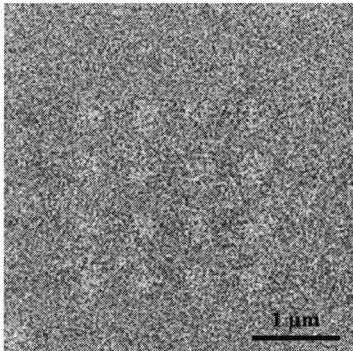

Figure 3: The LFM image after DPN.

After ferri/ferrocyanide etching, the Raman spectra of Au/Cr/Si substrate have been measured. The Raman shift on the Au/Cr/Si substrate is 1563, 1704 cm^{-1}, respectively. The 1704 cm^{-1} peak position of Raman spectrum is the C-O double bond stretching vibration band, and it has been enhanced. The C-O double bond is the tail group of the MHA. Other low intensity peak of Raman shift at 1400~1600 cm^{-1} is the MHA long alkyl chain of CH$_3$ deformation.

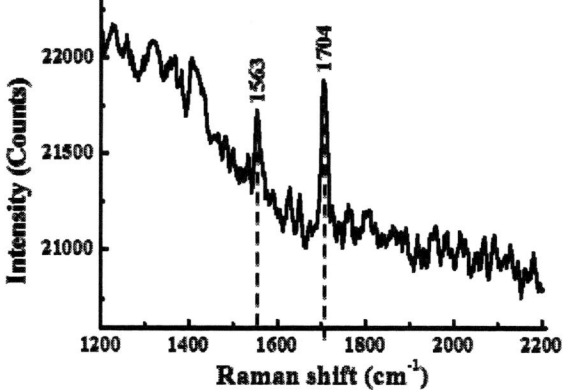

Figure 4: The Raman spectra of MHA on Au/Cr/Si substrate.

Conclusion

We have successfully created the SERS (Au/Cr/Si) substrates combining with dip-pen nanolithography technique and wet chemical etching. Experimental results show that the surface of Au layer is very smooth with Ra = 1.42 nm. The surface roughness of Au/Cr/Si substrate is suitable for the DPN experiment. The average contact angle of DPN substrate is 83.9°±2.2°. The 1704 cm^{-1} peak position of Raman spectrum is the C-O double bond stretching vibration band, and it has been enhanced. The DPN technique supplies nanostructure of gold pattern and it is effective and stable to fabricate the SERS-active substrate which can be applied in biomedical field, such as the protein chip and microorganisms' discrimination chip.

Acknowledgments

The authors would like to thank the Taipei Medical University Hospital for financially supporting this research under contract 99TMU-TMUH-03-1.

References

1. K. Kneipp, H. Kneipp and J. Kneipp, *Accounts of chemical research*, **39**, 443 (2006).
2. K. J. H. Cotton T. M. , Chumanov G. D., *J. Raman Spectrosc*, **22**, 729 (1991).
3. R. P. V. D. D.L. Jeanmaire, *Electroanal. Chem*, **84**, 1 (1977).
4. H. P. Fleischmann M, McQuillan AJ, *Chem. Phys*, **26**, 163 (1974).
5. C. L. Haynes, A. D. McFarland and R. P. Van Duyne, *Analytical chemistry*, **77**, 338a (2005).
6. R. J. C. Brown and M. J. T. Milton, *J Raman Spectrosc*, **39**, 1313 (2008).
7. A. D. Ormonde, E. C. M. Hicks, J. Castillo and R. P. Van Duyne, *Langmuir : the ACS journal of surfaces and colloids*, **20**, 6927 (2004).
8. S. Foteinopoulou, J. P. Vigneron and C. Vandenbem, *Optics express*, **15**, 4253 (2007).
9. N. A. Abu Hatab, J. M. Oran and M. J. Sepaniak, *ACS nano*, **2**, 377 (2008).
10. R. D. Piner, J. Zhu, F. Xu, S. Hong and C. A. Mirkin, *Science*, **283**, 661 (1999).
11. K. B. Lee, S. J. Park, C. A. Mirkin, J. C. Smith and M. Mrksich, *Science*, **295**, 1702 (2002).
12. A. Ivanisevic and C. A. Mirkin, *Journal of the American Chemical Society*, **123**, 7887 (2001).
13. H. Zhang, S. W. Chung and C. A. Mirkin, *Nano letters*, **3**, 43 (2003).
14. H. Zhang, Z. Li and C. A. Mirkin, *Advanced Materials*, **14**, 1472 (2002).
15. H. Zhang and C. A. Mirkin, *Chem Mater*, **16**, 1480 (2004).
16. S. R. J, D. J. A, I. Eleanore, O. Jenifer, R. Sergy, L. Tom, L. Tom, N. Mike and N. Mike, *Progress in biomedical optics and imaging*, **720703**, 1 (2009).
17. X. X. Han, B. Zhao and Y. Ozaki, *Analytical and bioanalytical chemistry*, **394**, 1719 (2009).
18. D. Lee, Y. J. Choe, M. Lee, D. H. Jeong and S. R. Paik, *Langmuir : the ACS journal of surfaces and colloids*, **27**, 12782 (2011).
19. Z. S. Li, W. D. Ruan, W. Song, X. X. Xue, Z. Mao, W. Ji and B. Zhao, *Spectrochim Acta A*, **82**, 456 (2011).
20. T.-Y. Liu, K.-T. Tsai, H.-H. Wang, Y. Chen, Y.-H. Chen, Y.-C. Chao, H.-H. Chang, C.-H. Lin, J.-K. Wang and Y.-L. Wang, *nature communication* **2358** (2011).
21. W. R. Premasiri, D. T. Moir, M. S. Klempner, N. Krieger, G. Jones and L. D. Ziegler, *Journal of Physical Chemistry B*, **109**, 312 (2005).
22. R. M. Jarvis and R. Goodacre, *Analytical chemistry*, **76**, 40 (2004).

Application of Commercial Manufacturing Methods to Mixed-Potential NO$_x$ Sensors

Cortney R. Kreller[a], Praveen K. Sekhar[b], Wenxia Li[c], Ponnusamy Palanisamy[c], Eric L. Brosha[a], Rangachary Mukundan[a], and Fernando H. Garzon[a]

[a] Los Alamos National Laboratory (LANL), Sensors and Electrochemical Devices Group, Los Alamos, New Mexico 87545
[b] Nanomaterials and Sensors Laboratory, Department of Electrical Engineering, Washington State University Vancouver, Vancouver, WA 98686
[c] ESL ElectroScience, 416 East Church Road, King of Prussia, PA 19406

> Commercial manufacturing methods have been used to fabricate a planar, self-heated, tape-cast, mixed-potential NO$_x$ sensor for application in vehicle on-board emission control systems. The device consists of dense La$_{0.8}$Sr$_{0.2}$CrO$_3$ (LSCrO) and Pt electrodes and a porous YSZ electrolyte on one side of a dense ceramic substrate and a Pt-heater with independent leads on the backside of the substrate. These planar mixed-potential sensors are capable of selectively detecting ppm levels of NO$_x$ when operated at a positive bias. Additionally, the device has yielded stable performance under months of testing as a result of the stable morphology of the electrode/electrolyte/gas phase interface.

Introduction

The development of low-cost, sensitive, and robust nitrogen oxide (NO$_x$) sensors will improve fuel economy and lower emissions of lean-burn gasoline and diesel engines. Lean-burn engines improve fuel efficiency and reduce exhaust hydrocarbon emissions. NO$_x$ sensors with sufficiently fast response time may be used to directly monitor the lean-burn combustion as part of an On Board Diagnostics (OBD) system to help optimize efficiency, analogous to the use of O$_2$-sensors to help maintain stoichiometric operation of gasoline engines. While operating in lean-burn mode improves fuel efficiency, it can also lead to increased NO$_x$ emissions as a result of increased operating temperature. The ongoing development of NO$_x$ storage catalysts (NSC) and Selective Catalyst Reduction (SCR) technologies requires a closed feedback loop with a sensitive and robust NO$_x$ sensor to improve NO$_x$ reduction efficiency and meet U.S. EPA emission standards (1).

Mixed-potential sensors are electrochemical devices that measure the non-Nernstian potential of a mixture of gases, where the mixed-potential is fixed by the rates of different electrochemical reactions occurring simultaneously at an electrode/electrolyte interface (2, 3). Employment of electrode materials that possess varying electro-catalytic activities towards the redox half reactions has been shown to further increase the mixed-potential response. Mixed-potential sensors for the selective detection of CO, HCs, H$_2$ and NO$_x$ have all been extensively investigated however, the commercialization of this technology has been hindered by poor reproducibility and stability (4, 5). Because the mixed-potential response is highly dependent on the kinetics of the electrochemical reactions occurring at each electrode, highly stable electro-active areas are required to maintain a constant sensor response over the lifetime of the device. Additionally, device sensitivity is improved by minimizing the amount of heterogeneous reaction occurring prior to the analyte gas reaching the electrochemical interface (6, 7). To address these issues, the Electrochemical Sensors and Devices Group at Los Alamos National Laboratory has developed a patented sensor design that incorporates dense electrodes and porous electrolytes in bulk, thin film, and tape cast form (8). By using dense electrodes, heterogeneous catalysis is reduced, increasing sensitivity, and the increased morphological stability of dense electrodes yields a robust electrochemical interface, increasing lifetime durability.

Based on the success of the laboratory bulk and thin film sensors, LANL has recently developed a commercial platform base for planar, self-heated, tape-cast, NO_x mixed-potential sensors in collaboration with Electro-Science Laboratories (ESL, King of Prussia, PA). As described in detail previously, the manufacturing methods developed by ESL can be readily scaled to cost-effective manufacturing methods already employed in the manufacturing of planar O_2 λ-sensors (9). In the present NO_x sensor, the working electrode, $La_{0.8}Sr_{0.2}CrO_3$ (LSCrO), is known to be a poor catalyst for the oxygen reduction reaction whereas the Pt counter electrode is known to possess high catalytic activity. Operated at open circuit, this combination of electrode materials serves as a sensor for reducing gases, such as hydrocarbons and ammonia (7, 10). However, the sensor can be transformed to a total NO_x sensor by appropriate tuning of the current bias and operating temperature (11-13).

As reported earlier, the initial prototype of the tape-cast sensors developed by ESL yielded an unbiased sensor response similar to that of bulk device analogs (9). However, under bias, these devices exhibited reduced NO_x sensitivity and significant drift over time. Post-mortem Environmental Scanning Electron Microscopy (ESEM) and X-ray computed tomography (XCT) revealed buried defects and cracks in the YSZ at the working electrode interface and a coarse Pt morphology that would be expected to change over time. Based on these initial tests, ESL adjusted the Pt ink formulation, the substrate composition, and the process design to address these limitations while preserving the desirable attribute of forming the device platform using a single firing. This paper reports on the behavior of the second generation of the mixed-potential commercial sensor prototypes.

Experimental

Based on the mixed-potential sensor design concept developed at LANL, ESL ElectroScience designed and developed a High Temperature Co-fired Ceramic (HTCC) structure and substrate and the required screen-printing patterns for the various layers based on commercial manufacturing technology (9). The two sides of the sensor prototype are shown in Figure 1. A Pt-heater with independent leads is printed on the backside of the ceramic substrate. A standard heater ink composition was used and the required heater resistance was obtained by modifying the heater pattern design. An insulating heater overcoat was applied to reduce the amount of ambient heat loss and provide a more uniform distribution of heat across the substrate. The sensing component is printed on the opposite side of the ceramic substrate and consists of screen-printed

LSCrO- working and Pt-counter electrodes coated with a porous 3 mol%-YSZ electrolyte layer.

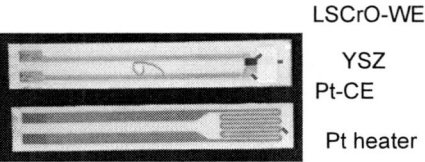

LSCrO-WE

YSZ
Pt-CE

Pt heater

Figure 1: ESL prototype sensor

In order to address the limitations found during testing of the initial prototypes, ESL adjusted the Pt ink formulation, the substrate composition, and the process design to address these limitations while preserving the desirable attribute of forming the device platform using a single firing. In order to optimize the working electrode and electrolyte materials properties, a post firing step was incorporated after the heater integrated substrate with Pt electrode was made by a single firing (co-sintering).

The sensors were mounted into commercial Ford/Visteon automotive packaging for flat plate O_2 sensors with 4 metal clips for making electrical connection to each of the two sensor and heater leads. The assembly was placed into a specially prepared quartz tube with a sealed end that placed the sensor electrodes in the test gas stream. A rubber stopper was used to hold the apparatus in place and seal the gas environment. In this "cold-wall" set-up, the Pt-heater is the only source of heat supplied to the sensor components. The sensor leads were connected to a Keithley 2400 source meter with the Pt electrode connected to the positive terminal. Under positive bias, current flowed from the Pt electrode to the LSCrO electrode. Two terminal measurements were performed at a fixed flow rate of 500 ccm with 10% O_2/ balance N_2 used as the base gas. The flow rates of the test gas and base gas were modified to set the concentration of the test gas while maintaining a total flow rate of 500 ccm.

Figure 2: SEM micrograph of Pt (left) and LSCrO (right) electrodes and porous YSZ electrolyte overlayer (bottom).

Results and Discussion

An SEM micrograph of LSCrO/YSZ/Pt sensing element is shown in Figure 2. Both the Pt electrode, shown on the left, and the LSCrO electrode, shown on the right, are well adhered to the ceramic substrate and appear to have clean, defect free surfaces. Additionally, the morphology at the electrode interface is well defined with no obvious defects or cracks.

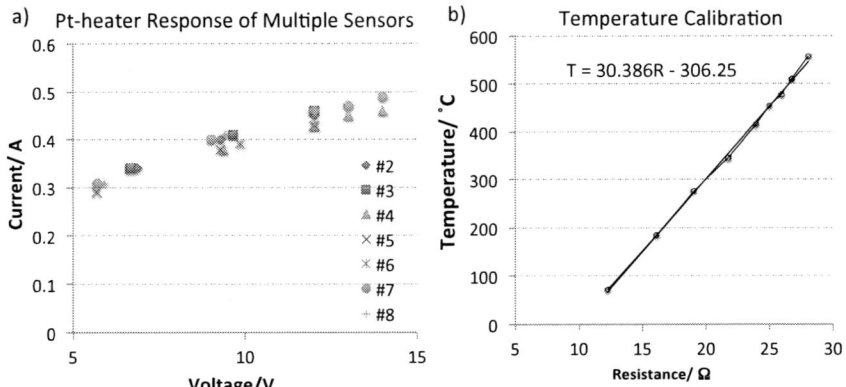

Figure 3: a) I-V response of Pt-heaters of multiple sensors and b) temperature calibration corresponding to Pt-heater resistance.

A total of 10 sensors were made by ESL in the second generation of commercial sensor prototypes. Eight sensors were chosen at random to examine the sensor-to-sensor variability in the I-V characteristics of the Pt-heater. As shown in Figure 3a, the Pt-heater resistance is highly reproducible between the devices with an average resistance at 12V of $26.95 \pm 0.92 \ \Omega$. Calibration of the Pt-heater and substrate platform was performed by adhering a K-type thermocouple to the sensing element and measuring the surface temperature and Pt-heater resistance as a function of applied voltage. As shown in Figure 3b, the sensing element surface temperature scales linearly with the Pt-heater resistance. The heater resistance was found to be constant after extended testing at varying applied voltages (data not shown). Current work is ongoing a) to verify that the temperature on the sensing element side of the substrate is also unchanged with time and b) to examine heater durability with increased number of temperature cycles.

Previous work on LSCrO/YSZ/Pt sensors in bulk or thin film configurations has demonstrated that when operated at open circuit, this combination of electrode materials serves as a sensor for reducing gases, such as hydrocarbons and ammonia (10). However, the sensor can be transformed to a total NO_x sensor by appropriate tuning of the current bias and operating temperature(11). Figure 4 shows the response of one commercial sensor prototype with the heater voltage maintained at 12 V to 100 ppm of each NO, NO_2, NH_3, C_3H_6, and C_3H_8 in a base gas of $10\%O_2/N_2$. Similar to the bulk device analogs, when the commercial sensor prototype is operated at open circuit, it yields a positive voltage response for the hydrocarbons and ammonia, and a negative voltage response to

NO_2. With the addition of a positive bias of 0.2 μA, the sensor yields a negative voltage response to both NO and NO_2, and is relatively insensitive to NH_3, C_3H_6, and C_3H_8.

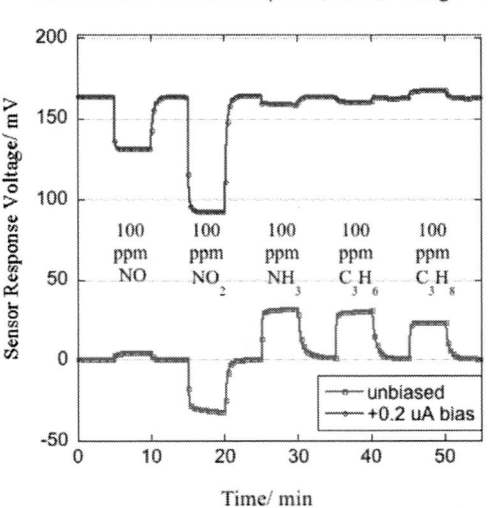

Figure 4: Sensor response to 100 ppm each NO, NO_2, NH_3, C_3H_6, and C_3H_8 in $10\%O_2/N_2$ base gas at open circuit and under +0.2 μA bias with an applied heater voltage of 12V.

The sensitivity of the prototype commercial sensor to NO is illustrated in Figure 5. At both unbiased and biased conditions, the sensor is capable of detecting NO concentrations as low as 5 ppm. Additionally, it is evident by comparing the biased and unbiased sensor response shown in Figure 5c that applying a +0.2μA bias increases the sensitivity of the sensor to NO by nearly an order of magnitude.

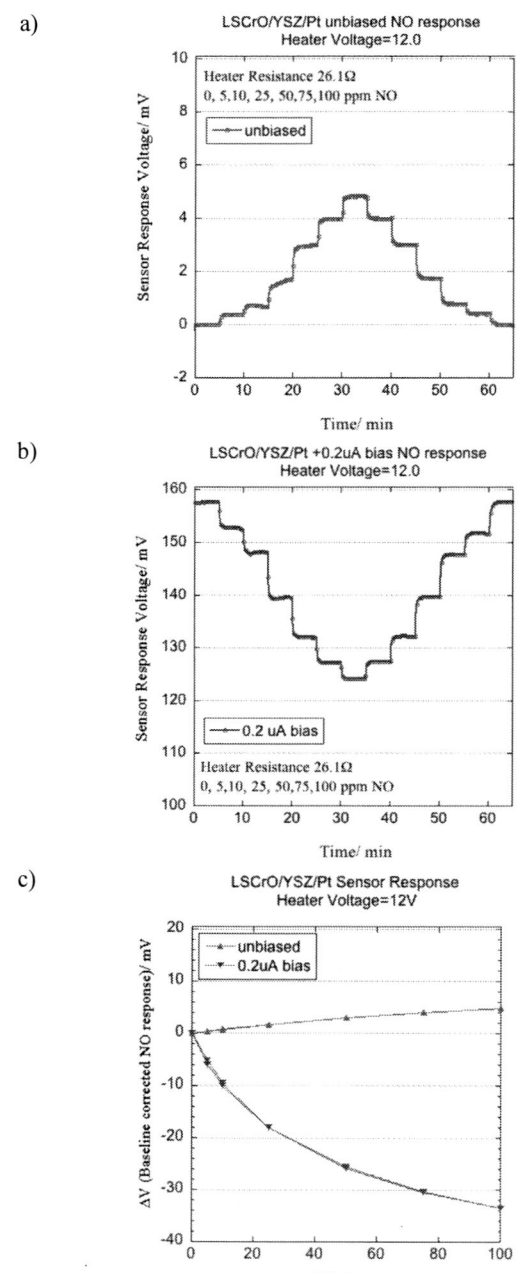

Figure 5: Sensor response to 0,5,10,25,50,75, and 100 ppm NO with (a) no bias, (b) +0.2 μA bias and (c) baseline corrected sensor response

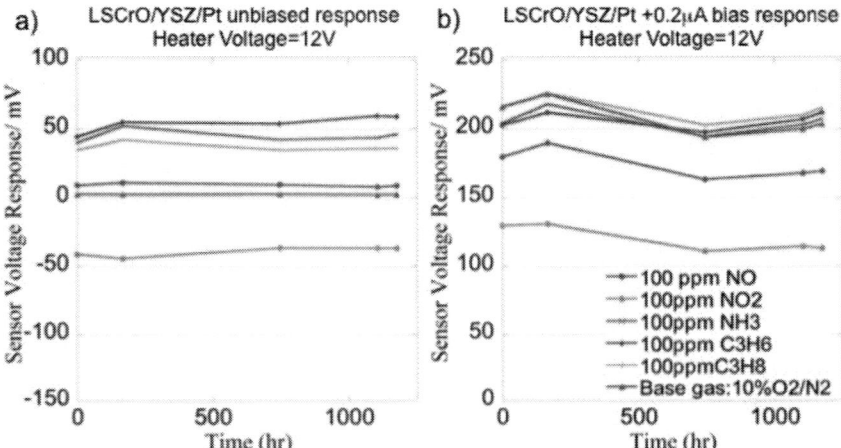

Figure 6: Sensor voltage response stability over 50 days of testing at open circuit and under 0.2 μA bias.

The stability of the response of a single sensor over 1200 hours of operation is shown in Figure 6. After an initial break in period of ~50 hours, the sensor response in both the biased and unbiased mode is relatively stable. The fluctuations that are observed in the sensor response shown in Figure 6 may be a result of changes in ambient temperature and thus fluctuations in the temperature of the inlet gas stream. The set-up used to operate this sensor was a "cold-wall" set-up where the sensor was encased in a closed end quartz tube to maintain a closed-gas environment, not necessarily a constant temperature environment. Future work will examine the sensitivity of the sensor response to small (±10°C) changes in gas temperature using a flow-through tube furnace. Further development of the commercial sensor prototypes will incorporate a Resistance Temperature Detector (RTD) in the center of the substrate in order to provide active feedback control of the temperature near the sensing element surface.

The sensor response stability shown in Figure 6 is a significant improvement over the first generation of the commercial sensor prototypes where a monotonic drift in performance was immediately observed in the biased mode as a function of time (9). In the first generation of prototypes, 3D x-ray tomography revealed buried cracks at the electrode/electrolyte interfaces and a roughening of the Pt morphology with device aging. Post-mortem X-ray tomography of the sensor tested over 1200 hours gave no evidence of defect formation or changes in morphology, indicating that the new fabrication processes implemented by ESL successfully addressed the limitations of the first generation of sensor prototypes.

Conclusions

A prototype planar, mixed-potential sensor for NO_x detection was fabricated using commercial manufacturing techniques. The second generation of the ESL tape-cast sensors has shown significant improvement over the first generation. The ratio of interfacial to bulk resistance and the electrochemically active area have been preserved so that the device sensitivity to a given analyte gas may be controlled by the application of an appropriate current bias. These devices are capable of selectively detecting ppm levels of NO_x when operated at a positive bias. The device response has been shown to be stable in both biased and unbiased operation for up to ~1200 hours, indicating that the manufacturing modifications incorporated by ESL have improved the durability of the sensor.

Acknowledgments

The research was funded by the US DOE, EERE, Vehicle Technology Programs. The authors wish to thank Technology Development Manager Roland Gravel.

References

1. Regulatory Announcement, EPA 420-F-08-004, Office of Transportation and Air Quality, **2010**.
2. J. W. Fergus, *Journal of Solid State Electrochemistry* **15**, 971-984 (2010).
3. F. H. Garzon, R. Mukundan, E. L. Brosha, *Solid State Ionics* **136-137**, 633-638 (2000).
4. W. J. Fleming, *Journal of The Electrochemical Society* **124**, 21-28 (1977).
5. P. Pasierb, M. Rekas, *Journal of Solid State Electrochemistry* **13**, 3-25 (2008).
6. P. K. Sekhar, E. L. Brosha, R. Mukundan, M. A. Nelson, D. Toracco, F. H. Garzon, *Solid State Ionics* **181**, 947-953 (2010).
7. R. Mukundan, E. L. Brosha, F. H. Garzon, *Journal of The Electrochemical Society* **150**, H279 (2003).
8. R. Mukundan, E. L. Brosha, F. H. Garzon, in *US Patent # 7,575,709, Vol. US Patent # 7,575,709*.
9. P. K. Sekhar, E. L. Brosha, R. Mukundan, W. Li, M. A. Nelson, P. Palanisamy, F. H. Garzon, *Sensors and Actuators B: Chemical* **144**, 112-119 (2010).
10. E. L. Brosha, R. Mukundan, R. Lujan, F. H. Garzon, *Sensors and Actuators B: Chemical* **119**, 398-408 (2006).
11. R. Mukundan, K. Teranishi, E. L. Brosha, F. H. Garzon, *Electrochemical and Solid-State Letters* **10**, J26 (2007).
12. D. L. West, F. C. Montgomery, T. R. Armstrong, *Journal of The Electrochemical Society* **153**, H23 (2006).
13. N. Miura, G. Lu, M. Ono, N. Yamazoe, *Solid State Ionics* **117**, 283-290 (1999).

Research on filter materials for LP gas sensors

Masakazu.Sai[a], Kazuya.Shinnishi[a], Kazunari.Kaneyasu[a],
Toshiya.Suzuki[b], and Masato.Takeuchi[b]

[a]Figaro Engineering Inc.1-5-11, Senbanishi, Mino, Osaka 562-8505, Japan
[b]Osaka Prefecture University1-1, Gakuen-cho, Naka-ku, Sakai, Osaka 599-8531, Japan

It is well known that semiconductor gas sensor characteristics may change when exposed to siloxane vapor. In this study, we examined filter materials to improve the siloxane durability of the LP gas sensor. Adsorption isotherms of octamethylcyclotetra-siloxane (D4) at 298 K were measured for each filter materials. As a result, γ-Al$_2$O$_3$ showed high performance in adsorption amount. However, the sensor with mixed filter materials of MPR-PEG (Mordenite zeolite modified with polyethylene glycol) and γ-Al$_2$O$_3$ was more stable after the D4 durability test than the sensors with the single filter materials. The γ-Al$_2$O$_3$ mixed with MOR-PEG maintains capacity for D4.

Introduction

Metal oxide semiconductor (MOS) gas sensor has features such as competitiveness in costs, compact size, ease of application circuit design and long-term performance stability. MOS gas sensors are used for various applications, for example: gas detectors and alarms; indoor air quality control; breath alcohol testers; and VOC monitors.
Residential gas detectors in Japanese market necessitate the product lifetime of at least 5 years. Therefore, it is important to ensure a long-term stability of gas sensors. Some gases in indoor environment may change characteristics of gas sensors, causing false alarms of gas detectors. For example, silicone putty, cosmetics, shampoos etc. contain siloxane compounds. It is well known that gas sensor characteristics may change when exposed to siloxane vapor.[1] Moreover, even low concentration of siloxane compounds dramatically damages gas sensors.[2] In order to solve this problem, filters are often used to reduce interference gases which may have bad effects to gas sensors. Activated carbon is known to be a good candidate to remove siloxane vapor.[3] However, carbon materials, which have high affinity to propane and butane, cannot be applied for LP gas sensors. Therefore, we currently use Mordenite zeolite which is granulated for improved fluidity using polyethylene glycol as a binder (MOR-PEG). However, the MOR-PEG adsorbent does not have a high enough performance to remove siloxane. In this study, we examined filter materials for LP gas sensors which show high resistance characteristics to siloxane vapor. Filter material examined is activated alumina, which is commercially available at low cost.

Experimental

We examined octamethylcyclotetrasiloxane (D4) adsorption isotherm and specific surface area (SSA) of each filter material. SSA was measured by a BET method using a N_2 adsorption isotherm. The pretreatment condition that was used was: evacuation at 673 K for 1.5 hours; treatment with oxygen at 673 K for 1.5 hours; evacuation at 373 K for 2.0 hours. Since the MOR-PEG contains polymer binder, only this sample was treated at 323 K.

D4 durability test was evaluated using the sensor as shown in figure 1. The gas sensing material was SnO_2 containing Pd of 1.5% by weight, with which a film of approximately 80 µm thickness was formed by screen-printing. The working temperature of the sensor element was 703 K. Filter materials were placed between the sensing element and the gas inlet. During the durability test, the sensor was placed in the test gas containing D4 of 60 ppm, and the gas sensing properties were examined in the other test chamber with clean air every 24 hours.

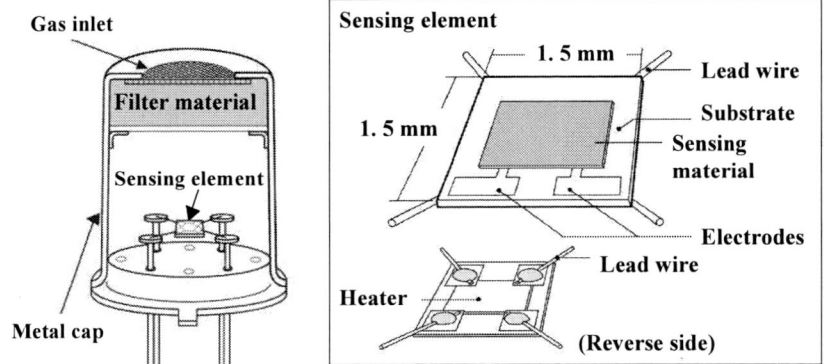

Figure 1. Sensor structure

Result and Discussion

Figure 2 shows the D4 adsorption isotherm and SSA. γ and θ-Al_2O_3 adsorbed larger amount of D4 than the MOR-PEG. Adsorption amount of D4 was independent on the SSA of adsorbents. In addition, correlation was not observed between acid-base surface properties of the γ-Al_2O_3 and the amount of D4 adsorption.

Figure 3 shows results of the D4 durability test. The sensor with the mixed filter materials (b) was more stable in property than the sensors with the single filter materials (a),(c). This result did not correspond to the D4 adsorption isotherms. Since the γ-Al_2O_3 surface is hydrophilic, water vapor coexisted in the test chamber might depress the D4 adsorption. However, the MOR-PEG containing a highly hydrophilic polymer binder may work as an absorber for water vapor. As a result, the γ-Al_2O_3 mixed with MOR-PEG maintains the adsorption capacity for D4.

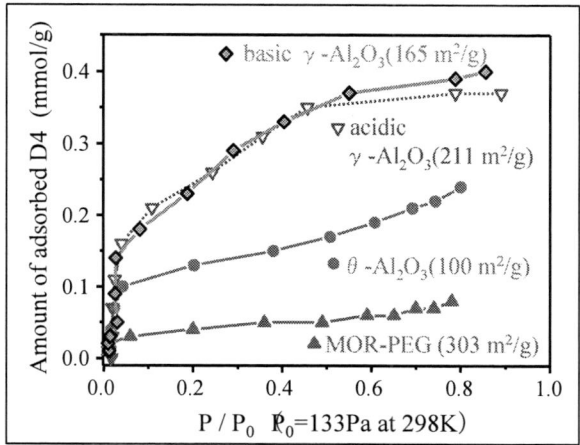

Figure 2. D4 adsorption isotherm and SSA of each filter material.

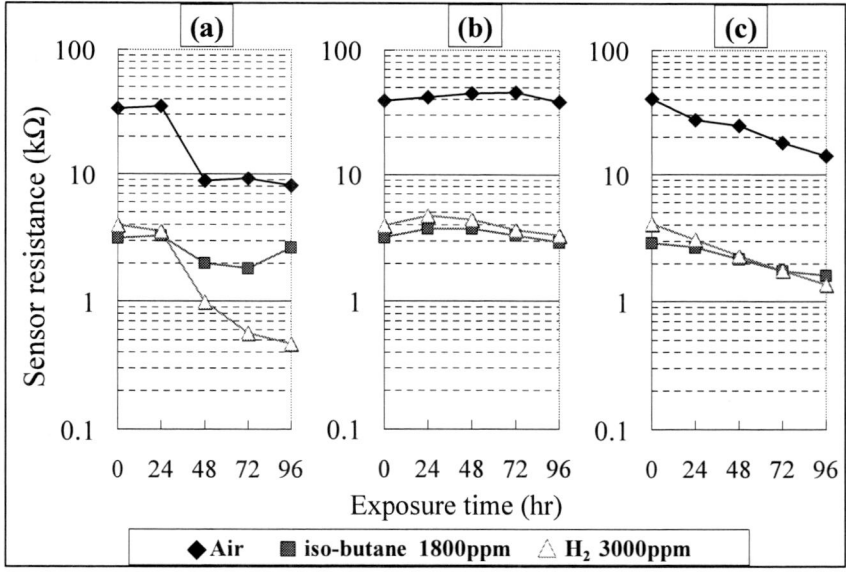

Figure 3. Results of D4 durability test using the three different filter materials.
(a) MOR-PEG 75mg
(b) MOR-PEG 37.5mg + basic γ-Al$_2$O$_3$ 37.5mg
(c) basic γ-Al$_2$O$_3$ 75mg

References

1. A. Katsuki, K. Fukui, *Sensors and Actuators B,* Vol.52, 30-37 (1998)
2. Japanese Patent No.3901602 (2007)
3. G. Busca et al., *Energy & Fuels,* Vol.23, 4156-4159 (2009)

CHAPTER 3

CHEMICAL SENSORS POSTER SESSION

Characterization and Electrochemical Response of Sonogel Carbon Electrode Modified with Nanostructured TiO2 and ZrO2 Film to Detect Common Neurotransmitters

S. K. Lunsford, M. K. Hughes and P. K. Nguyen

Department of Chemistry, Wright State University, Dayton Ohio 45435, USA

A new modified carbon electrode was developed by coating with nanostructured titanium dioxide and zirconium dioxide film and mixed film on top of graphite carbon electrode to detect common neurotransmitters such as 3,4-Dihydroxy-L-phenyl-alanine (L-DOPA) and dopamine. Cyclic Voltammetry (CV) was the method for the detection with these new sonogel modified electrodes, which showed good reversibility and reproducibility. This study proved that the mixture of TiO_2 and ZrO_2 is promising and could be expanded due to their good thermal stability, good conductivity and chemical inertness.

Introduction

In the pharmaceutical and clinical field there is need for nifty electrochemical sensors for in-situ detection of neurotransmitters since the nervous system can be altered by a change in concentration in the human body. Presently there is awareness in utilizing sonogel carbon conducting matrix as the conducting electrode due to the benefits such as high conductivity, relative chemical inertness, and modifiable chemical and biological modification. Our research group has been involved in the development of a new class of metal oxides (titanium dioxide, zirconium dioxide and mixtures of metal oxides). These specific nanoscale metal oxides have shown to be good for adsorption of neurotransmitters, good conductivity, and chemical inertness. This paper will describe the novel modified nanoscale metal oxides synthesized to detect common neurotransmitters such as L-DOPA. [1-4]

Experimental part

Techniques and Instrumentation

Synthesis of TiO_2, ZrO_2 and Mixture (TiO_2 + ZrO_2)

Titanium (IV) isopropoxide 97% (Sigma-Aldrich) and zirconium (IV) propoxide 70 % wt in 1-propoamol (Aldrich), deionized water, acetylacetone (Sigma-Aldrich), and ethanol (absolute ,200 proof, 99.5%) (Sigma-Aldrich) were used for the preparation of each solution. In mixture preparation, 2:4:1:62 was zirconium (IV) propoxide, titanium (IV) isopropoxide, water, acetylacetone and ethanol.

Construction of Bare Electrodes

A 0.5 mm O.D., 12.5 cm long copper wire was inserted into a capillary tube, (Sutter Instrument, 0.69 mm I.D., 1.2 mm O.D., 10 cm long, borosilicate glass both ends open) as the wire was exposed by 2.5 cm only on one end of the tube, the wire functioned as an electrical contact component. 0.7 g of graphite powder (Alfa-Aesar 99.0% 7-10 micron) was added 0.3 g of silicon oil, Xiameter PMX-200 Silicon fluid 100 cs (Aldrich) with magnetic chip in vial. This carbon paste of vial was placed on electro stir pan for smoothed carbon paste about 5 minutes, which becomes a homogenous carbon paste. Before the measurement, the modified electrode was smoothed on a piece of transparent paper to get a uniform, smooth, fresh surface. Previously made capillary tube with copper wire core was dipped firmly into the graphite packing about 0.7 cm of the dipping end of the tube.

Solution of TiO_2 and ZrO_2 and mixture ($TiO_2 + ZrO_2$) were under vigorous stirring. The one made above of a bare carbon electrode were dipped into the TiO_2 and ZrO_2 and mixture ($TiO_2 + ZrO_2$) separately coating for 3 seconds. After dry for 5 minutes, separately coating again with the same methods 4 times more. Then the electrode was heated at 230^0C for 20 minutes and cooled in room temperature.

Results and Discussion

In this work, ZrO_2, TiO_2, and sol-gel Mixture ($ZrO_2 + TiO_2$) were used as nanoporous electrodes to study the electron transfer of L-DOPA and dopamine in order to establish a new generation of chemical sensors. The detection of L-DOPA and dopamine were improved significantly due to the modified, ZrO_2 and TiO_2, nanoscale carbon hand-built electrodes compared to the bare carbon electrode response as illustrated in Table 1. The L-DOPA and dopamine detection with the modified nanoscale metal oxide carbon electrodes illustrate reversible behavior as compared to the bare carbon electrode response with only one anodic peak detected, thus not reversible behavior. These modified nanoscale materials have made the detection of L-DOPA and dopamine overall to be more reversible and give a quicker electron transfer response as observed by cyclic voltammogram data. [5]

The results in this paper are extended to the use of a novel Mixture from previous lab studies in the past where using simple modified sol-gel carbon electrode of TiO_2 or a modified sol-gel carbon electrode of ZrO_2 to detect the 1,2-dihydroxybenzne structures (dopamine and L-DOPA). Therefore the Mixture in Figure 1., has illustrated the optimized an improved separation and detection of the 1,2-dihydroxbenzenes.

In Figure 1, there is no need for prior separation of L-DOPA and dopamine with the modified sol-gel Mixture carbon electrode when analyzing by cyclic voltammetry, there are 4 peaks (two oxidation and two reductions peaks). With the bare carbon electrode in Figure 2., there is a problem attempting to detect L-DOPA and dopamine simultaneously, only two peaks are observed thus interference problem is observed.

Therefore these modified sol-gel electrodes are showing that good novel sol-gel sensors were developed to detect common neurotransmitters, which exhibited reversible behavior.

L-DOPA	E-pa (mV)	E-pc (mV)	Δ E
Bare Carbon	677.6	-	-
ZrO₂	759.1	429.0	330.1
TiO₂	606.0	501.5	104.5
Dopamine	E pa (mV)	E pc (mV)	Δ E
Bare Carbon	668.7	-	
ZrO₂	620.9	444.8	176.1
TiO₂	755.2	277.6	477.6

Table 1. Cyclic Voltammetry Data: Reversible Behavior between L-DOPA and Dopamine with bare carbon, ZrO₂, and TiO₂.

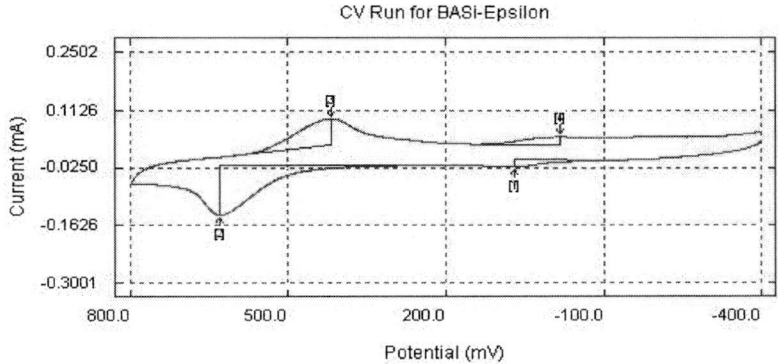

Figure 1. Modified Mixture (ZrO2 + TiO2) with L-DOPA and Dopamine in 0.1 M Sulfuric acid

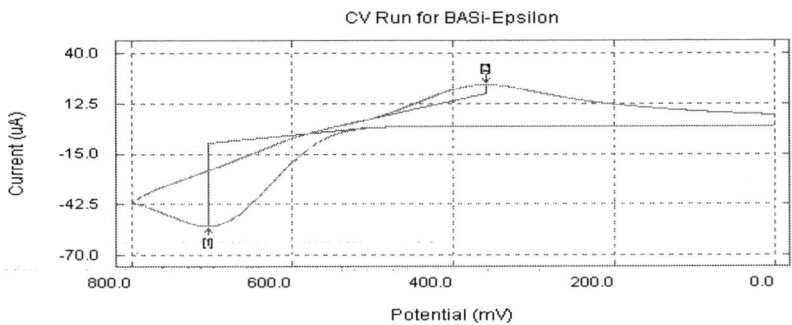

Figure 2. Bare L-DOPA and Dopamine in in 0.1 M Sulfuric acid

Acknowledgments

This work was encouraged and extended from previous sensor work carried out in the past by Dr. Dionysios D. Dionysiou from the University of Cincinnati, Department of Civil and Environmental Engineering.

References

1. H.B. Mark, N.F. Atta, Y.L. Ma, L.L. Petticrew, H. Zimmer, Y. Shi, S.K. Lunsford, J.F. Rubinson, and A. Galal, *Bioelectrochem. Bioeng.* **38**, 229 (1995).
2. N.F. Atta, I. Marawi, K.L. Petticrew, H. Zimmer, H.B. Mark, Jr., and A. Galal, *J. Electroanal. Chem.*, **408**, 47 (1996).
3. S. K. Lunsford, H. Choi, J. Stinson, and A. Yeary, *Talanta*, **73**, 172 (2007).
4. M.D.P.T. Sotomayor, A. A. Tanaka, and L.T. Kubota, Anal. Chim. Acta **455**, 215 (2002).
5. S.Q. Liu, J.J. Xu, and H.Y. Chen, Bioelectrochem. **57**, 149 (2002).

Functionalization of Pyrolyzed Carbon Micro Structures for Bio-nanoelectronics
Platforms

Mieko Hirabayashi[a], B. Mehta[a], A. Khosla[b], and S. Kassegne*[b]

[a] MEMS Research Lab, Department of Mechanical Engineering,
San Diego State University, San Diego, CA 92182, USA.
*Email: kassegne@mail.sdsu.edu
[b] School of Engineering Science, Simon Fraser University,
Burnaby, Canada. V5A 1S6

In this study, the surface-treatment of pyrolyzed carbon
microelectrodes - which are otherwise chemically inert - for
improving their attachment chemistry with double-stranded DNA
molecular wires as part of a bio-nanoelectronics platform is
investigated. Pyrolyzed carbon microelectrodes were fabricated
using standard negative lithography procedures with SU-8 (10)
negative-tone photoresist on a silicon wafer. These microelectrode
structures were then pyrolyzed and converted to a form of
conductive carbon that we refer to as PSU-8. Functionalization of
the resulting pyrolyzed structures was done using oxygen plasma
etching and the results confirmed with Fourier Transform Infrared
Spectroscopy (FTIR). Post-pyrolysis analysis using Electron
Dispersion X-ray Spectroscopy (EDS) showed a decrease in
oxygen content after pyrolysis and higher oxygen concentrations at
the edges and location of defects. FTIR results confirmed the
presence of carboxyl and hydroxyl groups in both untreated
pyrolyzed carbon (PSU-8) and plasma-treated PSU-8 structures.

Introduction

The well-known Moore's law that predicts that standard transistor capacities in integrated
circuits (IC) double approximately every two years continues to hold true (1). This
prediction has consistently been accurate since 1975 driven mostly by increasing the
density of silicon transistors, i.e., increasing the number of transistors per unit area by
both decreasing the size of each transistor and increasing the number of transistors used
in a given area. However, at some density, silicon will no longer be capable of stable
computing (2). If increasing computing speed can no longer be feasible by just increasing
transistor density, it is natural that other types of technology should be considered.

Carbon-based solutions to the silicon problems are currently under investigation in
several different realms varying from electronics chips to solar cells. In solar cells, for
example, such organic solutions have attracted attention because they are cheaper, easier
to manufacture, and have other properties that may have benefits in specific applications
such as transparency and flexibility (3). The main advantages offered by carbon are its
abundant availability (fourth most abundant element), excellent electrochemical stability,
good thermal and electrical conductivities, and excellent response to chemical surface

treatments that favor DNA attachment chemistry, particularly through electrokinetic means (4). DNA-based nanoelectronics involving metal electrodes has recently been demonstrated with either bare DNA strands (5), or metalized DNA (6-7), or DNA as a template or carrier for nanoparticles (8-14).

Some organic electronics being explored currently include DNA wires, DNA transistors, DNA memory, and DNA computers (2, 5). The current study focuses on the functionalization of a negative photoresist-derived conductive carbon electrode known as Pyrolyzed SU-8 (PSU-8) or Carbon-MEMS (C-MEMS) (15-16). The advantage to this type of carbon over carbon nanotubes or glassy carbon, which have similar properties to PSU-8, is that PSU-8 can be custom patterned for a specific application on the micron level using standard photolithography procedures. The purpose of the following study is to demonstrate that the chemical structure of PSU-8 is similar enough to glassy carbon and carbon nanotubes that oxidation methods like plasma treatment work just as well on PSU-8 for functionalization.

Methods

Generating the proper bio-nanoelectronics platform for functionalization involved first fabricating microelectrode using photolithography methods, then pyrolyzing the microelectrode to alter the structure of the photoresist from a non-conductive polymer, SU-8, to conductive PSU-8, then finally treating the microelectrode with oxygen plasma. Analysis of structure was done using FTIR and EDS. Imaging was done using light microscopes and SEM.

Photolithography

A silicon wafer was prepared for lithography by first washing with water, acetone, isopropyl alcohol, and water, drying with an air gun, and heating for 2 minutes at 200°C. Once the wafer had cooled, it was centered on the chuck of spin-coater and a layer of SU-8 was deposited. The SU-8 was spin-coated at 3000 rpm for 45 seconds, removed from the chuck, and placed on a hotplate at 45°C for soft baking. The temperature was increased to 95°C over a period of 40 minutes and then held at 95°C for 5 minutes. After soft baking, the wafer was removed from the hotplate and allowed to cool for 30 minutes. The wafer was set up under the UV lamp with the patterned mask placed over it and exposed for 30 seconds at 10mW/cm^2. Post baking was done on a hot-plate at 50°C to 100°C over a period of 30 minutes followed by stripping of none cross-linked polymer. For visual inspection and for ensuring proper development of the microelectrodes, imaging was done using 2-D and 3-D lenses on a Hirox light microscope.

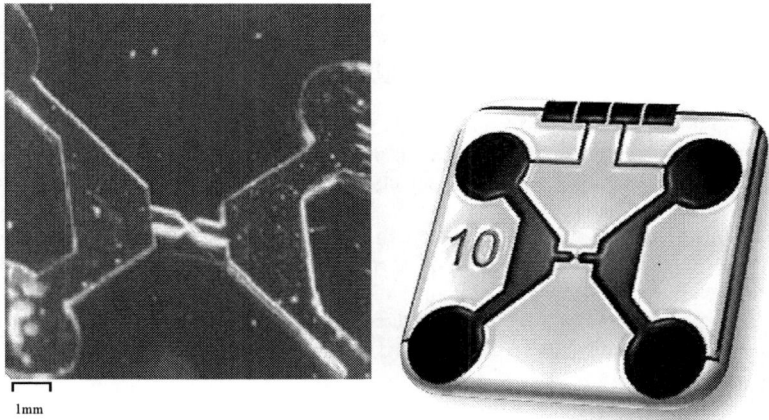

Figure 1. 3-Dimensional Hirox image of an SU-8 Electrode

Pyrolysis

Rate of heating, the amount of nitrogen flow, and the substrate, were optimized in the pyrolysis process to achieve microelectrodes with high conductivity and a strong bond between the substrate and the conductive material (12). Best results were found when SU-8-(10) structures on a silicon wafer with a silicon dioxide coating wafer were heated under continuous nitrogen flow of at least 2L/min and at the heating protocol described in Table I.

Table I: Pyrolysis Ramping Rate

Temperature	Time
Room-700°C	1.5 hours
700-900°C	1.5 hours
900-1000°C	1.5 hours
1000°C	1.5 hours
1000-900°C	1.5 hours
900-700°C	1 hour
700 °C -room	Auto off

Oxygen Plasma Etching

In order to functionalize the carbon, electrodes were exposed to oxygen plasma at a vacuum of 0.6 and 0.75 Torr at various times and powers as shown in Table II.

Table II: Plasma Treatment Parameters

Time (s)	Power (W)
60	25
7	50
60	50
7	75
60	75

Results

Electron Dispersion X-ray Spectroscopy (EDS)

Phillips FEI Quanta[TM] 450 Scanning Electron Microscope, EDS detector, and Oxford INCA software , were used to perform carbon/oxygen (C/O) weight percentage analysis for carbon and imaging for possible defects. Pure SU-8 samples before pyrolysis on a silicon and bare silicon were used as controls. Three PSU-8 electrodes were used in analysis. Two smooth, defect free sections on each electrode were chosen for EDS analysis. C/O weight percentages for carbon and error were calculated in INCA software. The six weight percentages were averaged and compared to a sample of SU-8 in Figure 3. The increase in the C/O weight percentage implies that the amount of oxygen in the sample decreased, which, as seen in Figure 3, is expected.

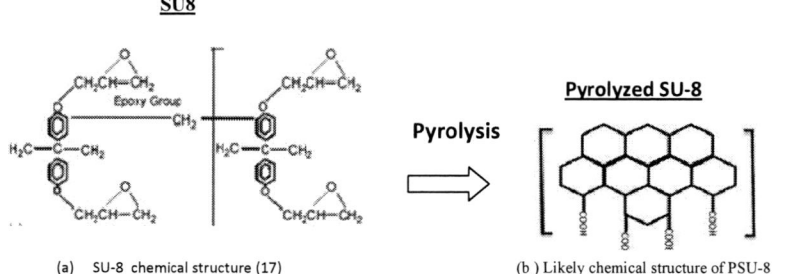

(a) SU-8 chemical structure (17)   (b) Likely chemical structure of PSU-8

Figure 2. Chemical structure of SU-8 before pyrolysis (17) and likely structure of PSU-8 after pyrolysis.

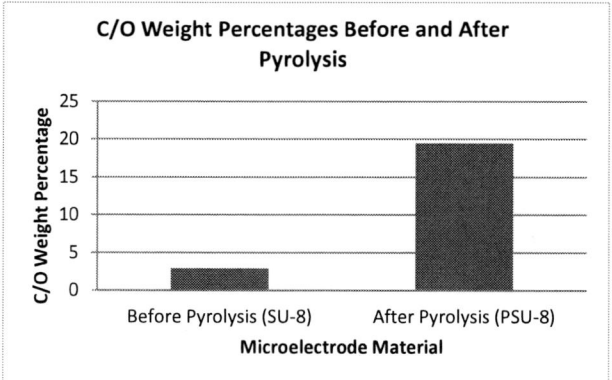

Figure 3. Graph of the change in C/O weight percentage of carbon after pyrolysis. Increase in C/O weight percentage indicates a decrease in oxygen. The y-axis represents change in weight percentage of carbon. The x-axis represents the two cases, i.e., before and after pyrolysis.

In addition to analyzing the C/O weight percentage of carbon at the smooth surface of the samples, additional data was taken at the edges and the defects of the electrodes to determine if defects and edge groups were more prevalent in these patterned microelectrodes similar to carbon nanotubes (5). Figure 4 shows that C/O weight percentages at the edges (n=6, two from each electrode) and defects (n=2, two separate electrodes) are slightly smaller than at the smooth surface indicating that there may be more carboxyl groups present along the edges and defects much like the carbon nanotube structure. The high standard deviation in the edge analysis is likely due to the wide variation in the location of the edges where data was taken. In addition, since the edge is typically slightly rounded, some of the edge spectrum could have been from a side wall rather than the edge, creating a larger variability in the data.

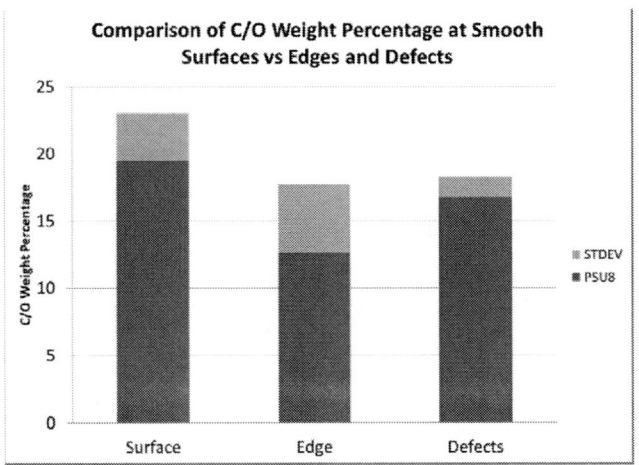

Figure 4. Comparison of C/O weight percentages at smooth surfaces versus edges and defects. Weight percentages were found to be smaller at edges and defects, indicating an increase in oxygen in these areas.

Fourier Transform Infrared Spectroscopy (FTIR)

Thermo Nicolet® FTIR was used to collect FTIR spectra and OMNIC Spectra software was used to process data. For each sample, the PSU-8 was removed from the silicon using a spatula and pressed into a KBr pellet. Background was collected before each sample. Software was used to calculate percent transmittance data to correct baseline and final data was processed in a spreadsheet.

Figure 5 shows FTIR results for treated and untreated PSU-8. All samples, including the control, show peaks at about 3450 and 1650 cm^{-1}. Peaks at about 3450 indicate hydroxyl (–OH) bond and peaks at about 1650 indicate carbon-oxygen bond (C=O) (6). We submit that, taken together, these two peaks imply the presence of carboxyl groups. However, because the EDS analysis shows that carboxyl groups may exist at edges and defects of untreated pyrolyzed sample, the FTIR profile may be picking up the carboxyl groups from the end and defect groups in the control. Therefore, FTIR may only be useful in confirming that the oxygen present in the sample is due the carboxyl rather than other bond, but cannot be used to confirm that plasma treatment increase the concentration of carboxyl groups.

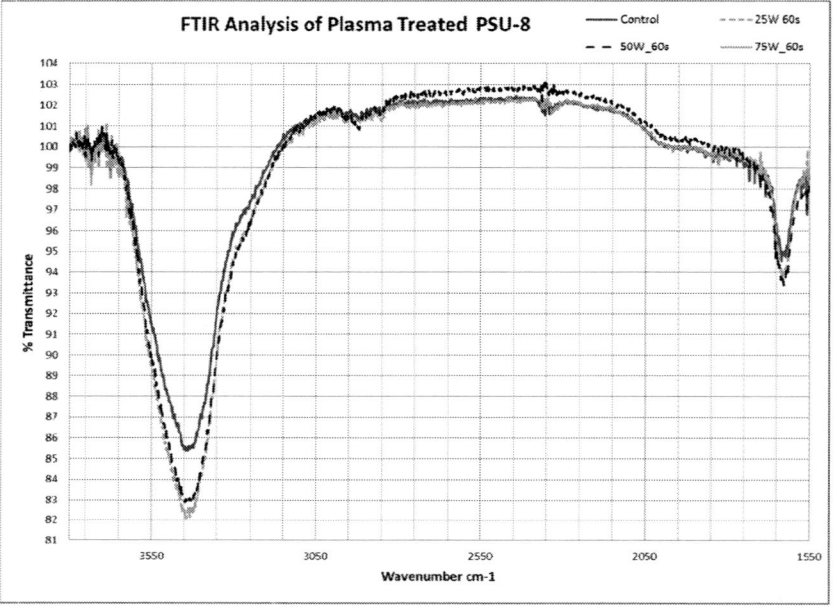

Figure 5. FTIR of PSU8 samples. Sample labeled 'control' is untreated PSU8. Other samples indicate the power and time of the oxygen plasma treatment.

Conclusions

Initial EDS analysis indicates, as expected, that oxygen content in SU8 decreases after pyrolysis when converted to PSU-8. Further analysis of PSU-8 electrodes shows an increase in oxygen weight percentage at the edges and defects of the electrodes. FTIR analysis shows peaks at 3450 and 1650 cm^{-1} indicating the likely presence of carboxyl groups in the PSU8 sample. While the FTIR proves the presence of carboxyl groups, there was little difference between the control and the plasma treated samples. However, because untreated samples were found to have increase oxygen content at edges and defects, the carboxyl found in the control samples was likely due to the edge and defect carboxyl groups. Therefore, in order to determine if plasma treatment is indeed successful at increasing carboxyl group concentration, another EDS analysis on the surface after plasma treatment should be done.

Acknowledgments

The authors would like to acknowledge San Diego State University Electron Microscope facilities and Dr. Steve Barlow for allowing us access to the SEM for our EDS analysis.

References

1. Moore, G.E. (1965). "*Cramming more components onto integrated circuits*", Electronics Magazine. p. 4.
2. Hassa, S., IEEE Computer Science Soc., 560-561, 2010.
3. Gunes, S., Neugebauer, H., and Sariciftci, N., Chem. Rev., 1324-1338, 107(2007).
4. Wang, C., Jia, G., Taherabadi, L.H., and M.J. Madou. (2005). "*A novel method for the fabrication of high-aspect ratio C-MEMS structures*". Journal of microelectromechanical systems, 14(2), 348.
5. Berlin Y.A.; Burin A.L.; Ratner M.A., '*DNA as a molecular wire, Super lattices and Microstructures*', Volume 28, Number 4, October 2000 , pp. 241-252(12).
6. Braun et al 1998Nature 391 775.
7. Richter et al 2001 Appl. Phys. Lett. 78 536.
8. Csaki, A., G. Maubach, D. Born, W. Fritzsche, Institute for Physical High Technology, Jena, Germany '*DNA-Based Construction for Nanoelectronics*'.
9. Mirkin et al 1996, Nature, 382 607.
10. R. P. Andreas, S. Datta, D. B. Janes, C. P. Kubiak, and R. Reifenberger ,The Handbook of Nanostructured Materials and Nano-technology, edited by (Academic Press, 1998).
11. Liu, Q., Wang, L., Frutos, A., Condon, A., Corn, R., and Smith, M., Nature, 403 pages 175-179, 2000.
12. Hamon, M., Hu, H., Bhowmik, P., Niyogi, S., Zhao, B., Itkis, M., and Haddon, R., Chem. Physics Lett. , **8-12**,347(2001).
13. Skoog, D., Holler, S., and Crouch, S., *Principles of Instrumental Analysis*, Sixth Ed., p. 461, Thomson Corporation, Canada (2007).
14. Joshi, M., Kal, N., Lal, R., Ramgopal, V., and Mukherji, S., '*Biosensors and Bioelectronics*', **2429–2435**, 22(2007).
15. Park, B., Zaouk, R., Wang, C., and Madou, M, "*A Case for Fractal Electrodes in Electrochemical Applications*", J. Electrochem. Soc., Volume 154, Issue 2, pp. P1-P5 (2007).
16. Park, B., Taherabadi, L., Wang, C., Zoval, J., and Madou, M., "Electrical Properties and Shrinkage of Carbonized Photoresist Films and the Implications for C-MEMS Devices in Conductive Media", J. Electrochem. Soc. 152, J136, 05.
17. M. Joshi et. al. Biosensors and Bioelectronics, 2007, 22,2429–2435.

332

ECS Transactions, 50 (12) 333-338 (2012)
©The Electrochemical Society

Self-Assembled Monolayers of Oligonucleotides as Receptor Layers for Mercury Ion Sensor

Ł. Górski[a], R. Ziółkowski[a], E. Malinowska[a]

[a] Faculty of Chemistry, Department of Microbioanalytics Warsaw University of Technology, Warsaw, Poland

> The feasibility of using gold electrodes modified with short-chain, thymine-rich, ss-DNA oligonucleotides for determination of mercury cation is examined. The methylene blue was used as a redox marker for analytical signal generation. Biosensor response was based on the difference in electrochemical signal before and after subjecting it to sample containing Hg^{2+} ion. The lower detection limit of 3 nmol L^{-1} for Hg^{2+} was observed, together with good selectivity against common metal cations.

Introduction

Sensors show many advantages over other analytical methods, including low cost, fast response, ability to perform continuous measurements directly in a sample matrix, and ease of miniaturization (1). As a consequence, numerous sensor types were employed for environmental analysis, industrial quality control and clinical diagnostics (2). Among various types of sensors, electrochemical transduction methods offer some important advantages, including lower detection limit, ease of miniaturization in terms of size and power consumption, low operational costs, good reproducibility and fast response (3). In recent years, electrochemical biosensors came to a special prominence. The receptor layers of these devices are fabricated of recognition elements of biological origin, such as enzymes, nucleic acids, antibodies or whole cells (4).

DNA-based biosensors are important for the detection of specific nucleic acid sequences, which could be important for diagnosis of certain genetic and infectious diseases or for tracing genetically modified foods (5). However, the scope of analytes that can be detected using DNA sensors is much broader, including metal ions, simple organic compounds, oligosaccharides, peptides, proteins and even cells or bacteria (6). For the determination of such analytes, functional nucleic acids, including aptamers and DNAzymes, are usually employed (7).

The interactions of metal ions with nucleic acids have been studied for many years, mainly due to the possible toxicity and cancerogenicity of certain heavy metal ions, depending to some extent on their affinity to DNA in living organisms (8). Metal ion binding to DNA is usually based on the Coulomb interactions between metal cations and negatively charged DNA residues. However, specific metal binding is also possible via the phosphate groups of the DNA backbone and the electron donor atoms of the bases.

Recently, the binding of metal ions to nucleic acids was employed for the construction of sensors for the determination of metal ions. In some of these devices, random DNA sequences are used in the receptor layer. In this case, the selectivity of resulting sensor depends mainly on electrostatic metal ion - DNA interactions. However, some specific oligonucleotide sequences show high selectivity towards certain metal ion.

333

Among them, T-T mismatch that interacts strongly with Hg^{2+} ion is of special importance (9). Electrochemical mercury biosensors employing thymine-rich oligonucleotides typically rely on conformational change of the oligonucleotide upon the forming of thymine–Hg^{2+}–thymine complex. Alternatively, hybridization of two partly-complementary strands or denaturation of double helix might be the source of analytical signal.

Experimental

Apparatus

Voltammetric measurements were performed with a CHI 660A and CHI 1040A electrochemical workstations (CH Instruments, USA). The experiments were carried out with a conventional three-electrode system consisting of a gold disk electrode (CH Instruments, USA), a gold wire auxiliary electrode and an Ag/AgCl/1.0 mol L^{-1} KCl reference electrode. All potentials are reported versus Ag/AgCl reference electrode at room temperature. The cyclic voltammetry was conducted at a sweep rate of 100 mV s^{-1} while the square wave voltammetry was performed at a pulse amplitude of 25 mV, increment of 4 mV and a frequency of 15 Hz.

Reagents

Tris-HCl, methylene blue, mercury(II) chloride, KH_2PO_4, NaOH, NaCl, EDTA, 6-mercapto-1-hexanol (MCH) were purchased from Aldrich Chemicals. H_2SO_4, HCl and H_2O_2 were purchased from POCh, Poland. All reagents were used without further purification. All solutions were prepared with Milli-Q water. Milli-Q water and all aqueous buffer solutions were sterilized using an autoclave. Deoxyoligonucleotides were purchased from Genomed Sp. Z o.o., Poland.

The base sequences were as follows (all used probes were thiolated DNA oligonucleotides):

T: 5′ -SH-$(CH_2)_6$- TTT TTT TTT TTT TTT -3′   (100% thymine bases)

A: 5′ -SH-$(CH_2)_6$- AAA AAA AAA AAA AAA -3′   (0% thymine bases)

AT: 5′ -SH-$(CH_2)_6$- CGA CTG TGA ATT CGT -3′   (33.3% thymine bases)

Oligonucleotide stock solutions were prepared with 10 mmol L^{-1} Tris–HCl, (pH 7.5) and stored in a −20° C freezer before use.

Solutions

The following solutions were prepared: piranha solution (H_2O_2/H_2SO_4; 3:1), immobilization buffer solution (1 mol L^{-1} KH_2PO_4, pH 4.5), Tris-HCl solution (50 mmol L^{-1}), mercaptohexanol solution (4 µmol L^{-1} in immobilization buffer solution), $HgCl_2$ solution in 50 mmol L^{-1} Tris-HCl solution containing 50 µmol L^{-1} methylene blue. The electrochemical measurements were carried out in 50 mmol L^{-1} Tris-HCl containing 50 µmol L^{-1} methylene blue and $HgCl_2$.

Reagents

Before any electrochemical experiments, the gold electrode was polished with alumina powder of grain sizes from 1 to 0.05 μm (CH Instruments, USA). Next, the electrode was washed with water and sonicated for 5 min in demineralized water at 400° C. Next, the piranha solution was dropped on the working gold disk electrode and incubated for 1 min. After removing the solution, the electrode was again washed with demineralized water. The last step of electrode cleaning was its voltammetric cycling in 50 mol L^{-1} Tris-HCl solution (pH 3.0), until the CV characteristic for a clean gold was obtained.

The DNA recognition monolayer was prepared as described in (10). Briefly, after the electrode cleaning, the 4 μmol L^{-1} solution of thiolated ssDNA in 1 mol L^{-1} KH$_2$PO$_4$ (pH 4.5) was dropped on the gold working disc electrode. The ssDNA immobilization was performed for 120 min. Then the solution was removed and the electrode was washed with 1 mol L^{-1} KH$_2$PO$_4$ (pH 4.5). Then electrode was incubated in the 4 μmol L^{-1} solution of MCH for 60 min. Finally, the solution was removed and the electrode was washed with immobilization buffer solution and the electrochemical measurements were conducted.

Results and discussion

The experiments were carried out to evaluate the possibility of developing a DNA-based electrochemical biosensor dedicated to determination of mercury ion. According to the literature, mercury ions strongly interacts with thymine bases creating thymine-Hg^{2+}-thymine complex. This interaction influences the general charge of DNA receptor monolayer, decreasing its negative value. This in turn should reduce the number of methylene blue redox marker particles occurring in the vicinity of the electrode surface. The electrochemical signal value of MB redox reaction should be proportional to above mentioned changes and to the mercury ion concentration in the sample solution.

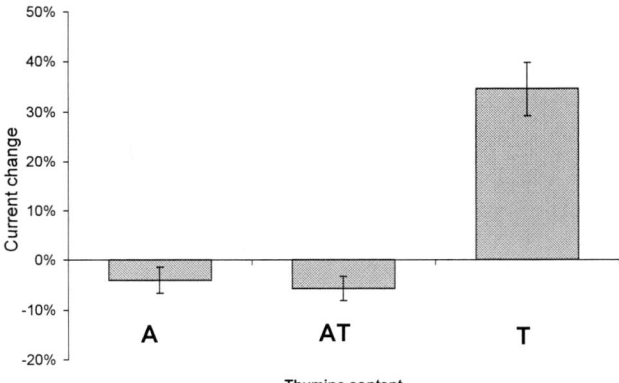

Figure 1. Influence of probe sequence on biosensor response. Experiments were conducted for 5 μM L^{-1} Hg^{2+} concentration.

To confirm the interactions between thymine bases and mercury ion, as well as the usefulness of proposed mechanism for determination of mercury ion, three different sequences of 15 nucleotides were tested. They differ in thymine base content: **T** (100% thymine bases), **A** (0% thymine bases) and **AT** (33.3% thymine bases). As it can be seen in Fig. 1, the significant change of analytical signal (MB redox current change) was obtained only for sequence which consisted exclusively of thymine bases. In the case of two other strands, the recorded response probably results from methylene blue adsorbed on the gold electrode (increasing MB current signal after second incubation) (11). Therefore, it can be concluded that thymine-rich oligonucleotides show strong interaction with mercury ion. It can be postulated, based on literature accounts, that thymine- Hg^{2+}-thymine complexes are formed within the receptor layer. Most probably, these thymine molecules originate from two different, parallel oligonucleotide strands from the monolayer.

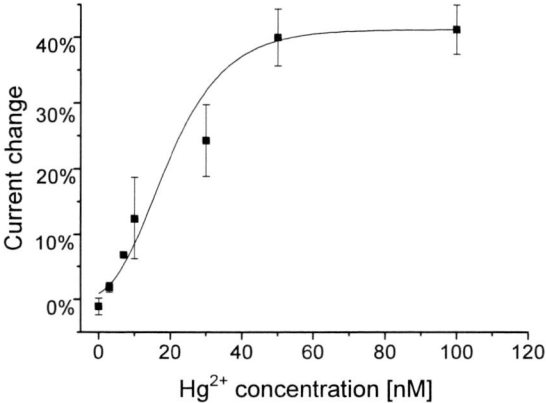

Figure 2.   Calibration curve towards Hg^{2+} ion for gold electrode modified with oligonucleotide **T** (5′ -SH-(CH$_2$)$_6$- TTT TTT TTT TTT TTT -3′).

After preliminary experiments, the calibration curve was prepared for sensor with a receptor layer consisting of **T** oligonucleotides. As it is shown in Fig. 2, together with increasing Hg^{2+} ion concentration, the biosensor response also increases. This results in the detection range from 3 to 50 nM L^{-1} of Hg^{2+} ions. The lower detection limit for proposed method (3 nM L^{-1}) is below maximum contaminant level for inorganic mercury in drinking water, set by US EPA (0.002 mg L^{-1}, ca. 10 nM L^{-1}) (12).

The last part of characterization of prepared biosensor was to evaluate its Hg^{2+} selectivity against some common metal cations (UO_2^{2+}; Ca^{2+}; Pb^{2+}; Cu^{2+}; Fe^{3+}; Mg^{2+}; Cd^{2+}). As it can be seen in Fig. 3, the strongest biosensor response was obtained for Hg^{2+} ions (34.6%). The strongest interference was observed for uranyl ion (11.5%). This can be explained by strong interactions with phosphate moieties present in sugar-phosphate backbone of DNA. The response to other tested metal cations is negligible, indicating high selectivity of proposed sensor towards mercury ion.

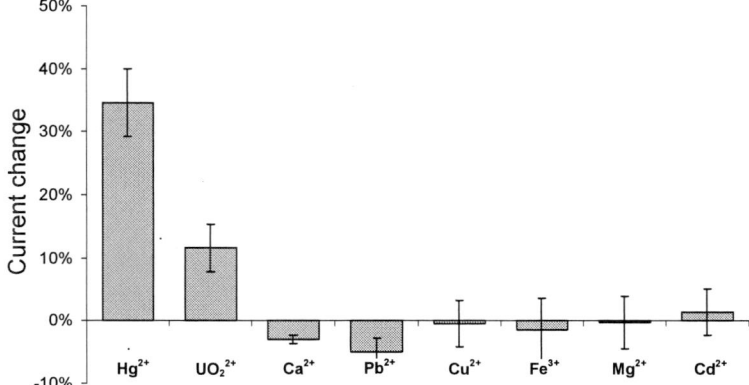

Figure 3. Selectivity of gold electrode modified with oligonucleotide **T** ($5'$ -SH-$(CH_2)_6$-TTT TTT TTT TTT TTT -$3'$) All cations were at 5 $\mu M\ L^{-1}$ concentration.

Conclusions

It has been shown herein that gold electrodes modified with short-chain ss-DNA oligonucleotides can be used for trace-level voltammetric determination of mercury cation. Proposed ssDNA recognition layer functions via strong interaction between thymine bases and Hg^{2+} ion. External redox marker, namely methylene blue, was employed for electrochemical measurements. Due to the competition between Hg^{2+} and methylene blue, current signal decreases proportionally to the analyte concentration. Proposed ssDNA-modified electrodes showed good selectivity towards Hg^{2+}, against some common metal cations, with UO_2^{2+} cation being the only one serious interferent.

The remarkably low detection limit of 3 $nmol\cdot L^{-1}$ for Hg^{2+} ions was achieved using proposed biosensor. This, together with high selectivity towards mercury ion, makes possible the application of the developed method for analysis of Hg^{2+} in real water samples.

Acknowledgments

This work was financed by the Polish Ministry of Science and Higher Education within a framework of the Operational Programme – Innovative Economy, Priority I, Action 1.3, Sub-Action 1.3.1, Project No. POIG.01.03.01-00-014/08-00.

References

1. R. Renneberg and F. Lisdat, *Biosensing for the 21st Century*, Springer, New York (2007).
2. S. Andreescu and O.A. Sadik, *Pure Appl. Chem.*, **76**, 861 (2004).
3. T. Hianik and J. Wang, *Electroanal.*, **21**, 1223 (2009).
4. J.P. Chambers, B.P. Arulanandam, L.L. Matta, A. Weis and J.J. Valdes, *Curr. Issues Mol. Biol.*, **10**, 1 (2008).

5. A. Sassolas, B.D. Leca-Bouvier and L.J. Blum, *Chem. Rev.*, **108**, 109 (2008).
6. J. Liu, Z. Cao andY. Lu, *Chem. Rev.*, **109**, 1948 (2009).
7. C. Teller and I. Willner, *Curr. Opin. Biotech.*, **21**, 376 (2010).
8. J. Anastassopoulou, *J. Mol. Struct.*, **651–653,** 19 (2003).
9. A. Ono and H. Togashi, *Angew. Chem. Int. Ed.*, **43**(33), 4300 (2004).
10. R. Ziółkowski, Ł. Górski, M. Zaborowski and E. Malinowska, *Bioelectrochemistry,* **80**(1), 31 (2010)
11. T. Sagara, H. Kawamura and N. Nakashima, *Langmuir,* **12**(17), 4253 (1996).
12. US Environmental Protection Agency, *National Primary Drinking Water Regulations,* EPA 816-F-09-004 (2009).

Development of Highly-Sensitive Electrochemical Measurement System on Dry Chemistry Using Ionic Liquid

Satoshi Arimoto, Makoto Takahashi, Akihito Kamei, and Toshihiko Yoshioka

R&D Bioscience Technology Development Office, Panasonic Corporation
3-4 Hikaridai, Seika-cho, Soraku-gun, Kyoto 619-0237, Japan

It is necessary to detect some disease marker molecules in body fluid in order to find out the disease in an early stage. Most of conventional chemical sensors seem to have problems in terms of sensitivity. We have focused on substitutional stripping voltammetry (SSV), which is highly-sensitive electrochemical measurement technique and been developing a sensor. However, it is impossible for conventional SSV to be applied for sensor without any treatment due to the complexities that SSV requires salt bridge and two electrochemical cells.
Hydrophobic ionic liquid which is room-temperature molten salt was mixed with polymer. The obtained gel possessed ionic conductivity, and could replace the auxiliary electrolyte solution and the salt bridge for SSV. We have succeeded to conduct SSV in a simplified system by use of the gel.

Introduction

In an aging society, the growth of medical spending has been a serious problem. In order to discover some diseases at an early stage, it has been required to develop new technologies that allow detection of a slight amount of disease markers in biological fluids. Most of conventional chemical sensors seem to have problems in terms of sensitivity. SSV is one of the most desirable high sensitive measurements (1, 2). However, the common SSV measurement system requires an auxiliary solution and a salt bridge in liquid system (as shown in Figure 1). This complexity of the measurement system interrupts an application of SSV for introduction to a compact sensing device.

Aiming to achieve the chemical sensor that employs the SSV in dry condition (i.e. solid system), room temperature ionic liquid (RTIL) mixed with polymer was introduced. RTIL is a salt that can keep molten state even in room temperature. The first RTIL is $AlCl_3$-ethylpyridinium bromide mixture having a melting eutectic point of -40 °C in 1951 (3). Since an air- and water-stable RTIL of 1-ethyl-3-methylimidazolium tetrafluoroborate was reported in 1992 (4), many kinds of RTILs have been synthesized and studied extensively because of their several attractive features such as negligible vapor pressure at room temperature, low-combustibility, ionic conductivity and wide electrochemical potential window. Due to these properties, some technologies have been developed in many fields (5, 6).

In this paper, we introduce a new simplified SSV measurement system using a gel made from RTIL (hereinafter called stripping gel). The stripping gel was prepared by mixing the hydrophobic RTIL with polymer. In addition, the proposed system kept high sensitivity of the SSV.

Experimental

Preparation of the Stripping Gel

50 mg of methyl methacrylate polymer (available from Wako Pure Chemical Industries, Ltd., Mw = 100,000) was dissolved in 1 ml of acetone by ultrasonic wave on ice cooling in a closed container. 50 μl of 1-buthyl-3-methylimidazolium bis(trifluoro methanesulfonyl)amide (available from TOKYO CHEMICAL INDUSTRY CO., LTD.) containing 10 mM of 1-buthyl-3-methylimidazolium chloride (available from TOKYO CHEMICAL INDUSTRY CO., LTD.) was added to the acetone solution and stirred well. The acetone solution was dropped on the silver electrode surface. The acetone was evaporated to form the gel-coated electrode in a vacuum condition.

Electrochemical Measurement

Cyclic voltammetry was conducted using a typical electrochemical cell. Counter and reference electrodes were Pt coil and Ag/AgCl (sat. KCl), respectively. All electrode potentials in this paper will be presented versus this reference electrode. The electrolyte solution was phosphate buffered saline containing some concentrations of ferrocenecarboxylic acid (Fc^{2+}-COOH). All electrochemical measurements were carried out by use of the potentiostat (BAS Co., Ltd, ALS-832C) at room temperature.

The comb-shaped working electrodes were Au (Line / Space = 2 / 2 um). The gel-coated electrode and comb-shaped electrodes were attached to the insulating substrate from the back side, as shown in Figure 2(a), and used for SSV. The measurement was conducted in two steps. The detailed procedure is described in the Results and Discussion.

Results and Discussion

Characterization of the Stripping Gel

In order to confirm that the electrochemical reaction could be carried out even in the stripping gel, two types of gel-coated electrode was prepared. Figure 3 illustrates cyclic voltammograms taken in the phosphate buffered saline in the presence (a) and the absence (b) of 10 mM 1-buthyl-3-methylimidazolium chloride in 1-buthyl-3-methyl-imidazolium bis(trifluoromethanesulfonyl)amide. The scan rate was 10 mV/s. Clear redox current peaks around 0.01 and -0.18 V were observed (Figure 3(a)). These current peaks were derived from oxidation and reduction of silver / silver chloride. The result also implies that the stripping gel behaved as a solid electrolyte. In addition ionic conductivity was maintained through the interface of the stripping gel and the phosphate buffered saline.

The ability of the stripping gel toward reduction of ferrocenecarboxylic acid (oxidant, Fc^{3+}-COOH) was also confirmed. Figure 4 shows cyclic voltammograms obtained with the comb-shaped working electrodes in phosphate buffered saline containing 100 μM of Fc^{2+}-COOH at the scan rate of 10 mV/s. Here, the comb-shaped working electrodes were composed of the first working electrode (generator) and the second working electrode (corrector). In the case that cyclic voltammetry was carried out with only the generator, clear redox current of Fc^{2+}-COOH / Fc^{3+}-COOH was observed (Figure 4c). Application of 0 V to the corrector, redox cycle of Fc^{2+}-COOH / Fc^{3+}-COOH was developed between the generator and the corrector. As a result, the oxidation current was amplified about tenfold (Figure 4b). Instead of application of 0V to the corrector, the corrector was connected to the gel-coated electrode, which was immersed in the same electrolyte solution. The gel-coated electrode used here was obtained by preparation of the stripping gel on the silver electrode (ϕ=3 mm). The amplification similar to the case of application of 0V was confirmed as shown in Figure 4a. This result means that the electrons generated on the gel-coated electrode caused by the oxidation of Ag to AgCl were spent for reduction of Fc^{3+}-COOH on the corrector.

Simplified SSV Using the Stripping Gel

The sample solution was dropped on the substrate so as to cover one of the comb-shaped working electrodes, reference electrode, counter electrode and the gel-coated electrode. SSV measurement was conducted in two steps by use of the system shown in Figure 2(b) (1, 2). In the step 1, each switch was connected to the respective terminal A. 0.4 V was applied to the generator for 600 seconds, resulting in oxidation of Fc^{2+}-COOH. Then, the corrector was electrically connected to the gel-coated electrode. A redox cycle of Fc^{2+}-COOH / Fc^{3+}-COOH was developed. During that time, silver chloride was deposited on the gel-coated electrode. The charge amount of deposition of silver chloride corresponds to that of reduction of Fc^{3+}-COOH on the corrector.

In the step 2, each switch was connected to the respective terminal B. The gel-coated electrode was swept from -0.10 to -0.30V at the scan rate of 5 mV/s. This sweep caused electrolysis of the silver chloride which had been deposited on the silver during the step 1. The generated chloride ion was dissolved into the gel. The concentration of Fc^{2+}-COOH was quantified from the amount of the current flowed during the step 2. Figure 5 shows the amount of current charge during step 2 as a function of the concentration of Fc^{2+}-COOH. Even 10^{-12}M of Fc^{2+}-COOH in phosphate buffered saline could be detected, although there was considerable measurement error.

Conclusion

We succeeded to develop a simplified electrochemical measurement system for quantifying a chemical substance contained in a sample solution at a significantly low concentration. This paper presented effective utilization of some features of RTILs. Replacing both the auxiliary solution and the salt bridge with the gel made from ionic liquid (shown in Figure 1), it has been achieved to simplify the measurement system of SSV. Further investigations aiming to develop a dry sensor using this technique are currently underway.

References

1. T. Horiuchi, O. Niwa, M. Morita and H. Tabei, *Anal. Chem.*, **64**, 3206 (1992).
2. T. Horiuchi, O. Niwa, M. Morita and H. Tabei, *Denki Kagaku*, **60**, 1130 (1992).
3. F. H. Hurley and T. P. Wier, *J. Electrochem. Soc.*, **98**, 203 (1951).
4. J. S. Wilkes and M. J. Zaworotko, *Chem. Commun.*, 965 (1992).
5. F. J. M. Rutten, H. Tadesse and P. Licence, *Angew. Chem. Int. Ed.* **46**(22), 4163 (2007).
6. T. Torimoto, T. Tsuda, K. Okazaki and S. Kuwabata, Adv. Mater., 22(11), 1196 (2010).

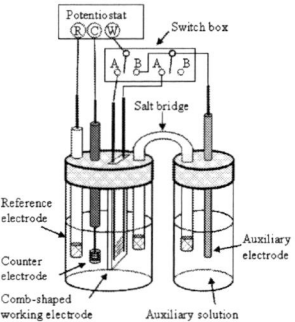

Figure 1. Schematic illustration of a conventional system of SSV.

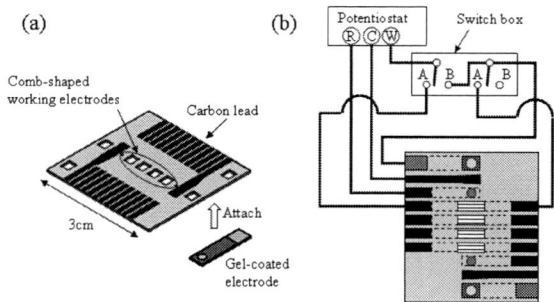

Figure 2. Schematic illustration of (a) a substrate and (b) a system of simplified SSV.

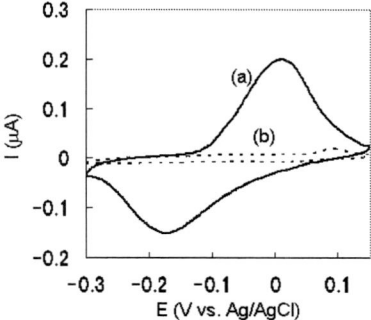

Figure 3. Cyclic voltammograms taken in the phosphate buffered saline in the presence (a) and the absence (b) of 10 mM 1-buthyl-3-methylimidazolium chloride.

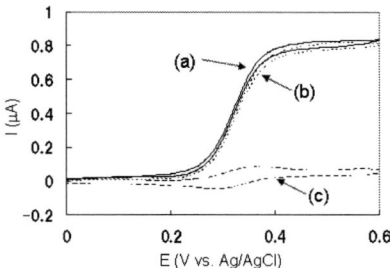

Figure 4. Cyclic voltammograms obtained with the comb-shaped working electrodes in phosphate buffered saline containing 100 μM of Fc^{2+}-COOH. The corrector was connected to (a) the gel-coated electrode, (b) the potentiostat, and (c) was in a floating state.

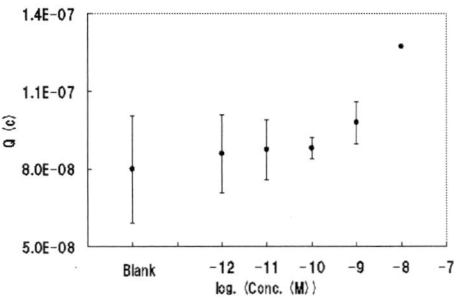

Figure 5. Plots of the amount of current charge during step 2 as a function of the concentration of Fc^{2+}-COOH.

344

Layer-By-Layer Catalytic Interface for Electrochemical Detection of Multiple Substrates Featuring Bio-Functionalized Carbon Nanotubes

J. S. Kirsch, X. Yang, and A. L. Simonian

Department of Materials Engineering, Auburn University, AL 36849

Built upon previously gained knowledge, a novel, layer by layer assembled carbon nanotube sensor for detection of two analytes is discussed. The main idea is to functionalize the carbon nanotubes with different enzymes and bio-polymers to build a multi-functional series of interfaces, which are sensitive to their respective substrates. Organophosphorus hydrolase and glucose oxidase should be separately, covalently immobilized onto multi-walled carbon nanotubes (MWNT-OPH and MWNT-GOx) for catalytic hydrolysis of the model organophosphate, paraoxon (PX) and catalytic oxidation of β-D-glucose. The catalytic response of the different layers was characterized by amperometry with redox mediator. While the sensor to simultaneously detect both glucose and paraoxon is not inherently practical, the experiment shows the possibility for co-interfacing multiple enzymes on the same platform. This may be very important to allow for the detection of several substrates, through the use of numerous biocatalysts on separate layers. Additionally, this approach might be applied for systems where several biocatalysts are assembled to work in a consecutive mode and the product of previous reaction might be a substrate for the next one.

Introduction

Layer-by-layer assembly is a nanofabrication technique which has the advantages of being versatile, relatively inexpensive, and easy, while allowing a bottom-up approach to thin film preparation. It is characterized by the addition of oppositely charged layers consisting of cationic or anionic species. The layers are built up by exposing the charged surface of the substrate/film to the anionic/cationic species in liquid form, which then adsorbs onto the substrate/film to form a layer through electrostatic attraction. Biomaterials, especially proteins which are water-soluble, are well suited for film development using this technique, due to the presence of charged sites at their surface (1-15). Recently, it has been the focus of many research groups for its application in biosensors, due to the inherent advantages of LbL assembly. One such advantage is the open structure of the films allowing for interpenetration of the layers by liquid media, which has been demonstrated to enhance the signal for single analyte systems (2, 3, 5, 8-10, 12-15). Thus, it may be advantageous to build a multi-analyte system, wherein multiple substrates could be simultaneously detected on the same platform. The layer-by-layer assembly technique is perfectly suited to this application, due to the ease of fabrication and the versatility in choosing the layer constituents (1-5, 7-15). The layer-by-layer assembly technique also has the added advantage of sensor renewability, i.e. after significant loss of sensor activity, the surface can be easily renewed by depositing a new final layer (9).

Carbon nanotubes are a popular choice for nanostructure scaffolding, sensors, and LbL assembly due to their excellent mechanical and electrical properties, and biocompatibility (7, 9-13, 16-23). In addition, carbon nanotubes may be well suited for application in LbL assemblies for their interpenetration and the highly developed porous structures between for fast diffusion of ions (7). Immobilization of biomolecules, such as proteins, onto carbon nanotubes has been widely demonstrated to enhance the stability and lifetime of proteins/enzymes, while also favorably impacting electron transfer during enzymatic catalysis of chemical compounds (9, 10, 12, 16-19, 22). However, in order for these systems to become useful, there are several challenges that must be overcome. Carbon nanotubes must be debundled and well dispersed and enzymatic activity must be preserved during the immobilization process (9, 10, 12, 17, 22, 23). Acid-treatment of the carbon nanotubes not only assists their dispersion in aqueous media, but also provides functional groups on the ends and on side wall defects, which can be exploited for the immobilization of biomolecules (9, 10, 12, 17, 22, 24).

3,4-ethylenedioxythiophene (EDOT) is a conductive monomeric compound which is polymerized to poly(3,4-ethylenedioxythiophene) (PEDOT) when a potential is applied (20, 25-27). It can therefore be used as an electrode coating for different applications, such as entrapment of enzymes and in anti-fouling from phenolic compounds (20, 26). The use of surfactants allows this monomer to be dispersed in aqueous solutions (20, 25). The enhanced electrical properties given to the coating when carbon nanotubes are added have been demonstrated, as well as the biocompatibility of the film (20). This has the distinct advantage of enabling an easily repeatable and controllable coating on the electrode surface, and has the possibility for catalytic properties (20, 26). It may be possible, depending on the permeability and thickness to use this as a replacement for the traditional electrostatic LbL process, which has very slow deposition/adsorption times (6). It may also be useful as a precursor layer to protect the solid electrode surface, and as a charged support for additional electrostatic layers.

Organophosphorus hydrolase (OPH) is a well studied enzyme actively investigated and engineered at Texas A&M University (TAMU), which catalyzes the hydrolysis of several types of organophosphate (OP) neurotoxins, and showing preferential activity towards P-O bonds (Eq. 1) (9, 22, 28-34). Organophosphate neurotoxins are typically used as pesticides and in chemical warfare agents, due to their acute toxicity and ability to irreversibly bind to and inhibit the acetylcholine esterase (AChE) neurotransmitter (35, 36). Traditionally AChE is used as a biorecognition element in biosensors to detect the presence of inhibitors; however, such systems are non-discriminate and can be inhibited by many different compounds such as heavy metals, phenols, antibiotics, etc (31, 37-39). Thus, there is a need for discriminate detection and determination of OP neurotoxins.

[1]

Glucose oxidase (GOx) is another well studied enzyme that is typically used in glucose biosensors for monitoring of β-D-glucose, especially for diabetes (2-5, 13-16, 18,

19, 21, 40, 41). With the concern of the recent rise in diabetes worldwide, considerable research has been focused on easy, fast, and specific detection of glucose in the bloodstream. GOx is a redox enzyme dependent on the cofactor FAD+ in its active site; GOx loses all activity without its cofactor FAD, and for electrochemical detection of glucose, GOx relies on an oxygen-depleted redox mediator (Eq.2-4) (14, 41). This is due to the location of the active site, which is deep inside the center of this enzyme, and is electrically insulated by the protein's structure (2). Immobilization of GOx onto carbon nanotube scaffolding has been demonstrated to exhibit direct electron transfer (DET) due to the electrical properties of the carbon nanotubes (13, 16, 18, 19, 21, 42).

Layer-by-layer assembly has also potentially allowed for "mediator-free" detection through "electrochemical wiring" of the enzymes with redox-modified polymers and protein entrapment (2-5, 15, 40).

$$glucose + GOD_{(Ox)} \rightarrow gluconolactone + GOD_{(Red)} \qquad [2]$$

$$GOD_{(Red)} + 2Fecp_2R^+ \rightarrow GOD_{(Ox)} + 2Fecp_2R + 2H^+ \qquad [3]$$

$$2Fecp_2R \rightleftarrows 2Fecp_2R^+ + 2e^- \qquad [4]$$

The objective of this research is to demonstrate multi-substrate biosensing through the use of two unrelated enzymes with different kinetic mechanisms which catalyze the hydrolysis/oxidation of two completely different substrates (paraoxon and glucose) in the layer-by-layer assembly technique. In order for this to be accomplished, both OPH (positively charged) and GOx (negatively charged) must be covalently attached to the carbon nanotubes and an electrochemical sensor must be fabricated using these constituents. Both enzymes must show activity to their respective substrate after immobilization onto the carbon nanotubes, and after sensor fabrication. Previous work has demonstrated that PEI and DNA, after immobilization onto multiwalled carbon nanotubes (MWNT), make an ideal cushioning support/interface between the electrode and the bio-catalytic interface (9). Through this demonstration, it is hoped that there will be a greater understanding of how the structure of the layers, the pH of the working buffer, and the enzyme loading affects the multilayer permeability and catalysis of the substrate, as well as the sensor response, when multiple analytes/substrates are present.

Materials & Methods

Materials:

(MWNTs (purity 95%, length 1-5 μm, diameter 30 +\- 10 nm) prepared by a chemical vapor deposition (CVD) process were purchased from Nanolabs. Organophosphorus hydrolase (OPH) was generously provided by Dr. James Wild and his research group from TAMU. Paraoxon purchased from ChemService,Inc, dissolved in 25 mL DI water and allowed to stir at 4°C for 72 hours. Glucose oxidase (GOx) from Aspergillus niger, DNA (lyophilized salmon sperm salt), N-hydroxysuccinimide (NHS), N-ethyl-N-(3-dimethylaminopropyl)carbodiimide hydrochloride (EDC), polyethyleneimine (PEI), ferrocene carboxylic acid (FCA), 2-(N-morpholino)ethanesulfonic acid (MES), N-cyclohexyl-2-aminoethanesulfonic acid

(CHES), and phosphate buffered saline (PBS) were all obtained from Sigma-Aldrich. DI water was obtained from Millipore Direct-Q 5 Water system (resistivity, 18MΩ cm^{-2}).

Instruments:

Cyclic voltammetry and amperometric measurements were performed using an electrochemical analyzer CHI 660 (CH Instruments, Austin, TX) connected to computer running a software package. A three-electrode electrochemical cell configuration was used, consisting of the modified glassy carbon (GC) electrode (2mm diameter) working electrode, with Ag/AgCl (3 M KCl) and platinum disk electrode respectively serving as the reference and counter electrodes, in a homemade 20 mL capacity batch cell. The Unicell system was used for flow mode applications consisting of a glassy carbon (2mm diameter) working electrode and a counter and reference electrode purchased from BASi. The working electrode in both cases was the same. All measurements were carried out at room temperature (25°C) unless otherwise stated.

Oxidation of Nanotubes:

The 50 mg of carbon nanotubes were suspended in a mixture of 150 mL of H2SO4 and 50 mL HNO3, and sonicated in a Branson bath sonicator for 6 hours (9). The black solution was then added to 800 mL of DI water and filtered through 0.2 μm polycarbonate filter membrane (Watman) and re-suspended in 1 L of water and subsequently filtered, and this was repeated two additional times (to near-neutral pH). A black grease-like substance was obtained, and then dried in a vacuum oven (on the filter paper) at 80°C overnight. A thin, black paper-like material was peeled off the filter paper and placed into a scintillation vial for storage.

Immobilization of OPH:

2 mg of the acid treated CNT were suspended in 5 mL of DI water with a tip sonicator (20% Amplitude, 1 hr) in an ice bath to prevent heating of the sample (9). The result was an optically homogenous black solution. 500 μL of DI Water and 500 μL of 500 mM pH 4.7 MES buffer were added to the dispersed nanotubes. 2.3 mL of a 50 mg/mL solution of NHS and 1.2 mL of a 10 mg/mL solution of EDC were added to the buffered nanotube solution, an allowed to stir for 30 min (22). The resulting solution was centrifuged at 13,200 RPM for 30 min and the supernatant was discarded to remove the excess EDC/NHS. The nanotubes were re-dispersed in 9 mL of MES buffer pH 6.7 and then 1 mL of 0.2 mg/mL OPH in 20 mM CHES pH 8.6 was added. This solution was placed on a platform shaker at its lowest speed and allowed to shake overnight at 4°C. The solution was then centrifuged at 13,200 RPM for 30 min, the supernatant was taken, and the nanotubes re-dispersed in CHES buffer three times, to remove unbound protein (9). The final supernatant was taken to check for activity, and if active, the centrifugation process was repeated until no activity was present. This solution was concentrated by dispersing the final CNT in 1 mL of CHES buffer.

Immobilization of GOx (under progress):

To obtain active CNT-GOx solution, a similar procedure to the one above should be used. In addition, to check for permeability, the apo-enzyme (GOx without FAD) should also be immobilized onto the CNT.

Preparation of PEI/DNA:

2 mg Acid treated MWNTs were added to PEI (1 mg/mL) and sonicated with a tip sonicator in an ice bath (20% Amplitude, 1 hr) to disperse the nanotubes. The resulting solution was centrifuged at 13,200 RPM for 30 min to reduce the amount of unbound polymer. 0.1 wt% solution of DNA was dissolved in DI water at 35 °C with a stir bar for 45 min. 0.1 wt% MWNT was added to the DNA solution and sonicated with a tip sonicator in an ice bath (20% Amplitude, 1 hr). The resulting solution was centrifuged at 13,200 RPM for 30 min to reduce the amount of unbound DNA (9, 10).

Preparation of LBL Assemblies:

A glassy carbon electrode was polished thoroughly with 1 μm, 0.1 μm, and 0.05 μm alumina electrode polish (Beuhler), and washed with DI water. The electrode was placed in a beaker with DI water and sonicated in the bath sonicator for 15 min. The electrode was then treated in 1 M NaOH for 5 minutes at 1.2V to produce a negatively charged surface. MWNT-PEI and MWNT-DNA were alternatively adsorbed onto the surface of the electrode by a small drop of the solution on top of the electrode, for a consistent amount of time. The drop was then washed off with DI water and then dried with pure nitrogen gas. Two bi-layers of MWNT-PEI/MWNT-DNA (total of four layers) were built. MWNT-OPH was then adsorbed using the same time and procedure, alternating with MWNT-GOx. After the first four, odd numbered layers were MWNT-OPH and even numbered layers were MWNT-GOx. Sensors were kept refrigerated until use, and each sensor was allowed to sit overnight in refrigerated conditions before use.

Results and Discussion

Using UV-Vis spectroscopy it was confirmed for free enzyme, supernatant and MWNT-OPH solutions, that there is very little activity in the supernatant, and that there is activity for MWNT-OPH, albeit reduced from free enzyme (Fig 1). Final concentration of CHES (pH 8.3) in each cuvette was 20mM and the reference was taken with respect to the enzyme. The reaction was initiated with increasing concentrations of paraoxon. Table 1 is a comparison of the V_{max} and K_M of each catalysis agent.

TABLE I. Comparison of Catalytic Rates of Each Agent

Agent	V_{max} (k_{cat}/s)	K_m (μmol)
WT-OPH	8873	0.0658
MWNT-OPH	2306	0.044
Supernatant	237.8	0.0395

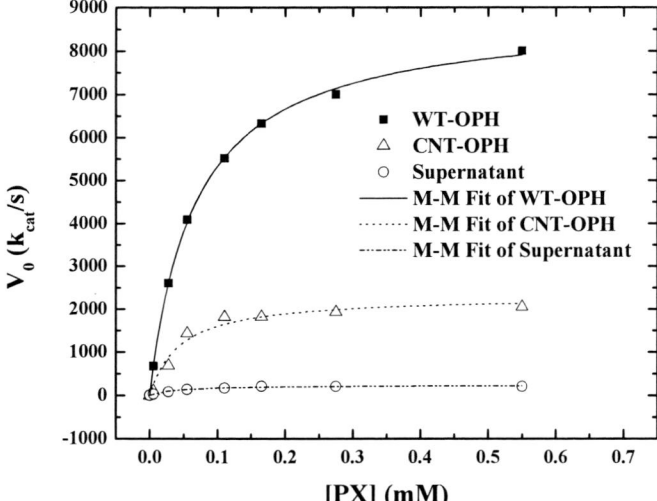

Figure 1. Michaelis Menton plot of free enzyme (WT-OPH), immobilized enzyme (CNT-OPH), and the supernatant drawn off of the CNT-OPH solution. The increased activity of the CNT-OPH over the supernatant confirms immobilization of OPH onto the MWNT.

Electrochemistry:

Cyclic voltammetry shows that p-nitrophenol is oxidized at 0.95V (Fig. 2). It was observed in amperometry that at 0.95V there was a peak for injections of paraoxon with a linear calibration, for sensors with MWNT-OPH (Fig. 3 & 4). The terminal layer of that particular sensor was MWNT-OPH, so there is no barrier to the enzyme that was immobilized onto the multiwalled carbon nanotubes. For adsorption time, it is shown (Fig.5) that increasing the amount of time for adsorption also increases catalytic response for one layer of OPH underneath one layer of GOx, corresponding to thickness of the layers. As layer thickness increases, so does the amount of enzyme available for catalysis of the substrate; in this case, there is more OPH available to cause hydrolysis of paraoxon. One would expect that at higher adsorption times however, that the thickness of the layer will eventually impede the ability for the substrate to reach the lower catalytic layers, and further layer adsorption would be futile. Therefore it is essential to determine the most optimal adsorption times, as well as the amount of enzyme bound to each of the carbon nanotubes.

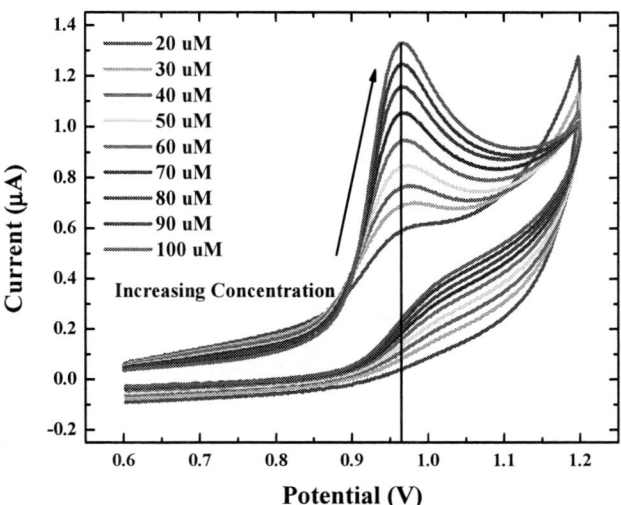

Figure 2. CV of PNP on bare glassy carbon electrode in batch mode cell. PNP was added to the solution in increasing concentration and the solution was stirred for several seconds before each CV. Scan rate was 100 mV/s from 600 mV to 1200 mV, in 10 mM PBS pH 7.4.

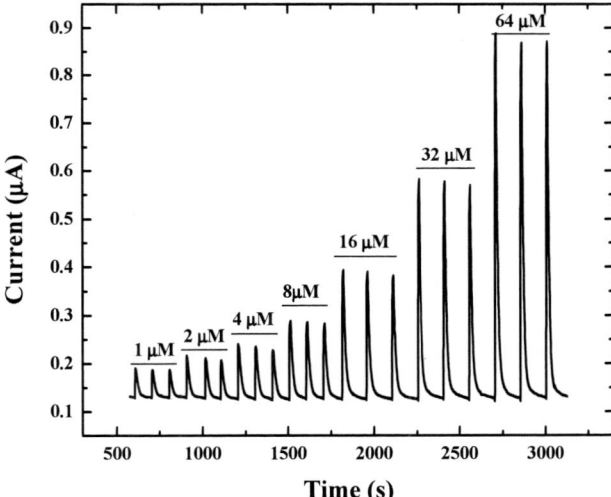

Figure 3. Amperometric response of a seven layer sensor (MWNT-PEI/MWNT-DNA)$_2$(MWNT-OPH/MWNT-GOx)(MWNT-OPH) to serial injections of increasing concentrations of paraoxon. Voltage was constant at 0.95V and flow rate was 20 mL/hr, with a buffer of 10 mM PBS pH 7.4.

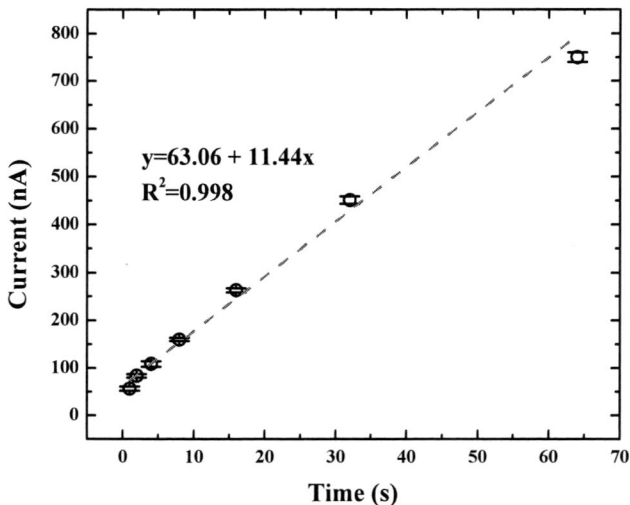

Figure 4. Amperometric calibration curve for the seven layer sensor shown above. The limit of detection is 260 nM paraoxon, calculated by multiplying the standard deviation of the baseline by three and dividing by the slope of the calibration curve.

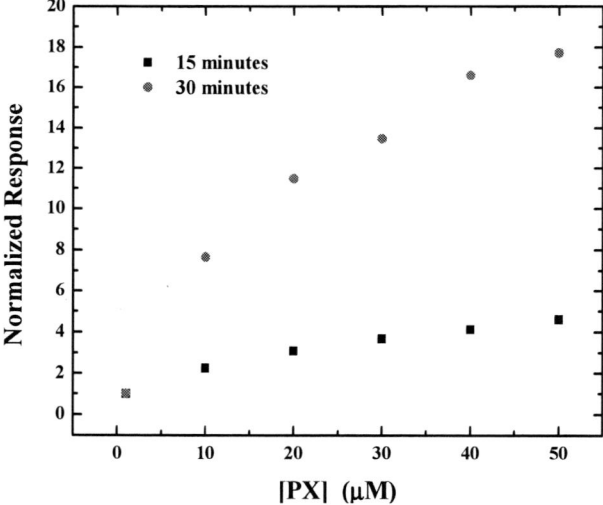

Figure 5. Comparison of the Amperometric response to paraoxon from two 8 layer sensors (MWNT-PEI/MWNT-DNA)$_2$(MWNT-OPH/MWNT-GOx)$_2$ with different adsorption times. The response is normalized to the first concentration. The difference may be explained through layer thickness, with the longer adsorption time allowing for a thicker layer. It is important to note that the catalysis layer is underneath a layer of MWNT-GOx which has no catalytic ability towards paraoxon.

What is expected to be obtained/demonstrated:

- Successful Immobilization of GOx
- Activity of GOx layers
- Layer thickness affect on catalytic activity (varying the adsorption time)
- Effect of pH on the system
- Permeability of substrate into the layers
- Sensor longevity/renewability
- Effect of SWNT vs. MWNT
- Viability of MWNT/PEDOT as precursor layer
- Viability of MWNT/PEDOT/Enzyme as multi-layer assembly

Acknowledgements

The authors would like to thank the AUDFS Center for their financial support, grant USDA-20053439415674A. This material was based on work which supported ALS by the National Science Foundation, while working at the Foundation. The views expressed in this article are those of the authors, and do not necessarily reflect the official policy or position of the NSF. The authors would also like to thank Dr. Wild's research group for providing the OPH and for their assistance with it when necessary.

References

1. K. Ariga, J. P. Hill and Q. Ji, *Phys. Chem. Chem. Phys.*, **9**, 2319 (2007).
2. E. J. Calvo, R. Etchenique, L. Pietrasanta, A. Wolosiuk and C. Danilowicz, *Analytical chemistry*, **73**, 1161 (2001).
3. E. J. Calvo and A. Wolosiuk, *ChemPhysChem*, **6**, 43 (2005).
4. V. Flexer, E. S. Forzani, E. J. Calvo, S. J. LudueÃ±a and L. I. Pietrasanta, *Analytical chemistry*, **78**, 399 (2006).
5. J. Hodak, R. Etchenique, E. J. Calvo, K. Singhal and P. N. Bartlett, *Langmuir*, **13**, 2708 (1997).
6. C. Jiang and V. V. Tsukruk, *Advanced Materials*, **18**, 829 (2006).
7. S. W. Lee, B. S. Kim, S. Chen, Y. Shao-Horn and P. T. Hammond, *Journal of the American Chemical Society*, **131**, 671 (2008).
8. Y. Lee, I. Stanish, V. Rastogi, T. Cheng and A. Singh, *Langmuir*, **19**, 1330 (2003).
9. S. Mantha, V. A. Pedrosa, E. V. Olsen, V. A. Davis and A. L. Simonian, *Langmuir* (2010).
10. D. Nepal, S. Balasubramanian, A. L. Simonian and V. A. Davis, *Nano letters*, **8**, 1896 (2008).
11. M. Olek, J. Ostrander, S. Jurga, H. MÃ¶hwald, N. Kotov, K. Kempa and M. Giersig, *Nano letters*, **4**, 1889 (2004).
12. V. A. Pedrosa, T. Gnanaprakasa, S. Balasubramanian, E. V. Olsen, V. A. Davis and A. L. Simonian, *Electrochemistry Communications*, **11**, 1401 (2009).
13. T. W. Tsai, G. Heckert, L. F. Neves, Y. Tan, D. Y. Kao, R. G. Harrison, D. E. Resasco and D. W. Schmidtke, *Analytical chemistry*, **81**, 7917 (2009).
14. W. Zhang, Y. Huang, H. Dai, X. Wang, C. Fan and G. Li, *Analytical biochemistry*, **329**, 85 (2004).
15. W. Zhao, J. J. Xu, C. G. Shi and H. Y. Chen, *Langmuir*, **21**, 9630 (2005).
16. C. Cai and J. Chen, *Analytical biochemistry*, **332**, 75 (2004).
17. Y. Gao and I. Kyratzis, *Bioconjugate chemistry*, **19**, 1945 (2008).
18. A. Guiseppi-Elie, C. Lei and R. H. Baughman, *Nanotechnology*, **13**, 559 (2002).
19. Y. Lin, F. Lu, Y. Tu and Z. Ren, *Nano letters*, **4**, 191 (2004).
20. M. Liu, Y. Wen, D. Li, H. He, J. Xu, C. Liu, R. Yue, B. Lu and G. Liu, *Journal of Applied Polymer Science* (2011).
21. Y. Liu, M. Wang, F. Zhao, Z. Xu and S. Dong, *Biosensors and Bioelectronics*, **21**, 984 (2005).
22. V. A. Pedrosa, S. Paliwal, S. Balasubramanian, D. Nepal, V. Davis, J. Wild, E. Ramanculov and A. Simonian, *Colloids and Surfaces B: Biointerfaces*, **77**, 69 (2010).
23. V. A. Sinani, M. K. Gheith, A. A. Yaroslavov, A. A. Rakhnyanskaya, K. Sun, A. A. Mamedov, J. P. Wicksted and N. A. Kotov, *J. Am. Chem. Soc*, **127**, 3463 (2005).
24. Z. Wang, M. D. Shirley, S. T. Meikle, R. L. D. Whitby and S. V. Mikhalovsky, *Carbon*, **47**, 73 (2009).
25. S. H. Cho, H. J. Lee, Y. Ko and S. M. Park, *The Journal of Physical Chemistry C* (2011).
26. V. Serafin, L. Agui, P. Yanez-Sedeno and J. M. Pingarron, *Journal of Electroanalytical Chemistry*, **656**, 152 (2011).
27. M. S. Yavuz, G. C. Jensen, D. P. Penaloza, T. A. P. Seery, S. A. Pendergraph, J. F. Rusling and G. A. Sotzing, *Langmuir*, **25**, 13120 (2009).

28. M. M. Benning, J. M. Kuo, F. M. Raushel and H. M. Holden, *Biochemistry*, **33**, 15001 (1994).
29. D. P. Dumas, S. R. Caldwell, J. R. Wild and F. M. Raushel, *Journal of Biological Chemistry*, **264**, 19659 (1989).
30. J. K. Grimsley, J. M. Scholtz, C. N. Pace and J. R. Wild, *Biochemistry*, **36**, 14366 (1997).
31. K. Lai, N. J. Stolowich and J. R. Wild, *Archives of biochemistry and biophysics*, **318**, 59 (1995).
32. T. E. Reeves, M. E. Wales, J. K. Grimsley, P. Li, D. M. Cerasoli and J. R. Wild, *Protein Engineering Design and Selection*, **21**, 405 (2008).
33. R. D. Richins, A. Mulchandani and W. Chen, *Biotechnology and bioengineering*, **69**, 591 (2000).
34. K. Y. Wong and J. Gao, *Biochemistry*, **46**, 13352 (2007).
35. T. A. Slotkin, E. D. Levin and F. J. Seidler, *Environmental health perspectives*, **114**, 746 (2006).
36. F. Worek, M. Koller, H. Thiermann and L. Szinicz, *Toxicology*, **214**, 182 (2005).
37. M. D. Luque de Castro and M. C. Herrera, *Biosensors and Bioelectronics*, **18**, 279 (2003).
38. S. Rodriguez-Mozaz, M. J. Alda, M. P. Marco and D. Barcelo, *Talanta*, **65**, 291 (2005).
39. S. Rodriguez-Mozaz, M. J. Lopez de Alda and D. Barcelo, *Analytical and bioanalytical chemistry*, **386**, 1025 (2006).
40. E. J. Calvo, C. Danilowicz and A. Wolosiuk, *Journal of the American Chemical Society*, **124**, 2452 (2002).
41. A. E. G. Cass, G. Davis, G. D. Francis, H. A. O. Hill, W. J. Aston, I. J. Higgins, E. V. Plotkin, L. D. L. Scott and A. P. F. Turner, *Analytical chemistry*, **56**, 667 (1984).
42. T. O. Tran, E. G. Lammert, J. Chen, S. A. Merchant, D. B. Brunski, J. C. Keay, M. B. Johnson, D. T. Glatzhofer and D. W. Schmidtke, *Langmuir*, **27**, 6201 (2011).

Chalcogenide Glass Chemical Sensor for Cadmium Detection in Industrial Environment

M. Milochova, M. Kassem, and E. Bychkov

Laboratory of Physical Chemistry of Atmosphere, University of Littoral, 59140 Dunkerque, France

Advanced chemical sensors for cadmium detection based on chalcogenide glasses have been developed for continuous or quasi-continuous *in situ* monitoring of industrial atmospheric particles containing heavy-metals. Analytical performance of the sensors and a laboratory prototype of the monitoring system are presented.

Introduction

Industrial atmospheric particles emitted by a metallurgical plant contain heavy metals (lead, cadmium, arsenic, etc.). The emissions are controlled on a regular basis, in most cases once a month. To reduce the environmental risk and possible impact on the neighboring population, a more frequent or continuous monitoring is required. The existing control procedure includes particle collection on a filter with subsequent filter mineralization and analysis, e.g. using ICP-MS. This procedure cannot be transformed into a continuous or quasi-continuous *in situ* monitoring. Recently, we have proposed an alternative technique including collection and analysis modules adapted for in situ measurements (1). The collection module (Figure 1) allows direct dissolution of heavy metals in a specific solvent. The analysis module is capable of quasi-continuous monitoring of dissolved heavy metals using advanced chemical sensors based on chalcogenide glasses (2,3).

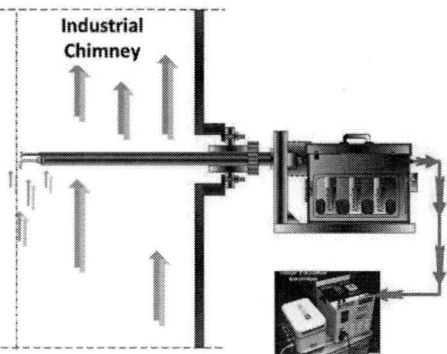

Figure 1. A schematic representation of the developed collection module and the collection procedure for an industrial chimney. Instead of using industrial particles collection on a filter with subsequent filter mineralization and analysis, the atmospheric particles are directly solubilized in a collection module allowing continuous monitoring.

Lead chemical sensors with improved selectivity, long-term stability and low detection limit have been developed for the analysis module (4). In this paper we report the R&D results for cadmium sensors.

Experimental Details

Glass Synthesis

$(CdSe)_x(AgI)_{0.5-x/2}(As_2Se_3)_{0.5-x/2}$ samples (x = 0.05, 0.10, 0.15, 0.20) were prepared using the appropriate proportions of previously synthesized CdSe, AgI and As_2Se_3. The purified elements were weighed in correct proportions using a balance of precision $\pm$ 0.1 mg to give a total sample mass of 3 g. They were sealed under vacuum (10^{-6} mbar) in a cleaned silica tube (8 mm ID and 1 mm wall thickness), heated slowly in a rocking furnace to 850 °C at a rate of 5 K min^{-1}, maintained at this temperature for 24 hours, and cooled down to 650 °C before quenching in cold salt/water mixture. Further details of glass synthesis are published elsewhere (5).

Sensor Preparation

Sensor membranes of 5-6 mm in diameter and 2-4 mm thick were cut from the melt and polished. Silver was sputtered on the back side of the membrane, and a metallic wire was attached to it by a silver microadhesive. The membranes were then sealed into PVC tubes (Figure 2). Fourteen chemical sensors with different membrane compositions were prepared for potentiometric measurements.

Figure 2. A chalcogenide glass and a chemical sensor on its basis (1).

Potentiometric Measurements

The potentiometric measurements have been carried out using the following electrochemical cell:

Ag,AgCl | KCl (sat.) || KNO$_3$ (0.1 M) || Cd(NO$_3$)$_2$, KNO$_3$ (0.1 M) | Cd Sensor [1]

Other details of potentiometric measurements and sensors tests are given elsewhere (1-3).

Experimental Results and Discussion

Cadmium Sensitivity

Sensor sensitivity to Cd^{2+} ions was studied in cadmium nitrate solutions in the concentration domain between 10^{-7} M and 10^{-3} M. Figure 3 shows typical calibration curves in Cd(NO$_3$)$_2$ solutions for a new cadmium sensor measured continuously over a two week time period. The response slope S is sub-Nernstian (24-26 mV/pCd vs. S_0 = 29.6 mV/pCd at 298K):

$$E = E^0 + RT/2F \ln a_{Cd}^{2+}, \qquad [2]$$

where S_0 = 2.303 $RT/2F$, a_{Cd}^{2+} is the thermodynamic activity of Cd^{2+} ions, R, T and F have the usual meaning.

The sensor response was found to be stable and reproducible ($\pm$ 4-5 mV for the standard potential E^0). The low detection limit is about 3 10^{-7} M or $\approx$30 ppb.

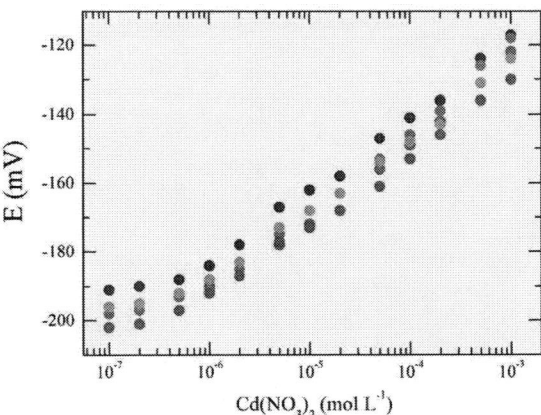

Figure 3. Typical calibration curves for a typical new cadmium sensor measured over a two weeks' time period: day1, day 2, day5, day10, day14.

Selectivity in Standard Solutions

The sensors appear to be highly selective in the presence of alkali, alkali-earth and most heavy-metal cations. Nevertheless, Cu^{2+} and Pb^{2+} ionic species exhibit strong interfering effect. Selectivity coefficients of the new cadmium sensors are shown in

Tables I and II measured using a constant concentration of interfering cation or constant concentration of Cd^{2+} ion, respectively.

The Nikolsky-Eisenman equation [3] was used to calculate the selectivity coefficient $K_{Cd^{2+},M^{Z+}}$ (6):

$$E = E^0 + RT/2F \ln (a_{Cd}^{2+} + K_{Cd^{2+},M^{Z+}} \, a_M^{Z+}),$$ [3]

where a_M^{Z+} is the thermodynamic activity of interfering M^{Z+} ions. Lower numerical value of the selectivity coefficient corresponds to better selectivity of the potentiometric chemical sensor.

TABLE I. Selectivity coefficients for cadmium sensors in the presence of some alkali, alkali-earth and heavy-metal cations measured using a constant concentration of interfering cation.

Interfering Ion	Interfering Ion Concentration	Selectivity Coefficient
Potassium	10^{-1} M	3.8×10^{-7}
	10^{-2} M	6.0×10^{-7}
	10^{-3} M	1.1×10^{-6}
Calcium	1 M	1.0×10^{-6}
Nickel	10^{-1} M	1.6×10^{-3}

TABLE II. Selectivity coefficients for cadmium sensors in the presence of lead and copper cations measured using a constant concentration of cadmium ions.

Interfering Ion	Cadmium Ion Concentration	Selectivity Coefficient
Lead	10^{-1} M	1×10^2
	10^{-2} M	2×10^2
	10^{-3} M	3×10^2
	10^{-4} M	5×10^2
Copper	10^{-1} M	2×10^3
	10^{-2} M	4×10^2
	10^{-3} M	8×10^2
	10^{-4} M	2×10^4

pH influence

The optimal pH range for sensor measurements depends on cadmium ion concentration in the solution. At high pH, the precipitation of cadmium hydroxide causes a decrease of the sensor potential. In acid solutions, the H^+ ions affect the sensor response (Figure 4).

The optimal pH range appears to be $6 \leq pH \leq 8$ for diluted cadmium concentrations but shifts to more acid solutions, $4 \leq pH \leq 7.5$, with increasing $Cd(NO_3)_2$.

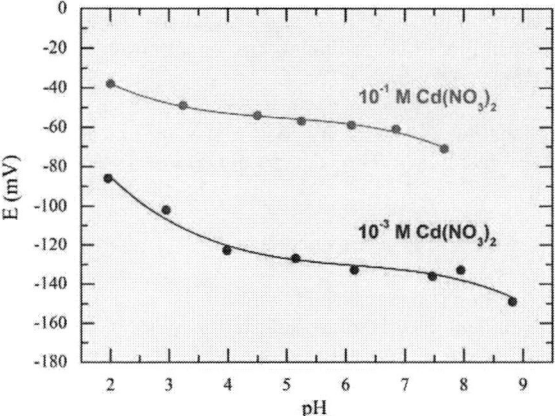

Figure 4. Cadmium sensor response in $Cd(NO_3)_2$ solutions as a function of pH. A potential decrease at pH > 7 is caused by precipitation of cadmium hydroxide. The H^+ ions affect the sensor response in acid solutions.

Reproducibility of Sensor Potentials

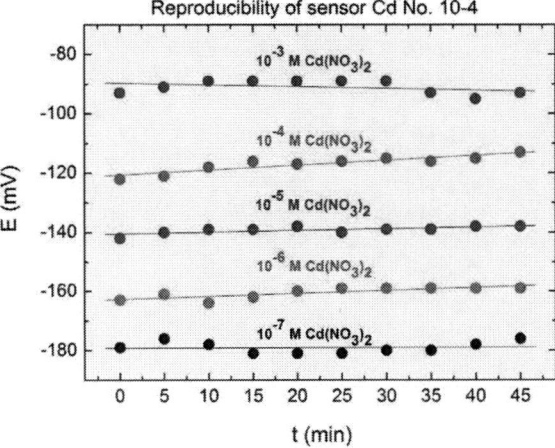

Figure 5. Typical reproducibility of cadmium sensor in $Cd(NO_3)_2$ solutions as a function of measurement time with an intermediate storage of the sensor in open air.

Potential reproducibility is an important sensor parameter for continuous *in situ* measurements. The reproducibility was studied using a series of consecutive sensor measurements in standard solutions with an intermediate storage of the sensors in open air (2). Typical reproducibility of the cadmium sensors is shown in Figure 5 for different $Cd(NO_3)_2$ concentrations.

The sensor reproducibility is reasonably good: $\pm$ 2-4 mV and essentially invariant in diluted and concentrated cadmium nitrate concentrations.

Conclusions

New $CdSe-AgI-As_2Se_3$ chemical sensors exhibit high sensitivity and low detection limit sufficient for cadmium detection in industrial environment, i.e., for cadmium monitoring in case of industrial atmospheric particles. The sensors show also high selectivity in the presence of alkali, alkali-earth and many heavy-metal cations but remain rather sensitive to copper and lead interfering species. Reasonably good potential reproducibility and long-term stability are promising for in situ applications; however, further studies are needed to ensure optimized and reliable sensor performance.

Acknowledgments

This work was supported by the EU Interreg IVA 2-Seas program (the CleanTech project).

References

1. M. Milochova and E. Bychkov, *J. Phys. Soc. Japan*, **79**, Suppl. A, 173 (2010).
2. M. Miloshova, D. Baltes, and E. Bychkov, *Water Science Technology*, **47**, 135 (2002).
3. E. Bychkov, Yu. Tveryanovich, and Yu. Vlasov, in *Applications of chalcogenide glasses*, R. Fairman and B. Ushkov, Editors, p. 103, Semi-conductors and Semimetals Series, Vol. 80, Elsevier, New York – London, (2004).
4. M. Milochova and E. Bychkov, *French Patent No. 06 08805*, registered on October 6, 2006, issued on January 9, 2009: BO de la propriété industrielle n° 09/02 (publ. No. 2 906 804).
5. M. Kassem *et al*, *Solid State Ionics*, **181**, 466 (2010).
6. W.E. Morf, *The principles of ion selective electrodes and of membrane transport*, Akadémiai Kiado, Budapest (1981).

ECS Transactions, 50 (12) 363-368 (2012)
©The Electrochemical Society

Electrochemical Pump Consisting of Cu^{2+}-Poly(acrylic acid) Gel

K. Takada, N. Yamamura, A. Hayashi, T. Yasui, and A. Yuchi

Graduate School of Engineering, Department of Material Science and Engineering,
Nagoya Institute of Technology, Gokiso, Nagoya 466-8555, Japan

An electrochemical pump consisting of poly(acrylic acid) incorporating Cu^{2+} ion was developed. The pump discharges and charges liquid at the level of a few hundred of picoliter per second upon expansion and contraction of the gel induced by reduction and oxidation of the copper, respectively. The pump discharged ca. 5 μl of water at −1.3 V vs. Ag/AgCl for 3 h, whereas charged ca. 1.2 μl at +1.0 V for 3 h. At constant currents of −4.4 and +4.4 μA cm^{-2}, the volumes of water discharged and charged were found to be nearly proportional to the charge at ca. 0.10 cm^3 C^{-1} for both reduction and oxidation processes, except for an initial stage of the electrolysis. Further, upon repeated galvanostatic electrolyses, the same volume-charge slopes were obtained. These results indicate a reversibility and reproducibility of the pump.

Introduction

Micropumps have attracted much attention accompanying widespread use of nano- and micro-channels for microanalysis and microsynthesis. Micropumps can mainly be classified into mechanical and non-mechanical ones according to the presence or absence of a movable part. The former includes piezoelectric (1), electrostatic (2), and thermo pneumatic pumps (3), and the latter includes electrokinetic (4) and magnetohydrodynamic pumps (5). In addition to these, there have recently been reported a micropumps driven by chemical reaction. The driving forces are chemically generated gas (6) and electrochemically generated gas (7). Desirable features for micropumps are a simple constitution, facile control of flow rate, low energy consumption. The above mentioned pumps, except for an electrokinetic pump, have relatively complicated structure. An electrokinetic pump has a simple structure, but requires large energy (~kV) to operate. An electrostatic pump also needs high voltage (80 – 130 V).

Recently, micropumps consisting of a polymer gel have been developed. The driving force of the pump is a volume change of a gel induced by changes in temperature (8) and pH (9). In addition to those stimuli, a polymer gel is known to show abrupt volume change in response to light, solvent composition, and voltage. Since the volume change can be induced by a relatively small change in the stimuli, a pump based on a polymer gel can operate at low energy. In addition, the gel pump can be downsized to molecular level.

Poly(acrylic acid) gel exhibits sharp volume change upon changes in Cu^{2+} concentration. At higher concentration the gel shrinks due to formation of coordination and/or electrostatic bonds between carboxylate groups of the gel and Cu^{2+} ion. We have

363

previously demonstrated that the volume of a poly(acrylic acid) gel incorporating Cu^{2+} ion can be controlled by electrochemical redox of copper (10). The gel expanded upon reduction Cu^{2+} resulting from dissociation of the bond between Cu^{2+} and carboxylate, whereas collapsed upon oxidation of Cu^0 resulting from formation of the bond. On the basis of this, we have also demonstrated that a film electrode modified with poly(acrylic acid) gel containing Cu^{2+} ion works as an actuator (10). In the present study, by utilizing the volume change of the gel, we developed an electrochemical pump consisting of a Au pipe as a cylinder and poly(acrylic acid) gel containing Cu^{2+} as a piston, which discharges and charges liquid at the level of a few hundred of picoliter per second.

Experimental

Materials and Apparatus

Acrylic acid (reagent grade, Wako Pure Chemical Industries) was distilled prior to use. All other reagents were of at least reagent grade quality and were used without further purification. Aqueous solutions were prepared with distilled-deionized water. Electrochemical experiments were carried out with a Hokuto HZ-5000 potentiostat. All measurements were performed at room temperature (25 ± 1 °C).

Preparation of Pump

Poly(acrylic acid) gels were prepared by radical polymerization of acrylic acid (700 mM) and N,N'-methylenebis(acrylamide) (7 mM) with ammonium peroxodisulfate (3.5 mM) and N,N,N',N'-tetramethylethylenediamine (8 µM) as an initiator and an accelerator, respectively. A schematic illustration of a pump is shown in Figure 1. The gel (typically 8 µl) serving as a piston was synthesized in the upper side of a Au pipe (1 mm i.d., 14 mm length), which serves as a cylinder, under nitrogen at 60°C for 3 h. The pipe with the gel was then set in a lab-made holder (1.5 mm i.d., polycarbonate) and the upper end was covered with a Au mesh (100 mesh) to prevent the gel from expanding to the upper direction. The Au mesh was fixed with another holder, which also works as a solution reservoir for gel expansion. The gel was brought into contact with a 0.1 M $Cu(NO_3)_2$ aqueous solution containing 1 mM $NaClO_4$ from the lower side of the pipe for 3 h to incorporate Cu^{2+} ion. The gel was then conditioned with a 1 mM $NaClO_4$ solution for at least 2 days. An U-shaped Teflon tube (0.5 mm i.d.) partially filled with water was attached to the lower side of the holder to monitor the water level, which varies with contraction and expansion of the gel. The Au pipe served as a working electrode, and a Ag/AgCl and a coiled Pt wire set in the upper reservoir were used as reference and counter electrodes, respectively.

Results and Discussion

Electrochemical Properties of Pump

A pump consisting of Cu^{2+}-poly(acrylic acid) gel is expected to expel liquid when Cu^{2+} is reduced to Cu^0 as the gel (piston) expands, whereas to take up liquid when Cu^0 is re-oxidized to Cu^{2+} as it contracts. In order to characterize electrochemical properties as well as a performance of the pump, cyclic voltammetry was performed. Figure 2 shows CV and changes in water volume discharged/charged from the pump in contact with a 1

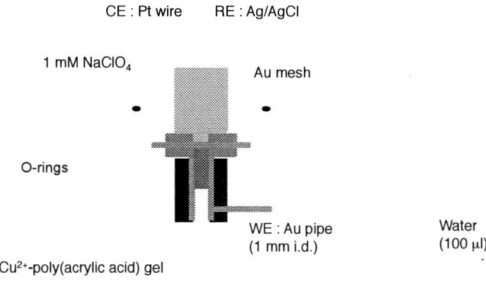

Figure 1. Schematic illustration of Cu^{2+}-poly(acrylic acid) gel pump.

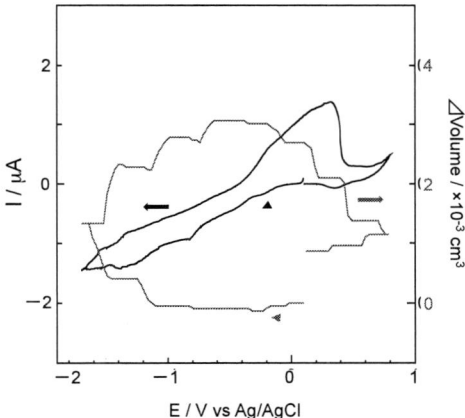

Figure 2. Cyclic voltammogram and changes in volume of Cu^{2+}-poly(acrylic acid) gel pump in contact with 1 mM $NaClO_4$ solution at 0.1 mV s^{-1}.

mM $NaClO_4$ solution at the upper side at a scan rate of 0.1 mV s^{-1}. Here, the changes in water volume was defined as a difference from that at an open circuit. An appearance of peaks corresponding to Cu^{2+} reduction at –1.0 V vs. Ag/AgCl and Cu^0 oxidation at +0.2 V indicates that copper incorporated in the gel which is plugged in a Au pipe at low supporting electrolyte concentration retains electrochemical activity. The water level of the pump remained constant over the potential range from +0.1 to –1.0 V. Upon further negative going scan to –1.7 V and reversed positive scan to –0.6 V the pump appeared to discharge ca. 3.1 μl of water. Since the discharge was evident at the potential where the reduction of Cu^{2+} takes place, it was attributed to expansion of the gel upon the reduction. During the rest of the anodic scan, the pump did not take up water up to –0.2 V. It started to take up at –0.2 V and continued to +0.1 V after the reversal of potential scan direction at +0.8 V (totally 2.2 μl). Similar to the reduction, the charging of the pump started with oxidation of Cu^0, it can be concluded as a manifestation of collapse of the gel upon the oxidation. In a successive potential cycling similar results for both current and changes in water volume (3.9 µl for reduction and 2.8 µl for oxidation) were observed. These results indicate that the pump can reversibly discharge/charge liquid.

Electrochemical Control of Pump

On the basis of the CV, a potentiostatic control of the pump was carried out at −1.3 and +1.0 V for 3 h each. At −1.3 V the pump discharged ca. 5 µl of water, whereas charged ca. 1.2 µl. These results are consistent with those of CV. Further the fact that a similar experiment without Cu^{2+} in a gel showed virtually no discharge/charge supports the mechanism of the pump proposed. The pump with Cu^{2+}, however, appeared to take about 50 min to start expelling water when it was reduced. This will be discussed later. In addition, the volume charged upon oxidation was smaller than that discharged upon reduction, indicating that the Cu-poly(acrylic acid) gel did not shrink to its original volume. This is likely because that water incorporated in the gel is difficult to return back to the reservoir against gravitational force upon re-oxidation, and/or that the gel expanded to a lower direction upon reduction and then the upper part of the gel collapsed upon re-oxidation.

Galvanostatic control of the pump with Cu^{2+} was also carried out at −4.4 µA cm^{-2} for reduction and at +4.4 µA cm^{-2} for oxidation (Figure 3). On a first run a volume of water discharged at −4.4 µA cm^{-2} for 210 min appeared to be ca. 1.4 µl, whereas that charged at +4.4 µA cm^{-2} for 144 min was ca. 1.0 µl. On a second run those at −4.4 µA cm^{-2} for 240 min and +4.4 µA cm^{-2} for 180 min were 1.4 and 1.1 µl, respectively. The charges consumed during those processes as well as the volumes discharged/charged are summarized in Table 1. Based on these results, the change in volume per charge was also evaluated. This factor is one of the important ones for practical applications because it is facile if a pump can be controlled by a charge. The changes in a volume per charge were calculated to be 0.11 cm^3 C^{-1} for both reduction and oxidation on the first run and 0.10 cm^3 C^{-1} on the second run. These results clearly demonstrate that the pump can be controlled by a charge and has excellent reversibility and reproducibility upon potential switching. Flow rates of the pump were calculated to be in the order of 0.1 nl s^{-1} for both reduction and oxidation.

The pump under the galvanostatic control was found to take ca. 5 x 10^{-3} C to start discharging upon reduction. This delay would reflect a friction between a gel and a Au pipe, which prevents the gel from expanding. This postulation is further supported by the fact that the charge required to start expelling appeared to be the same as that for the potentiostatic control described above. Since time consumed to start for the galvanostatic method (ca. 50 min) was found to be different form that for the potentiostatic one (ca. 30 min), water transport into a gel to expand it is not the process causing the delay.

In order to improve the initial stage of the expelling process to be proportional to a charge, an attempt to reduce the friction was made by modifying inside of the Au pipe with 3-mercapto-1-propanesulfonic acid. Since 3-mercapto-1-propanesulfonic acid has a negative charge, it is expected to repulsively interact with carboxylate of the gel. The pump modified with 3-mercapto-1-propanesulfonate started to discharging right after an application of +0.6 V and discharged ca. 1.8 µl of water for 3 h, and took up 0.2 µl at

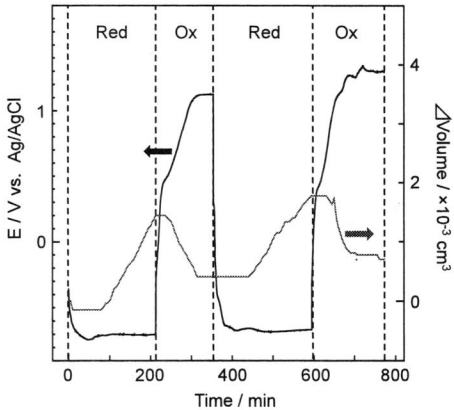

Figure 3. Galvanostatic control of Cu^{2+}-poly(acrylic acid) gel pump in contact with 1 mM $NaClO_4$ solution at −4.4 and at +4.4 μA cm^{-2}.

TABLE I. Charge, Δvolume, and Δvolume per charge of Cu^{2+}-poly(acrylic acid) gel pump actuated by galvanostatic electlolysis

	ΔVolume / cm^3		Charge / C		ΔVolume per Charge / $cm^3\ C^{-1}$	
	Red[a]	Ox[b]	Red[a]	Ox[b]	Red[a]	Ox[b]
Run1	1.4×10^{-3}	1.0×10^{-3}	1.3×10^{-2}	8.6×10^{-3}	0.11	0.11
Run2	1.4×10^{-3}	1.1×10^{-3}	1.4×10^{-2}	1.1×10^{-2}	0.10	0.10

a: −4.4 μA cm^{-2}, b: +4.4 μA cm^{-2}

−1.3 V for 3 h. A plot of changes in volume vs. charge for the reduction process appeared to be linear (Figure 4). The most notable thing is that the charge spent to start expelling was found to be ca. 0.7 mC, which is much smaller than that of the unmodified pump (3.2 mC). These results clearly indicate the effectivity of the chemical modification of the Au pipe to improve the disproportionality at the initial stage.

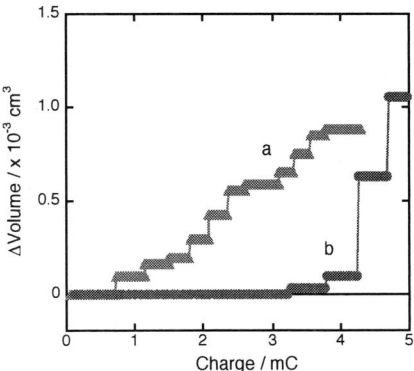

Figure 4. Relationship between charge and volume of water expelled from the pump (a) modified with 3-mercapto-1-propanesulfonate and (b) without modification.

Conclusions

An electrochemical pump consisting of Cu^{2+}-poly(acrylic acid) gel was developed on the basis of expansion and contraction of the gel. The pump expelled and took up water at rate of a few hundred of picoliter per second upon galvanostatic reduction at -4.4 and oxidation at $+4.4$ μA cm^{-2}, respectively. The water volumes discharged and charged were found to be proportional to the electric charge for both reduction and oxidation, except for the initial stage of the electrolysis, indicating excellent reversibility. The delay to start discharging at the initial stage of the reduction was mainly ascribed to a friction between the gel and the Au pipe. This delay was improved by modification of the Au pipe with 3-mercapto-1-propanesulfonate, which decreases the friction by electrostatic repulsion.

Acknowledgments

This work was supported in part by Grant-in-Aid for Scientific Research (No. 23510145) from Japan Society for the Promotion of Science.

References

1. S. Shoji, S. Nakagawa, and M. Esashi, *Sens. Actuators A*, **21**, 189 (1990).
2. J. Xie, J. Shih, Q. Lin, B. Yang, and Y. Tai, *Lab Chip*, **4**, 495 (2004).
3. A. Tuantranont, W. Mamanee, T. Lomas, N. Porntheerapat, N. V. Afzulpurkar, and A. Wisitsoraat, *IEEE Int. Conference on Nanotechnology*, **7**, 1203 (2007).
4. P. Wang, Z. Chen, and H.-C. Chang, *Sens. Actuators B,* **113**, 500 (2006).
5. A. Homsy, S. Koster, J. C. T. Eijkel, A. Berg, F. Lucklum, E. Verpoorte, and N. F. de Rooij, *Lab Chip*, **5**, 466 (2005).
6. Y. H. Choi, S. U. Son, and S. S. Leekoi, *Sens. Actuators A*, **111**, 8 (2004).
7. D. A. Ateya, A. A. Shah, and S. Z. Hua, *Rev. Sci. Inst.,* **75**, 915 (2004).
8. A. Richter, S. Klatt, G. Paschew, and C. Klenke, *Lab Chip*, **9**, 613 (2009).
9. D. T. Eddington and D. J. Beebe, *J. Microelectromech. Sys.,* **13**, 586 (2004).
10. K. Takada, N. Tanaka, and T. Tatsuma, *J. Electroanal. Chem.*, **585**, 120 (2005).

Superoxide anion radical sensor using GC electrode modified with heparin/PEDOT and polymerized iron porphyrin

Ryo Matsuoka[a], Takeshi Kondo[a,b], Makoto Yuasa[a,b]

[a]Department of Pure and Applied Chemistry, Faculty of Science and Technology, Tokyo University of Science, Noda, Chiba 278-8510, Japan
[b]Research Institute for Science and Technology, Tokyo University of Science, Noda, Chiba 278-8510, Japan

Abstract

Blood coagulation on the electrode surface is a serious problem for a superoxide anion radical ($O_2^-\cdot$) sensor because it causes deterioration of the current signal especially *in vivo*. In order to solve this problem, we have developed iron porphyrin-based $O_2^-\cdot$ sensors immobilizing heparin, which is an anti-thrombotic compound, on the surface. In the present study, we immobilized heparin with a polyethylenedioxythiophene (PEDOT) matrix on poly[iron tetrakis(3-thienyl)porphyrin]-modified glassy carbon (GC) electrode, and the stability of the modified sensor to blood coagulation was investigated. The PEDOT matrix was found to be more effective for stable electrochemical detection of $O_2^-\cdot$ after immersion in a fibrinogen/thrombin solution for two weeks, while serious electrode fouling due to blood clot formation was occurred at a sensor using a polypyrrole matrix after immersion in the solution for one week.

1. Introduction

Reactive oxygen species (ROS) derived from $O_2^-\cdot$ are believed to be essential for normal cellular growth and homeostasis (1). However, excess ROS may cause an oxidative stress condition, which can cause various diseases such as cancer, arteriosclerosis and inflammation in a living organism by the strong radical toxicity. Therefore, *in vivo* measurement of $O_2^-\cdot$ concentration should be very important to clarify the relationship between $O_2^-\cdot$ and various diseases. Especially, development of a rapid *in vivo* $O_2^-\cdot$ analysis has been desired.

For $O_2^-\cdot$ detection, absorption spectrophotometry such as cytochrome *c* reduction method and nitroblue tetrazolium (NBT) reduction method, chemiluminescence and electron spin resonance (ESR) such as luminol method have been employed conventionally (2). However, these methods are not useful for *in vivo* and real-time $O_2^-\cdot$ measurement. On the other hand, some electrochemical methods using electrodes modified with superoxide dismutase (SOD) and cytochrome *c* are useful for a real-time detection. However, problems in stability and reproducibility are often raised when enzymes are used for surface modification of electrodes (3-5).

In order to obtain a stable $O_2^-\cdot$ sensing system to a long-term use, we have developed electrodes modified with iron porphyrin complex, which is the active center of cytochrome c, as an electrocatalyst (6, 7). The $O_2^-\cdot$ sensor was fabricated by the electropolymerization of iron tetrakis(3-thienyl)porphyrin (FeT3ThP) on a GC electrode, and *in vivo* quantitative measurement of $O_2^-\cdot$ was successful at this electrode (8). However, blood clot formation on the electrode surface caused a significant decrease in sensitivity during use.

In our previous study, we immobilized anti-thrombotic heparin with polypyrrole (PPy) matrix on the iron porphyrin-modified $O_2^-\cdot$ sensor surface (7). Heparin is generated in liver and small intestine and is an acidic mucopolysaccharide with a mixture of various molecular weight ranging from 3,000 to 35,000 (average molecular weight is 12,000) with a sulfated glucosamine and glucuronic acid repeating structure (9-18). In addition, heparin is widely used in the clinical field as a fast-acting anticoagulant (10-12). In this study, we used polyethylenedioxythiophene (PEDOT) for a matrix to immobilize heparin because it is expected to be more difficult to be oxidized, hence more stable than PPy matrix (19-21). PEDOT/heparin-poly(FeT3ThP)-modified GC electrode was found to be useful for detection of $O_2^-\cdot$ without formation of a fibrin layer in a fibrinogen/thrombin solution (Fig. 1).

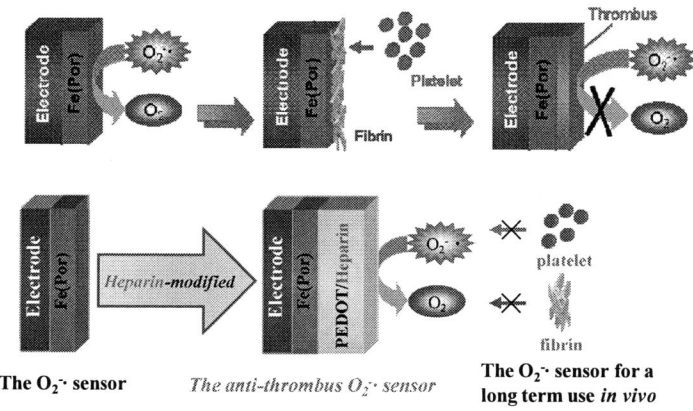

Fig. 1 Structure of PEDOT/heparin-modified $O_2^-\cdot$ sensor.

2. Materials and Methods

Preparation of poly(FeT3ThP)-modified GC electrode

FeT3ThP was synthesized by the procedure described in our previous report (7, 8). FeT3ThP and tetrabutyl ammonium perchlorate (TBAP) were dissolved in CH_2Cl_2. 1-methyl imidazole as an axial ligand of the porphyrin was added to the solution to prepare a monomer solution. The monomer solution was added to dichloromethane containing TBAP to prepare a solution for electropolymerization. The solution was put in a two-chamber three-electrode electrochemical cell, and poly(FeT3ThP) was formed on a needle-type GC electrode (Fig. 2). Electropolymerization was performed by potential cycling from 0 to +2.0 vs. Ag/Ag^+. Potential sweep rate was 50 mV/s (7). After rinsing with dichloromethane, poly(FeT3ThP)-modified GC electrode was obtained.

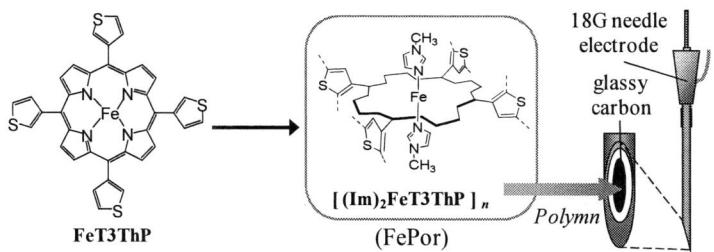

Fig. 2 Preparation of poly(FeT3ThP)-modified GC electrode.

Preparation of PEDOT/heparin-poly(FeT3ThP)-modified GC electrode

Heparin (Sigma-Aldrich) was dissolved in 0.5 mL pure water (1000 U/mL), followed by the addition of EDOT (Sigma-Aldrich) in the solution to be 5.0 mM. Electropolymerization of a PEDOT/heparin film on the electrode surface was carried out with a two-chamber three-electrode electrochemical cell assembly, using the poly(FeT3ThP)-modified electrode as a working electrode. Potential cycling was performed in the heparin/EDOT solution with a potential range from 0 to +2.0 V vs. Ag/AgCl, potential sweep rate of 50 mV/s, and a cycle number of 10 (22, 23). The heparin can be immobilized in the PEDOT film by electrostatic interaction between anionic heparin and cationic PEDOT (Fig. 3). After rinsing with water, a PEDOT/heparin-poly(FeT3ThP)-modified GC electrode was obtained.

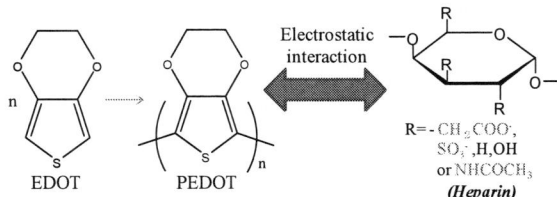

Fig. 3 Elecrostatic interaction between PEDOT and heparin.

Electrochemical $O_2^-\cdot$ detection

For evaluation of a superoxide anion radical sensor, a xanthine (XAN)/xanthine oxidase (XOD) system was used for in situ $O_2^-\cdot$ generation (7). 0.1 mM XAN-containing phosphate buffered saline (PBS) was flowed to an electrochemical flow cell at a flow rate of 1 mL/min, and amperometric current was recorded with a two-electrode system at a potential of +0.5 V vs. stainless steel (SUS 304). After the current became constant, XOD-containing PBS solution (0, 5, 10, 20 and 50 mU/mL) was added to the XAN-containing PBS solution to generate a certain concentration of $O_2^-\cdot$.

Anti-thrombotic examination of the sensors

To estimate the anti-thrombotic properties of a sensor, fibrin formation after immersion of the sensor in a fibrinogen (3 mg)/thrombin (1 U) solution was observed. Since insoluble fibrin can be formed from fibrinogen by the activation of thrombin at a blood clots formation process, thrombus formation should be suppressed by inhibiting the activation of thrombin (15). O_2^- detection current was recorded after immersion in a fibrinogen/thrombin solution to estimate anti-thrombic activity of the sensor (7).

3. Results and Discussions

Preparation of poly(FeT3ThP)-modified GC electrode

Fabrication of a FeT3ThP-modified GC electrode was confirmed by differential pulse voltmmetry (DPV) in PBS. A voltammetric peak was observed for the Fe(II/III) redox reaction at around -0.1 V vs. Ag/AgCl indicating the presence of iron porphyrins (7). In addition, electrochemical detection of $O_2^-\cdot$ at thef sensor was examined using a XAN/XOD system for *in situ* $O_2^-\cdot$ generation. Current increase upon XOD concentration (ΔI) is known to be calculated to $O_2^-\cdot$ concentration ($[O_2^-\cdot]$) (8),

$$[O_2^-\cdot] = \sqrt{{x}/{3.0}} \times 10^{-6} \qquad [1]$$

Where x is XOD concentration. The calibration curve (ΔI vs. $[O_2^-\cdot]$, Fig. 4) showed a linear relationship with the $[O_2^-\cdot]$ range from 0 to 1.3 µM. , indicating successful detection of $O_2^-\cdot$ at the poly(FeT3ThP)-modified electrode (8).

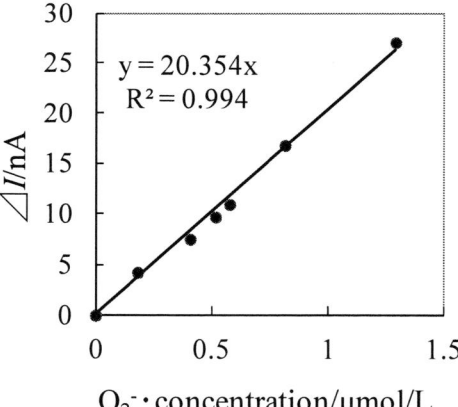

Fig. 4 Calibration curve for $O_2^-\cdot$ at a poly(FeT3ThP)-modified GC electrode. Electrode potential was 0.5 V vs. stainless steel.

$O_2^-\cdot$ detection at a PEDOT/heparin-poly(FeT3ThP)-modified GC electrode

Electrochemical detection of $O_2^-\cdot$ at the PEDOT/heparin-poly(FeT3ThP)-modified GC electrode was examined using a XAN/XOD system for *in situ* $O_2^-\cdot$ generation. The calibration curve was found to be linear in the $O_2^-\cdot$ concentration range from 0 to 1.3 μM. The slope of the calibration curve was found to be similar to that obtained at a poly(FeT3ThP)-modified GC electrode (Fig. 4), indicating that the PEDOT/heparin layer on the surface did not interrupt the access of $O_2^-\cdot$ to the active site of the electrocatalyst (iron porphyrins). Thus, this electrode can be used for quantitative detection of $O_2^-\cdot$ (7).

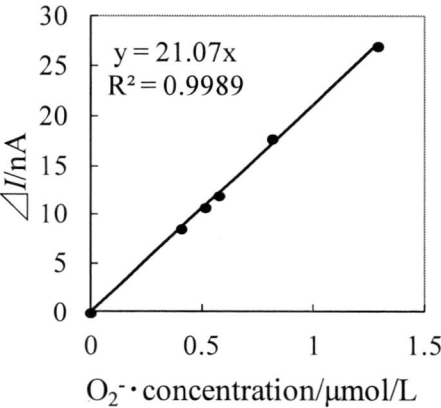

Fig. 5 Calibration curve for $O_2^-\cdot$ at a PEDOT/heparin-poly(FeT3ThP)-modified GC electrode. Electrode potential was 0.5 V vs. stainless steel.

Anti-thrombotic evaluation of a PEDOT/heparin-poly(FeT3ThP)-modified GC electrode

As an example of anti-thrombotic test, the modified electrodes were immersed in a fibrinogen/thrombin solution containing antithrombin (0.135 mg/mL), which is a blood coagulation factor inhibitor. This antithrombin concentration was half of that in the physiological condition, enabling an accelerated test for blood clot formation on the electrode surface (17, 18, 20). For a poly(FeT3ThP)-modified GC electrode, almost no current response for $O_2^-\cdot$ detection at 1 mM was observed after the electrode was immersed in the fibrinogen/thrombin solution for 30 min due to electrode fouling by blood clot formation. In the case of a PPy/heparin-poly(FeT3ThP)-modified GC electrode, the current response could be observed when the immersion time in the fibrinogen/thrombin solution within 5 h. However, PEDOT/heparin-poly(FeT3ThP)-modified GC electrode showed excellent durability to the fibrin formation by the immersion at least within 24 h (Fig. 6). Figure 7 shows scanning electron microscopy (SEM) images of PEDOT/heparin-poly(FeT3ThP)-modified and poly(FeT3Th)-modified GC electrode surfaces before and after immersion in the fibrinogen/thrombin solution for 1 h. The result demonstrates clearly that the PODOT/heparin layer suppressed the formation of a fibrin layer.

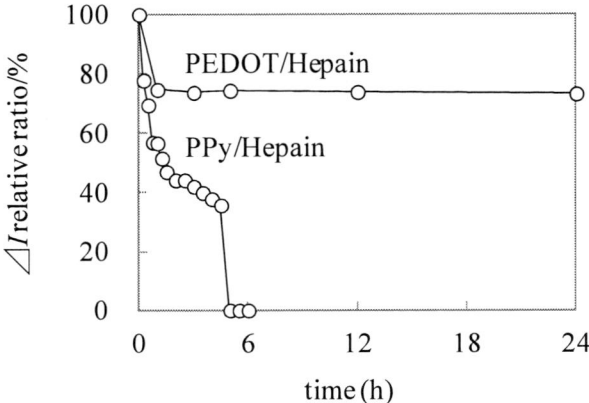

Fig. 6 Relative current response for $O_2^-\cdot$ detection at PPy/heparin- and PEDOT/heparin-poly(FeT3ThP)-modified GC electrodes as a function of immersion time in the fibrinogen/thrombin solution containing 0.135 mg/mL antithrombin. $O_2^-\cdot$ concentration was 1mM. Electrode potential was +0.5 V vs. stainless steel.

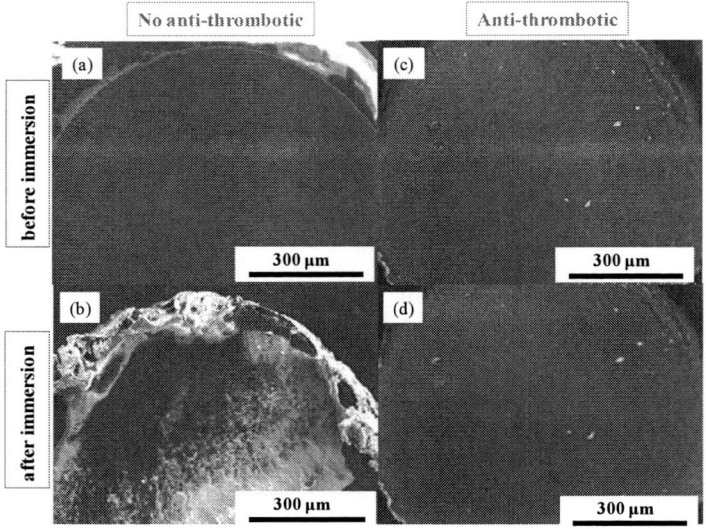

Fig. 7 SEM images of (a and b) poly(FeT3ThP)-modified and (c and d) PEDOT/heparin-poly(FeT3ThP)-modified GC electrode surfaces (a and c) before and (b and d) after immersion in a fibrinogen/thrombin solution for 1 h.

Finally, we have examined the anti-thrombotic activity of the heparin-modified sensors using a fibrinogen/thrombin solution containing antithrombin at a physiological concentration (0.27 mg/mL) (Fig. 8). The trend of the current response change to immersion time was similar to that in an accelerated test (Fig. 6), and the PEDOT/heparin-poly(FeT3ThP)-modified GC electrode was found to be durable to blood clot formation for at least two weeks.

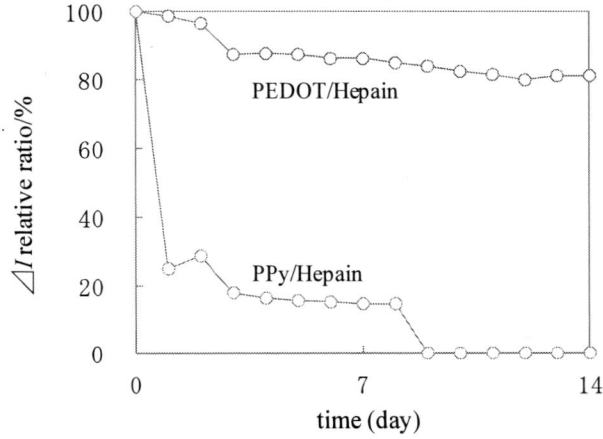

Fig. 8 Relative current response for $O_2^-\cdot$ detection at PPy/heparin- and PEDOT/heparin-poly(FeT3ThP)-modified GC electrodes as a function of immersion time in the fibrinogen/thrombin solution containing 0.27 mg/mL antithrombin. $O_2^-\cdot$ concentration was 1mM. Electrode potential was +0.5 V vs. stainless steel. The current response at the PPy/heparin-poly(FeT3ThP)-modified GC electrode was found to decrease remarkably with one day. This should be due to removal of heparin associated with oxidative degradation of PPy matrix.

4. Conclusion

PEDOT/heparin-poly(FeT3ThP)-modified GC electrode was prepared by electropolymerization of FeT3ThP and EDOT/heparin. Current response for $O_2^-\cdot$ detection at the PEDOT/heparin-poly(FeT3ThP)-modified GC electrode was found to be similar to that at a poly(FeT2ThP)-modified GC electrode. The calibration curve showed a linear relationship in the $O_2^-\cdot$ concentration range from 0 to 1.3 μM. The PEDOT/heparin-poly(FeT3ThP)-modified GC electrode was found to be durable to blood clot formation in fibirinogen/thrombin solution containing antithrombin at a physiological concentration (0.27 mg/mL) for at least two weeks. Thus, the PEDOT/heparin-poly(FeT3ThP)-modified GC electrode should be useful for a real-time and *in vivo* quantitative analysis of $O_2^-\cdot$.

References

1) Y. Ando, S. Imamura, Y.M. Hong, M.K. Owada, T. Kakunaga and R. Kannagi, *Biochem. Life Sci. Adv.*, **7**, 259, (1988).

2) F. Lisdat, B.Ge.E. Ehrentreich-Förster, R. Reszka, and F.W. Scheller, *Anal. Chem.*, **71**, 1359, (1997).

3) J.M. Cooper, K.R.Greenough, and C.J.McNeil, *J.Electroanal.Chem.*, **347**, 267, (1993).

4) K.T.Toomas, T.T.Alexey, A.M.Edward, W.Hillhouse, P.Manning, C.J.McNeil, *Free Radical Biology & Medicine.*, **25**, 973, (1998).

5) K.V. Gobi, F. Mizutani, *J. Electroanal Chem.*, **484**, 172, (2000).

6) M. Yuasa and K. Oyaizu, *Curr. Org. Chem.*, 9, 1685, (1991).

7) M. Yuasa K. Oyaizu, A. Yamaguchi, M. Ishikawa, K. Eguchi, T. Kobayashi Y. Toyoda and S. Tsutsui, *Polymn. Adv. Technol.*, **16**, 616, (2005).

8) M. Yuasa K. Oyaizu, A. Yamaguchi, M. Ishikawa, K. Eguchi, T. Kobayashi Y. Toyoda and S. Tsutsui, *Polymn. Adv. Technol.*, **16**, 287, (2005).

9) J.A. Huntington, *Biochem. Sci.*, **31**, 8 (2006).

10) A.Dementiev, J.Dobó P.G.Gettins, *J. Biol.Chem.*, **281**, 3452 (2006).

11) C.R. Ricketts, *Biochem.*, **51**, 129 (1952).

12) I. Bjork, U. Lindahl, *Mol. Cell. Biochem.*, **48**, 161 (1982).

13) J.E. Jorpes, H. Bostrom, V. J. Mutt, *J. Biol. Chem.*, **183**, 607 (1950).

14) B. Casu, *Adv. Carbohydro. Chem. Biochem.*, **43**, 51 (1985).

15) K. Salminista, K. Lidholt, U. Lindahl, *FASEB J.*, **10**, 1270 (1996).

16) D.M. Tollefen, *Adv. Exp. Med. Biol.*, **425**, 35 (1997).

17) D.M. Tollefen, *Arch. Pathol.. Lab. Med.*, **126**, 1394 (2002).

18) D. O'Keeffe, S.T. Olson, N. Gasiunas, J. Gallagher, T.P. Baglin, and J.A. Huntington, *J. Biol. Chem.*, **279**, 50267 (2004).

19) T. Skotheim, R. Elsenbaumer, J. Reynolds, (eds), *Handbook of Conducting Polymers, Marcel Dekker, Inc.*, (1988).

20) L. Groenendaal, F. Jonas, D. Freitag, H. Pielartzik, J.R. Reynolds, *Adv. Mater.*, **12**, 481 (2000).

21) L. Groenendaal, G. Zotti, P. Aybert, S.M. Waygright, J.R. Reynolds, *Adv. Mater.*, **15**, 855 (2003).

22) M.D. Levi, E. Lankri, Y. Gofer, D. Aurbach, T. Otero, *J.Electrochem. Soc.*, **149**, E204, (2002).

23) A.J. Bard, F.-R.F. Fan, H.S. White, B.L. Wheeler, *J. Am. Chem. Soc.*, **102**, 5442, (1980).

High Sensitive Amperometric Detection of Glucose using Conductive DLC
Electrode in Higher Potential Region

Kensuke Honda, Akira Nakahara, Hiroshi Naragino, Kohsuke Yoshinaga

Graduate school of Science and Engineering, Yamaguchi University
1677-1, Yoshida, Yamaguchi-shi, Yamaguchi 753-8512, Japan

The possibility of a high sensitive amperometric detection of
glucose using electrochemically-generated hydroxyl radicals at N-
doped DLC was examined. Flow injection analytical system using
N-doped DLC as an electrode of electrochemical detector showed
higher performance of glucose detection in a potential region of
oxygen evolution (>3.6 V vs. Ag|AgCl). Linear dynamic range
was 5 orders of magnitude (1 μM to 10 mM) and theoretical
detection limit (S/N = 3) was 1.37 mM at 3.6 V vs. Ag|AgCl. In
addition to glucose, linear relations between values of response
current and of sample concentration were observed in molecules
easily oxidized by OH radicals (ascorbic acid, 2-propanol, ethyl
acetate and ethyl acetoacetate). The oxidation of glucose mediated
by OH radicals generated at N-doped DLC in higher potential
region is included in the mechanism of amperometric glucose
detection at N-doped DLC.

Introduction

Recently, conductive diamond like carbon films fabricated by doping nitrogen atoms
in hydrogenated amorphous carbon (N-doped DLC) have been gaining much attention as
an alternative electrode material to boron-doped diamond (BDD) for ideal polarizable
properties (wide working potential range), rapid deposition, and a property that the film
can be deposited on any substrate (1). Oxygen evolution is occurred in higher potential
region at N-doped DLC as well as that at BDD.

Higher overpotential for O_2 evolution was caused because interactions between
adsorbed hydroxyl radicals (M(•OH)) and electrode surfaces were weak and adsorbed
hydroxyl radical discharged directly to O_2 without forming higher oxides (M-O).
Adsorbed hydroxyl radicals can mediate oxidation of organic molecules at BDD and at
N-doped DLC (composed of sp^3 carbons)(2, 3). Electroanalyses of molecules with
higher oxidation potential can be achieved by amperometric detection of products
resulting from oxidation mediated by electrogenerated OH radicals at N-doped DLC.

In vitro measurement of glucose in blood is of great importance in a clinical diagnosis
of diabetes mellitus (hyperglycemia) characterized by high glucose level (> 7.0 mM).
Continuous monitoring of glucose is necessary to control and treat diabetes. The most
commonly used commercial glucose sensors are platinum-printed electrode-based sensor
(Glucose oxidase is immobilized). Most of the sensors are for single-use due to
adsorption of reaction product on platinum surface.

In our present work, the method of amperometric detection of glucose through oxidation mediated electrogenerated OH radical has been developed to realize an electroanalytical assay for continuous glucose monitoring.

Experimental

Nitrogen-doped a-C:H (N-doped DLC) films were prepared using a SAMCO model BP-10 microwave-assisted plasma chemical vapor deposition (CVD) system (SAMCO Co., Ltd.) (The same procedure described in Ref. [1] was applied). Films were deposited on p-Si (111) substrates (5×10^{-3} Ω cm, SUMCO Co., Ltd.) after in-situ sputter cleaning with argon ion (150 W for 15 minutes). Vapor at a flow rate of 5 mL/min was introduced into the evacuated reaction chamber from degassed acetonitrile (carbon and nitrogen source) kept at 50℃ by heating. N-doped a-C:H films of 1.5 µm thickness were grown by applying r. f. power (13.56 MHz) on the cathode which held Si substrates on a quartz liner. The deposition time was 40 minutes. The chamber pressure was adjusted at 10 Pa, and the stage temperature was set at 250℃ during deposition. DLC films with 1.72 (semimetal) Ω cm (values were obtained by four point probe methods) of resistivity could be synthesized by applying 300 W of RF power.

The resulting N-doped DLC membrane was introduced into flow cell in amperometric detector in flow injection analysis (FIA) system (Figure 1). The FIA system was consisted of a micro-L pump (GL Sciences, PU-611C), an auto sampler (Spark-Holland, Triathlon), an electrochemical amperometric (EC) detector (GL Sciences, ED703). The mobile phase using FIA was 0.1 M H_3PO_4 and flow rate was set at 0.25 ml min^{-1}. Injection volume was 25 µl.

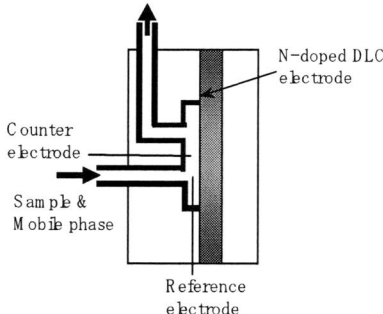

Figure 1. Configuration of flow cell with N-DLC electrode in FIA system.

Results and Discussion

Cyclic Voltammetry

The oxidation of glucose was examined at N-doped DLC electrode. Figure 2 (A) shows the cyclic voltammograms of 5mM glucose in 0.1 M H_3PO_4. Well-defined oxidation peak of glucose was not observed in figure 2 (A). The difference between cyclic voltammograms of glucose oxidation and back ground voltammograms is shown in figure 2 (B). The anodic current linearly increased from 1.8 V vs. Ag|AgCl with increasing potential. These results indicate that glucose was not directly oxidized at N-doped DLC surface and anodic reaction relating glucose oxidation that was accelerated with increasing rate of oxygen evolution occurred at N-doped DLC surface.

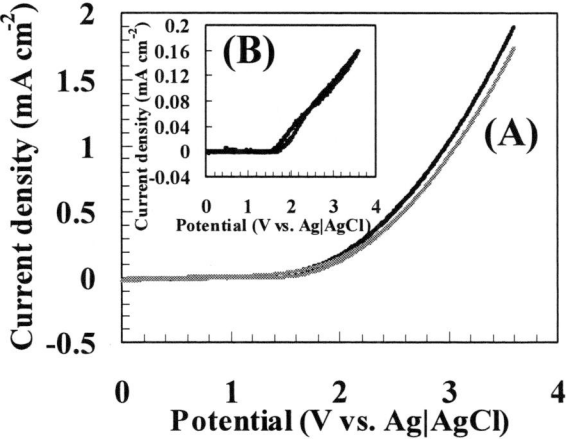

Figure 2. (A) Cyclic voltammograms of 5mM glucose (black line) in 0.1 M H_3PO_4 (grayline). (B) I-E curve of current differences between CV of glucose oxidation and back ground CV.

Flow Injection Analysis

Figure 3 shows the time profile of response current of FIA system obtained at N-doped DLC electrode for 20-μl injection of 5 mM glucose at 3.6 V vs. Ag|AgCl. Figure 4 shows hydrodynamic voltammograms obtained at N-doped DLC electrode for 20-μl injections of 5 mM glucose at pH 1. Faradaic current starts to increase at ca. 2 V vs. Ag|AgCl and reaches a plateau at ca. 3.8 V. Peak current exhibits sigmoidal shape. Back ground current was ca 4 μA (Figure 3), which was attributable to O_2 evolution. This current response was not changed before and after the injection of solution of mobile phase. The peak currents, therefore, were derived from direct oxidation of compounds related to glucose. These results suggest a possibility of amperometric detection of glucose at N-doped DLC in higher potential region. From this data, the potential for amperometric detection was set at 3.6 V vs. Ag|AgCl.

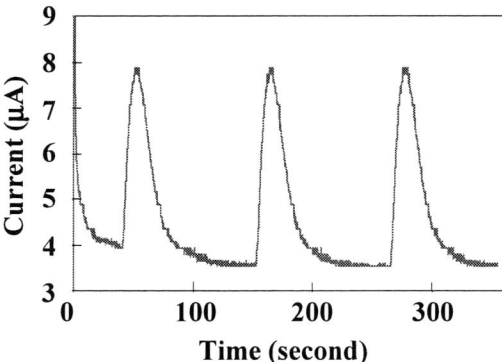

Figure 3. Time profile of response current of FIA at N-doped DLC for 5 mM glucose at 3.6 V vs. Ag|AgCl.

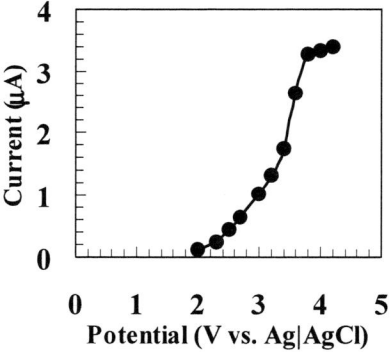

Figure 4. Hydrodynamic voltammograms for 5 mM glucose at N-doped DLC electrode.

Dependence of Glucose Concentration

The relation between amperometric current at 3.6 V and glucose concentration was examined. Amperometric currents show linear relation with glucose concentration in a range from 1 µM to 10 mM (linear dynamic range), as shown in figure 5. The theoretical detection limit (S/N = 3) can be 1.37 mM. The average concentration of glucose in blood (3 to 8 mM) is within the linear dynamic range in figure 5. This result indicates that analytical method using N-doped DLC has satisfactory accuracy for the analysis of glucose.

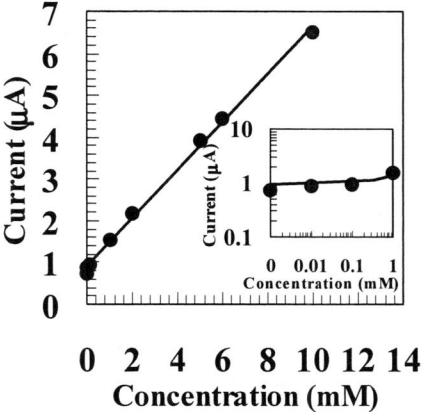

Figure 5. Analytical curve for currents vs. glucose concentration.

Reproducibility

The reproducibility was checked by multiple injections of 5 mM glucose at intervals of 1 minute. A highly reproducible response with a variety of peaks less than 10 % (n =20) was observed as shown in figure 6. It suggests that N-doped DLC is expected to be a promising electrode material for practical application of electrochemical detection of glucose.

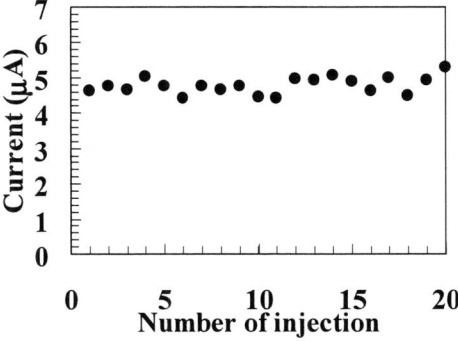

Figure 6. Response current of FIA peak heights for 5 mM glucose.

Mechanism of Glucose Detection and Measurements for Various Molecules

The mechanism of glucose detection at N-doped DLC at higher potential region was examined. Results of CV measurement and FIA analysis suggest that amperometric current was caused by the oxidation of products that resulted from glucose oxidation mediated by electrogenerated OH radicals at N-doped DLC.

OH radicals have been reported to be generated through water electrolysis at N-doped DLC and boron-doped diamond and not in organic electrolyte even in higher potential region. Response current for 5 mM glucose at 3.6 V vs. Ag|AgCl was not observed at N-doped DLC films by using 1 M $LiClO_4$ in acetonitrile as a mobile phase. On the contrary, in FIA measurement, response currents for 5 mM uric acid that was directly oxidized at N-doped DLC were observed in both 0.1 M H_3PO_4 aqueous solution and 1 M $LiClO_4$ in acetonitrile. Hence, aqueous electrolyte is necessary to detect glucose at N-doped DLC in higher potential region.

Response currents for 20-μl injections of 5 mM glucose in the potential region from 1.0 V to 3.6 V were not observed at glassy carbon and polycrystalline Pt electrodes installed in FIA system. At active electrodes like Pt, the absorbed hydroxyl radical ((M(•OH)) forms higher oxide (M-O) and oxygen evolution occurs through this higher oxide because there is a strong interaction between hydroxyl radicals (that forms on the electrode surface at initial step of water discharge) and the electrode surface. Hence, at polycrystalline Pt and glassy carbon (active) electrodes, the oxidation of organic molecules mediated by •OH hardly occurs, resulting in no response in FIA measurement. These results also support that anodic current for glucose injection at N-doped DLC was caused by oxidation of the product after glucose oxidation mediated by •OH.

Oxidation of molecules by hydroxyl radicals has been reported to occur through hydrogen abstraction reaction by •OH (3), and α-hydrogen is easily abstracted by •OH. In FIA measurement, response currents for molecules consist of hydrogen, which has been reported to be hardly abstracted by hydroxyl radicals (acetic acid, methanol and ethanol) (4), could not be observed at N-doped DLC films at 3.6 V.

However, in the case of four molecules (ascorbic acid, 2-propanol, ethyl acetate and ethyl acetoacetate) that have been reported to occur hydrogen abstraction reaction easily by OH radicals (4, 5) or have higher •OH scavenging ability (6), linear relations between response current and concentration were observed. The linear dynamic ranges and the sensitivity to four molecules were summarized in Table 1. Sensitivities to glucose and ascorbic acid were higher by ca. 2 ~ 3 orders of magnitude than those to 2-propanol, ethyl acetate, and ethyl acetoacetate. This trend is roughly consistent with •OH scavenging ability and the order of the rates of hydrogen abstraction reaction occurred by •OH.

Hence, the oxidation of glucose mediated by electrogenerated OH radicals at the surface is included in the mechanism of amperometric glucose detection at N-doped DLC.

TABLE I. The results of FIA measurement for various molecules at N-doped DLC.

Molecules	Linear dynamic range	Sensitivity
Glucose	1 μM - 10 mM	0.574 μA/mM
Ascorbic acid	100 μM - 10 mM	2.720 μA/mM
2-Propanol	40 mM - 200 mM	6.90 nA/mM
Ethyl acetate	140 mM - 200 mM	6.60 nA/mM
Ethyl acetoacetate	80 mM - 200 mM	12.0 nA/mM

Summary

In this study, flow injection analytical system for a high sensitive amperometric detection of glucose could be realize by using electrochemically-generated OH radicals at N-doped DLC in high potential region (>3.6 V vs. Ag|AgCl). The linear dynamic range of glucose detection was 5 orders of magnitude (1 μM to 10 mM) and the theoretical detection limit (S/N = 3) was 1.37 mM at 3.6 V vs. Ag|AgCl. In addition to glucose, amperometric detection for molecules easily oxidized by OH radicals (ascorbic acid, 2-propanol, ethyl acetate and ethyl acetoacetate) could be possible. It is clarified that N-doped DLC electrode has potential application of electrochemical assay as a high sensitive glucose detection in higher potential region.

It was clarified that electrogenerated hydroxyl radicals play an important role in amperometric detection of glucose at N-doped DLC in high potential region. This FIA system showed higher amperometric response to ascorbic acid that is included in human blood. High response current for ascorbic acid (and easily oxidized molecules by OH radicals) is one of the problems (interferences) that should be fixed in order to realize an electroanalytical assay for selective and continuous glucose monitoring. In order to enhance sensitivity to glucose compared with that to coexisting molecules, it is necessary to make clear the origin of anodic current for glucose injection in higher potential region (e.g. products of glucose oxidation occurred by OH radicals) and the mechanism of amperometric detection of molecules (active for •OH oxidation like ascorbic acid) coexisting in blood. Improvement of performance of FIA system with N-doped DLC to detect glucose selectively in mixed media (like blood) is now in progress by enhancing sensitivity to glucose.

Acknowledgments

A part of this work was supported by Grant-in-Aid for Scientific Research from the Ministry of Education, Culture, Science and Technology of Japan (No. 23656242).

References

1. Y. Tanaka, M. Furuta, Y. Katsuki, K. Kuriyama, R. Kuwabara, A. Fujishima and K. Honda, *Electrochimica Acta*, **56** (3), 1172 (2011).
2. B. Marselli, J. Gcia-Gomez, P.-A. Michaud, M. A. Rodrigo and Ch. Comninellis, *J. Electrochem. Soc.*, **150**(3), D79 (2003).
3. K. Honda, Y. Yamaguchi, Y. Yamanaka, M. Yoshimatsu, Y. Fukuda, A. Fujishima, *Electrochim. Acta*, **51**(4), 588 (2005).
4. L. Nelson, O. Rattigan, R. Neavyn, H. Sidebottom, *International Journal of Chemical Kinetics*, **22**, 1111 (1990).
5. I. M. Campbell, P. E. Parkinson, *Chem. Pharm. Lett.*, **53**(2) 385 (1978).
6. T. Tomita, M. Kashima, Y. Tsujimoto, *Chem. Pharm. Bull.*, **48**(3) 330 (2000).

New Application of Produced Pigment from Bacteria to Detect of Ammonia in Combination with Flow Injection for Ammonia Analysis

Y. Iida[a] and I. Satoh[b]

[a] Department of Applied Bioscience, Kanagawa Institute of Technology, 1030 Shimo-ogino, Atsugi, Kanagwa, Japan
[b] Department of Applied Chemistry, Kanagawa Institute of Technology, 1030 Shimo-ogino, Atsugi, Kanagwa, Japan

A novel detection method for ammonium ions based on a spectrophotometric FIA (Flow Injection Analysis) with use of pyocyanine as a coloring reagent in combination with micro fluidic gas diffusion device was developed in this study.

A blue pigment produced by microorganisms was isolated, characterized and then, applied to an FIA (Flow Injection Analysis) as a coloring reagent followed by investigation of the properties.

The pigment was separated by a silica-gel column chromatography, and the molecular weight was determined by EI-MS. The structure was analyzed by ^{1}H-NMR. The isolated compound was found to be identical to pyocyanine. The pigment-producing microorganism was identified as Pseudomonas sp. from 16S DNA analysis. Noting that the property of the pyocyanine of which absorbance depended on the pH shift, the compound as a coloring reagent was applied to an FIA for determinaiton of ammonium ions.

Introduction

The technique of analytical chemistry is essential for determination in various fields such as medical, food and environmental and so on, and new techniques are developed energetically[1-3]. Flow injection analysis (FIA) is one of the analytical techniques, and many techniques of chemical sensor including enzyme sensor are applied as a FIA[4-5].

There are many enzymes which produce NH_3 or CO_2 molecules in their catalytic reactions such as many kinds of amino-acid oxidases or decarboxylases. By using these enzymes as a recognition element, the biosensing system for their substrates could be easily developed in combination with a gas-diffusion unit. A miniaturization of the gas-diffusion unit will take advantage over the sensitivity because the surface area of the gas-diffusion tubing against flow volume is increased as the diameter of the gas-diffusion tubing decreases. In our previous studies[6-8], we try to development of those miniaturized gas-diffusion devices. In addition to the gas-diffusion device, development of the new reagent for detection of NH_3 or CO_2 molecules is an important issue for the performance of the sensing system and the reagent promised to applicable to combination with the enzymes which produced those gas (NH_3 or CO_2).

In our laboratory, during the screening of microbial which produce some enzyme, some bacteria which produce a pigment was obtained. And the color of the pigment was changed by pH changing. Therefore, the pigment will be a new coloring agent for indicator of pH shift. In this study, the pigment was identified followed by isolation and was applied to determination of ammonia as a coloring agent.

Experimental

Materials, Chemicals and General procedure

All reagents used in this study were commercially available and of analytical grade. Ultrapure water with a resistivity of 18.2 MΩ–cm was obtained from an EQG-3S system (Nippon Millipore K. K., Tokyo), and used in all procedures. Wakogel C-300 silica gel (Wako Pure Chemical Industries, Osaka) was used for the column chromatography. F_{254} silica gel-coated glass plate (Merck, Darmastadt, Germany) was used for thin layer chromatography (TLC). ^{1}H-NMR (300 MHz), ^{13}C-NMR (125 MHz), and 2D-NMR spectra were measured with JMN-LA300 spectrometer (JEOL Ltd., Tokyo) in a CDCl$_3$ solution. Chemical shifts were given in δ (ppm) values relative to tetramethyl silane (TMS) as an internal standard. Gas chromatography/mass spectrometry (GC-MS) spectra were measured with a GCMS-QP1000A spectrometer (Shimadzu Co.) (column: DB-1 column, J & W Scientific, 15 m x 0.25 mm; line: He flow at 1 mL min^{-1}; column temperature profile: 50 ℃ for 1 min, increasing from 50-250 ℃ at 10 ℃ min^{-1}, 250 ℃ for 10 min; electron potential: 70 eV). FT-IR spectrum was obtained by using of FT/IR 410 spectrometer (JASCO Co.). Reverse phase high performance liquid chromatography (HPLC) was performed on a ODS-3 column (5 μm, 4.6 x 150 mm : GL Sciences) in a JASCO system (JASCO Co.). Detection was carried out with an UV detector at 254 nm (UV-920, JASCO Co.).

Culture of pigment producing bacteria and isolation of pigment

The bacteria which produce blue pigment was pre-cultured in the sterilized LB broth at 37℃ for 12 h as a pre-incubation. After pre-incubation, The microorganisms was inoculated on a LB agar by platinum loop and incubate it at 37℃ for 48 hours. After incubation, the agar was soaked in chloroform (1 L) for 24 hours followed by the removing the microorganisms by using bacteria spreader. And, the solvent was evaporated under reduced pressure. The extracted material (0.127 g) was dissolved in H$_2$O (100 mL) and extracted with three 100 mL portions of n-hexane (Hex), three 100 mL portions of chloroform (CHCl$_3$). The evaporation of each fraction resulted dried matter 20 mg (H$_2$O), 10 mg (Hex), 91 mg (CHCl$_3$), respectively. Among these 3 fractions, the CHCl$_3$ fraction showed the deep blue color, and therefore this fraction was further purified by the silica gel column chromatography (3.0 x 25 cm). When the column was eluted with CHCl$_3$-MeOH (20 : 1), six distinct fractions were obtained. Among these 6 fractions, Fr-B showed the blue color and the weight was 20 mg. This fraction was applied on the reverse phase HPLC, and eluted with H$_2$O-MeOH with its ratio being varied from 80 : 20 to 0 : 100 during 10 min-elution at a flow rate 1.0 mL min^{-1}. The detector of HPLC was UV/Vis and the wave length was 310 nm. Finally 14 mg dried matter was obtained as the purified matter.

Molecular Analysis Data of the Purified Matter.

IR v_{max} 2925, 1631, 1604 cm^{-1}; EIMS: m/z = 211 (M$^+$), 212 (M+H)$^+$, 196 (M-(CH3+N$_2$))$^+$; ^{1}H-NMR (400 MHz, CDCl$_3$+CD$_3$OD) δ 4.98 (s, 3H) , 7.55 (d, J = 8.5, 1H) , 8.04 (d, J = 8.5, 1H) , 8.23 (t, J = 8.2, 1H) , 8.40 (t, J = 8.5, 1H) , 8.50 (t, J = 8.2, 1H) , 8.73 (d, J = 8.2, 1H) , 8.76 (d, J = 8.2, 1H)

Identification of the bacteria

The bacteria was identified by API 20 NE (Sysmex Co.) and by 16 S ribosomal RNA gene analysis.

Flow system and procedure

A schematic diagram of the flow system is shown in Fig. 1. The system was assembled with two double-plunger pumps (Sanuki DM3M-2044, DMX-2000, Sanuki Industry Co., Ltd., Tokyo), a rotary injection valve with a 100 µl sample loop, the immobilized acid urease column (300 µl) with a water-jacket, a gas-diffusion unit, a UV/VIS detector (UV-970, JASCO Corp., Tokyo) with a quartz flow-through cell (volume 32 µl, light-path length 10 mm), and a pen recorder (Multi-Pen Recorder; type R-62M3, Rikadenki Kogyo Co. Ltd., Tokyo). The temperature around the gas diffusion unit was regulated with a constant temperature bath (F·I·A Instruments Co., Ltd.).

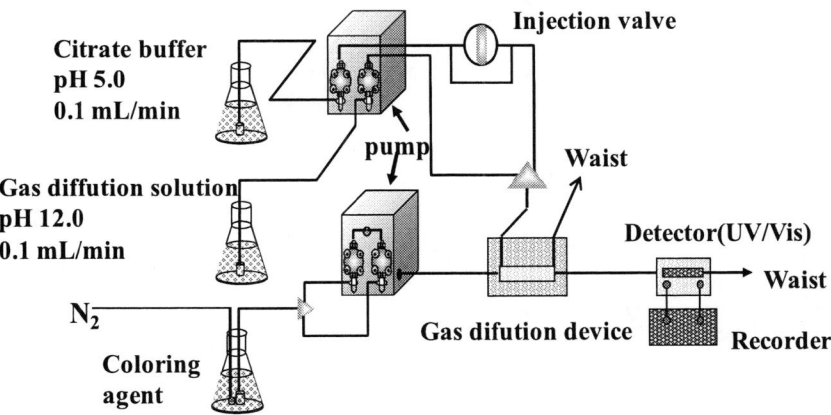

Fig.1 Schematic diagram of the FIA system

Citrate buffer (50 mM, pH 5.0) as the carrier solution (0.1 ml min^{-1}) was successively pumped through the system. Sample solutions were introduced into the system via the rotary injection valve. Ammonium ions formed in the enzymatic hydrolysis of urea were converted to gaseous ammonia molecules by mixing with the strongly alkaline buffer (gas-diffusion buffer: 100 mM sodium phosphate, pH 12.0), and the mixed solution (0.2 ml min^{-1}) was transferred to the gas-diffusion unit consisting of a double tubing structure. The absorbance of Thymol Blue flowing streams in the PTFE tubing was varied by gaseous ammonia diffusion across the PTFE tubing, and subsequent increase in absorbance at 596 nm attributable to the reaction was successively monitored by a flow-through type of a UV/VIS detector and displayed on the pen recorder. The coloring reagent solution (Thymol Blue solution; 0.15 mM, pH 8.4, 0.1 ml min^{-1}, Bacteria produced pigment; 200 μM, pH 4.2, 0.2 ml min^{-1}) was passed with a wet nitrogen streaming (120 ml min^{-1}) into the reservoir.

Results and Discussion

Characterization of the pigment

From the results of GC-MS and ^{1}H-NMR, H-H and C-H COSY NMR spectra, the pigment produced by the microorganism was identified as a pyocyanine ($C_{13}H_{10}N_2O$).

Identification of the microorganism

From the results of API 20 NE test, the microbial was speculated as a *Pseudomonas aeruginosa* (98 %). And from the results of 16S ribosomal RNA gene was also indicated as a *Pseudomonas aeruginosa* (98.4 %).

Application of the pigment for ammonia detection

To evaluate detection properties of this method, ammonium chloride solutions were injected into the sensing system without the enzyme column. And the properties of obtained pigment as a coloring reagent was compared with Thymol Blue. An absorbance at 596 nm (ΔA) of Thymol Blue solution and 280 nm of the solution of the pigment was changed due to ammonia diffusion across the PTFE tubing in the gas diffusion unit. As shown in Fig. 2, a good linear relationship between the concentrations of ammonium chloride solutions and the peak height (changes in absorbance) was obtained. Therefore, optimum conditions of this system was researched in further studies.

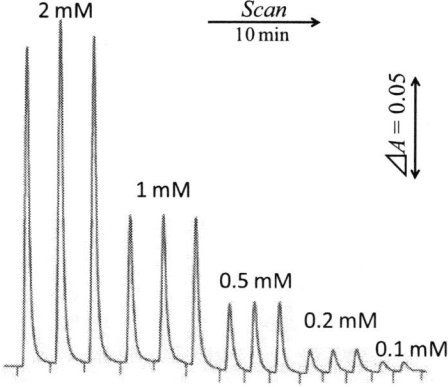

Fig.2 Response curves to injection of ammonium chloride standards with various concentrations.

Effect of the flow rate on the responses. The influence of flow rate on the response of this ammonia detection system was investigated. Various combinations of flow rate of carrier, gas diffusion reagent and pyocyanine solution were investigated.

As shown in Fig.3, the responses depended on the flow rate and when the flow rate was fast, the tendency for a reaction to be low was observed (the flow rate of each solution used in this study was shown in Table 2.). The lowest response was obtained when the flow rate of each solution was 0.2 mL/min. On the other hand, the highest response was observed when the flow rate of each solution was 0.05 mL/min. However, 0.05 mL/min of flow rate of any solution made measurement time long, 0.1 mL/min of flow rate of each solution was decided to research in further investigation.

Table 1. Combination of flow rate of each solution used in this study.

	Carrier (mL/min)	Gas diffusion (mL/min)	pyosianin (mL/min)
A	0.2	0.2	0.2
B	0.2	0.2	0.1
C	0.2	0.2	0.05
D	0.1	0.1	0.2
E	0.1	0.1	0.1
F	0.1	0.1	0.05
G	0.05	0.05	0.1
H	0.05	0.05	0.05

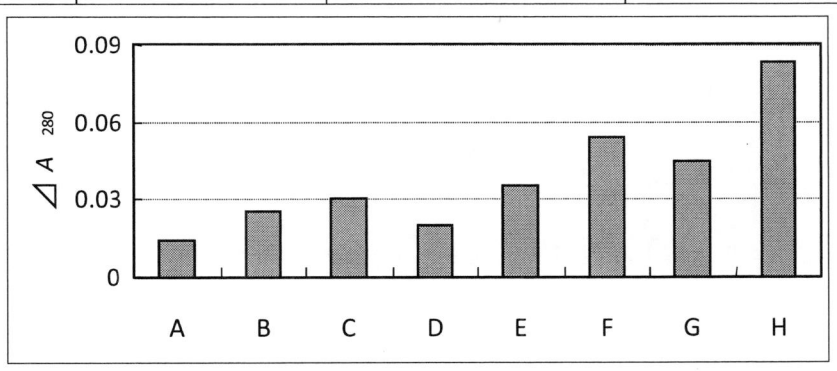

Fig. 3 Effect of the flow rate on the response of ammonia detection.

Effect of the concentration of citric acid in the coloring reagent on the response.
The response of this system will depend on the concentration of citric acid used as a buffer in the coloring reagent (pyocyanine solution) because a principle of this method measures the pH shift and the pH of coloring reagent is influenced by CO_2 in the air.

Therefore, the influence of buffer reagent on the measuring was observed and the results was shown in Fig. 4. The data showed that if the concentration of citric acid become high, the response will become low. And when no citric acid contained solution was used as a coloring, highest response was obtained.

In usual, a buffer agent such as citric acid, phosphate and so on will need as a buffer to prevent stability of base line from CO_2 or some interfering compounds which influence on pH shift. However, in the case of this study, no interfering effect from CO_2 was observed. We consider the reason that solubility of CO_2 in methanol is low in comparison with that of in water. Therefore the base line was stable in no buffer containing reagent and so, the coloring reagent (no buffer containing pyocyanine solution) was used in further research.

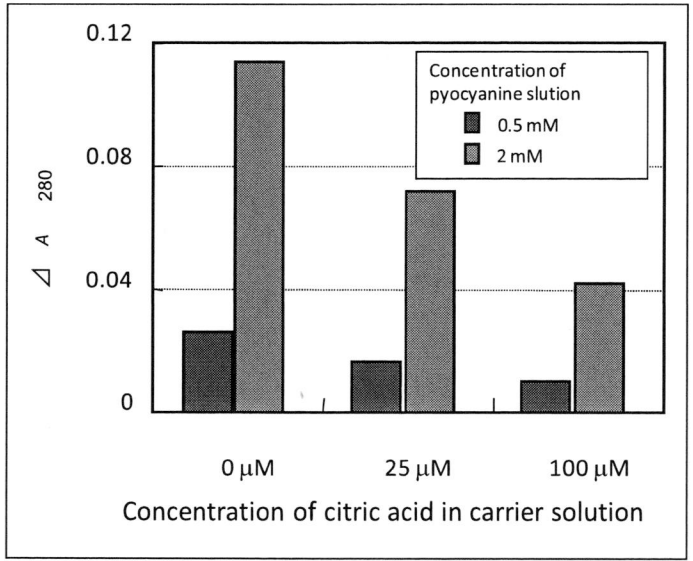

Fig.4 Effect of the concentration of citric acid in the carrier solution on the response.

Effect of the concentration of the pyocyanine on the response. To evaluate response of the system to the concentration of pyocyanine was observed. Although the peak become large when the concentration of a coloring agent is higher, the response of 100 μM is not double in comparison to that of 50 μM. It means that the concentration of pyocyanine will become saturate, therefore, 50 μM coloring reagent was used in further research.

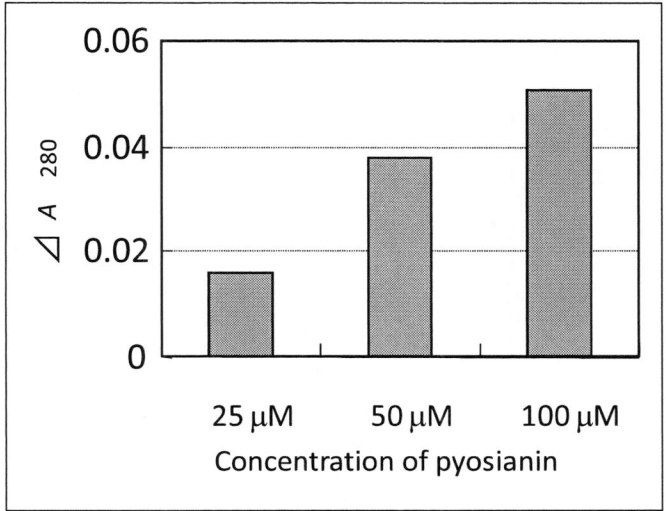

Fig. 5   Effect of the concentration of the pyosianine on the response

Application to the determination of ammonia. The optimized condition for determination of ammonium ion was decided, various concentrations of ammonium chloride solutions were further injected and measured by this system. As shown in Fig. 6, a good linear relationship between the concentrations of ammonium chloride solutions and the peak height (changes in absorbance) was obtained and the good linearity was obtained in a range of 15.6 μM – 2.0 mM. And the coefficient was calculated to be 0.995. The response to ammonium chloride was obtained within 3 min. A relative standard deviation of change in absorbance (ΔA) at the each concentration of ammonium chloride was about 2.8 % ($n = 5$). On the other hand, when the result of using of conventional Thymol Blue as a coloring agent was also shown in the Fig. 6. The peak was lower than that of pyocyanine. It means that the sensitivity of pyocyanine as a coloring reagent was higher than that of Thymol blue. The reason why pyocyanine has higher sensitivity might be the transition pH range. Thymol Blue transitions from yellow to blue at pH 8.0–9.6. On the other hand, pyocyanine transitions from red to blue at pH 4.2–6.1. Therefore, the influence of pH shift by ammonia on the pyocyanine is larger than that of pH shift by ammonia on the thymol blue.

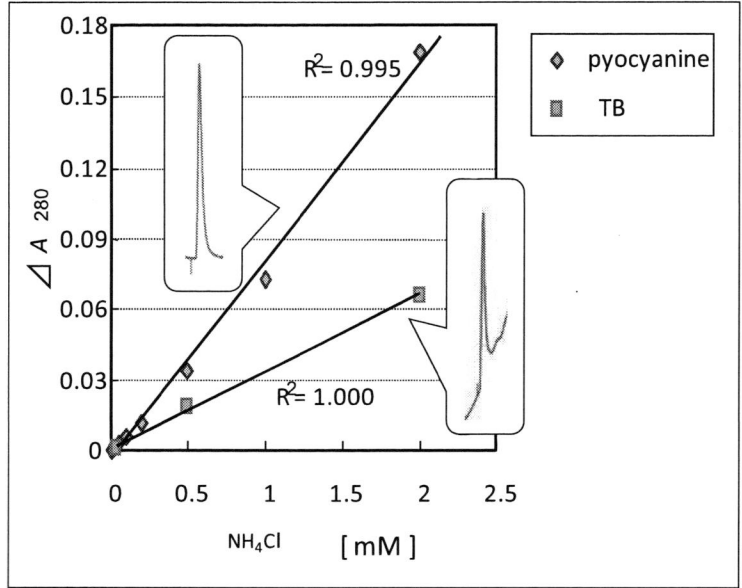

Fig. 6 Calibration curve for ammonia measurement and comparison with conventional coloring reagent.

References

1. X. Xia, Y. Yibin, *Food Rev. Int.*, **27**(3), 300 (2011).
2. W. Jerome,Jr., L. Barry, C. Ray, K. Mel, *Anal. Chem.*, **83**(12), 4557 (2011).
3. H. Supaporn Kradtap, G. Kate, *Anal. Lett.*, 44(1-3), 483 (2011).
4. A.-H. Nabil N., S. Bahruddin, *Anal. Lett.*, 44, 13 (2011).
5. P. Paula C. A. G., Lucia M., Saraiva M. F. S., LIMA Jose L. F. C. *Anal. Lett.*, 44(1-3), 374 (2011).
6. Y. Iida, Y. Chiba, K. Matsumoto, T. Noda, I. Satoh, *Desalication and Water Treatment,* **17**, 1 (2010).
7. Y. Iida, M. Ikeda, M. Aoto, I. Satoh, *Talanta*, **64**, 1278 (2004).
8. Y. Iida, Y. Suganuma, K. Matsumoto, I. Satoh, *Analytical Sciences,* **22**, 173 (2006).

ECS Transactions, 50 (12) 393-399 (2012)
©The Electrochemical Society

Electrochemical Immunosensor for Diagnosis of Parasitical Human Diseases

C. A. Erdmann, J. Inaba, A. G. Viana, C. A. Pessoa, K. Wohnarth and J. R. Garcia

Chemistry Department - State University of Ponta Grossa Av. Gal. Carlos Cavalcanti,
4748, Uvaranas, Ponta Grossa-PR – 84030-900, BRAZIL

> In this paper the preparation and electrochemical characterization
> of an immunosensor for Chagas disease is described. The results
> obtained indicating the possibility of modification of ITO surface
> by formation of an LbL film constructed by the intercalation of
> PVS as polyanion and AgNP-SiPy$^+$ as polycation. The LbL films
> were then modified by the incorporation of Chagas ATG on theirs
> surface. These modified LbL films showed a pronounced
> impedance change when in the presence of the Chagas ATB
> indicating that this kind of film could be used in a device for
> diagnosis of Chagas disease.

Introduction

Fast and precise diagnosis of infectious and parasitical disease is crucial to guide
authorities to correct actions in the case of epidemics situation, since the immediate
identification of the infection focus or parasite source is essential for an efficient reaction.
On the other hand, for an individual infected, delays in a diagnosis can harm the
treatment necessary and result in an aggravation of the illness. In developing countries
this situation is aggravated by the lack of financial resources applied to the detection and
prevention of infectious and parasitical disease due to the fact that the majority of the
technologies used in the diagnosis have a high cost, using complex systems and a
requirement of great physical infrastructure for their implementation (1).

In this context immunoassays play a central role for diagnosis in clinical and
biological sampling. Due to the complex nature of clinical samples, immunoassays
typically employ a structure that facilitates an interaction between antigen (ATG) and
antibody (ATB) in order to increase specificity and sensitivity. Recently, a variety of
nanomaterials, many of them electroactive, possessing size dimensions perfect for
devices miniaturization have been shown to have utility in the development of new
immunoassays (2-6).

Electrochemical immunosensors based on the specific reaction of the ATG and ATB
with electrochemical transduction have been a remarkable subject for clinical diagnosis
(7-8) due to the possibility of producing methods feasible for use *in vitro* and *in vivo*,
with low cost, small size, and short response time. These methods involve the
immobilization of an ATB component on the electrode transducer and the use of an
electrochemical active substance for signal generation. Then, the sensor can be exposed
directly to a biological fluids; if ATG is present, an ATG – ATB complex will be formed.
This formation will induce changes of physical and chemical characteristics in the film on
the electrode surface, and that change can be then detected. Despite the transduction

393

principle, the coupling of the molecular recognition receptor to the transducer surface is of key importance for the control and to improve characteristics of the sensor (9-10).

Aiming at to contribute for the development of diagnosis methodology for endemic parasitical diseases in Brazil the major objective of this work is the preparation of electrochemical imunosensors for diagnosis of neglected parasitical disease as Chagas, leishmaniasis and schistosomiasis. At this paper the preparation and electrochemical characterization of an immunosensor for chagas disease is described.

Experimental

All chemicals used in the experiments describe were of analytical grade and they were used without previous purification and solutions were prepared with water from a Millipore Milli-Qplus system. The $AgNO_3$, the $NaBH_4$ and the anionic polyelectrolyte poly(vinylsulfate) (PVS) were acquired from Sigma. The cationic polyelectrolyte silsesquioxane 3-n-propylpyridinium chloride ($SiPy^+Cl^-$), was prepared according to a method described elsewhere (11). The biomolecules (*Trypanosoma* ATG and ATB) used were extract from an Elisa test kit used in qualitative and semi-quantitative Chagas diagnosis from ANALISA (Brazil).

The Chagas disease sensor was produced by immobilization of the ATG on to a modified conducting glass substrate (ITO - indium doped tin oxide film). This immobilization was carried out using the Layer-by-Layer (LbL) procedure. In this methodology bilayers of a polycation and a polyanion are adsorbed over the substarate. Each of these bilayers will be formed by a layer of polyvinyl sulfonate (PVS^-) and a metallic layer of nanoparticles of silver (Ag) incorporated into the inorganic matrix of $SiPy^+$. The use of metallic nanoparticles is justified for the necessity of the introduction of transducers that could increase the level of the electrochemical signal to enhance the quantification of the antibody presence in the body fluids.

Synthesis of the Silver Nanoparticles

The silver nanoparticles were prepared using the $SiPy^+Cl^-$ as nanoreactor and stabilizer. This hybrid composite ($AgNP-SiPy^+$) was obtained by chemical reduction of the $AgNO_3$ in solutions contained different concentration of $SiPy^+Cl^-$ (3 g L^{-1} and 4 g L^{-1}), the chemical reduction was carried out using $NaBH_4$. After addition of the reducing agent the reaction mixture was remained over magnetic agitation for 1 hour. It was prepared $AgNP-SiPy^+$ using different amounts of $AgNO_3$ (2% and 5% *m:m* related to the $SiPy^+Cl^-$ total mass) and $NaBH_4$ (excess of 10% and 100% related to the molar quantity need to reducing the added Ag). The nanoparticles dispersion was characterized by UV-visible measurements in a spectrometer from VARIAN, model Cary 50. The nanoparticle size distribution was determinate by Dynamic Light Scattering Analysis (DLS) using an equipment from Micro Trac, model Zetatrac.

LbL films assemblage

For assembling the LbL films used in the sensors the PVS was used as polyanion and the composite AgNP-SiPy$^+$ was used as polycation, thus forming a bilayer over ITO. In this process the PVS concentration was maintained in 3 g L^{-1} and the AgNP-SiPy$^+$ dispersion used was chosen as the one with smaller particles. The films growth was followed by UV-visible measurements using the same spectrometer used in the AgNP-SiPy$^+$ dispersion characterization. After that the morphological characterization of the LbL films was evaluated by Atomic Force Microscopy (AFM). The AFM measurements were carried out on films surfaces with 2, 4, 6, 8, 10, 14 bilayers and on clean ITO. These measurements were performed with a SHIMADZU equipment, model SPM 9600 in the non-contact mode.

Subsequent to verification of the satisfactory formation of the LbL films for the modification of the ITO substrate the ATG for the Chagas disease was incorporate into films with different bilayers numbers by immersing these films in a solution containing the Chagas ATG during 2 h. After this the films were immersed in a solution containing BSA (Bovine Serum Albumin) to form a kindly environment for the ATG protein and also to block any free active adsorptive sites.

Electrochemical Measureaments

The electrochemical measurements were carried out using a three electrode electrochemical cell (10 mL of capacity), using the LbL as working electrode, a platinum electrode as auxiliary electrode and a Ag/AgCl$_{(sat.)}$ as reference. These measurements were performed in an AUTOLAB Potentiostat/Galvanostat, model PGSTAT100 combined with the frequency analyzer module FRA2. Cyclic Voltammetry analyses were performed intent to identify the AgNP-SiPy$^+$ incorporation at the films and impedance measurements were performed before and after immersion of the modified films in a solution containing the Chagas ATB to evaluated changes on response of the films in these different environment.

Results and Discussions

As described above the formation of the AgNP-SiPy$^+$ was performed using different conditions. The variation in the SiPy$^+$Cl$^-$ concentration does not produce changes on the AgNP-SiPy$^+$ obtained, however the changes on the Ag and on reducing agent quantity influence the nanoparticles as demonstrated on Table I.

TABLE I. Conditions used for AgNP-SiPy$^+$ dispersion formation using a SiPy$^+$Cl$^-$ concentration of 3 g L^{-1}.

Silver Quantity (%)[*]	NaBH$_4$ Quantity (%)[**]	Medium Particle Size (nm)
2	10	21
2	100	35
5	10	225
5	100	280

[*] - 2% and 5% m:m related to the SiPy$^+$Cl$^-$ total mass
[**] - excess of 10% and 100% related to the molar quantity need to reducing the added Ag

The UV-visible spectrum for the AgNP-SiPy$^+$ dispersion produced with the conditions described in the first line of the Table I is showed on Figue 1-A. It could be seem in this figure that the dispersion presented a higher intensity band in *ca.* 250 nm assigned to the SiPy$^+$Cl$^-$ polymer and a brod low intensity band in *ca.* 400 nm assigned to the plasmonic band of the AgNP-SiPy$^+$. Due to the fact that the nanoparticles obtained in the conditions described in the first line of the Table I this dispersion was used to LbL films formation.

The Figure 1-B shows the UV-visible spectra obtained by the LbL films formation with different numbers of bilayers. This figure shows that the growth of the film is constant until the formation of the 20th bilayer. It is interesting to mention that this kinetic measurement was made for different substrate immersion time intervals (1 min to 10 min) producing the same results. It indicates that the kinetics of the mass transfers is high.

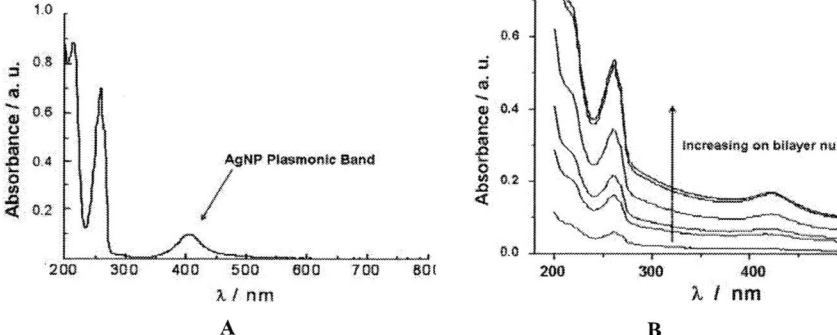

Figure 1. UV-visible spectra for **A)** Silver nanoparticles dispersion obtained with the conditions as listed on the first line of Table I; **B)** LbL films obtained by the bilayer formation with PVS (3 g L^{-1}) and AgNP-SiPy$^+$ (condition listed on the first line of Table I) obtained with films grown with different numbers of bilayers.

The LbL films produced with different numbers of bilayers was submitted to AFM analyses and these measurements produced the thickness values shown in Table II. The thickness of the LbL film with just 1 bilayer is lower than that obtained for the clean ITO surface. It is due to the fact that the first layer fills the depression on the ITO surface. After that the thickness increases faster for each bilayer until reaching the 20th one, supporting the behavior observed with the UV-visible measurements.

TABLE II. Thickness of the LbL films obtained with different numbers of bilayers for 1 μm^2 and for 5 μm^2.

Bilayer Number	Thickness in 1 μm^2 (nm)	Thickness in 5 μm^2 (nm)
0	9.93	16.08
1	4.00	13.25
5	17.43	20.34
10	35.04	41.20
15	90.49	169.12
20	107.28	260.44
Thickness by layer	**4.98**	**9.88**

The electrochemical behavior of the LbL films was followed by cyclic voltammetry measurements. This characterization is exemplified in Figure 2 that brings the voltammograms for the films obtained with one and two bilayers. As could be seem in this figure the capacitive current for the LbL film with 2 bilayers is higher than those values for the 1 bilayer film. This fact is consistent with a higher surface area caused by the increase in mass transferred into the film. Additionally for the 2 bilayer film a pronounced Faradaic current peak was observed with Epa in *ca.* 0.13 V assigned to the process of nanoparticle surface oxidation. With the increase in numbers of bilayers the films start to present a resistive behavior component and due to this fact the films with two bilayers were chosen for the sensor preparation.

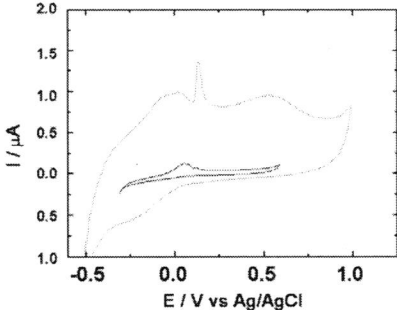

Figure 2. Voltammograms for the films obtained with: Dark Gray – 1 bilayer and Light Gray – 2 bilayers, obtained in a PBS buffer with pH = 4.8 and 100 mV s^{-1}.

Figure 3 brings the Nyquist plot that compares the impedance change for LbL films obtained with 2 bilayers without ATG modification and containing ATG in the absence and in presence of the Chagas ATB. It could be seen from this Figure that there is a profound change on the impedance behavior of the films and the presence of the ATG on the film surface give selectivity to the sensor for the ATB presence.

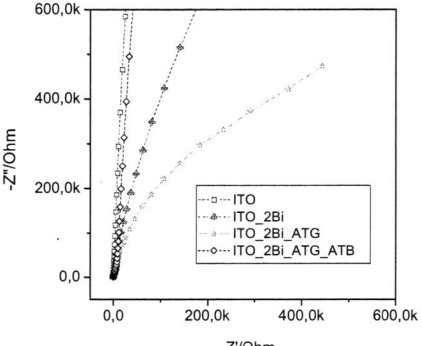

Figure 3: Nyquist plot for LbL films obtained with 2 LbL bilayers modified with ATG and the impedance response for films without modification in the presence of the Chagas ATB.

The results obtained indicate the possibility for the modification of ITO surface by formation of an LbL film constructed by the intercalation of PVS as polyanion and AgNP-SiPy$^+$ as polycation. The LbL films were than modified by the incorporation of Chagas ATB on their surface. These modified LbL films showed a pronounced impedance change when in the presence of the Chagas ATB indicating that this kind of film could be used in a device for diagnosis of Chagas disease.

Conclusions

The results obtained indicating the possibility of the modification of ITO surface by formation of an LbL film constructed by the intercalation of PVS as polyanion and AgNP-SiPy$^+$ as polycation. The LbL films were than modified by the incorporation of Chagas ATG on theirs surface. These modified LbL films showed a pronounced impedance change when in the presence of the Chagas ATB indicating that this kind of film could be used in a device for diagnosis of Chagas disease.

Acknowledgments

We would like to thank the following funding institutions from Brazil: CNPq, CAPES and Fundação Araucária (SUS – 08/2009) for financial support.

References

1. F.B. Diniz, R.R. Ueta, *et al.*, *Biosensors and Bioelectronics,* **19**, 79, (2003).
2. F. N. Crespilho, V. Zucolotto, et al., *Eletrochemistry Science*, **1**, 194, (2006).
3. L. Ping, M. Ligen, et al., *Microchimica Acta*, v.**171**p.297, (2010).
4. Y. Gushikem, E. V. Benvenutti, et al., *Pure Applied Chemistry*, **80**, 1593, (2008)

5. R. V. S. Alfaya, Y. Gushikem, *Journal of Colloid Interface Science*, **313**, 438, (1999).
6. S. T. Fujiwara, C. A. Pessoa, et al., *Analytical Letters*, v.**35**, p.1117, (2002).
7. J Wang, B Tian and K.R Rogers, *Anal. Chem.*, **70**, 1682, (1998).
8. G.D Liu, Z.Y Wu, et al., *Anal. Chem.*, **73**, 3219, (2001).
9. .S Babkina, E.P Medyantseva, et al., *Anal. Chem.*, **68**, 3827(1996).
10. C Fernández-Sánchez, A Costa-García, *Anal. Chim. Acta*, **402**, 119, (1999).
11. Y. Gushikem, R. V. S. Alfaya, A. A. S. Alfaya, *Br. Pat., INPI 9803053-1*, (1998).

Preparation of Fine Implantable Needle Type Biosensors for Blood Vessel Glucose Monitoring

K. Edagawa and M. Yasuzawa

Department of Chemical Science and Technology,
The University of Tokushima, Tokushima 770-8506, JAPAN

Fine needle type glucose sensors with variety of outer membrane were prepared and their glucose sensor properties were evaluated by in vitro measurement in phosphate buffer solution and horse serum, and in vivo measurement using the veins of rats and rabbits. Extra-outer membrane materials of polyglutamic acid (PGA) and heparin, which provide hydrophilic surface were introduced in addition to polyurethane/polydimethylsiloxane composite outer membrane. Glucose sensors without extra-outer film and with PGA extra-outer film showed glucose response on the day of implantation, while clear response was not obtained on the second day of implantation due to the adsorption of thrombus on the sensor surface. On the other hand, the sensor with heparin outer membrane provided satisfactory glucose response also on the second day.

Introduction

Biosensors are used in various fields, such as medical treatment, food, and environment. Especially, the blood glucose degree meter used for self-monitoring of blood glucose (SMBG) is indispensable to a diabetic's health care administration. It is well known that keeping good control of the blood glucose degree can prevent the onset and progression of serious diabetes complications. Therefore, it is important to accurately recognize the blood glucose degree and provide appropriate treatments, such as insulin therapy. Moreover, blood glucose managements are now well recognized to be of significantly important in variety of medical treatment especially surgical operation, since drugs and stresses may cause insulin resistance, which lead to an acute hyperglycemia state even in non-diabetes patient (1). In order to prevent hyperglycemia, excess amount of insulin are injected to the patient. However, insulin resistance is a passing phenomenon, which continue for a certain time and it can suddenly reduce its effect to provide hypoglycemia. When the hypoglycemic state of less than 30 mg/dl remain for a while, damage of the brain can be presented and may also lose one's life. Therefore, rapid recognition of hypoglycemia is essential for effective treatment.

There are several continuous glucose monitoring system (CGMS) in the market and it is getting popular as an effective equipment that lower physical and mental load of diabetes patient on glucose measurement. However, it is well known that the glucose level of subcutaneous tissue differ significantly at the time of rapid glucose concentration change. Therefore, it can be possible that the direct blood glucose measurement is required for the case of serious insulin resistance, while the access to the blood vessel increase the risk of serious infection.

In this study, fine needle type glucose sensors with variety of outer membrane were prepared in order to evaluate their glucose sensor properties in blood vessel. Extra-outer membrane materials γ-polyglutamic acid (PGA) and heparin, which provide hydrophilic surface were introduced in addition to polyurethane(PU)/polydimethylsiloxane (PDMS) composite outer membrane. The efficiencies of outer membrane were evaluated by in vitro measurement in phosphate buffer solution and horse serum, and in vivo measurement using the veins of rats and rabbits.

Experimental

Reagents and Chemicals

Glucose oxidase (GOx) (244 U/mg, purified from Aspergillus niger) was purchased from Biozyme laboratories. γ-Polyglutamic acid (Mw = 800000 – 1000000) was kindly supplied from Nippon Poly-Glu. Cellulose acetate was obtained from Kishida Chemical. Polyurethane (Tecoflex) was purchased from Thermedics Inc (Woburn, MA). Polydimethylsiloxane (MED-4211) was purchased from Nusil (Carpenteria, CA). Nafion® (perfluoronated ion-exchange polymer 5% wt. in a mixture of lower aliphatic alcohols and water) was purchased from Aldrich. Horse serum containing 0.55 mmol dm^{-3} glucose was purchased from Tissue Culture Biologicals (Los Alamitos, CA). Platinum-iridium (0.10 mm in diameter, Pt 90%-Ir 10%) wire was purchased from Nilaco. The insulation tube of polyimide (0.12 mm i.d., 0.16 mm o.d.) was purchased from Furukawa Electric. All other reagents were of analytical grade, and were used without further purification.

Preparation of Fine Needle-Type Glucose Sensor

The schematic illustration of the glucose sensor is shown in Fig. 1. The sensor fabrication was carried out according to methods of Wilson et al. (2,3), which consists of three main steps: permselective inner-film preparation, enzyme immobilization, and biocompatible outer-film construction. The preparation of the permselective inner film was prepared by alternate three-times coating of both Nafion and cellulose acetate using Nafion and cellulose acetate solutions. Immobilization of enzyme (glucose oxidase) was performed by cross-linking method using bovine serum albumin and glutaraldehyde. The outer film was prepared using a tetrahydrofuran solution containing polyurethane and polydimethylsiloxane. Extra-outer film of PGA was prepared by passing the electrode through a wire loop, which has a membrane of PGA. Heparin extra-outer film was prepared by coating heparin film on polyethylene imine film.

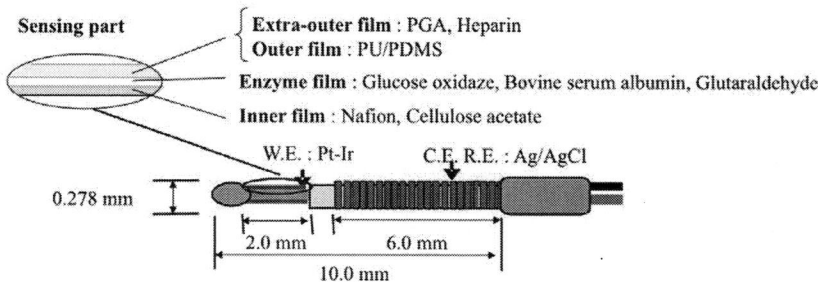

Figure 1. Schematic illustration of fine needle-type glucose sensor.

Glucose Sensor Response Measurement

The amperometric responses of the prepared electrodes to glucose were examined at 40°C in a 0.1 mol dm^{-3} phosphate buffer solution of pH 7.4 containing 0.1 mol dm^{-3} NaCl and horse serum containing 0.55 mmol dm^{-3} glucose, by measuring the electrooxidation current at a potential of 0.6 V (*vs.* Ag/AgCl) for hydrogen peroxide detection. Amperometric measurements were performed with a Potentiostat Model 3104 (Pinnacle Technology Inc.).

The calibration of the sensor was carried out by adding increasing amounts of glucose to the measuring solution. The current was measured at the plateau (steady-state response), and was related to the concentration of the analyte.

The sensor response was also evaluated by in vivo measurement using rats and rabbits. The sensor was implanted in the veins of rats and rabbits under anesthesia. Glucose and insulin were injected in order to provide variation of glucose level. Periodic blood glucose was also measured using commercially available glucose meter. Conversion of sensor response current to glucose level was performed using one-point calibration method (4). Blood glucose value around 100 mg dL^{-1}, which was obtained at a stable unchanged sensor response, was used for calibration (rectifying points). When the measurement was completed, the sensors remained in place either disconnected or connected to the wireless potentiostat 3102 (Pinnacle Technology Inc.).

Results and Discussion

In vitro Measurement

Figure 2 shows typical calibration curves of the electrodes prepared without extra-outer film and heparin film. The response current increased with increasing concentration of glucose up to 22.4 mmol dm^{-3}, while the linear relationship between the glucose concentration somewhat wider with heparin extra-outer film. The sensitivity of the sensors prepared was approximately same even after the modification of extra-outer film.

In vivo Measurement

Glucose sensors with PU/PDMS outer film, which provided satisfactory performance on the monitoring in subcutaneous tissue, showed response to glucose on the day of implantation, while the response reduced as time passed and it did not show clear glucose response after 18 hours. Adsorption of the thrombus was observed on the surface of the sensor after the removal from blood vessel. On the other hand, the sensor with heparin outer membrane provided satisfactory glucose response also in the second day (Fig. 3). However, the introduction of PGA outer film did not show clear improvement, and no response was confirmed after two days of implantation.

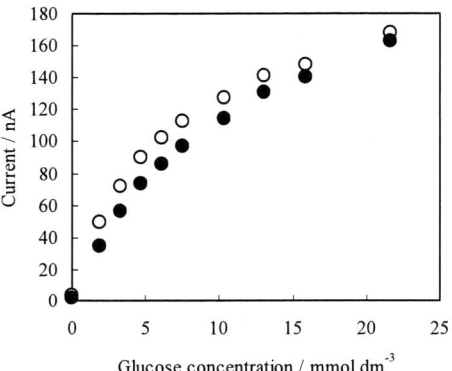

Figure 2. Typical calibration curves of electrodes without extra-outer film (open circle) and with heparin extra-outer film (closed circle) measured in a 0.1 mol dm^{-3} phosphate buffer solution (pH 7.4) at 40°C.

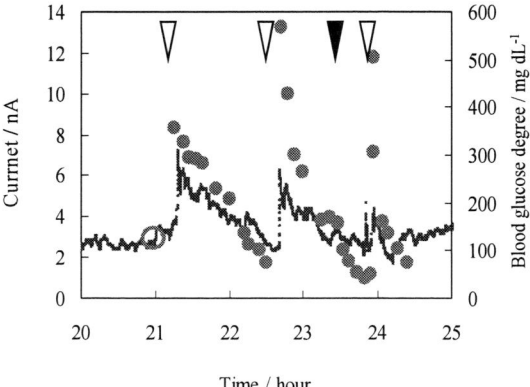

Time / hour

Figure 3. Comparison of the glucose sensor response (solid line) and blood glucose measured using commercial glucose meter (solid circle) with time. Glucose sensor was inserted in blood vessel for 2 days. White and black arrows indicate the injection of glucose and insulin, respectively.

Conclusion

Fine needle type glucose sensors with PU/PDMS outer membrane and with extra-outer membranes of PGA and heparin were prepared and implanted in blood vessel for continuous glucose monitoring. Glucose sensors with PU/PDMS outer membrane and PGA extra-outer membrane provided good response on the day of implantation, while their function were lost on the second day. Glucose sensors with heparin extra-outer membrane provided good glucose response even after two days of implantation. This indicate that the sensor with heparin extra-outer membrane may have a potential to provide immediate notice of a transient insulin resistance recovery, and reduce the risk of presenting serious hypoglycemic state, which may cause critical illness.

Acknowledgments

The authors are grateful to Nippon Poly-Glu Co. Ltd., Japan for providing γ-polyglutamic acids.

References

1. R. P. Dellinger,M. M. Levy, J. M. Carlet, et al, *Crit Care* Med **36**, 296 (2008).
2. D. S. Bindra, Y. Zhang, G. S. Wilson, D. R. Thevenot, D. Moatti, and G. Reach, *Anal Chem*, **63**, 1692 (**1991**).
3. B. Aussedat, M. Dupire-Angel, R. Gifford, J. C. Klein, G. S. Wilson, and G. Reach, *Am. J. Physiol. Endorinol. Metab.*, **278**, 716 (2000).
4. C. Choleau, J. C. Klein, G. Reach, B. Aussedat, V. Demaria-Pesce, G. S. Wilson, R. Gifford, W. K. Ward, *Biosens Bioelectron* **17**, 647 (2002).

406

CHAPTER 4

MICRO/NANOFABRICATION

408

ECS Transactions, 50 (12) 409-412 (2012)
©The Electrochemical Society

Micro-systems and Nanotechnologies in ELISA and Droplet Generation Applications

C. H. Yeh and Y. C. Lin*

Department of Engineering Science, National Cheng Kung University, Tainan, Taiwan

Micro-systems and nanotechnologies were successfully applied for ELISA detection and droplet generation. Under the AuNPs as a label of antibody in the electro-microchip, the detection limit for the protein A is 1 ng/mL. Then, the developed microfluidic chips can spontaneously generate the droplets of four various concentrations in different microchannels and generate stable and uniform emulsions that are less than 5μm in diameter by using the electro-spray phenomenon.

Introduction

Immunoassays, which are based on the specific recognition of an antibody by a corresponding antigen, are generally quantified by measuring the specific activity of a label. Based on the material used for protein labeling, they are classified as radioimmunoassay [1], enzyme-linked immunosorbent assay (ELISA) [2], fluorescent dyes immunoassay [3], chemiluminescence immunoassay [4], metalloimmunoassay [5], etc. The immunogold labeling technique (colloid gold-labeled proteins) is commonly used because the high electron density of the gold nanoparticles exhibits a dark brown spot under an electron microscope, resulting in a distinct image for labeling. Metallic nanoparticles (e.g. gold and silver colloids) have recently been successfully used as labels technology because of their easily controllable size, unique optical and electrical properties, and high biocompatibility with antibodies, proteins, RNA, and DNA. This study presents a newly immunoassay that uses an electro-microchip to detect the immuno-reaction signal, metal nanoparticles (AuNPs, AgNPs, and PtNPs) as a label of antibody and as a catalyst for silver precipitation, and the silver enhancement reaction to magnify the detection signal [6, 7]. The metal nanoparticles were introduced into the electro-microchip by the specific binding of the antibodies-ANPs conjugates and then were coupled with silver enhancement to produce black spots of silver metal. The silver precipitation constructs a bridge between two electrodes of the electro-microchip allowing electrons to pass. The microfluidic device which integrated the flow-focusing geometry provides the approach to generate the water-in-oil emulsions and oil-in-water emulsions in liquid-liquid systems [8]. The biomaterial emulsions most widely display the carriers for the immobilization of enzymes and proteins are the same as the controlled release of drugs. Because the microcapsule size and distribution have influence on clearance rate from the body and ultimately determine the drug dosage, it is important to control the size of the uniform biomaterial microspheres and narrow the size distribution. Because of the excellent properties, such as biocompatibility, biodegradable, and non-toxicity, bio-polymers which included alginate, chitosan, gelatin, and PLGA have been widely used the biomedical carrier for drug delivery. These microfluidic chips have the single flow-focusing geometry, multi flow-focusing geometry, and electro-spraying integration [9, 10].

409

Materials and Methods

The electro-microchip and developed microfluidic chips were designed by using AutoCAD® 2008 software, and these chips were fabricated by using the micro-electro-mechanical systems (MEMS) technology. For the sandwich immunoassay, the 1st antibodies are immobilized on the glass slides, and the antigens are then bound to the 1st antibodies. Excess unbound antigens are removed by washing with doubly distilled water. When added to the chip, the gold-conjugated 2nd antibodies bind to the antigens which are already bound to the 1st antibodies, resulting in the formation of a sandwich-like complex, as shown in Fig. 1. In the silver enhancement process, the silver ions in the silver enhancement solution are reduced by the promotion of the nanogold, and a large number of silver particles are precipitated. These silver precipitations construct a "bridge" between the two electrodes of the electro-microchip allowing electrons to pass, which further decreases the impedance. The variation of impedance can be easily measured with a commercial LCR meter.

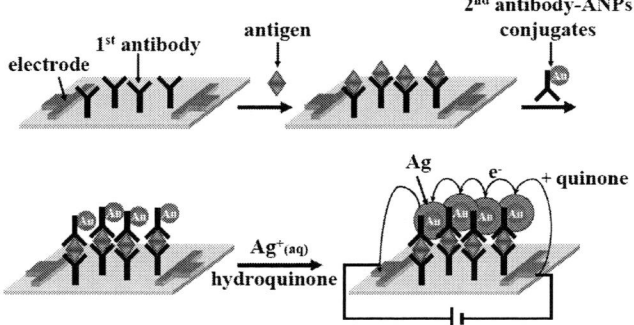

Fig. 1 Schematic illustrations of antibody-ANP conjugate recognition and signal amplification with silver enhancement: sandwich immunoassay (three-layer format).

For the droplet generation, the micro-mixers and tree-like microchannel network produced droplets of four various concentrations in different microchannels, and the flow-focusing device was used to spontaneously assemble for water-in-oil droplets in each microchannel, as shown in Fig. 2a. In order to generate four uniform sized droplets which had the concentration gradient distribution by using the flow-focusing devices, the widths of bifurcate channels (X and Y in Fig. 2a) were adjusted to obtain the same flow velocity in microchannels. Then, in order to decrease the droplet size, an electro-spraying microfluidic chip was integrated with a hydrodynamic flow-focusing function and parallel electrodes to generate the electro-spray phenomenon under microfluidic conditions, as shown in Fig. 2b. The electro-spray condition (Taylor cone) was generated in the developed microfluidic chip with a high electric field, and this can be used to produce emulsions with smaller diameters.

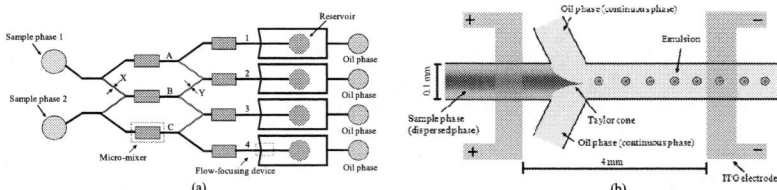

(a) (b)

Fig. 2 (a) The uniform droplet size with different concentrations generated in the every microchannel. (b) Schematic of the electro-spray microfluidic chip. The microchannel was 100 µm in width, and the parallel ITO electrode gap size was 4 mm.

Results and discussions

In the experimental results for ELISA applications, under the various gap size (from 20 µm to 200 µm) tests, the 20 µm electrode gap of the electro-microchip had the best sensitivity and its detection time was the shortest. The sandwich immunoassay was used for quantitative analysis. For example, under the AuNPs as a label of antibody, the detection limit is 1 ng/mL of antigen (protein A) with 10 µg/mL of 1st antibody (IgG) immobilized on slides, as shown in Fig. 3. The results of the proposed immunoassay are faster than those of conventional ELISA.

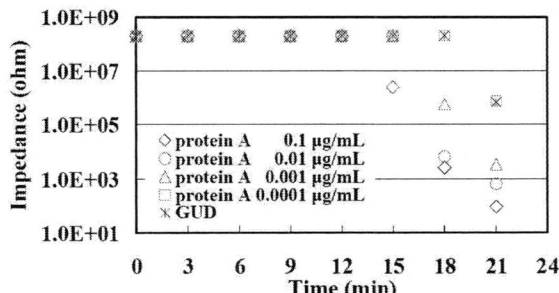

Fig. 3 10 µg/mL concentration of the first antibody and then 0.1~0.001 µg/mL of antigen. The results show that the detection limit is 1 ng/mL.

In the experimental results for droplet generation applications, the micro-mixers and tree-like microchannel network produced droplets of four various concentrations in different microchannels, and the flow-focusing device was used to spontaneously assemble for water-in-oil droplets in each microchannel (Fig. 4). In order to decrease the droplet size, an electro-spraying microfluidic chip was integrated with a hydrodynamic flow-focusing function and parallel electrodes to generate the electro-spray phenomenon under microfluidic conditions. The Taylor cone can then generate stable and uniform emulsions that are less than 5 µm in diameter.

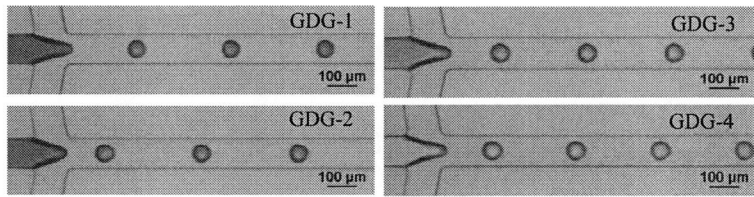

Fig. 4 Water droplets with different trypan blue concentrations generated in the individual microchannel. The droplet size is 55 μm under the 2 μL/min of the sample phase flow rate and the 10 μL/min of the oil phase flow rate.

Conclusions

The developed ELISA detection is easy to perform, only a small amount of the reagent is required (as little as 40 μL sample protein per well), and it is fast (within 30 min), convenient, and low-cost. This approach has many potential uses in protein microarray research and clinical diagnosis. Then, these microfluidic chips have advantages of the generation of uniform droplets, multi-droplets with a concentration gradient distribution, the multi-droplets with a concentration gradient distribution on one step, and the scale-up production of droplets.

Acknowledgments

The authors would like to thank the Center for Micro/Nano Science and Technology, National Cheng Kung University, Tainan, Taiwan, R.O.C. for access to equipment and technical support. This research was partially supported by the Southern Taiwan Science Park Administration (STSPA), Taiwan, R.O.C. under contract no.100CB02. Funding from the Ministry of Education and the National Science Council of Taiwan, R.O.C. under grants NSC 100-2221-E-006-050- and 99-2221-E-006-203-MY3 are gratefully acknowledged. This paper is also partially supported by "Aim for the Top University Plan" of the National Chiao Tung University and Ministry of Education, Taiwan, R.O.C.

References

[1] M. Bonamico, C. Tiberti, A. Picarelli, P. Mariani, D. Rossi, E. Cipolletta, M. Greco, M. D. Tola, L. Sabbatella, B. Carabba, F. M. Magliocca, P. Strisciuglio and U. D. Mario, Gastroenterology **96**, 1536-1540 (2001).
[2] J. S. Rossier and H. H. Girault, Lab Chip **1**, 153-157(2001).
[3] W. C. W. Chan and S. M. Nie, Science **281**, 2016-2018 (1998).
[4] T. Tanaka and T. Matsunaga, Anal. Chem. **72**, 3518-3522 (2000).
[5] M. Dequaire, C. Degrand and B. Limoges, Anal. Chem. **72** , 5521-5528 (2000).
[6] C.-H. Yeh, H.-H. Huang, T.-C. Chang, H.-P. Lin, and Y.-C. Lin, Biosens. Bioelectron. **24**, 1661-1666 (2009).
[7] C.-H. Yeh, W.-T. Chen, T.-C. Chang, H.-P. Lin, and Y-C. Lin, Sens. Actuator B-Chem. **139**, 387-393 (2009).
[8] S. L. Anna, N. Bontoux, and H. A. Stone, Applied Physics Letters **82**, 364-366 (2003).
[9] C.-H. Yeh, Y.-C. Chen, and Y.-C. Lin, Microfluid. Nanofluid. **11**, 245-253 (2011).
[10] C.-H. Yeh, M.-H. Lee, and Y.-C. Lin, Microfluid. Nanofluid. **12**, 475-484 (2012).

Wafer Scale Processing of Plasmonic Nanoslit Arrays in 200mm CMOS Fab Environment

K. Malachowski, R. Verbeeck, T. Dupont, C. Chen, Y. Li, S. Musa, T. Stakenborg, D. Sabuncuoglu Tezcan, and P. Van Dorpe

Imec, Kapeldreef 75, 3001 Leuven, Belgium

Plasmonic nanoslits have great potential for single molecule applications. We report a wafer scale process for these structures using process steps compatible with a standard CMOS fab environment. This process allows a large scale fabrication of designed nanoslits with extremely small gap sizes and lengths tuned to exhibit optical resonances. Moreover, adjacent grating nano-antennas were successfully implemented, generating strong and localized electric fields in the nanoslit. These slits have practical applications in surface enhanced Raman spectroscopy-based molecular sensing and plasmonic tweezers.

Introduction

Nanofluidic systems based on solid-state nanopores, i.e. a two-reservoir system containing ionic solutions separated by a membrane containing a single nanopore, have been studied widely (1, 2). By monitoring the ionic current through the nanopore, the size and charge of molecules or small particles translocating through the pore can be determined as they generate current spikes or dips. Hence, such systems show a promise for the analysis of biological entities, such as bacteria, viruses, proteins or DNA. In particular, DNA sequencing has been put forward as an important application domain (3). Specifically for DNA strand sequencing, it has been shown theoretically that the different bases can be distinguished while the DNA strand is translocating through a pore. Combining nanopores with other techniques that locally monitor the molecular properties can pave the way for single molecule diagnostics.

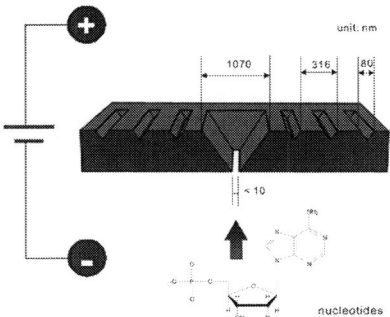

Figure 1 Conceptual visualization of a nanofluidic system based on solid state nanopore for DNA strand sequencing.

Surface enhanced Raman scattering (SERS) is a technique that allows the optical probing and identification of single molecules near metallic nanostructures exhibiting resonantly enhanced local electric fields at visible frequencies. While this general principle has been shown for a myriad of nanostructures, single molecule performance is usually observed in random nanoparticle aggregates exhibiting localized surface plasmon resonances, with so called hot spots for interparticle gaps of ~ 1 nm (4). However, the experimental reproducibility of this kind of naturally-formed hot spots is a serious problem. Therefore, we designed the combination of an engineered artificial plasmonic nanostructure with large local field enhancements and a solid state nanopore for molecular confinement as shown in Figure 1. We have shown that a gold nanoslit supports strongly localized plasmon resonances with a high spatial resolution and large SERS enhancement factors (5, 6).

Requirement specification

Nanoslit-cavity

The device is conceptually shown in Figure 1, and consists of a metal coated inverse pyramidal cavity with a narrow nanoslit at the cavity bottom. We have shown earlier that this can be fabricated in a lab-mode by anisotropically etching a silicon-on-insulator wafer on both the top and the back surface and subsequently coating it with metal (in our case this is usually gold). The metal sidewall length is tuned to optimize the coupling of far-field radiation with a specific wavelength to the local plasmonic mode in the nanoslit. For an excitation wavelength of 785 nm, which is commonly used for Au nanostructures, the slit length is around 700 nm (5).

The width and the length of the nanoslit are the most critical parameters. The local field enhancement increases with orders of magnitude for a decreasing gap size. We have previously shown that the nanoslit acts as a highly confined surface plasmon waveguide (7) and that for slit widths around 10 nm and slit lengths between 50 and 100 nm the first order Fabry-Perot resonances can be excited in the visible range. In this way the electromagnetic field enhancement can increase with more than one order of magnitude compared to non-resonant nanoslits. For an excitation wavelength of 785 nm, the optimal slit length is 90 nm.

Gratings

The field enhancement, and consequently the SERS enhancement factor can be increased even more by adding reflective structures next to the cavities (8). This avoids the loss of electromagnetic energy by surface plasmons propagating on the flat metal surface. Efficient reflection can be achieved using Bragg grating mirrors, which show optimized reflection if the pitch of the grating corresponds to $\lambda_{SPP}/2$, where λ_{SPP} is the surface plasmon wavelength on the flat metal surface. The grating antennas are conceptually shown in Figure 1. In this particular case, for the triangular gratings and an excitation wavelength of 785 nm, the designed grating pitch is 316 nm, with groove widths of 80 nm. Obviously, the distance between the cavity and the gratings is very important to avoid destructive interference and is set at 325 nm. The presence of this kind of gratings can supply a stronger field enhancement (5).

Fabrication process

Having the detailed requirements in mind, a translation into a process flow can be considered. The main requirement for the plasmonic nanoslit array is the dimension of the metal nanoslit in the range of around 10 nm × 90 nm. Considering that the metal coating causes shrinkage of the original nanopore dimensions, the target dimension of the silicon nanoslit is 80 nm × 160 nm. To reach repeatable dimensions across a 200 mm wafer a buried etch stop layer in form of silicon dioxide and an epi silicon layer is needed. Furthermore, to ensure that the process steps can be integrated into a CMOS compatible fab environment, silicon on insulator (SOI) wafers with 1 μm buried oxide thickness and a 700 nm thick silicon epi layer were chosen. The nanoslit is designed as a part of a nanofluidic system, requiring wafer back side processing to generate a channel through the wafer thickness. The whole process integration scheme shown in Figure 2 for two nanoslit structures is therefore divided into two main parts containing front and back side processing and are detailed described further in text.

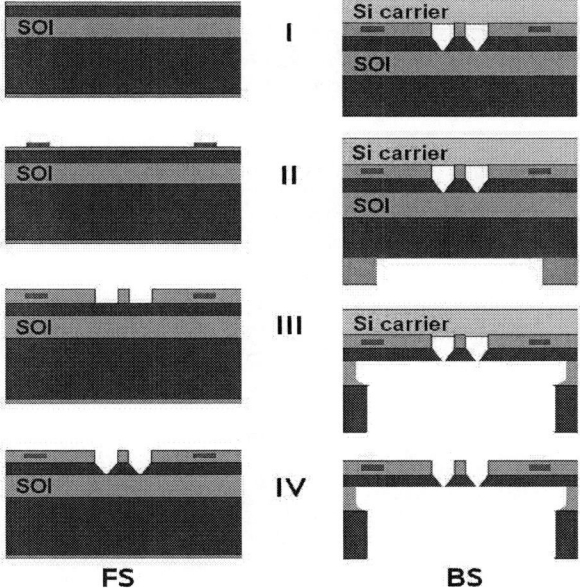

Figure 2 Schematic flow diagram for CMOS compatible solid state nanoslit processing. Wafer front side (FS) process flow shown in left column and back side (BS) in right column. FS: I) initial oxide deposition; II) metal markers deposition; III) hard mask patterning; IV) slit etch. BS: I) temporary Si carrier bonding and thinning; II) lithography; III) deep Si etch; IV) Si carrier removal.

Wafer front side processing

Using SOI wafers with (100) epi crystal orientation, the anisotropic etching behavior of KOH is used to create a nanoslit on top of the buried oxide layer, which acts as an etch stop. The angle towards the (111) crystal orientation of 54.7 degree in combination with the thickness of the epi layer enables an exact hard mask dimension calculation to define the slit dimensions using a photolithography mask. For a simulated slit dimension, e.g. of 80 nm × 130 nm by using a 700 nm silicon epi wafer, the hard mask opening should be around 1.07 μm x 1.15 μm. Thinner silicon epi layers down to 200 nm were investigated giving a disadvantage in device performance and analytical characterization of produced structures. As shown in Figure 3 FS IV the buried oxide is used as a stopping layer. Wafer batch processing was done in a wet bench with a 30% KOH solution and a temperature of 50 °C resulting in a process time of 6 minutes to etch the 700 nm epi layer. A short dip of the wafers in a diluted HF solution of about 20 seconds before KOH etch was introduced to avoid any presence of a native oxide layer, which would prevent the silicon from uniform etching. The silicon dioxide hard mask was deposited using a CVD tool at 400 °C with a thickness of 50 nm. As shown in Figure 3 FS III, the hard mask was patterned using a photo resist and 193 nm lithography followed by fluorine-based dry etching. Furthermore, Bragg mirror reflection gratings for loss reduction of the surface plasmons were build into the hard mask and etched at the same time as the slit. Due to dimension differences of the gratings (~100 nm) and the slit (~1 μm) a double-exposure approach using two separate reticles for the slit and gratings were used. Both features were printed into the same resist and commonly developed.

Due to further processing steps on the wafer back side some metal markers (TiN) were introduced as shown in Figure 3 FS II. These markers were used for the back side lithography alignment and are not part of the active front side structure.

All process steps are fully compatible with CMOS environment and combined to reach repeatable nanostructures on a 200 mm wafer. A top SEM view of processed nanoslit including plasmonic nanostructures can be found in Figure 3.

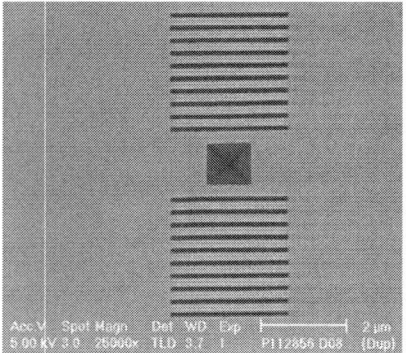

Figure 3   Top SEM view on solid state nanoslit with dimension of 80 nm x 160 nm etched into (100) silicon wafer (right) and in combination with plasmonic antennas for local field enhancement (left).

The main driver to reach repeatable structures in nanometer range is the structuring of the oxide hard mask and its dimension. We found that the best way for fine tuning of the dimension is achieved during the HF dip before KOH etching. Before wafer batch processing we measured therefore 2 to 4 exemplary wafers to adjust the HF dip time for the whole batch. The etch performance in KOH solution was checked on several wafers and is exemplary displayed for wafer D12 and D16 in Figure 4. The smallest feature of around 80 nm can be measured on all the wafers with a small error coming partially from processing and the measurement itself on SEM.

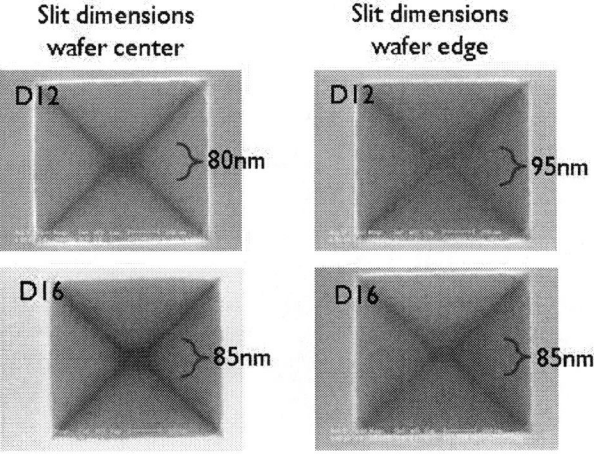

Figure 4 Top SEM view on solid state nanoslits captured from 2 different wafers (D12 and D16) and 2 different positions on the wafer (center and edge) reflecting a good dimension uniformity and wafer-to-wafer repeatability.

This integration exercise enables a reliable processing of nanoslits in different configurations. A cross-sectional build up of the initial concept drawing and the processed solid-state nanoslit is displayed in Figure 5.

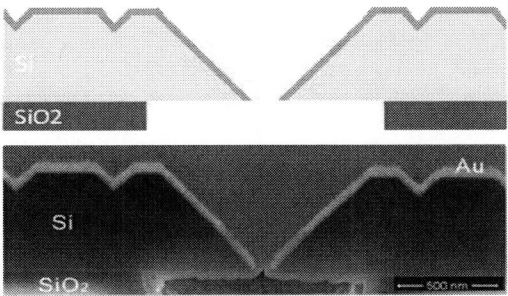

Figure 5 A cross-section view from the concept (upper drawing) towards a fabricated nanoslit (bottom SEM picture).

Wafer back side processing

The target final device thickness is 200 µm, therefore a carrier wafer was temporary bonded to the wafer front side to enable the necessary process steps on the back side using standard equipments without a risk of wafer breakage. Thus, glue from Brewer Science is spun-on the device wafer with a thickness of 16 µm and bonded to a standard 200 mm silicon wafer with 725 µm thickness as a carrier, as shown in Figure 3 BS I. In the same Figure the thinning of the device wafer down to 200 µm by mechanical grinding is displayed. For the subsequent back side lithography initially a polishing step using CMP was required to enable a reliable alignment to the front side markers. Further optimizations of the grinding and litho steps enabled a reliable back side alignment without CMP. Due to the amount of silicon to be etched a 40 µm thick positive photoresist AZ 40XT from MicroChemicals was used to withstand the BOSCH process for 50 minutes as shown in Figure BS II. Figure 6 shows an infra red image used during the alignment procedure, where the front side metal markers including numbering are clearly visible as well as the grinding marks after wafer thinning. In the same picture a hole in the photoresist of 70 µm in diameter is visible.

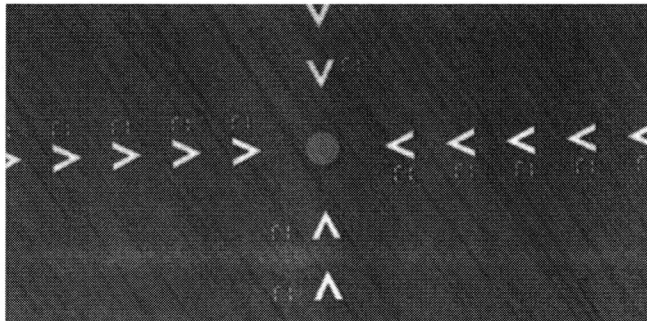

Figure 6   IR back side alignment on lithography step of a 200 µm thin device wafer. Front side metal markers (arrows) and a 70 µm diameter hole (picture middle) in developed photo resist is visible as well as grinding marks from thinning process.

Deep silicon etch of holes with 70 µm in diameter using the Bosch process is used to open up the nanoslits from the wafer backside. The etching time to reach the buried oxide needed to be adjusted on dummy wafers using SEM cross-sections. Due to the extreme depth of the holes and its small opening an optical inspection was not reliable enough to analyze the etching depth. A faster alternative to the SEM cross-section method was found in opening the front side membrane after deep silicon etch using ultrasonic excitation as shown in Figure 7. It was found that the buried oxide layer is etched very slowly compared to silicon allowing a significant process window and thus over etch amount to compensate the silicon thickness non-uniformity over the wafer. The notching depth as shown in Figure 7 after deep silicon etch is not important for the nanoslit functionality.

For complete opening of the micro fluidic channel a buffered HF solution bath is further used to remove the last remaining layer, which is the buried oxide from the SOI and to expose the nanoslit as shown in Figure 3 BS III. For the subsequent temporary carrier separation a high temperature process around 220 °C using slide de-bonding approach is applied. The carrier wafer can be reused after remaining glue cleaning. During the slide de-bonding procedure a significant amount of the glue material is squeezed into the nanoslits. A special sequence for cleaning was therefore developed without any use of ultrasonic excitation, which would result in a damage of the membranes as shown in Figure 7.

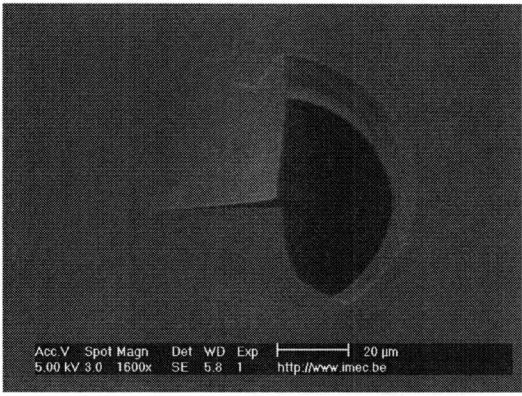

Figure 7   Top SEM view of broken front side SOI and silicon epi membrane after ultrasonic excitation. Notching around the 70 μm hole becomes visible.

Due to contamination restrictions for gold coatings, the thin gold layer was applied on die level but could be basically done on wafer-level as well. The 200 μm thick wafer was singulated using an UV blue tape and conventional blade dicing. The process conditions need to be adjusted to avoid membrane breakage, which are mainly caused by the water jet application during the dicing process itself and the subsequent water jet cleaning.

Sample evaluation

Generally, nanostructures optimized for SERS are first evaluated by measuring the Raman signal from a self-assembled monolayer (SAM) of Raman active molecules on the gold surface. 4-aminothiophenol is one of the commonly used SAM chemicals for SERS. Through the gold-thiol bonds (Au-S), the SAM molecules will be strongly absorbed on the surface of the gold layer. Indeed, we show in Figure 8 that the nanoslit generates strong signals from this monolayer of molecules (5). For single molecule investigations it is important that the hot spot generating the signal is indeed very localized. To demonstrate this high spatial resolution, we developed an electron beam induced deposition (EBID) method to directly place carbonaceous nanoparticles as the Raman analytes into the nanoslit. The bright dots in the TEM cross-section images of Figure 8 (B) are these selectively deposited particles, with a diameter of around 20 nm. From the resultant SERS spectra, we can clearly see that the hot spots are only localized inside the slit, rather than at the side wall or on the top edge.

Meanwhile, this experiment also supplies the most direct evidence for the intrinsic localized properties of SERS (6). This result is very important for SERS measurements of DNA translocating through the nanopore. When the chip is loaded into the flow cell for DNA nucleotides identification, only the molecules inside the slit will be excited and generate a measurable SERS spectrum.

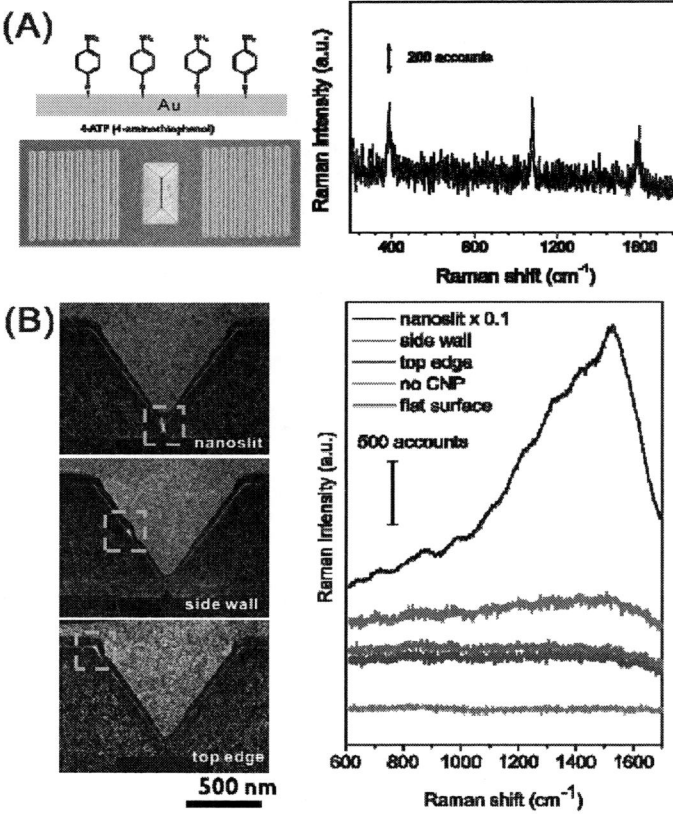

Figure 8 A) SAM coated nanoslit sample with a clear SERS spectrum; B) SERS spectra from carbonaceous nanoparticles (bright dots) placed at different locations indicating localized hot spots inside the nanoslit.

Another interesting application of such structures is plasmonic trapping of deep-subwavelength objects. At the resonant wavelength of the nanoslit, the strong field gradient generated near the hot spots result in a strong plasmonic (optical) force. This new technology enables to trap small nano-objects, at low incident intensity, without damaging the objects, as would be the case for conventional optical trapping of objects in that size range. An example was shown in trapping polystyrene beads with a diameter at around 20 nm (9). As shown in Figure 9, the bead(s) can be trapped in a single- and a double- states. The trapping mechanism increases the chance of trapping and analyzing single nano-objects or molecules.

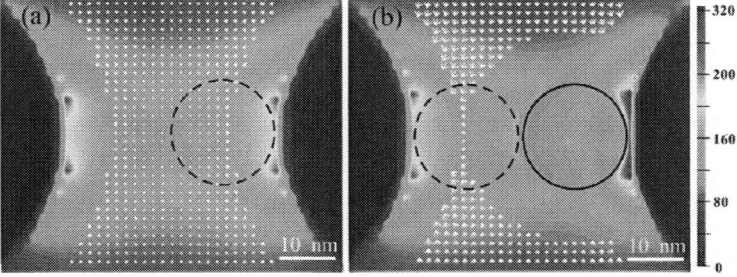

Figure 9 Plasmonic trapping of polystyrene beads. Numerical simulated optical field profiles and the force during the trapping: (a) single-bead trapping; (b) double-bead trapping. The white arrows represent the direction and the relative intensity of the plasmonic force on the coming bead.

Conclusions

The wafer-level manufacturing of nanoslit arrays using a 200 mm CMOS environment has been discussed and an array with repeatable feature dimension formed on a wafer is shown in Figure 10. The front side cavity and the gratings are anisotropically etched by 30% KOH solution, while the back deep cavity is etched by BOSCH dry etching. After removing the stop oxide layer and the deposition of gold/titanium layers on the front side, the plasmonic slit on a freestanding membrane is prepared. This kind of structures has shown promising performance on applications as the SERS-based molecular detection and the plasmonic trapping of nanoscaled dielectric objects. Especially, the combination of the repeatable wafer-level processing of fluidic nanoslit arrays and the SERS technology has a promising feasibility towards the parallel DNA sequencing.

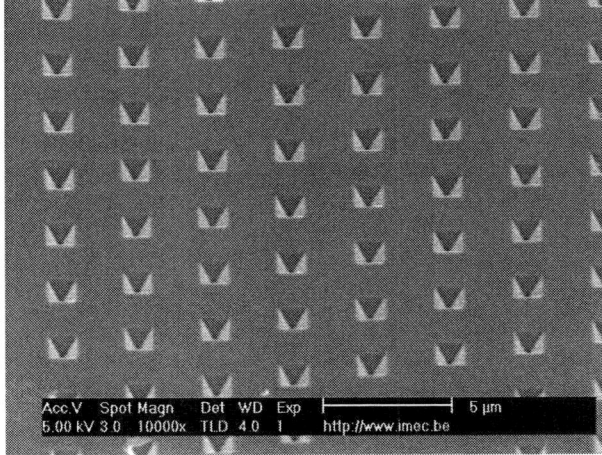

Figure 10 Top view of plasmonic nanoslit arrays processed on 200 mm wafer level.

Acknowledgments

The authors would like to thank John Slabbekoorn and Myriam Moelants for their support on lithography and Nina Tutunjyan and Danny Goossens for their support on dry etching. Pol Van Dorpe and Chang Chen acknowledge financial support from the FWO-Flanders.

References

1. Dekker, C. Solid-state nanopores, *Nat. Nanotechnol.*, **2** (4), p. 209-215 (2007).
2. Branton, D., et al. The potential and challenges of nanopore sequencing, *Nat. Biotechnol,.* **26** (10), p. 1146-1153 (2008).
3. Clarke, J., et al. Continuous base identification for single-molecule nanopore DNA sequencing, *Nat. Nanotechnol.*, **4** (4), p. 265-270 (2009).
4. Kneipp, K., et al. Detection and identification of a single DNA base molecule using surface-enhanced Raman scattering (SERS), *Phys. Rev. E.*, **57** (6), p. 6281-6284, (1998).
5. Chen, C., et al. Focusing plasmons in nanoslits for surface enhanced Raman scattering, *Small*, **5** (24), p. 2876-2882, (2009).
6. Chen, C., et al. Direct evidence of high spatial localization of hot spots in surface enhanced Raman scattering, *Angew. Chem. Int. Ed.*, **48** (52), p. 9932-9935 (2009).
7. Chen, C., et al. Highly confined surface plasmon polariton resonances in rectangular nanopore cavities. *Phys. Status Solidi-R.*, **4** (10), p. 247-249 (2010).
8. Chen, C., et al. Groove-gratings to optimize the electric field enhancement in a plasmonic nanoslit-cavity. *J. Appl. Phys.* **108** (3), 034319 (2010).
9. Chen, C., et al. Enhanced optical trapping and arrangement of nano-objects in a plasmonic nanocavity. *Nano Lett.*, **12** (1), p. 125-132 (2012).

ECS Transactions, 50 (12) 423-434 (2012)
©The Electrochemical Society

Tunable Young's Modulus in Carbon MEMS using Graphene-based Stiffeners

C. M. Washburn[a], T. N. Lambert[a], J. Blecke[a], D. Davis[a], P.S. Finnegan[a], B.G. Hance[a], D. R. Wheeler[a], T.E. Beechem[a], T.M. Alam[a], M.T. Brumbach[a], and J.M. Strong[b]

[a] Sandia National Laboratories, P.O. Box 5800, Albuquerque, NM 87185
[b] LMATA, 4209 Balloon Park Rd. NE, Albuquerque, NM 87109

> Carbon composite micro-electromechanical systems (C-MEMS)
> incorporating 2 wt.% graphene stiffeners show a 65% increase in
> Young's modulus and 11% increase in conductivity. An improved
> reduced graphene oxide (iRGO), is blended into pyrolytic carbon
> beams prepared for resonant frequency testing. Designed around a
> 10:1 (length: width) aspect ratio, the linearity of wt.% iRGO in the
> cantilevers as a function of resonant frequencies is evaluated. The
> collection of the 1st through 3rd bending modes using laser doppler
> velocimetery (LDV) of the graphene filled cantilevers shows an
> increase in frequency response with nanomaterial loading (wt.%).
> A model was developed using the 3-bending modes and correlated
> with cross sectional geometry and density to extract a Young's
> modulus.

Introduction

Carbon composites incorporating graphene nanomaterial stiffeners, drives new materials and devices into micro-electromechanical system (MEMS) development to improve dynamic range, sensitivity, lifetime, and functionality when compared to current state of the art MEMS technology. Proposed carbon composite structure are possible replacements for single crystal/metal MEMS beams, diaphrams, struts etc. at a fraction of the expense. A carbon composite material also renders a device to be less prone to stiction under high G-force loading, and have tremendous resilience under extreme mechanical deformation and shock.

The pyrolysis of photo-patternable materials and the basic micro-electromechanical properties related to pyrolytic carbon materials and resonator device has been demonstrated by George Whitesides *et al.* (1). Since then, Marc J. Madou *et al.*(2) and Richard L. McCreery *et al.*(3) have developed carbon on carbon approaches toward carbon MEMS, and high surface area electrochemical sensors, along with carbon for anode/cathode materials for Li-ion battery applications. Groups at Sandia have also demonstrated pyrolyzed carbon's potential bio-applications by electrochemically placing nano-materials on its' surface. (4).

In this paper, improved reduced graphene oxide (iRGO) nanomaterial is dispersed into a photo-active polymer matrix using commercial novalac polymer sets for carbon composite MEMS patterning and device development. High temperature pyrolysis converts the iRGO sheet and photo-active polymer into an amorphous carbon composite with disordered graphene sheets coupled into the matrix. Finally, an all carbon pattern is then released from the supporting substrate for testing. iRGO was synthesized following a recent literature report (5) and was characterized using Powder X-ray diffraction (PXRD), Raman spectroscopy and elemental analysis.

423

A MEMS single cantilever beam provides the basis for testing and evaluating carbon composite cantilevers. Graphene nanomaterial and engineered resonators have been shown (6,7) to directly impact the Young's modulus and conductivity of the final material. Furthermore, we present evidence of graphene structure improving mechanical and electrical properties of pyrolytic carbon structures, suggesting unique carbon-carbon bonding arrangements due to the introduction of graphene which is still being understood.

Graphene Synthesis and Characterization

Synthesis

Reduced graphene oxide (iRGO) was synthesized via the chemical reduction of improved graphene oxide (iGO)(7). Approximately 600 mg of iGO was dispersed in deionized water (600 mL) and divided into two equal portions. These mixtures were stirred for approximately 24 hours followed by bath sonication for 1 hour. Immediately following sonication, the two portions were combined into a poly(methylpentene) Nalgene bottle (1 L capacity) and $N_2H_4 \cdot H_2O$ (3 mL) was added with stirring. The bottle was sealed, and the reaction was heated to 85°C in an oil bath for 24 hours with continuous stirring. Upon completion, the reaction was removed from the heat, cooled, and black solids were isolated by filtration through a coarse glass fritted funnel. The solid product was washed with deionized water (1 L), and methanol (500 mL) and dried at 60°C under vacuum. Additional aggregate solids that subsequently precipitated from the black filtrate were collected and washed as previously described. In general, the reaction yielded approximately 300 mg of iRGO.

Characterization Performed

Powder X-ray Diffraction (PXRD)

Samples were mounted as EtOH slurries directly onto a zero background holder purchased from The Gem Dugout, State College, PA 16803 and allowed to dry. Samples were scanned at a rate of 0.02° / 2(s) in the 2θ range of 5−80° on a Bruker D8 Advance diffractometer in Bragg-Brentano geometry with Cu Kα radiation and a diffracted beam graphite monochromator.

Raman Spectroscopy

Raman spectra were recorded using a Thermo Scientific Smart Raman DXR instrument with a DXR 633 nm laser with a high-resolution gradient from 150 cm^{-1} to 2100 cm^{-1}. Resulting data was analyzed using the Thermo Scientific software and re-plotted for presentation purposes using Kaleidagraph software.

Elemental Analysis

Elemental analyses were performed on a Perkin-Elmer 2400 CHN-S/O Elemental Analyzer.

Characterization Results
Powder X-ray diffraction (PXRD)

iGO forms a well-ordered layered structure, as indicated by a well-defined d_{001} peak in its powder X-ray diffraction (PXRD) pattern. The exact 2θ values can range from $8\text{-}12°$, depending on extent of hydration, and were determined to be $\sim 2\theta = 10.99°$ (correlates to a basal spacing of 8.08 Å) for the iGO samples prepared here, seen in Figure 1. The peak at $\sim 2\theta = 10\text{-}12°$ is lacking in the iRGO sample and the peak for the iRGO is observed at $2\theta = 23.78°$ indicating a basal separation of 3.70 Å, larger than that of graphite at 3.35 Å. The broadness of the (002) peak is also consistent with poor ordering along the stacking direction and indicative of a powder comprised of disordered graphene-like nanosheets. The broad 101/001 peak(s) at $2\theta = 42\text{-}44°$ suggests poor through plane alignment also consistent with a disordered material.

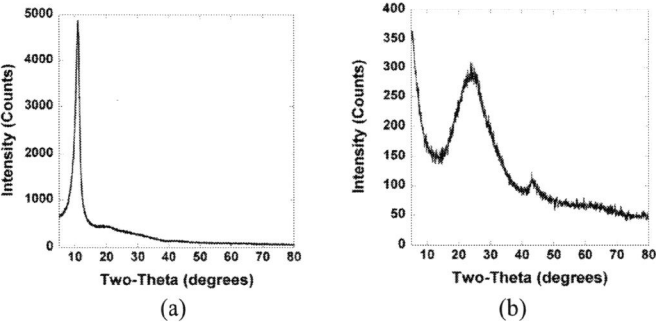

(a)   (b)

Figure 1. PXRD of the (a) GO with peak at 10.99 ° and (b) iRGO with peak at 24° indicating disordered graphene-like nanosheets.

Raman Spectroscopy

The chemical reduction of iGO to iRGO (7) was also confirmed with Raman spectroscopy. The iGO as prepared here exhibits D- and G-bands at 1311 and 1594 cm^{-1} with a D/G ratio of 1.02. Upon hydrazine reduction, the G-band shifts from 1594 to 1586 cm^{-1}, while the D-band shifts from 1311 to 1336 cm^{-1}, as seen in Figure 2 (a) and (b). A marked increase in the I_D/I_G ratio of is also observed (I_D/I_G = 1.5). Shifts to lower wavenumbers are expected upon reduction and the accompanying increase in I_D/I_G ratio for RGO has been explained by the presence of smaller but more numerous sp^2 domains in the carbon (5).

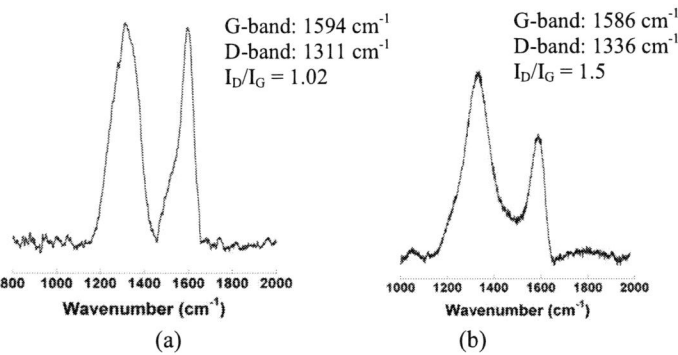

Figure 2 (a) and (b). Raman Spectra for (a) iGO before the conversion to (b) iRGO upon hydrazine reduction.

Elemental Analysis

GO and iRGO were also analyzed for their C, H, N elemental composition. Values for the iGO prepared here are: 41.8% C, 55.8% O and 2.4% H. The iRGO elemental composition was found to be: 69.4% C, 23.8% O, 1.3% H along with 5.6% N, as seen in Table I. The decrease in oxygen content is consistent with the chemical reduction of the graphene oxide. The presence on nitrogen in the iRGO is expected due to the N_2H_4 reducing agent (8).

Table I: Elemental analysis of graphene materials[a]

Sample	C	H	O	N	C/O (EA)
Graphite	98.1	NA	1.9	-	51.6
iGO	41.77	2.44	55.75	0.04	0.749
iRGO	69.40	1.27	23.75	5.57	2.92

[a] iGO = improved graphite oxide, iRGO = improved reduced graphene oxide (hydrazine method)

MEMS Design and Fabrication

Test Devices

The test devices used for this set of experiments are simple cantilevers of varying sizes that are affixed to bond pads, as shown in Figure 3(a) and (b). Both the cantilevers and the pads exist on the same layer and are comprised of the same material. Fillets at the base of the cantilevers serve to reduce mechanical stress at the joint between cantilever and pad. Several cantilevers of the same size share a single bond pad, and are placed along the edge of the pad with large spacing relative to cantilever width to minimize interference between devices.

All of the cantilevers, regardless of size, share the same thickness and a common aspect ratio of 10:1 length to width and a ratio of approximately 365:1 bond pad area to cantilever area. This aspect ratio was chosen to facilitate the successful release of the cantilevers while still maintaining effective adhesion of the bond pad to the substrate. The design is critical to the release method.

The cantilever and bond pad design allow for the bond pads to remain connected to the substrate after release. The bond pad is partially undercut during the release process; however, it remains connected to the silicon surface while the cantilevers are completely freed from the silicon surface.

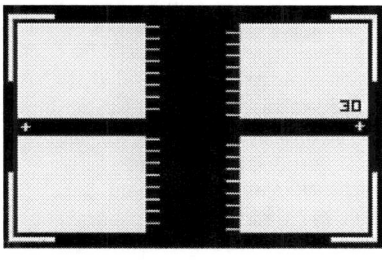 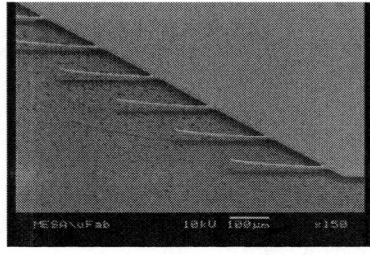

(a)   (b)

Figure 3 (a) and (b). (a) Describing the on mask layout of a 30 μm wide X 300 μm long beam and (b) the equivalent SEM image of the fabricated device

Device Fabrication

To fabricate the carbon composite MEMS (C-MEMS) devices, improved reduced graphene oxide (iRGO) is weighed and incorporated into photoresist AZ4330. The concentrations of iRGO used for this set of experiments were 0.5%, 1.0%, and 2.0 wt.%. The slurry is mixed for 2 hours with sonication to agitate and suspend the iRGO within the photoresist. The AZ4330 photoresist and iRGO mixture is then manually dispensed onto 100 mm n-type silicon wafer (10-100 ohm-cm) and spin casted at 4000 rpm for 30 seconds, followed by a soft bake at 90°C for 90 seconds on a hot plate. This produces a thin photoresist and iRGO film 3.4 μm thick. The pattern is then exposed onto the coated wafer using a Karl Suss MA-6 manual contact aligner for 30 seconds at a wavelength of 420nm and an intensity of 20mW/cm². The exposed wafer is submerged for 90 seconds with heavy agitation into AZ400K 1:4, a buffered KOH based developer leaving the cantilever pattern. The patterned resist is then cured by baking the wafer at 110°C for 10 minutes, followed by a deep ultra-violet (DUV) exposure in a Fusion F300 UV Curing System at 260 nm with an intensity of 160mW/cm², for 30 seconds. A final ramped bake from 100°C to 280°C at a ramp rate of approximately 10 degrees per minute completes the curing process. The resist cures by evolving the majority of the water, CO_2 and CO through decarboxilation and decarbonylation during the baking process. The wafer is then cleaved and baked in a reducing atmosphere of $H_2:N_2$ (5%:95%) in a Lindberg tube furnace at 1150°C, then soaked for 1 hour and allowed to cool to room temperature via natural convection.

After pyrolyzed carbon conversion, a xenon difluoride (XeF_2) silicon etch release process is performed to remove the silicon from under the graphene stiffened carbon MEMS device. Using an XACTIX e1 Series XeF_2 etch system releases the cantilevers from the substrate after 210 cycles. XeF_2 etchant is a highly selective isotropic etchant for silicon and undercuts the carbon cantilevers releasing them from the substrate, leaving them suspended above the silicon. The XeF_2 vapor is pulsed in cycles in the process chamber, resulting in highly efficient silicon etching. This isotropic etch allows large features to be undercut, etching laterally at nearly the same rate as its vertical etch.

Testing and Results

Thin Film Measurement Results

Four point probe measurements were taken on 2.5 cm x 2.5 cm quartz substrates purchased from GM Associates. The iRGO loading ranged from 0.0 wt.%, 0.5%, 1.0%, to 2.0 wt.% for these samples, along with one standard deviation of 5 recorded measurements. The substrates were processed through the same high temperature reducing atmosphere as the carbon beams to understand the conductivity as a function of iRGO loading. The measurement was performed on a 4-wire NI PXI-4070 Signature conductivity meter with a four point Signatone tungsten carbide probe tips at 1 millimeter spacing. To find the average conductivity (S/cm), five measurements on each sample are collected and supported with thickness verification using cross sectional SEM. The average film thickness ranged between 850 nm and 900 nm and had no bearing on loading. For the conductivity calculation, the 900 nm thickness was chosen and the results, shown in Figure 4, indicate an increase in conductivity with increased iRGO loadings, with an 11% increase in conductivity with 3.1% standard deviation for 2.0 wt% iRGO as compared to 0.0 wt.% pyrolytic carbon.

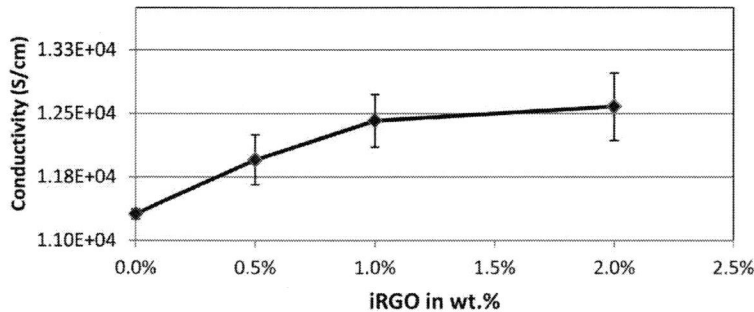

Figure 4. Describing the conductivity of pyrolyzed carbon verse iRGO (wt.%) on quartz substrates.

Laser Doppler Velocimeter (LDV) Methodoloy

Dynamic measurements were recorded using a scanning laser doppler velocimeter (LDV). A die containing the sample beams was adhered to a piezoelectric shaker on a sizeable mass. A pseudo-random input was used to excite the test pieces with a frequency range of 0 to 1.5 kHz. Single point LDV measurements at the tip of each beam were recorded as well as scans to verify the bending modes. The input voltage to the piezoelectric shaker from the signal generator was used as the reference signal.

A model of a cantilever and gradient-based optimization techniques were used to fit material and geometric parameters to the experimental data. The frequency of the ith bending mode for a homogenous, fixed-free cantilever beam can be written in closed-form, shown by Eq. [1], where L is the length of the beam, E is the Modulus of Elasticity, I is the area moment of inertia about the neutral axis, m is the mass per unit length and λ_i is a constant corresponding to the fixed-free boundary condition (8). The cross-section of each beam was determined by SEM to be wing-shaped and approximately 1.75 µm at the thickest location. Figure 5 shows an SEM image of a sample cross-section. A semi-elliptical cross-section was assumed for the model, described by Eq. [2] and Eq. [3], where A_c is the cross-sectional area I is the area moment of inertia about the centroid, a is two times the beam width and b is the maximum thickness at the center of the beam (8). While there is evidence that the beam tip curves up, the assumption is still made that the centroid of the beam lies on the neutral axis.

$$f_i = \frac{\lambda_i^2}{2\pi L^2}\left(\frac{EI}{m}\right)^{\frac{1}{2}} \qquad [1]$$

$$A_c = \frac{\pi ab}{2} \qquad [2]$$

$$I_{xc} = \frac{ab^3}{72\pi}(9\pi^2 - 64) \qquad [3]$$

Figure 5. SEM cross-section of a sample cantilever beam with 2.0 wt.% RGO

Laser Doppler Vibrometer (LDV) Results

Dynamic tests were performed on a small set of cantilever beams to determine how the amount of iRGO loading changes the frequency response of the system in order to gain insight into the effects of the iRGO on the Modulus of Elasticity. Fixed-free cantilever beams were evaluated in two different sizes, 20 μm X 200 μm and 30 μm X 300 μm long. The test pieces also varied by iRGO loading as samples were fabricated with 0.0%, 0.5%, 1.0% and 2.0 wt.% iRGO loading for both sizes.

Resonant frequency spectra in Figures 6 (a) and (b) illustrate the dynamic response of a 30 μm X 300 μm carbon beam as a function of iRGO loading at the first bending mode and second bending mode, respectively. Figure 7 describes the linear relationship of iRGO loading as a function of peak resonant mode from 0.0 wt.% to 2.0 wt.% with R^2 being 0.91 and 0.94 for a fit parameter. The nanomaterial loading in Figure 8 extends to demonstrate and compare engineered beams of a 10:1 aspect ratio have reproducible trends in mechanical rigidity due to the iRGO dispersion and carbonization. Our hypothesis that the mechanism for increased Young's Modulus is that graphene basal plane is providing a scaffold or template to adjust the surrounding amorphous carbon phases is still under investigation as to the actual mechanism.

To support these results, an investigation into beam-to-beam variation of 26 beams with a length of 300 μm and 1.0 wt.% iRGO loading were tested. The average frequency of the first bending mode, Figure 6(a), was found to be approximately 18.47 kHz with a standard deviation of 0.64 kHz.

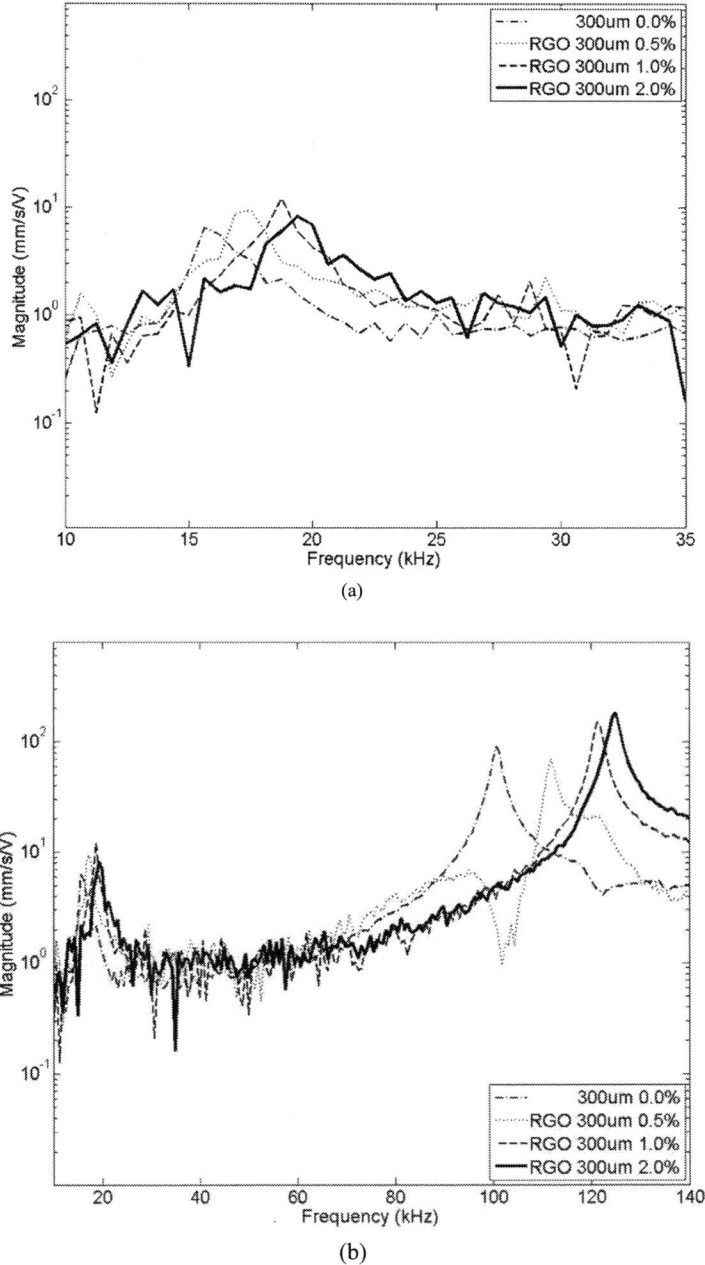

(a)

(b)

Figure 6. Resonant frequency response for 300 μm long beams for iRGO loadings of 0.0%, 0.5%, 1.0% and 2.0 wt.%, showing the frequency of the (a) 1st bending modes, and (b) the frequency shift of the 1st and 2nd bending modes responses.

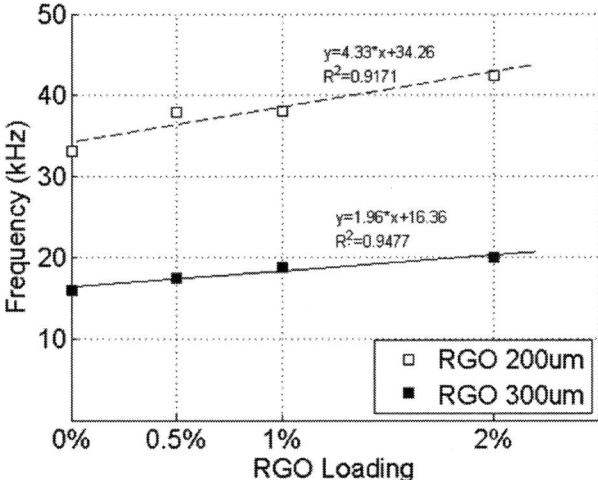

Figure 7. The change in the 1st resonant frequency mode for a 20 μm x 200 μm and 30 μm x 300 μm beam as a function of iRGO loading in wt.%.

The average frequency of the first 3 bending modes was also calculated. For the 200 μm long beam case, the frequencies of the first 3 bending modes increased by approximately 15-20% with the addition of 0.5% RGO loading. That value grows to 25-35% for 2.0% RGO loading. For the 300 μm long beam case, the frequencies of the first 3 bending modes were increased by approximately 8% for 0.5% iRGO loading and 25% for 2.0% iRGO loading.

Figure 8 establishes the relationship to Modulus of Elasticity computed from fitting the measured bending modes to a closed-form beam equation (9). Parameters chosen for modeling included a density of 1.4 g/cm^3 and maximum thickness at the center of the cross-section, 1.75 μm, with the length of the beam allowed to range from 200-210 μm due to undercut of the etch step in the fabrication process. The largest unknown, the Modulus of Elasticity, was allowed to range between 15 and 100 GPa. The cost function of the gradient-based optimization aimed to minimize the error between the computed natural frequency of the first 3 bending modes from the model and those recorded experimentally in a least-squared fashion. Parameters in the model were fit for 200 μm long beams of each iRGO loading. The optimized Modulus of Elasticity is plotted in Figure 8 and illustrates an increasing trend of moduli for increased iRGO loading. The optimization did not fit the frequencies of the first 3 bending modes perfectly so the percent error was computed. Using these frequency extremes as well as the nominal parameters from the optimization, a range of moduli was computed for each beam type. The maximum and minimum moduli for each beam are shown as the error bars (+/- 10 %) in Figure 8. This analysis indicates the Modulus of Elasticity increases from 41 GPa to 68 GPa, a 65% improvement, by the addition of 2.0 wt.% RGO loading.

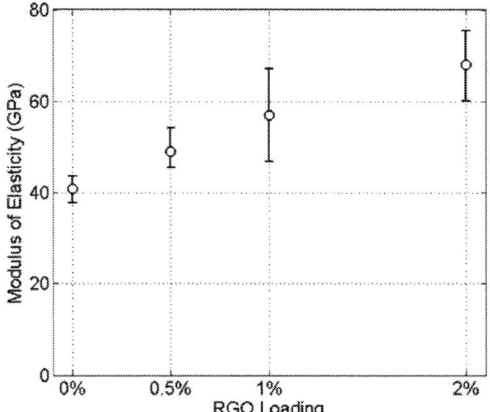

Figure 8. Estimated Modulus of Elasticity for beams loaded with iRGO by fitting the first 3 bending modes computed from a closed-form model of a fixed-free cantilever beam parameters.

Conclusion

The development of a carbon composite incorporating graphene nanomaterial as a stiffener has been demonstrated. The incorporation of graphene leads to an increase in conductivity and Modulus of Elasticity with increasing concentrations. Graphene as a template surface for ordering amorphous carbon during thermal treatment is suggested as a possible reason for these increases, although this still needs verification; however, the electromechanical data indicates this is a viable new approach for developing carbon-carbon composites using photo-sensitive materials as well new tunable device characteristics. In our limited experience, not all nanomaterial fillers behave similarly to graphene under the previously reported conditions (10). This provides the MEMS community with a new opportunity in materials and device design.

Acknowledgments

The authors would like to acknowledge the following individuals and funding sources for their contributions to make this a successful paper; Lee Massey, Christine Ford, Robert Bernstein, and Tedd Rohwer. Sandia is a multi-program laboratory operated by Sandia Corporation, a Lockheed Martin Company, for the United States Department of Energy's National Nuclear Security Administration under contract DE-AC04-94AL85000.

References

1. Whitesides, G.M.; Schueller, O.J.A; Brittain, S.T.; Marzolin, C. *Fabrication and Characterization of Glassy Carbon MEMS.* Chem. Mat. Vol. 9. 1399-1406. 1997.
2. Madou, M.; Wang, C.; Malladi, K.; *Fabrication of suspended carbon microstructures by e-beam writer and pyrolysis.* Carbon. Vol. 13. 2602-2607 2006.
3. McCreery, R.; Ranganathan, S. ; Majji, S.M.; Madou, M. *Photoresist-Derived Carbon for Microelectomechanical Systems and Electrochemical application.* Journal of The Electrochemical Society, Vol. 147. No.1. 277-282. 2000.
4. Burkel, D.B.; Washburn, C.M.; Raub, A.K.; Brueck, S.R.J.; Wheeler, D.R.; Brozik, S.M.; Polsky, R. *Lithographically Defined Porous Carbon Electrodes*, Small. Vol. 5. No. 24. 2792-2796. 2009
5. Marcano, D.C.; Kosynkin, D.V.; Berlin, J.M.; Sinitskii, A.; Sun, Z.; Slesarev, A.; Alemany, L.B.; Lu, W.; Tour, J.M. *Improved Synthesis of Graphene Oxide .* ACS Nano. Vol. 4, No. 8. 4806-4814. 2010.
6. Bunch, J.S.; Van der Zande, A.M; Verbridge, S.S.; Frank, I.W.; Tanenbaum, D.M.; Parpia, J.M.; Craighead, H.G.; McEuen, P.L. *Electromechanical Resonantors from Graphene Sheets.* Science. vol. 315. 490-493. 2007
7. Zalalutdinov, M.K; Robinson, J.T., Junkermeier, C.E.; Culbertson, J.C.; Reinecke, T.L.; Stine, R.; Sheehan, P.; Houston, B.H.; Snow, E.S. *Engineering graphene mechanical systems.* ACS Nano Letters. DOI: 10.1021/ nl3018059, 2012
8. Lambert, T.N.; Luhrs, C. C.; Chavez, C. A.; Wakeland, S.; Brumbach, M.T.; Alam, T.M. *Graphite oxide as a precursor for the synthesis of disordered graphenes using the aerosol through plasma method.* Carbon. Vol. 48. 4081-4089. 2010
9. Stankovich S, Dikin DA, Piner RD, Kohlhaas KA, Kleinhammes A, Jia Y, et al., *Synthesis of Graphene-Based Nanosheets Via Chemical Reduction of Exfoliated Graphite Oxide.* Carbon. Vol. 45. 1558-1565. 2007
10. Blevins, Robert D. 1979 *Formulas for natural frequency and mode shape* Van Nostrand Reinhold Co., New York.
11. Strong, J.M.; Washburn, C.M.; Wheeler, D.R. *Carbon MEMS Accelerometer*, COMSOL conference proceedings, Boston, MA, 2011.

ECS Transactions, 50 (12) 435-440 (2012)
©The Electrochemical Society

Residue-Free Dry Etching of Polymer Sacrificial Layer
for Microelectromechanical-System Device Fabrication

K. Takagahara[a], K. Ono[a], K. Kuwabara[a], T. Sakata[a], H. Ishii[b], Y. Sato[a] and Y. Jin[a]

[a] NTT Microsystem Integration Laboratories
3-1, Morinosato Wakamiya, Atsugi-shi, Kanagawa Pref., 243-0198 Japan
[b] Toyohashi University of Technology
1-1, Hibarigaoka, Tempaku, Toyohashi-shi, Aichi Pref., 441-8580 Japan

Residue-free dry etching of a photosensitive-polymer sacrificial layer using $O_2/CF_4/CO$-plasma exposure is described for the fabrication of microelectromechanical-system (MEMS) devices. Energy dispersive X-ray spectroscopy reveals that $O_2/CF_4/CO$-plasma exposure removes polymer sacrificial layers without leaving residue and severely damaging Au structures. Since the $O_2/CF_4/CO$-plasma exposure hardly damages Au structures, this process is applicable for the removal of sacrificial layers for MEMS with Au structures.

Introduction

The selection of sacrificial layer materials and a removal process of them is very important in microelectromechanical-system (MEMS) devices fabrication. Photosensitive polymers (1-6) are promising sacrificial layer materials because they can be directly patterned by conventional spin coating and photolithography. After dry-etching of polymer sacrificial layers by active oxygen species exposure, such as O_2-plasma exposure, there often remains silicon oxide (SiO_x) residue on the surfaces of the devices, which is derived from the oxidized silane coupling agent in the photosensitive polymers (7). Since this residue deteriorates the mechanical and electrical characteristics of the devices, it must be removed. The polymer and SiO_x residue are known to be simultaneously removed by O_2/CF_4-plasma exposure (8). However, this has a disadvantage in that the device surfaces are oxidized and eventually etched. Hence, there is a need to develop a method to completely remove the photosensitive-polymer sacrificial layers without oxidation and etching of the device surfaces. In this paper, polymer removal using $O_2/CF_4/CO$ plasmas is proposed for gold (Au) surfaces, which are often used in MEMS structures.

Reaction Mechanism

Photosensitive-polymer sacrificial layers are composed mainly of organic resins and silane coupling agents as adhesion promoters (9-13). The O_2/CF_4-plasma exposure is effective to remove the polymer sacrificial layers because it removes the organic resins as well as the silane coupling agents, which remain as SiO_x residue after O_2-plasma exposure. However, it is not applicable to the fabrication of MEMS using Au structures, because it etches not only the polymer sacrificial layers but also the Au structure.

435

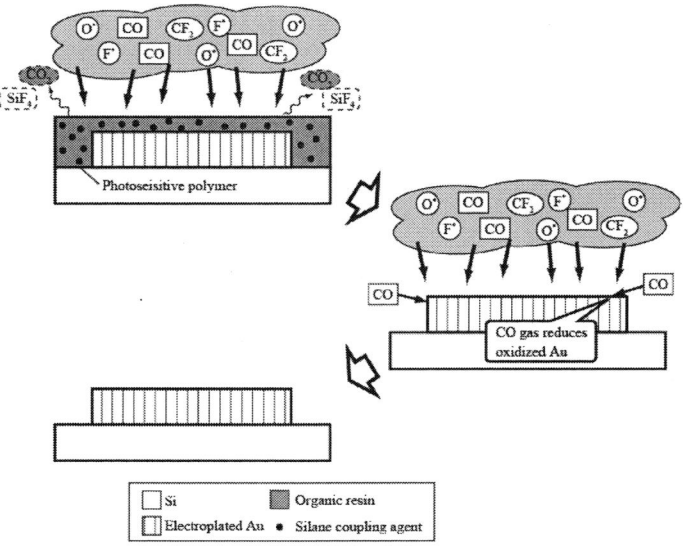

Figure 1. Schematic of photosensitive-polymer sacrificial layer removal by $O_2/CF_4/CO$-plasma exposure.

Au etching in O_2/CF_4 plasma is known to proceed in three steps (14): Au oxidation by oxygen radicals (O^*), binding of the Au oxide with $(CF_2)_n$ and subsequent desorption from the surface. In this reaction, producing Au oxide in the first step is necessary for Au etching to proceed because pure Au does not bind with $(CF_2)_n$.

Figure 1 shows a schematic of our approach to remove the photosensitive-polymer sacrificial layer using $O_2/CF_4/CO$-plasma exposure. Here, we added CO as a reduction gas to prevent Au oxidation in the first step of Au etching while maintaining the capability to remove the polymer and SiO_x residue. The $O_2/CF_4/CO$ plasma produces active species, such as O^*, fluorine radicals (F^*), fluorocarbon gases (ex. CF_2), etc. The organic resins and the silane coupling agents combine with active species to become carbon oxide gases and silicon fluoride gases and desorb from sample surface. Au structures combine with O^* to become Au oxide. The CO gas reduces the produced Au oxide immediately to pure Au. This oxidation-reduction cycle is repeated during the $O_2/CF_4/CO$-plasma exposure, and the Au etching reaction does not proceed to the binding of the Au oxide with $(CF_2)_n$. Therefore, the $O_2/CF_4/CO$-plasma exposure completely removes the photosensitive-polymer sacrificial layer without etching the Au structures.

Experimental

EDS analysis

We examined the chemical composition analysis of Au structure surfaces after polymer removal by O_2, O_2/CF_4, and $O_2/CF_4/CO$ plasma exposure. Samples of the 10-um-thick photosensitive-polymer layer (CRC8300; Sumitomo Bakelite Co., LTD.)

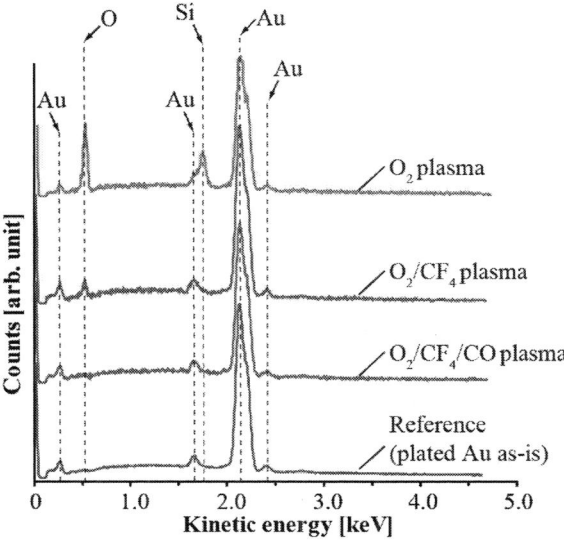

Figure 2. EDS spectra of the Au surfaces after plasmas exposure and reference.

formed on the Au structures were exposed to O_2, O_2/CF_4, and $O_2/CF_4/CO$ plasmas at 100°C and RF power of 200 W. Gas flows were 200, 180/20 and 180/20/80 sccm at the constant total pressure of 40 Pa in the etching chamber. Figure 2 shows the energy spectra of sample Au and reference (electroplated Au as-is) surfaces analyzed with an energy dispersive X-ray spectrometer (EDS). After the O_2-plasma exposure, peaks for silicon and oxygen appeared, which means that SiO_x residue remained on the Au surfaces. For O_2/CF_4 plasma, the silicon peak was absent and the oxygen peak remained. This means that the SiO_x residue was removed but Au was oxidized. On the other hand, in the case of $O_2/CF_4/CO$ plasmas, neither silicon nor oxygen peaks remained. This indicates that the photosensitive polymer was removed without SiO_x residue and the Au oxides.

Measurement of Au surface resistance

To investigate the effect of adding CO gas, we measured the variation of sheet resistance of Au surfaces. Au and adhesive titanium (Ti) films with 70- and 100-nm thickness were formed on the Si substrate by vacuum evaporation. Figure 3 shows the normalized sheet resistances measured after exposure to the $O_2/CF_4/CO$ plasmas for 3 minutes at 100°C and RF power of 200 W. CO flow was varied under the constant flow of O_2 and CF_4 (180 and 20 sccm) while the total pressure was kept at 40 Pa. At the CO flow of 0 sccm, the sheet resistance increased 1.4 times from that before exposure as a result of the Au oxidation and etching. The sheet resistance decreased and became closer to 1.0 as the CO flow increased. There are two possible reasons for the suppression in the resistance variation: one is that adding CO gas to the constant total pressure simply attenuates the reactive species produced in O_2/CF_4 plasmas; the other is that CO gas reduces Au oxide and stops the Au etching reaction as described above. To examine these possibilities, we measured the $O_2/CF_4/CO$-plasma-exposure-time dependence of the sheet

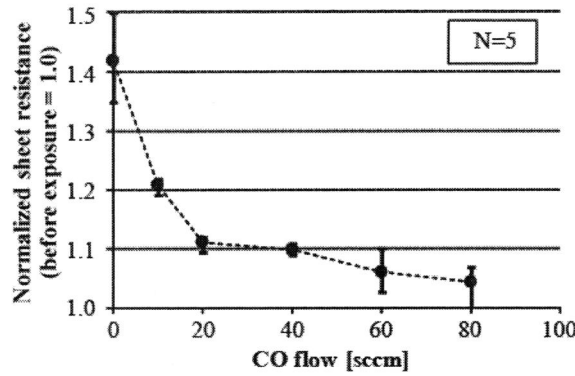

Figure 3. CO flow rate dependence of normalized sheet resistance.

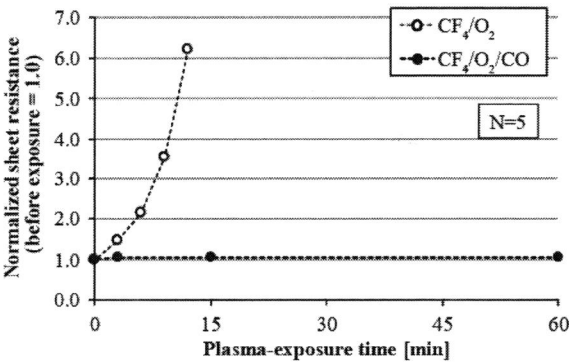

Figure 4. Plasma-exposure-time dependence of normalized sheet resistance.

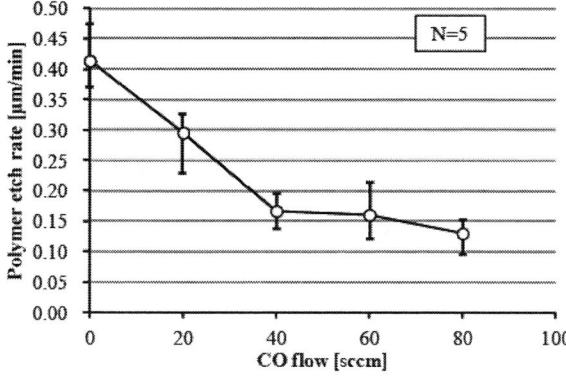

Figure 5. CO flow rate dependence of photosensitive-polymer etch rate.

resistance (Fig. 4). Although the sheet resistance increases with increasing O_2/CF_4-plasma-exposure time, lengthening the $O_2/CF_4/CO$-plasma-exposure time does not increase the sheet resistance. If CO gas simply attenuates the reactive species, lengthening the $O_2/CF_4/CO$-plasma-exposure time should increase the sheet resistance. Therefore, these results indicate that the $O_2/CF_4/CO$-plasma exposure can completely remove the photosensitive-polymer sacrificial layer without oxidizing and etching Au structures.

Application for sacrificial layer removal

Figure 5 shows the CO-flow-rate dependence of the photosensitive-polymer etch rate under the constant flow of O_2 and CF_4 (180 and 20 sccm) while the total pressure was kept at 40 Pa. At the CO flow of 0 sccm, the polymer etch rate is about 0.40 μm/min. The etch rate decreases as the CO flow increases. At the CO flow of 80 sccm, the polymer etch rate is about 0.13 μm/min, which is about 1/3 of that at 0 sccm. These results indicate that the etching time for photosensitive-polymer removal is longer for $O_2/CF_4/CO$-plasma exposure than for O_2/CF_4-plasma exposure. However, since lengthening the $O_2/CF_4/CO$-plasma-exposure time does not increase the sheet resistance of Au structures as shown in Fig. 4, this process is applicable for the removal of sacrificial layers for MEMS with Au structures, which often need an overetching.

Conclusion

We developed a residue-free dry etching process for photosensitive-polymer sacrificial layers using $O_2/CF_4/CO$-plasma exposure for the fabrication of MEMS devices. Adding CO gas reduces Au oxides, which is an intermediate in the Au-etching reaction. EDS spectroscopy reveals that $O_2/CF_4/CO$-plasma exposure removes polymer sacrificial layers without leaving residue and severely damaging Au structures. Measurement of Au sheet resistance also supports this result. Since the $O_2/CF_4/CO$-plasma exposure hardly damages Au structures, this process is applicable to the removal of photosensitive-polymer sacrificial layers.

Acknowledgement

The authors thank Messrs. K. Kudou, M. Yano and Dr. K. Machida of NTT Advanced Technology Corp. for their considerable support.

References

1. A. B. Frazier and M. G. Allen, *Proc. Int. Conf. Micro Electro Mechanical Systems*, 4 (1992).
2. P. P. Sergio, P. B. K. Linda and T. -C. N. Clark, *IEEE MTT-S Int. Microwave Symp. Digest*, 165 (2000).
3. P. Robert, D. Saias, C. Billardl, S. Boret, N. Sillon, C. Maeder-Pachurka, P.L. Charvet, G. Bouche, P. Ancey and P. Berruyer, *Proc. Int. Conf. Solid State Sensors, Actuators and Microsystems*, 1714 (2003).

4. H. U. Rahman, T. Hesketh and R. Ramer, *Proc. Int. Conf. Emerging Technologies (ICET)*, 116 (2008).
5. D. Molinero and L. Castaner, *Proc. Spanish Conf. Electron Devices*, 281 (2009).
6. B. A. Ganji and B. Y. Majlis, *IEEE Int. Conf. Semiconductor Electronics (ICSE)*, 267 (2010).
7. K. Takagahara, K. Kuwabara, T. Sakat and Y. Sato, *Proc. 23rd microprocesses and nanotechnology conference (MNC)*, 10C-3-3 (2010).
8. Y. Horiike and M. Shibagaki, *Jpn. J. Appl. Phys.*, **15**, 13 (1976).
9. K. L. Mittal, in *Silanes and other coupling agents*/2004, p. 439, V.S.P. Intl Science, Tokyo (2004).
10. T. Banba, E. Takeuchi, A. Tokoh and T. Takeda, *Proc. Electronic Components and Technology Conference*, 564 (1991).
11. Q. Yao, J. Qu, J. Wu and C. P. Wong, *Proc. Int. Symp. Advanced Packaging Materials*, 27 (1999).
12. M. B. Vincent, L. Meyers and C. P. Wong, *Proc. Int. Symp. Advanced Packaging Materials*, 49 (1998).
13. S. Luo and C. P. Wong, *IEEE Trans. Components and Packaging Technologies*, 38 (2001).
14. Y. Taniguchi and T. Shin-mura, *J. Vac. Sci. Technol. A*, **16**, 2042 (1998).

ECS Transactions, 50 (12) 441-445 (2012)
©The Electrochemical Society

Hydrodynamic Cell Enrichment in Double Spiral Microfluidic Channels

Jiashu Sun,[a] Mengmeng Li,[a] Chao Liu,[b] Guoqing Hu,[b] Xingyu Jiang[a]

[a]CAS Key Lab for Biological Effects of Nanomaterials and Nanosafety, National Center for NanoScience and Technology, Beijing, 100190, China.
[b]Institute of Mechanics, Chinese Academy of Sciences, 100190, China.

> Rapid cell separation and enrichment has attracted significant attention because of its important applications in cell biology studies and biomedical diagnostics. we have developed a compact microfluidic platform, composed of 6-loop double spiral microchannels, to separate cancer cells from blood cells. The curved geometry induces a Dean drag force acting on cells to compete with the inertial lift, resulting in different equilibrium positions for different sized cells. The separation chip is characterized with the mixture of 4T1 cells and the murine blood cells at different flow rates, with decent separation and enrichment efficiency.

Introduction

Rapid cell separation and concentration has attracted significant attention because of its important applications in cell biology studies and biomedical diagnostics (1). Traditional techniques used to separate cells by size include centrifuge and microfiltration. Other innovative techniques, *e.g.*, fluorescence-activated cell sorting (FACS), magnetically activated cell separation (MACS), acoustic cell filter, and cell affinity chromatography (CAC), have also been developed as more sophisticated tools to conduct cell separation based on cell properties. Conventional techniques used to separate different size cells include centrifuge and microfiltration, which are bulky, laborious and sometimes perform ineffectively in separating cells over a broad size range (2, 3).

Microfluidic methods, taking advantage of unique hydrodynamic effects under small length scales and low Reynolds numbers, allow precise and continuous cell separation in a compact format with throughput comparable to traditional techniques(4-6). A simple method is to construct micropost arrays that are separated from each other by different distances, serving as microsieves to separate different sized cells. Besides, the particular geometries of microfluidic systems will induce additional hydrodynamic forces, which could be adopted to separate cells by size. For example, using inertial lift and Dean drag forces in asymmetric curving channels, the randomly distributed particles of different sizes can be focused into several steams that are size-dependent, leading to equilibrium separation at a high flow rate The most remarkable feature of size-based passive microfluidic separation techniques is to eliminate the need of cell labeling, as well as applying any external forces except driving pressures (7). However, most passive-based cell sorters published so far have been focused on proof-of-concept demonstration by using specified particle mixtures or pre-conditioned biological samples, and hence lacking the biological sense.

441

Theory

Many factors, including geometries of channel, flow conditions, and properties of suspending liquid medium, influence the performance of the microfluidic particle/cell sorting and concentration devices. We developed a double spiral microchannel with 6-loop for each direction to particle/cell manipulation, as illustrated in Fig. 1(a). At a relatively high Reynolds number, a secondary cross-section flow (Dean flow) will develop in curved microchannels due to the action of centrifugal forces. Neutrally buoyant particles suspended in liquids are also subjected to drag force and lift force, which result in differential particle migration within the microchannel, as illustrated in Fig. 1(b). The equilibrium positions of particles with different sizes are determined by the competition between hydrodynamic (drag and lift) forces acting on the particles in high-velocity flows. To describe the motion of particles in closed microchannels, two dimensional Reynolds numbers are introduced here: the channel Reynolds number (Re_c) and the particle Reynolds number (Re_p),

$$\mathrm{Re}_C = \frac{\rho U_m D_h}{\mu}$$

$$\mathrm{Re}_P = \mathrm{Re}_C \frac{a^2}{D_h^2} = \frac{\rho U_m a^2}{\mu D_h}$$

[1]

At small particle Reynolds number $Re_p \ll 1$, the magnitude of the lift forces responsible for movement away from the channel center and walls, can be written as

$$F_Z = \frac{\mu^2}{\rho} \mathrm{Re}_P^2 \, f_C(\mathrm{Re}_C, x_C)$$

[2]

According Eq. [2], the particle size has a dramatic influence on this lateral force. In a straight channel, the lift coefficient vanishes at the axis of the channel, increases in magnitude with distance towards to the wall until a maximum is reached, and then decreases again. When the Reynolds number is increased, the equilibrium position of a particle in quadratic flow is moved towards the wall. The curvature of the present double-spiral microchannel creates additional transverse drag forces on particles and alters the position of flowing particles.

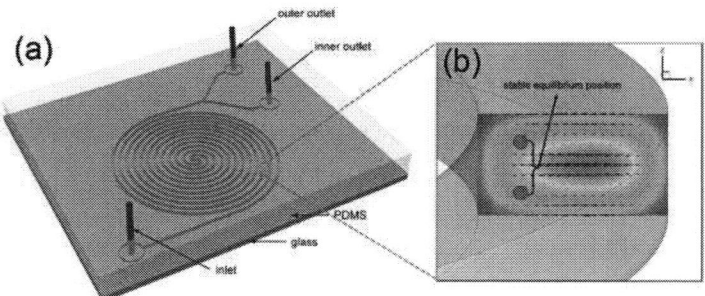

Figure 1. (a) Schematics of the microfluidic cell sorter containing 6-loop double spiral microchannels for particle/cell separation; (b) Illustration of the Dean flow inside a curved microchannel and the stable equilibrium positions of the particles.

Materials and methods

The microfluidic cell sorter for particle/cell separation consists of 6-loop double spiral microchannels with one inlet and two symmetric bifurcated outlets, as depicted in Fig. 1. The channel connected with the inlet first spirals in the counterclockwise direction to form 6 loops, changes its direction through a center junction, spirals out by another 6-loop clockwise channels, and ends with bifurcated outlets. The spiral microchannels are 150 μm wide and 80 μm deep, and the distance between two adjunction loops is 450 μm. The outlet towards to the double spiral channels is termed as the inner outlet, and the other is the outer outlet.

The microfluidic device is fabricated using standard soft-lithography techniques with SU8 master mold on a silicon substrate [8, 9]. After the PDMS slab is peeled from the mold, three ports at the inlet and outlets are punched through the PDMS with a flat tip needle. The PDMS slab is then bonded with a glass substrate (25 mm × 75 mm) post oxygen plasma treatment. Three plastic tubes are inserted through the ports and glued by the liquid PDMS on the top of device. The assembled device is finally placed into an oven at 70 °C for 30 min to cure the liquid PDMS and increase bonding.

In the experiment for separating cells, the channels were first filled with PBS, followed by loading the mixture solution of 4T1 cells and murine blood cells via the pump. The assembled device was mounted onto the stage of a Leica DMI 6000 microscope. Fluorescent streak images were obtained by a CCD camera with exposure time of 5 s and superimposed using Adobe Photoshop CS4.

Results and Discussion

We applied the microfluidic platform to sort and concentrate 4T1 cells from murine blood cells at a mixture concentration of 2×106 cells/ml in PBS. The average size of the 4T1 cells is around 15 μm in diameter, significantly larger than the surrounding murine blood cells (erythrocytes are 6 μm and peripheral blood lymphocytes are 7-10 μm). Due to this large differences in cell size, we can simply differentiate 4T1 cells from blood cells by analyzing the bright-field microscopic images. The initial concentration ratio between 4T1 and blood cells in the cell mixture before separation was 0.86%. After cells traversing through 6-loop double spiral channels at a certain flow rate, most 4T1 cells and partial of blood cells migrated to the inner outlet, and most blood cells and few 4T1 cells moved to the outer outlet, as seen in Fig. 2. By counting more than 1000 cells in the inner and outer outlets, we found that the 4T1 cells in the inner outlet were concentrated by 3.93-fold at the flow rate of 10 mL/hr, and 5.23-fold at 40 mL/hr (Fig. 2). At a throughput of 40 mL/hr, the normalized 4T1 concentration at the outer outlet is 0.165, which can be read from Figure 6. Given the initial 4T1 ratio that is 0.86 %, the ratio of 4T1 cells to the collected cell samples at the outer outlet is 0.14 %, and the ratio of blood cells equals to 99.86%. Similarly, based on the normalized 4T1 concentration ratio of 5.23, we can obtain that the ratio of 4T1 cells at the inner outlet is 4.5% and that of blood cells is 95.5%. At a throughput of 10 mL/hr, the normalized 4T1 concentration at the outer and the inner outlet is 0.44 and 3.93, respectively, yielding a purity of 4T1 cells to be 0.37 % and 3.38%.

Although the concentration ratio of tumor cells in the current design is considerably lower compared to immuno-binding methods or common microfluidic filtration methods, our passive microfluidic technique allows label-free and size-based cell separation with a high throughput. Moreover, simplicity of the system makes it suitable for preliminary separation of biological samples that have a size difference. More importantly, our passive cell separation technique allows easy retrieval of target cells that remain viable and intact, which could be used for further cell culture, gene analysis, and cancer prognosis. In order to enhance the separation efficiency and concentration ratio, designing of multiplex cascading of separation circuits or integration of a chip-based detection downstream to refine the collected cells should be attempted in future study.

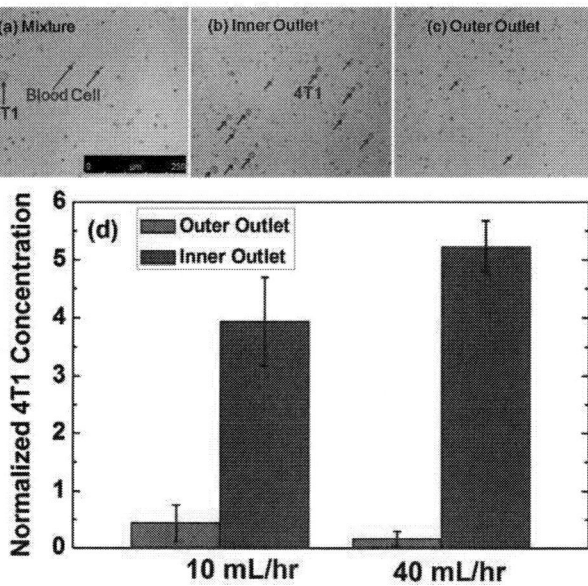

Figure 2. (a-c) Microscopic images for the mixture of 4T1 and blood cells, the inner outlet and the outer outlet at a flow rate of 40 mL/hr; (d) The normalized 4T1 concentration at the inner and outer outlets at two flow rates.

Acknowledgments

J.S. and X.J. acknowledge financial support from MOST (2011CB933201, 2009CB930001), NSFC (51105086, 21025520, 90813032), and CAS (KJCX2-YW-M15). G.H. acknowledges financial support from MOST (2011CB707604) and CAS(KJCX2-YW-H18).

References

1. D.R. Gossett, W.M. Weaver, A.J. Mach, S.C. Hur, H.T.K. Tse, W. Lee, H. Amini, D. Di Carlo, *Anal. Bioanal. Chem.,* **397,** 3249 (2010).
2. H. Tsutsui, C.M. Ho, *Mech. Res. Commun.,* **36,** 92 (2009).
3. P.G. Schiro, M. Zhao, J.S. Kuo, K.M. Koehler, D.E. Sabath, D.T. Chiu, *Angew Chem Int Ed Engl,* **51,** 1 (2012).
4. J. Sun, C.C. Stowers, E.M. Boczko, D. Li, *Lab Chip,* **10,** 2986 (2010).

5. B. Yuan, Y. Jin, Y. Sun, D. Wang, J.S. Sun, Z. Wang, W. Zhang, X.Y. Jiang, *Adv Mater,* **24,** 890 (2012).
6. K.H. Kang, Y. Kang, X. Xuan, D. Li, *Electrophoresis,* **27,** 694 (2006).
7. W.C. Lee, A.A.S. Bhagat, S. Huang, K.J. Van Vliet, J. Han, C.T. Lim, *Lab Chip,* **11,** 1359 (2011).
8. J.S. Sun, S.K. Vajandar, D.Y. Xu, Y.J. Kang, G.Q. Hu, D.Q. Li, D.Y. Li, *Microfluid. Nanofluid.,* **6,** 589 (2009).
9. J. Sun, Y. Gao, R.J. Isaacs, K.C. Boelte, C.P. Lin, E.M. Boczko, D. Li, *Anal Chem,* **84,** 2017 (2012).

CHAPTER 5

CANTILEVERS AND MICRODEVICES

448

Nano/Micro Patterned Phononic Crystals

Bongsang Kim[a], Janet Nguyen[a], Charles Reinke[a], Maryam Ziaei-Moayyed[a],
Ihab El-Kady[a,b], Drew Goettler[b], Mehmet Su[b], Zayd C. Leseman[b] and Roy H. Olsson III[a]

[a] Sandia National Laboratories, Albuquerque, New Mexico 08540, USA
[b] University of New Mexico, Albuquerque New Mexico 08901, USA

Recently, with the application of micro/nano machining
technologies, there have been immense strides in the research of
phononic crystals. This paper reviews basics of micro/nano
fabricated 2D phononic crystals, and discusses their promising
applications, particularly for RF signal processing and thermal
conductivity manipulation with several examples of Sandia
National Laboratories.

Introduction

Phononic crystals (PnCs) are artificial acoustic metamaterials constructed by periodic
inclusions of acoustic scatters in a host or matrix material to yield certain frequencies to
be completely reflected by the structures, therefore creating bandgaps in transmission
frequency response (1). In the past decade, phononic crystals of various scales have been
demonstrated ranging from hand assembled rubber balls in acoustically lossy materials
such as water and epoxy to micro/nano-scale fabricated 2D phononic crystals in the
frequency range of 10-5,000 MHz (2-11).

In recent years, micro/nano-machined phononic crystals are particularly gaining
attention. Benefiting from semiconductor-micromachining technologies leads to the
ability to fabricate repeatable structures and flexibility in lithographic patterns. This has
led to various experimentation on a wide variety of geometries and sizes (2-7, 10-12).
The ability to suspend lithographically defined 2-D phononic crystals above the substrate
using micromachining techniques has allowed for low loss phononic waveguides (2, 3, 11,
13) and cavities (5, 10, 14) to be demonstrated. Also, wide ranges of materials have been
attempted as the material sets for phononic crystal inclusions and the surrounding media,
particularly focused on creating phononic bandgaps at extremely high frequencies (2-5, 9,
11).

Various promising applications of micro/nano-machined phononic crystals have
been discussed and suggested. Phononic crystals operating as acoustic mirrors are
applicable to mechanically vibrating structures such as gyroscopes and microresonators
(15). Through strategic defect placements, phononic crystals can be used as acoustic
waveguides (2, 3, 11, 13) and cavity resonators with high quality factor and low insertion
loss within a small foot-print (5, 10, 14). Recently, phononic crystal based liquid sensors
and liquid manipulators have been demonstrated (8, 16). Also phononic crystals have
demonstrated the capability of acoustic focusing and negative refraction (17, 18) which
may potentially lead to miniaturization and performance enhancement in
ultrasound,nondestructive testing and other acoustic imaging applications.

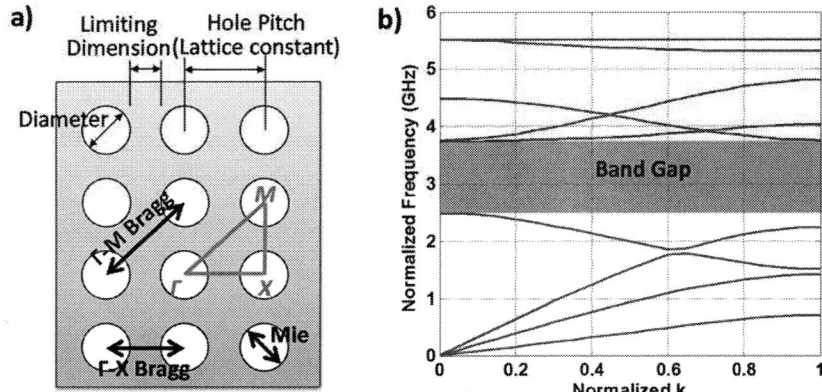

Figure 1. a) Schematic of a 2D square lattice phononic crystal. Bragg resonances are between neighboring holes and the Mie resonance is within the hole. b) An example of a dispersion relation of a 2D phononic crystal simulated by the PWE (plane-wave expansion) method. A band gap can be created by phononic crystals where the existence of phonons in a certain frequency range is prohibited.

Recently, the ability to manipulate material thermal conductivity in single crystal silicon using micro/nano-machined phononic crystals has been demonstrated (*19-21*). Unlike metals, in many semiconductor materials, heat conduction is dominated by phonons where the thermal energy is transferred by the vibration of the neighboring atomic lattices while electrical conduction occurs through electrons. For example, silicon thermal conduction takes place through phonons at GHz to THz ranges. Therefore, if only phonons can be independently manipulated through phononic crystals, it is possible to control material thermal conductivity without significantly altering the electrical conductivity.

Micro/Nano Patterned Phononic Crystal

Phononic crystals are artificial acoustic materials that alter wave propagation and phonon dispersion by periodic variation of acoustic properties. Phononic bandgaps can be formed by local resonance conditions within phononic crystals, forbidding acoustic wave propagation. For example, a 2D square lattice phononic crystal consisting of periodic inclusions shown in Figure 1a contains several resonances. Bragg resonances are between neighboring inclusions in the perpendicular and horizontal (*Γ-X*) directions or diagonal (*Γ-M*) direction. The Mie resonance is the resonance within the inclusion. When these local resonance frequencies approach each other, they interfere and wave propagation is impeded creating a band gap at that frequency as shown in the example case of Figure 1b. As the impedance mismatch between the inclusion and matrix increases the scattering at the interface increases typically resulting in a wider bandgap.

2D phononic crystals have been frequently demonstrated via micro/nano-machining techniques. Lithography-based micro/nano-machining techniques provide the capability to fabricate repeated micron-nanometer scale patterns with great precision and reliability. Fabrication process compatibility with CMOS IC and MEMS/NEMS is

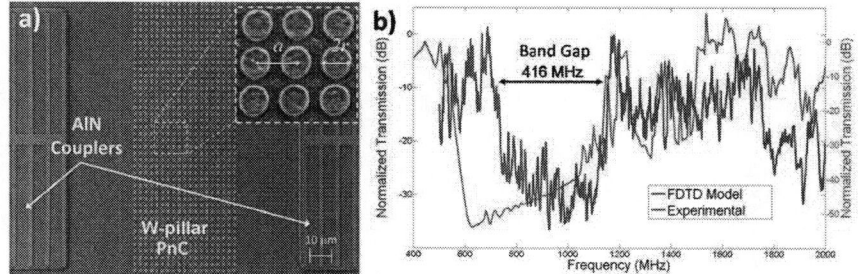

Figure 2. a) A phononic crystal constructed by tungsten pillar inclusions in a silicon dioxide matrix. This phononic crystal membrane is suspended from the substrate. b) FDTD (finite difference time domain) modeled and measured frequency response of the W-SiO$_2$ phononic crystal. A 416 MHz-wide band gap could be created at 1 GHz. These results are from (11)

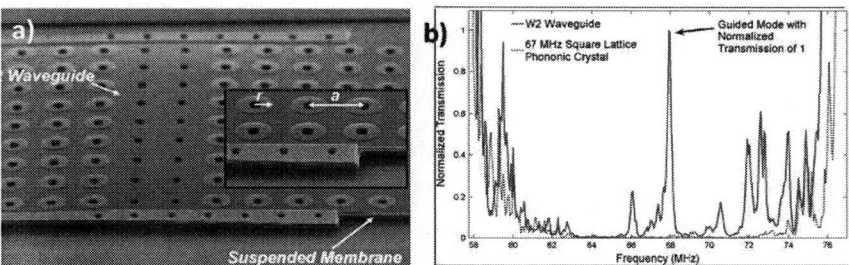

Figure 3. a) SEM image of W2 waveguide constructed by removing two rows of inclusions from a W-SiO$_2$ phononic crystal. b) Measured normalized transmission of the wave guide. Several guided modes exist inside the bandgap region including a dominant mode at 68 MHz with a normalized transmission of unity. These results are from (2).

another plus in the application point of view. At Sandia National Laboratories we have explored diverse material sets to create phononic crystals operating at extremely high frequencies using micro/nano-machining techniques (2, 3, 5, 15, 22). Particularly, using tungsten pillars in a silicon dioxide matrix, a 416 MHz wide fundamental bandgap could be realized at a 1 GHz center frequency as shown in Figure 2 (11). This solid-solid phononic crystal takes advantages of similar acoustic velocities between the inclusion and matrix material, therefore creating phononic bandgaps at higher frequencies becomes much easier with given lithography limitations compared to air-solid phononic crystals.

RF Applications

In phononic crystal structures, by removing or altering some inclusions, devices such as acoustic waveguides and resonant cavities can be built which can be used for RF signal processors.

Figure 3a shows an example of waveguide, which can route and bend acoustic signals. This structure was built by removing two rows of inclusions in a 8-period square

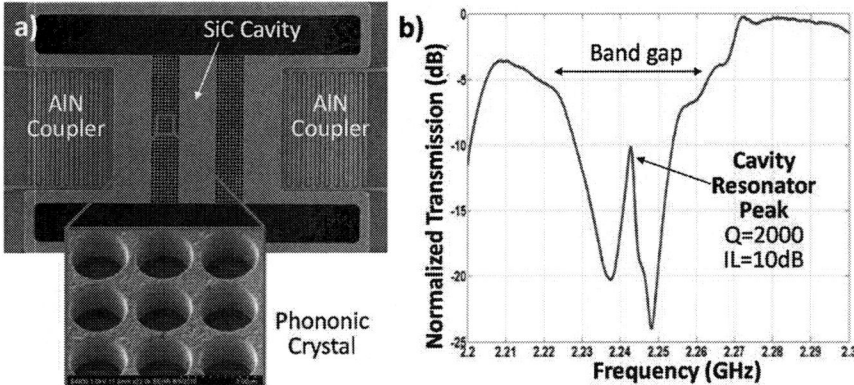

Figure 4. a) SEM image of a fabricated SiC phononic crystal cavity resonator. The cavity was surrounded by 5 layers of phononic crystal on each side. b) Measured frequency response of the SiC phononic crystal cavity resonator. The resonator was measured at 2.25 GHz resonant frequency with Q~2000 and IL=10dB. These results are from (5).

lattice tungsten-silicon dioxide phononic crystal (2). As can be seen in the measured frequency response plot of Figure 3b, this device presents several guided modes inside the bandgap with a dominant mode at 68 MHz with the normalized transmission of 1. This high transmission wave guide is due to the low loss materials forming the phononic crystal. In other devices (11)with a similar design, the normalized transmission loss increased to -10 dB at much higher frequency (GHz), which was because the acoustic energy had to be focused into the very narrow waveguide formed by removing only several rows of inclusions at such high frequencies. Such phononic crystal waveguides are useful to miniaturize delay elements commonly used in signal processing and delay line oscillators as was noted in similar concepts with photonic crystals (23).

Figure 4 shows a resonant cavity constructed by an air-SiC phononic crystal (5). With a minimum feature size of 200 nm, periodic through-holes were patterned with a lattice constant of 1.83 um in a SiC membrane, realizing a bandgap at a center frequency of 2.25 GHz. An acoustic cavity was prepared within the phononic crystal, resulting in a high Q resonator (Q~2000) with a small transmission loss (I.L.=10dB) at 2.25 GHz. Such high quality factor is achieved via the low loss of the SiC material. Phononic crystal based RF signal processors have an advantage in size over their micromechanical counter parts. For example, the above device occupies almost 10x smaller space compared to micromechanical resonators with similar performance (24). Also, by its nature, phononic crystal is easier to dealing with very wide bandwidth, whereas, micromechanical resonators have demonstrated better control over frequency tuning and other device performance at narrower bandwidth, therefore, these two technologies will be mutually complementary and supportive.

Demonstrated phononic-crystal based RF signal processors have used CMOS compatible micromachining techniques, therefore direct applications to many promising microsystems are possible which will lead to system-level miniaturization and performance enhancement in many RF communication devices.

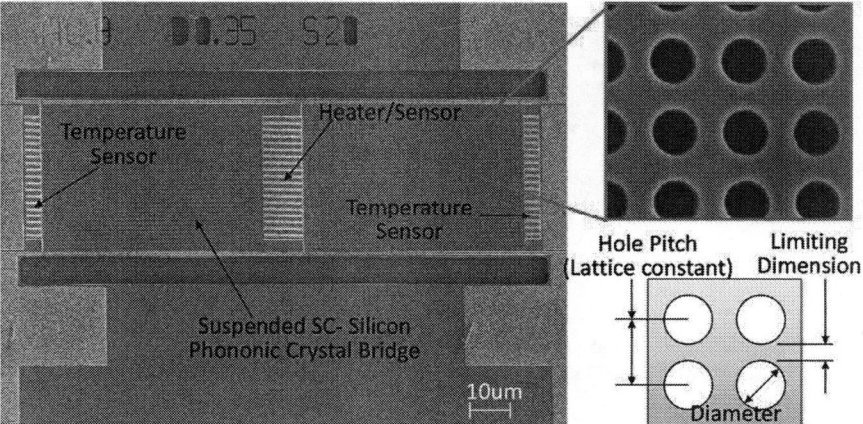

Figure 5. SEM image of phononic crystal thermal conductivity test structure (*21*). While heat is supplied at the suspended phononic crystal membrane center, the temperature gradient is measured to extract the phononic crystal thermal conductivity.

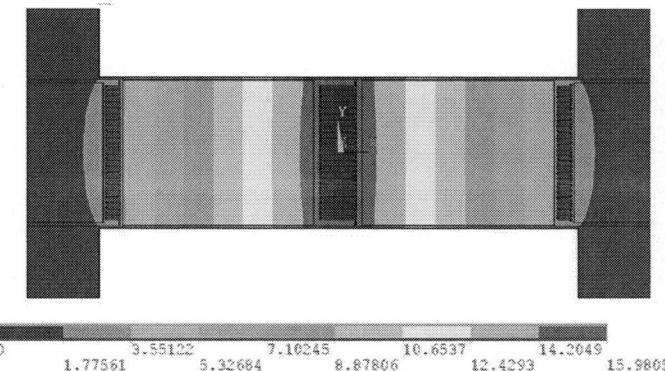

Figure 6. ANSYS finite simulated thermal model of the test structure in Figure 5.

Thermal Conductivity Manipulation

As mentioned in the introduction, in semiconductor materials, thermal conductivity can be controlled by manipulating phonon transport. Phononic crystals can effectively manipulate phonon transport in two ways to control material thermal conductivity. First, as the patterned inclusions approach each other, the size of thermal pathways is reduced comparable to the phonon mean-free-path, therefore heat transfer is significantly impacted as could be frequently observed in small-scale structures such as nano-wires and thin-film materials (*25, 26*). Secondly, through periodic variation of the acoustic properties, the phonon dispersion is modified which impacts the energy-carrying efficiency of phonons, therefore altering the thermal conductivity. Also, if a phononic bandgap is created at the frequencies responsible for heat conduction, heat transfer

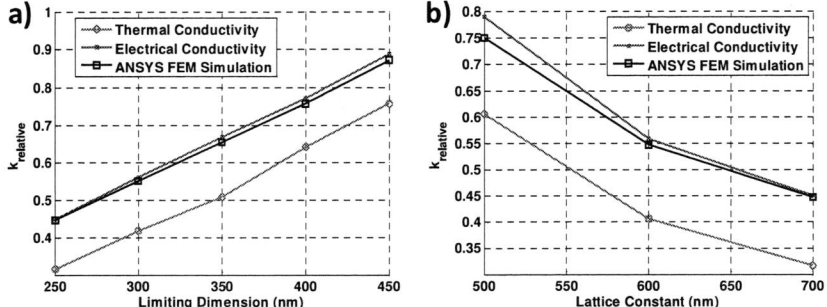

Figure 7. Measured relative thermal (red circles) and electrical (blue crosses) conductivities as well as ANSYS simulated volume reduction effects are marked (black squares). a) These data are of the samples with the same lattice constant of 700 nm with varying limiting dimensions. b) Data for samples with the same limiting dimension of 250 nm with varying lattice constants. These results are from (21).

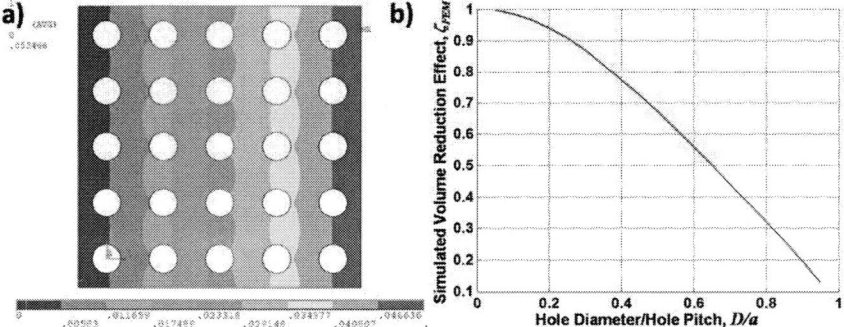

Figure 8. a) ANSYS finite element simulation of the effective conductivity reduction by introducing periodic through holes. b) Calibration curve of ANSYS simulated volume reduction effect factor, ζ_{FEM} vs. D/a. These results are from (21)

efficiency can be impacted. The former is enhancing incoherent phonon scattering and the latter is enhancing coherent phonon scattering.

Figure 5 shows SEM images of the test structure utilized to measure the impact of phononic crystals on silicon thermal conductivity (21). Using lithography techniques, periodic sub-micron through holes were patterned in a 500nm-thick lightly n-type doped (phosphorus 10^{16}) single crystal device layer of a SOI (silicon-on-insulator) wafer. Various combinations of lattice constants and hole diameters were designed, and after patterning holes, the oxide under the device was removed to release the phononic crystal membrane from the substrate. The device thermal conductivity was measured in a vacuum chamber while heat was supplied through an aluminum joule heater installed in the membrane center. The temperature gradient across the membrane was measured using pre-calibrated temperature sensors, and using the thermal models shown in Figure 6, the thermal conductivities of the phoninic crystal membranes were extracted. In parallel a

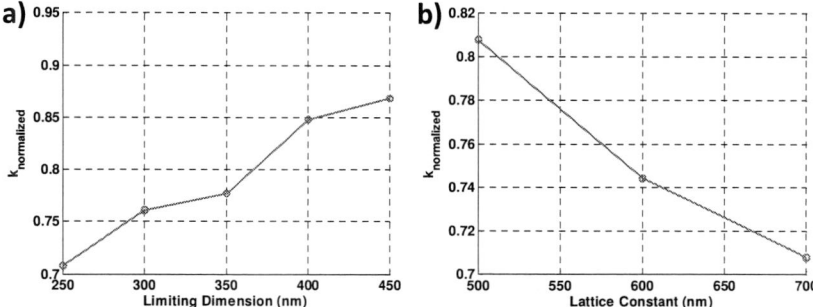

Figure 9. Measured normalized thermal conductivities. a) These data are from the samples with the same lattice constant of 700 nm and varying limiting dimensions. b) Data from the samples with the same limiting dimension of 250 nm with varying lattice constants. These results are from (21).

similar test structures were prepared to measure the electrical conductivity of phononic crystals. This set of devices had the identical phononic crystal patterning but were doped much higher (phosphorus 10^{19}).

Figure 7 shows the relative thermal conductivity and electrical conductivity from the measurement, which are defined as,

$$k_{relative} = k_m / k_{m,control} \qquad [1]$$

$$\sigma_{relative} = \sigma_m / \sigma_{m,control} \qquad [2]$$

where k_m and σ_m are the measured thermal and electrical conductivities respectively, and $k_{m,control}$ and $\sigma_{m,control}$ are the measured thermal and electrical conductivities of the control device which doesn't have any holes, but instead is just solid membrane. Figure 7a is the case of the same lattice constant with varying hole sizes and limiting dimensions. Figure 7b is in the case of the same limiting dimension but with varying lattice constancies and hole diameters. As can be seen, for all the samples with periodic holes there were significant reductions in both thermal and electrical conductivities. Particularly, the thermal conductivity was reduced much more than the electrical conductivity even with the same phononic crystal designs. However, much of these reductions must be from the effect of volume removal by introducing hollow through holes, therefore we have investigated the volume reduction effect using ANSYS simulation as shown in Figure 8a. This finite element simulation models only continuum effects such as Fourier's law in case of the thermal conductivity and Ohm's law in case of electrical conductivity, not any phonon or electron scattering effects, therefore only the relative ratio between the hole size and pitch is important. Figure 8b shows the calibration curve for the simulated volume reduction effect, ζ_{FEM}, with respect to the d/a (hole diameter/hole pitch) in the case of a square lattice of periodic holes, and these values are added in Figure 7. As can be seen, for all measured samples, the electrical conductivity was reduced to the same value as the volume reduction effect factor of the corresponding phononic crystal designs, however the thermal conductivities were reduced much further.

Here, we introduce normalized thermal conductivity, k_n, which is defined as,

$$k_n = \frac{k_{relative}}{\zeta_{FEM}} = \frac{k_m / k_{m,control}}{\zeta_{FEM}} \qquad [3]$$

This normalized thermal conductivity tells how much the thermal conductivity is further reduced beyond the contribution from the volume reduction effect. Figure 9 shows the reprocessed thermal conductivity data using equation [3]. As can be seen in Figure 9a, as the limiting dimension decreases, further reductions in normalized thermal conductivity can be observed which indicates incoherent scattering plays a significant role. However, the comparisons among the samples with the same limiting dimension shown in Figure 9b, also suggests that there is a clear differences in the normalized thermal conductivity. Particularly, as the lattice constant (hole pitch) increases, the thermal conductivity is further reduced. This indicates that also coherent scattering from the periodic hole placement may have contributed to the reduction in thermal conductivity. One hypothetical explanation is, as the lattice constant increases two Bragg resonant frequencies, $f_{\Gamma\text{-}X}$ and $f_{\Gamma\text{-}M}$, of the phononic crystals approach each other, resulting in more coherent scattering and further reductions in thermal conductivity. This phenomenon is currently under investigation using the PWE (plane wave extension) simulation method.

These devices were fabricated using only conventional lithography techniques, which not only opens direct applications to many micromachined devices without much modification to their original fabrication process but also allows for their repeatable mass production. Particularly, the ability to control material thermal conductivity without changing the electrical conductivity will improve the energy conversion efficiency when used in thermoelectric devices such as Peltier coolers or waste heat power generators.

Conclusions

Phononic crystals are an emerging research field. During the past decade, micro/nano-machining technologies have enabled repeated and reliable fabrication of phononic crystals operating at MHz to GHz frequency ranges. The ability to control acoustic wave propagation at such high frequencies together with the potential for integration with other micromachined devices such as CMOS circuits and MEMS devices is poised to impact many interesting applications ranging from RF signal processing and medical ultrasound to thermoelectric energy harvesters.

Acknowledgments

The authors would like to thank the staff of the Microelectronics Develop Laboratory at Sandia National Laboratories for their efforts to fabricate of the phononic crystal devices and Chris Nordquist and Mark Balance for use of RF and thermal characterization resources. Sandia National Laboratories is a multi-program laboratory managed and operated by Sandia Corporation, a wholly owned subsidiary of Lockheed Martin Corporation, for the U.S. Department of Energy's National Nuclear Security Administration under contract DE-AC04-4AL85000."

References

1.   F. R. Montero de Espinosa, E. Jiménez, M. Torres, Ultrasonic Band Gap in a Periodic Two-Dimensional Composite. *Physical Review Letters* **80**, 1208 (1998).

2.	R. H. Olsson III, I. El-Kady, Microfabricated phononic crystal devices and applications. *Measurement Science and Technology* **20**, 012002 (2009).

3.	R. H. Olsson III, I. F. El-Kady, M. F. Su, M. R. Tuck, J. G. Fleming, Microfabricated VHF acoustic crystals and waveguides. *Sensors and Actuators A: Physical* **145–146**, 87 (2008).

4.	T.-T. Wu, L.-C. Wu, Z.-G. Huang, Frequency band-gap measurement of two-dimensional air/silicon phononic crystals using layered slanted finger interdigital transducers. *Journal of Applied Physics* **97**, 094916 (2005).

5.	M. Ziaei-Moayyed, M. F. Su, C. Reinke, I. F. El-Kady, R. H. Olsson, in *Micro Electro Mechanical Systems (MEMS), 2011 IEEE 24th International Conference on.* (2011), pp. 1377-1381.

6.	S. Benchabane, A. Khelif, J. Y. Rauch, L. Robert, V. Laude, Evidence for complete surface wave band gap in a piezoelectric phononic crystal. *Physical Review E* **73**, 065601 (2006).

7.	N.-K. Kuo, C. Zuo, G. Piazza, Microscale inverse acoustic band gap structure in aluminum nitride. *Appl. Phys. Lett.* **95**, 093501 (2009).

8.	R. Lucklum, M. Ke, M. Zubtsov, Two-dimensional Phononic Crystal Sensor based on a Cavity Mode. *Sensors and Actuators B: Chemical,* (2012).

9.	S. Mohammadi, A. A. Eftekhar, A. Khelif, W. D. Hunt, A. Adibi, Evidence of large high frequency complete phononic band gaps in silicon phononic crystal plates. *Appl. Phys. Lett.* **92**, 221905 (2008).

10.	S. Mohammadi, A. A. Eftekhar, W. D. Hunt, A. Adibi, High-Q micromechanical resonators in a two-dimensional phononic crystal slab. *Appl. Phys. Lett.* **94**, 051906 (2009).

11.	R. H. Olsson III *et al.*, in *Ultrasonics Symposium (IUS), 2009 IEEE International.* (2009), pp. 1150-1153.

12.	H. Estrada *et al.*, Influence of lattice symmetry on ultrasound transmission through plates with subwavelength aperture arrays. *Appl. Phys. Lett.* **95**, 051906 (2009).

13.	A. Khelif, S. Mohammadi, A. A. Eftekhar, A. Adibi, B. Aoubiza, Acoustic confinement and waveguiding with a line-defect structure in phononic crystal slabs. *Journal of Applied Physics* **108**, 084515 (2010).

14.	F. Li, J. Liu, Y. Wu, The investigation of point defect modes of phononic crystal for high Q resonance. *Journal of Applied Physics* **109**, 124907 (2011).

15.	D. Goettler *et al.*, Realizing the frequency quality factor product limit in silicon via compact phononic crystal resonators. *Journal of Applied Physics* **108**, 084505 (2010).

16.	Y. Bourquin, R. Wilson, Y. Zhang, J. Reboud, J. M. Cooper, Phononic Crystals for Shaping Fluids. *Advanced Materials* **23**, 1458 (2011).

17.	S. Yang *et al.*, Focusing of Sound in a 3D Phononic Crystal. *Physical Review Letters* **93**, 024301 (2004).

18.	M. Ke *et al.*, Negative-refraction imaging with two-dimensional phononic crystals. *Physical Review B* **72**, 064306 (2005).

19.	J. Tang *et al.*, Holey Silicon as an Efficient Thermoelectric Material. *Nano Letters* **10**, 4279 (2010/10/13, 2010).

20.	P. E. Hopkins *et al.*, Reduction in the Thermal Conductivity of Single Crystalline Silicon by Phononic Crystal Patterning. *Nano Letters* **11**, 107 (2011/01/12, 2010).

21.	B. Kim *et al.*, in *MEMS 2012, 25th IEEE International Conference on Micro Electro Mechanical Systems.* (Paris, France, 2012), pp. 176-179.

22. C. M. Reinke, M. F. Su, R. H. I. Olsson, I. El-Kady, Realization of optimal bandgaps in solid-solid, solid-air, and hybrid solid-air-solid phononic crystal slabs. *Appl. Phys. Lett.* **98**, 061912 (2011).
23. T. F. Krauss, Slow light in photonic crystal waveguides. *Journal of Physics D: Applied Physics* **40**, 2666 (2007).
24. S. Gong, N.-K. Kuo, G. Piazza, GHz High-Q Lateral Overmoded Bulk Acoustic-Wave Resonators Using Epitaxial SiC Thin Film *Microelectromechanical Systems, Journal of* **21**, 253 (2012).
25. M. Asheghi, K. Kurabayashi, R. Kasnavi, K. Goodson, Thermal conduction in doped single-crystal silicon films. *Journal of applied physics* **91**, 5079 (2002).
26. A. I. Hochbaum *et al.*, Enhanced thermoelectric performance of rough silicon nanowires. *Nature* **451**, 163 (2008).

Photothermal Cantilever Deflection Spectroscopy

Seonghwan Kim[a], Dongkyu Lee[a], Rachel Thundat[a], Mehrdad Bagheri[a], Sangmin Jeon[b], and Thomas Thundat[a]

[a] Department of Chemical and Materials Engineering, University of Alberta, Edmonton, Alberta T6G 2V4, Canada
[b] Department of Chemical Engineering, Pohang University of Science and Technology (POSTECH), Pohang, Republic of Korea

A real-time technique that does not rely on chemical interfaces or biological receptors for molecular identification of picogram quantities of biomaterials such as DNA molecules in a high throughput fashion is described. This technique combines the extremely high sensitivity of microfabricated bi-material cantilever beams with the high selectivity of mid infrared (IR) spectroscopy to nanomechanically transduce the photon absorption-induced temperature variations of the molecules. Picogram amounts of target molecules were first adsorbed on the cantilever without using any receptors. Illuminating a bi-material cantilever sequentially with a mid-IR radiation results in photon absorption by the molecule at a certain wavelength, which results in a small temperature variation, and the resultant deflection of the bi-material cantilever. A plot of cantilever deflection as a function of an illuminating wavelength closely follows the IR absorption spectrum of the target molecules. We have used this technique to rapidly identify different DNA strands.

Introduction

Selective and sensitive detection of biomolecules rapidly without using any reagents using an inexpensive handheld device has immediate applications in many fields (1, 2). Currently available techniques, however, are time consuming and use expensive reagents and bulky equipment that requires trained personnel for operation. For example, a crime lab requires sophisticated and time consuming techniques such as electrophoresis to sequence DNA samples for comparison with DNA from suspects. However, if there can be a simple technique that can rapidly determine match versus non-match for the DNA strands, it will be possible to avoid the time consuming process of gel electrophoresis.

There exist many sensing techniques can be the basis for rapid, sensitive, and selective detection of molecules. For example, spectroscopic techniques such as Fourier Transform Infrared Spectroscopy (FTIR) have excellent selectivity for molecular detection based on vibrational characteristics of the molecules. The spectrum in the mid infrared (IR) region is known as the "molecular fingerprint regime" due to the uniqueness of the molecular vibrations in this region that are free from overtones. The IR spectroscopy in the mid-IR region can provide molecular identification rapidly. Though very selective, IR spectroscopy is not very sensitive and requires micro and milligram quantities of sample for analysis. Moreover the instrumentation methods for obtaining IR spectra are complex and bulky. Therefore, IR spectroscopy is routinely used for laboratory-based molecular identification.

Another technique that can detect chemicals and biomolecules in a rapid fashion is sensors based on micro-electro-mechanical-systems (MEMS) (3). Recent advances in microfabrication techniques have resulted in the development of many sensors. For example, microfabricated cantilever sensors are extremely sensitive sensors for molecular adsorption. Although cantilever sensors are extremely sensitive, they do not offer any chemical or biological selectivity. The selectivity in detection is obtained using bio-receptors (for example, antibodies) immobilized on the cantilever surface. Chemical and biological specificity is often achieved by immobilizing a monolayer of chemical or biological receptors on the cantilever surface. The selectivity, therefore, is determined by the selectivity of the receptor layer. Immobilizing receptors on the sensor surface in a reproducible manner is a challenge. Non-uniformity of the receptor coverage on the sensor surface results in large variation in sensor response. Possible random orientation of receptors and expense of receptors, and multiple steps involved in immobilizing receptors have been formidable challenges.

A single technique that can combine the selectivity of the IR spectroscopy with the unprecedented sensitivity of the microfabricated cantilever beams offers an ideal platform for developing portable and inexpensive chemical and biological sensors. This combined technique is based on photon-absorption-induced variation of thermal properties (photothermal effect) detected with microfabricated cantilever beams. This technique was originally proposed by Barnes et al. to identify surface species on a hydrogen-terminated amorphous Si cantilever (4, 5). Since then this technique of photothermal cantilever deflection spectroscopy (PCDS) has been demonstrated for limited number of chemical and biological materials such as explosives (6-8), DNA/RNA (9), and bacteria (10, 11).

Method

Bi-material Cantilevers

Microfabricated cantilevers, similar to those used in Atomic Force Microscopy (AFM), can detect extremely small forces in the range of pico-Newtons. These cantilevers are generally fabricated from silicon or silicon nitride by top-down micromachining methods and can be produced efficiently and affordably. Typical dimension of these cantilevers are around 200-500 microns in length, 40-90 microns in width and 0.5 - 1 micron in thickness. The spring constant of these cantilever are in the range of 0.03 - 1 N/m. Although the cantilevers have micrometer dimensions, their response are in nanometer-scale lends itself to their reference as nanomechanical transducers. These microcantilevers can be transformed into heat sensors using bimetallic effect by depositing a thin layer of metal by vacuum evaporation. When the cantilevers are made bimetallic by depositing a thin layer of metal on one side of the cantilever, they become extremely sensitive to temperature change. In this study, rectangular silicon cantilevers (CSC12-E) are obtained from MikroMasch USA (San Jose, CA). The dimension of each cantilever is 350 μm in length, 35 μm in width, 1 μm in thickness. The microcantilevers are cleaned by rinsing with acetone, ethanol, and UV ozone treatment then coated with 10 nm of chromium (adhesion layer) followed by 200 nm of gold using e-beam evaporator.

PCDS Technique

The PCDS technique is based on exposing a bi-material cantilever with adsorbed chemical or biological molecules to mid-IR light such that the wavelength of the light is scanned over a wavelength window. Absorption of IR light by the molecules on the cantilever excites the molecules into resonance producing very small changes in the temperature due to nonradiative decay of molecular vibration. Since the cantilever is bi-material, any change in the temperature results in cantilever bending. Therefore, the bi-material cantilever acts as a transducer for extremely small thermal changes occurring on the cantilever surface. When the cantilever with adsorbed material is exposed to different wavelengths in a sequential fashion, the bending of the cantilever will follow the IR absorption spectra of the adsorbed material. By using a differential technique, it is possible to detect and identify (molecular speciation) adsorbed molecules with high selectivity and sensitivity. Since the cantilever resonance frequency can be monitored before and after adsorption, the adsorbed mass can be determined with very high precision using variations in the resonance frequency.

Biomaterials

We have used synthetic single-stranded DNA (ssDNA) of 20 bases of adenine, thymine, cytosine, and guanine for our PCDS experiments. The initial concentration of each DNA solution was 10 mg/mL. These ssDNAs were deposited on the cantilever by inserting the cantilever into a micro glass capillary containing the DNA solution and slowly taking the cantilever out of the micro glass capillary.

Experimental Setup

The experimental setup for PCDS is shown in Figure 1. We have used an old Foxboro Miran 1A-CVF single beam spectrophotometer as the IR source for our PCDS setup. A chopped beam of IR light at 80 Hz is focused on the bi-material cantilever. The chopped light illuminates the cantilever, alternately heating and cooling the cantilever at a frequency equal to that of the chopper. When adsorbed molecules are resonantly excited, they create additional heating and thus additional cantilever bending. The IR wavelength is then scanned from 2.5 microns to 14.5 microns by a rotating filter wheel. If the adsorbed molecules on the cantilever absorb a particular wavelength, the cantilever bends more due to the generation of additional heat. This additional heat originates from the non-radiative decay associated with IR absorption by the sample. The cantilever bending is detected using a laser beam from a diode laser focused at the free end of the cantilever. The reflected light from the laser is directed to fall on a position sensitive detector (PSD). When the cantilever bends, the reflection of the laser shifts, and the PSD transmits a change in photo-induced voltage to a lock-in amplifier. The lock-in amplifier records the maximum amplitude of cantilever deflection during each heating and cooling cycle. The changes in the lock-in signal correspond to the degree of IR absorbance of the adsorbed material on the cantilever at a particular wavelength. Thus, a plot of the lock-in signal as a function of wavelength shows the IR spectrum of the material on the cantilever.

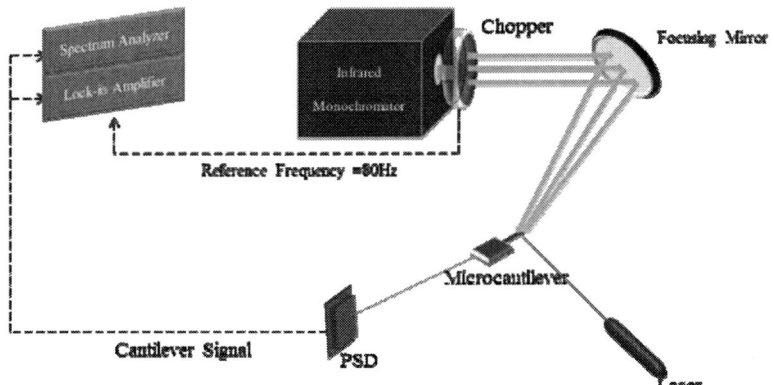

Figure 1. Schematic figure of the PCDS experimental setup.

Determination of Adsorbed Mass

In PCDS, the resonance frequency shift is also used to determine the mass of the adsorbed molecules. The relationship between adsorbed mass and resonant frequency shift is given by (12)

$$\Delta m = m_0 \left(\frac{\Delta E}{E_0} + 3\frac{\Delta t}{t_0} - 2\frac{\Delta f}{f_0} \right)$$

[1]

where Δm is the mass of the adsorbate and m_0 is the mass of the clean cantilever. E_0, t_0, and f_0 are the initial values of the Young's modulus, thickness and resonance frequency of the cantilever, respectively. ΔE, Δt, and Δf are the changes in the Young's modulus, thickness, and resonance frequency of the cantilever, respectively. In this experiment, the changes in thickness and Young's modulus are negligible and ssDNA molecules are considered uniformly adsorbed on the surface, then equation 1 can be simplified to

$$\Delta m = -2m_0 \frac{\Delta f}{f_0}$$

[2]

Using this equation, the adsorbed mass of ssDNA molecules on the cantilever is determined.

Results and Discussion

Figure 2 shows the IR spectra of the ssDNA of 20 bases of adenine, thymine, cytosine, and guanine. Each spectrum shown in this figure is an average of 3 different scans. The amount of ssDNA adsorbed on the surface was determined by measuring the resonance frequency before and after ssDNA adsorption. The amount of ssDNA varied between 2 and 6 nanograms. The PCDS is able to detect 100 picograms of adsorbed materials with a reasonable signal-to-noise ratio. On the other hand the conventional techniques would require milligrams to micrograms of the same material. Although the main prominent peaks are present at around the same energy location, they are slightly shifted in energy. These spectra clearly show that the different bases have finite different signatures. The DNA molecules are complex molecules containing thousands of molecules. Each of these molecules vibrates independently. However, these vibrations can be superimposed to form an overall spectrum. At present, there exists no theory that can predict the spectrum of a given DNA sequence or a cell. The observed spectrum is a superposition of spectral response of individual molecules. Since there are many different molecular groups present in a single DNA molecule, the observed spectrum is a combination all different spectrum. It should also be noted that even at the picogram level there are thousands of molecules on the cantilever surface. However, the spectrum of each strand will be identical since intermolecular interactions are negligible. In future it should be possible to make a library of PCDS IR absorbance spectra that could be used to identify the material on the cantilever.

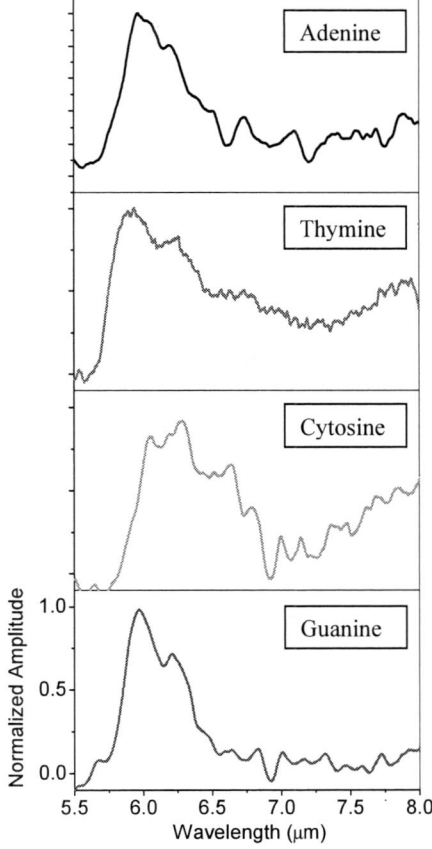

Figure 2 PCDS spectra of four different ssDNA of 20 bases of adenine, thymine, cytosine, and guanine.

Conclusions

We have demonstrated detection of ssDNA strands adsorbed on the bi-material cantilever using PCDS. These experiments show that the PCDS has the potential to increase efficiency for biological identification, for example, DNA molecules in a simple and rapid fashion. When material is attached to the cantilever array using the spotting method, the spectra obtained may be used to differentiate between DNA from various sources. The PCDS technique offers high sensitivity and selectivity. The technique is compatible with developing into a handheld device. With increased reliability of all aspects of this technique, the PCDS technique could be further applied to DNA and genetics work. In genomics research, there is increased interest in being able to differentiate between species or individuals based on small variations in their DNA sequences.

Acknowledgments

This research is supported by Canada Excellence Research Chairs program.

References

1. T. Thundat and A. Majumdar, in *Sensors and Sensing in Biology and Engineering*, F. G. Barth, J. A. C. Humphrey, and T. W. Secomb, Editors, p. 337, Springer Wein, New York (2003).
2. T. Vo-Dinh and B. Cullum, *Fresenius J. Anal. Chem.*, **366**, 540 (2000).
3. J. Fritz, *Analyst*, **133**, 855 (2008).
4. J. R. Barnes, R. J. Stephenson, M. E. Welland, Ch. Gerber, and J. K. Gimzewski, *Nature*, **372**, 79 (1994).
5. J. R. Barnes, R. J. Stephenson, C. N. Woodburn, S. J. O'Shea, M. E. Welland, T. Rayment, J. K. Gimzewski, and Ch. Gerber, *Rev. Sci. Instrum*, **65**, 3793 (1994).
6. P. G. Datskos, M. J. Sepaniak, C. A. Tipple, N. Lavrik, *Sens. Actuators B*, **76**, 393 (2001).
7. P. G. Datskos, S. Rajic, M. J. Sepaniak, N. Lavrik, C. A. Tipple, L. R. Senesac, and I. Datskou, *J. Vac. Sci. Technol. B*, **19**, 1173 (2001).
8. A. R. Krause, C. Van Neste, L. R. Senesac, T. Thundat, and E. Finot, *J. App. Phys.*, **103**, 094906 (2008).
9. E. T. Arakawa, N. V. Lavrik, S. Rajic, and P. G. Datskos, *Ultramicroscopy*, **97**, 459 (2003).
10. E. T. Arakawa, N. V. Lavrik, and P. G. Datskos, *Appl. Opt.* **42**, 1757 (2003).
11. A. Wig, E. T. Arakawa, A. Passian, T. L. Ferrell, and T. Thundat, *Sens. Actuators B*, **114**, 206 (2006).
12. M. Varshney, P. S. Waggoner, C. P. Tan, K. Aubin, R. A. Montagna, and H. G. Craighead, *Anal. Chem.*, **80**, 2141 (2008).

Development of Insulated Conductive AFM Probes for Experiments in Electrochemical Environment

Yexian Wu, Terunobu Akiyama, Peter van der Wal, Sebastian Gautsch, Nico de Rooij

Ecole Polytechnique Fédérale de Lausanne (EPFL),
The Sensors Actuators and Microsystems Laboratory (SAMLAB), Neuchâtel,
Switzerland

We present an insulated conductive atomic force microscope (AFM) probe with Au tip. The probe is based on a previously reported Pt_xSi_y-tip AFM probe. After the fabrication of the Pt_xSi_y-tip probe, a homogeneous gold cluster layer was grown exclusively on the conductive tip apex by feedback controlled electrochemical deposition of Au in a $HAuCl_4$ based electrolyte. The size of the gold clusters was around 2~5 nm in diameter. A controlled deposition of approximately 10 nm-thick gold layer was successfully achieved. A typical tip radius of curvature was 40 nm. A potential application of this probe is simultaneous characterization of mechanical and electrical properties of molecules.

Measuring mechanical properties of molecule junctions and simultaneous recording of their electric response is one of the key techniques to advance researches on molecular electronics. One of the promising experimental setups is based on atomic force microscope with a conductive probe (CP-AFM). A typical configuration is shown in Fig. 1. This setup enables to measure both a mechanical force applied on the tip and an electrical charge transport though the tip. To implement such a setup, there are, however, several technical difficulties. For example, a complex fabrication process is required to fabricate an insulated conductive probe. It is also challenging to attach an electrochemically controlled molecular switch in between the apex of the tip and the substrate under such AFM configuration. In this paper, we present a gold-coated insulated conductive AFM probe based on the previous work [1].

To efficiently attach a molecule, gold is a preferred metal, e.g., molecules with thiol groups can bind to gold surfaces [2]. However, the use of gold is often prohibited in a cleanroom environment. Only a few batch fabricated probes with gold tip has been reported [3]. Our probe, schematically shown in Fig. 1, was realized by micro fabrication and subsequent electrochemical deposition of gold. In an ideal setup, a molecular is attached between the apex of the tip and the substrate. If a voltage is applied, a current passing through the conductive tip can be measured and, at the same time, a force applied to the tip can be measured by deformation of the cantilever. It is important that the interconnection to the conductive tip of the cantilever is insulated all the way to the apex.

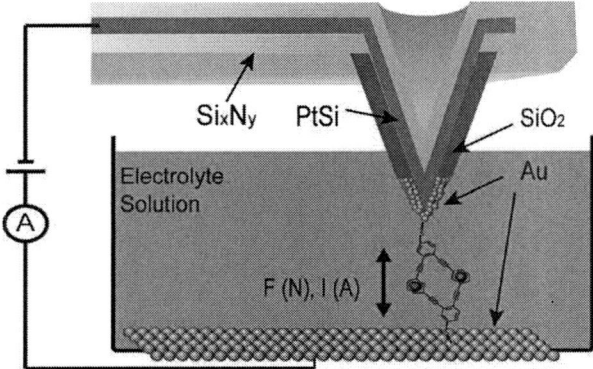

Figure 1. Schematic view of the insulated CP-AFM probe, which is intended to be used for mechanical and electrical measurements with molecular junctions.

The conductive probe in this study is based on a previously published platinum silicide probe [1]. These newly fabricated probes are shown in figure 2. The silicon nitride cantilever has an embedded electrical contact to the tip apex (Pt_xSi_y). The tip [Fig. 2(b)] located at the end of the Si_3N_4 cantilever, is covered with a passivation layer of silicon oxide. The passivation ensures that the sensing probe has a durable insulation. Fig. 2(c) shows a close-up view of the tip area, where the Pt_xSi_y apex with approximately 70 nm in diameter is protruding from the insulation.

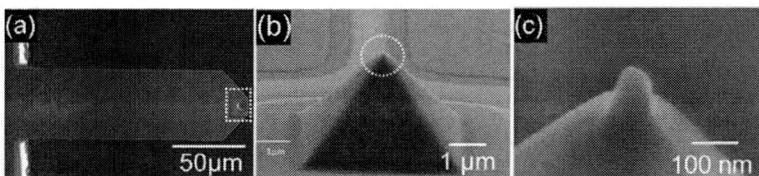

Figure 2. SEM image of (a) conductive probe, (b) the SiO_2 insulated Pt_xSi_y tip and (c) the exposed tip apex.

The post-processing of the insulated conductive probe was a controlled electrochemical deposition of gold. A schematic of the electrodeposition setup is shown in Fig. 3(a). The probe was immersed in a gold thiosulphate-sulphite solution, connected to a potentiostat in a 3-electrode setup and used in cyclic voltammetry mode. The deposition potential was swept between 0 V and -0.75 V (vs Ag/AgCl reference electrode). Fig. 2(b) shows the cyclic voltammogram of the probe. Gold deposition takes place between -0.55 V and -0.75 V, which indicates the voltage range for the following deposition experiments.

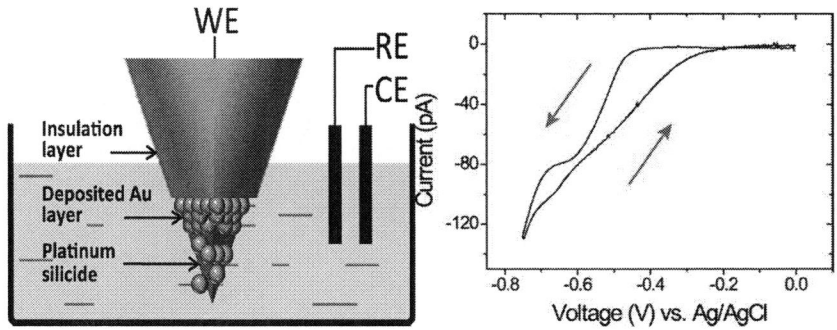

Figure 3. Schematic of electrochemical deposition on the Pt_xSi_y tip.

Results of the electrochemical deposition are shown in Figure 4. The original apex before deposition has an exposed area of 0.16 μm^2 on the tip apex. Fig. 4(b) and (c) show the apexes after cycling down to -0.55 V and -0.75 V respectively (one cycle). Fig. 4(d) shows the apex after cycling down to -0.75 V for two cycles. The cycling speed was 10mV/s. One can see that gold clusters were accumulated exclusively on the exposed platinum silicide apex. Both the deposition potential and the deposition time influenced significantly the deposition of gold clusters. The grain size increased when a higher deposition current was applied.

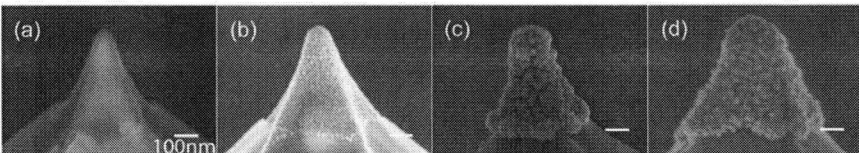

Figure 4. The opened CP-AFM tips: (a) No deposition; (b) cycling down to -0.55 V one cycle; (c) cycling down to -0.75V one cycle and (d) two cycles.

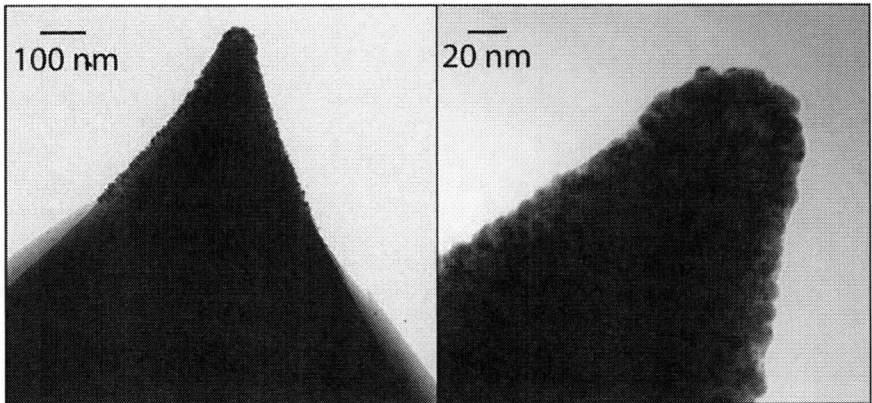

Figure 5. TEM images of modified gold coated CP-AFM tip.

Figure 5 shows TEM images of the probe in Fig. 4(b). In the left image, the light part on the side of the pyramid is the SiO_2 layer which is covering the Pt_xSi_y (darker part). Gold clusters are exclusively accumulated on the opened apex. The SiO_2-covered surface is clean. The size of the gold clusters is around 2~5 nm. The tip radius of curvature of 40 nm can be estimated from the right picture in Fig. 5. The thickness of the deposited layer is approximately 10 nm.

In summary, we developed a process to successfully deposit a 10 nm gold layer onto the Pt_xSi_y tip apex of an insulated conductive probe. A small and relatively sharp (~80 nm in diameter after gold deposition) conductive tip was obtained. The probe will be used for experiments with molecules in electrolyte solutions.

Acknowledgments

This work is financially supported by Swiss National Science Foundation (SNSF) via the Synergia project CRSII2 126969. We thank Wenjing Hong, Ilya Pobelov and Prof. Thomas Wandlowski from University of Bern for the joint work on electrochemical deposition.

References

1. T. Akiyama et al., *Japanese Journal of Applied Physics*, **43** (2004), p.3865-3867.
2. X. D. Cui et al., *Science*, **294** (2001), p.571-574.
3. P. Dobson et al., *Physical Chemistry Chemical Physics*, **8** (2006), p.3909-3914.

Characterization of Piezoresistive Microcantilever Sensors with Metal Organic Frameworks for the Detection of Volatile Organic Compounds

I. Ellern[a], A. Venkatasubramanian[a], J. H. Lee[b]
P. J. Hesketh[a], V. Stavila[c], M.D.Allendorf[c], A.L. Robinson[d]

[a]G.W. Woodruff School of Mechanical Engineering, Georgia Institute of Technology, Atlanta, GA, 30332
[b]Intel Corporation, Rio Rancho, NM, 87124
[c]Sandia National Laboratories, Livermore, CA, 94551
[d] Sandia National Laboratories, Albuquerque, NM, 87185

Metal-Organic frameworks (MOFs) with their high surface area, excellent chemical and thermal stability and analyte specific adsorption make them promising material for gas sensing/storage. In this paper, a vapor detection sensor based on HKUST-1 MOF is presented. HKUST-1 with the above mentioned properties makes it an ideal material to be used with strain based piezoresistive microcantilever sensor. Cantilever sensors which were fabricated, characterized and coated with HKUST-1 were exposed to different Volatile Organic Compounds (VOCs) including acetone, isopropanol, methanol and water at varying concentrations. From our experiments, we observe that the device is able to distinguish between the different analytes quantitatively based on their time constants for adsorption or desorption. The results show that acetone has the highest time constant while water has the lowest time constant.

Introduction

The need for a highly efficient molecular framework for applications in gas separation, sensing and storage has pushed the frontiers in the nanoporous materials research. Porous Metal-Organic Framework (MOF) has emerged as an important class of materials possessing many desirable properties expected in a molecular sieve such as tailorable permanent nanoporosity, complete desorption, high degree of chemical and thermal stability and analyte specific adsorption. A typical MOF consists of metal cations such as Zn (II) linked by anionic organic linkers groups such as carboxylates, yielding a rigid but open framework that can accommodate guest molecules. Adsorption of analyte molecules in these MOFs is governed by a number of mechanisms (1). In rigid MOFs, uptake is controlled primarily by adsorbate-pore surface interaction and steric interactions. In addition, however, some MOFs exhibit a degree of structural flexibility not observed in conventional recognition layers (1-8).

We recently demonstrated that this property can be used for chemical detection by strain-based transduction mechanisms (9). Thus, the suitability of MOFs with strain based chemical sensors like the piezoresistive microcantilever makes it an ideal candidate for chemical sensing. In this paper, we primarily focus on the results of adsorption of volatile organic compounds obtained on cantilever arrays coated with the well characterized HKUST-1 MOF. The HKUST-1 MOF was selected because of its ability to adsorb a number of species (10), high surface area and previously demonstrated suitability to strain based microcantilever sensor (9, 11). Through such a study we aim to expand the notional space of the analytes that can be used with these microcantilevers and hence obtain their adsorption properties using very small amounts of sample.

HKUST-1 has the structure of formula $Cu_3(BTC)_2(H_2O)_x$, comprises a binuclear Cu_2 paddlewheel unit (12). Its structure consists of two types of "cages" and two types of "windows" separating these cages. Large primary cages (13.2 and 11.1 Å in diameter) are interconnected by 9 Å windows of square cross section. The large cages are also connected to tetrahedral shaped secondary pockets of roughly 6 Å through triangular shaped windows of about 4.6 Å (3.5 Å in the hydrated form).

In this paper, we will briefly discuss the fabrication procedure of the microcantilever array, the MOF film deposition, the experimental setup and the characterization methods in the Methods section. We shall then focus on the response for water, methanol, isopropanol and acetone with the HKUST-1 MOF in the Results and Discussion sections.

Methods

I. Microcantilever Array Fabrication

N-doped piezoresistive microcantilever array sensors were fabricated using microfabrication techniques with dimensions 230 μm in length and 80 μm in width, as described in reference (11). The layer configuration of the device is shown in Figure 1; it was optimized for maximum response using COMSOL multiphysics modeling, the results of which will not be presented here as it is beyond the scope of the paper.

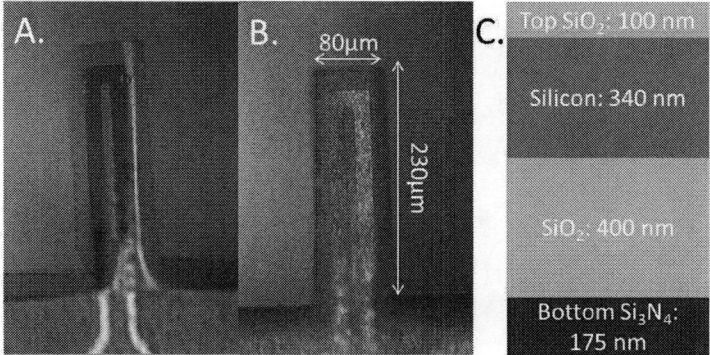

Figure 1.(A) Uncoated microcantilever sensor. **(B)** MOF coated microcantilever sensor.**(C)**Layer structureof the cantilever device.

II. Characterization

Post fabrication, each device was subjected to careful characterization by measuring the resistance of the cantilever array and collecting an optical image of the array as shown in Figure 1A. Subsequently, the devices were sent to Sandia National Laboratory for MOF film deposition. Upon receipt of the coated devices, the resistances were measured again to ensure that the device was not damaged during the deposition process and an optical image of the coated cantilever was obtained, as shown in Figure 1B. Finally, the device was wirebonded to the measurement package and loaded into the experimental setup. Device resistance was measured throughout the experimental cycle.

III. Cantilever Array Response Measurement Set up

Measurements were made in a custom test cell, where a Wheatstone bridge was used to obtain the cantilever response. Dry nitrogen was used as carrier gas and the analyte concentrations were regulated using a hydrator. A high flow rate mass flow controller (MFC1) was used in conjunction with a low flow rate MFC (MFC4) to introduce a range of diluted concentrations of Volatile Organic Compounds (VOCs) to the device chamber (Figure 2). The response was obtained by measuring the voltage across the Wheatstone bridge. The bridge consisted of two known resistors and two MOF coated microcantilevers in an arrangement which added response from the two devices.

All experiments were conducted at room temperature (23 °C) and at atmospheric pressure. MOF film was activated by removing adsorbed water by flow of nitrogen for 900s at 40 °C prior to beginning of the experiment. Higher temperatures are necessary to ensure complete removal of water, however, physical limitations such as the integrity of the self assembled monolayers (SAMs) coatings on gold and the physical integrity of the cantilever limit the temperature of activation. Our previous experience showed failure of the SAM coating at temperatures above 50 °C at atmospheric pressure. Temperatures exceeding 190 °C at low vacuum are required to completely dehydrate the MOF (13). Change in voltage was used to ensure a steady state was reached before introducing the first analyte.

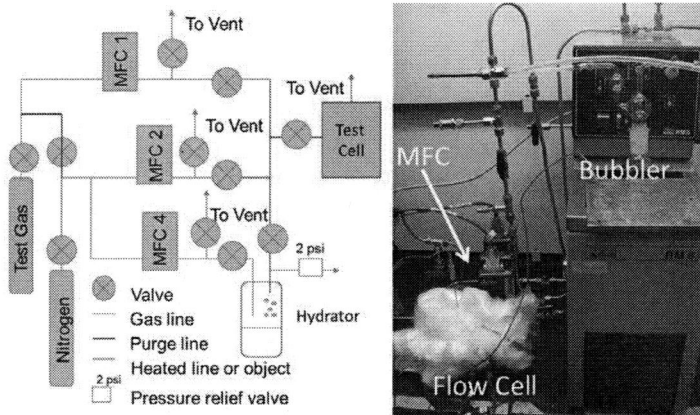

Figure 2. Three MFC's allow for testing gasses, VOC's and mixtures of both. MFC1 is dedicated to carrying purging gas at high flow rate, MFC2 for delivery gases and MFC4 for delivery of VOC's. Gas lines and the hydrator have the capability to be heated.

IV. HKUST-1 Thin film deposition

The MOF thin film was deposited by a procedure similar to that described in (11,14). Since the deposition procedure had been adequately validated before (11), only optical images of the MOF coated cantilever (Figure 1B) were obtained to ensure the presence of MOF film on the cantilevers. The MOF is coated onto one side of the beam.

Results and Discussion

Previously we presented our response for water, chloroform and hexane for the HKUST-1 MOF (15). Following the encouraging signs we have expanded our study to other VOCs including methanol, isopropanol and acetone and compared it with the adsorption measurements from water. In this section we will discuss our results.

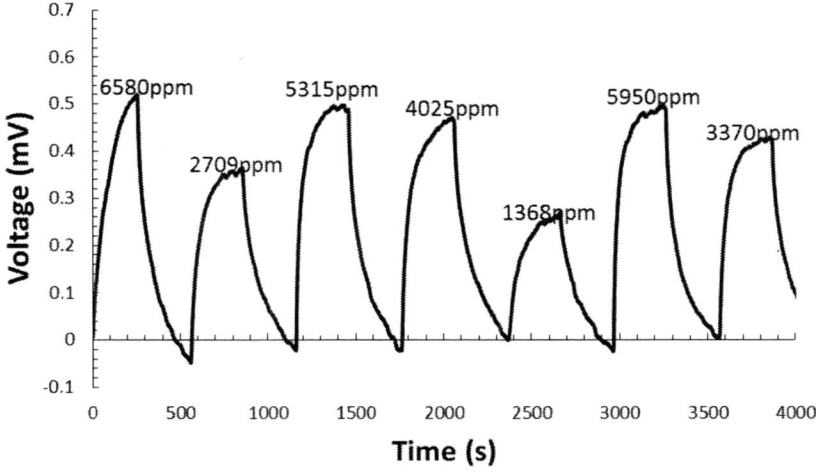

Figure 3. Response to methanol with HKUST-1 at 23°C.

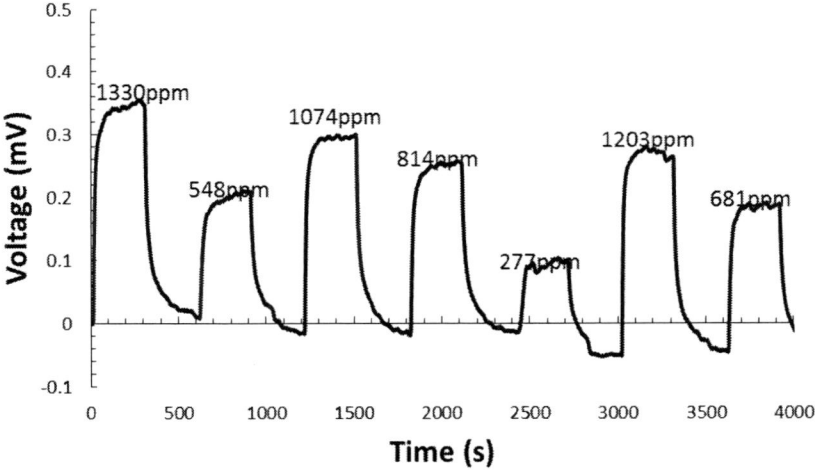

Figure 4.Response to water with HKUST-1 at 23 °C.

Figure 3 shows a sample of 4000 seconds of the transient response with a HKUST-1 coated cantilever for methanol in our experiment. Figure 4 shows similar response to water. Experiments started with introduction of analyte vapor into the flow cell (at 0 s) at a known concentration. The analyte vapor flow was halted once equilibrium was reached and nitrogen purge was started. Thus the cycle was repeated for different concentrations in a random order. This cyclic procedure remained same for all the analytes. From the data in Figures 3 and 4, it is evident that the sensor returns to a rough base line, suggesting reversibility may be occurring. From such transient response data for analytes including water, methanol, isopropyl alcohol and acetone, we have obtained the time constants for exposure and return to a dry nitrogen purge.

In this paper we have defined the time constant as the time taken to reach 63.2 % of the equilibrium value either during adsorption or desorption for approximately the same concentration. From Table I we see that the time constants vary significantly for each analyte for approximately the same concentration. We also observe that the desorption time constant is higher than the adsorption time constant. While the exact mechanism for this behavior is not known, we feel this may be due to the strong Van der Waals forces and hydrogen bonding between the analytes and the open MOF metal centers available for adsorption. However it should also be noted that the MOFs may not be fully dehydrated because of physical limitations imposed by the deposition procedure and the cantilever structure. To further understand the observed behavior of time constants between analytes, we must consider the complex geometry of the HKUST-1 MOF in conjunction with the physicochemical properties of the analytes listed in Table II.

Based on the geometry of the MOF and the physicochemical property of the analytes, the limiting dimension for transient adsorption/desorption process would be the size of the triangular shaped windows connecting the large primary cages with the tetrahedral shaped secondary pockets. Water being smaller than this triangular shaped windows (~4.6 Å) would pass through this passage with relative ease compared to methanol. The reason for higher time constants for methanol compared to water is currently unknown. Comparing this with the large analytes (isopropanol and acetone) whose kinetic diameters are larger than this window, would imply that these large analytes would mainly be adsorbed in the primary cages, hence resulting in lower adsorption. However as with methanol and water, the reason for the difference in time constants between isopropanol and acetone is not completely clear.

TABLE I. Time constants for HKUST-1 coated microcantilever sensors at 23 °C.

Analyte	Time Constant (s) (Adsorption)	Time Constant (s) (Desorption)	Concentration (ppm)
Water	14.1	24.6	2539
Methanol	44.8	86.7	2709
Isopropanol	17.5	19.6	2559
Acetone	56.7	81.3	2930

TABLE II. Physicochemical properties of fluids used in this study

Analyte	Kinetic Diameter (Å)	Dipole Moment (D)
Water	2.65 (16)	1.85 (16)
Methanol	3.8 (17)	1.7 (16)
Isopropanol	4.7 (18)	1.7 (19)
Acetone	4.7 (20)	2.9 (19)

Figure 5 shows the equilibrium response in terms of differential voltage versus analyte concentration for different analytes on the HKUST-1 coated cantilever. From the figure we observe that in general the vapor uptake is highest for water, followed by methanol, isopropanol and finally acetone. The reason for this behavior can be traced back to important physicochemical properties like kinetic diameter and dipole moment as listed in Table II. From Table II, we observe that as the kinetic diameter of the analyte increases, the equilibrium adsorption generally decreases. This is consistent with the fact that as the size of the analyte increases, fewer molecules can be accommodated into the pore framework.

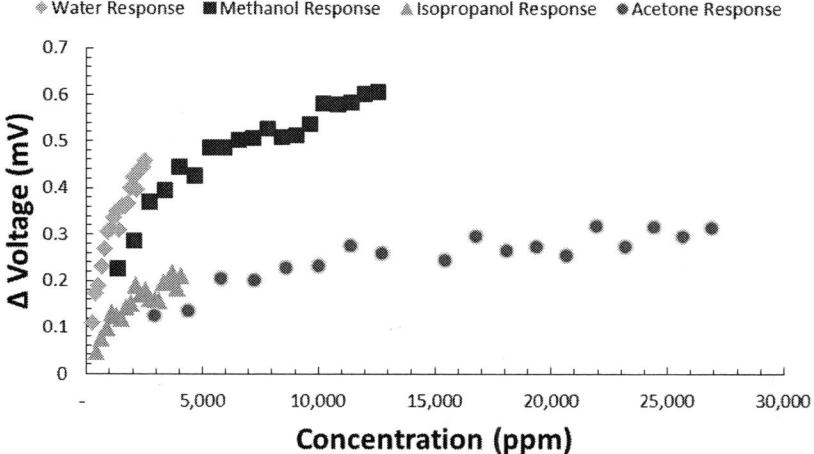

Figure 5. Response to VOCs with HKUST-1 coated cantilevers at 23 °C.

Conclusions

This work has demonstrated chemically induced strain based detection utilizing HKUST-1 MOF on microfabricated cantilever sensors. The characteristic response to four analytes, specifically, acetone, isopropanol, methanol and water provides some degree of discrimination based upon response time. From our experiments with different VOCs and water, we gained a fundamental insight into the adsorption process of these analytes and our preliminary experiments prove that it is possible to obtain quantitative parameters relevant to the adsorption process with very small amounts of MOF sample. However, we do agree that further improvements can be applied to our setup to obtain the thermodynamic parameters of adsorption. The compatibility of HKUST-1 MOF with strain based piezoresistive microcantilever sensors is encouraging for us to explore the detection of a mixture of analytes with an array of microcantilevers functionalized with different MOFs to take advantage of their respective chemical selectivity.

Acknowledgments

This work was supported by the Sandia Laboratory Directed Research and Development (LDRD) Program. The technical assistance of Gary Spinner at the Nanotechnology Research Center is gratefully acknowledged.

References

1. J. R. Li, R. J. Kuppler, H. C. Zhou, *Chem. Soc. Rev.*, **38**, 1477-1504(2009).
2. D. S. Coombes, F. Cora, C. Mellot-Draznieks, R. G. Bell, *J. Phys. Chem. C*, **113**, 544-552(2009).
3. T. Devic, P. Horcajada, C. Serre, F. Salles, G. Maurin, B. Moulin, D. Heurtaux, G. Clet, A. Vimont, J. M. Greneche, B. Le Ouay, F. Moreau, E. Magnier, Y. Filinchuk, J. Marrot, J. C. Lavalley, M. Daturi, G. Ferey, *J. Am. Chem. Soc.*, **132**, 1127-1136(2010).
4. A. J. Fletcher, K. M. Thomas, M. J. Rosseinsky, *J. Solid State Chem.*, **178**, 2491-2510(2005).
5. P. K. Thallapally, *Abstr. Pap. Am. Chem. Soc.*, **238** (2009).
6. K. Uemura, R. Matsuda, S. Kitagawa, *J. Solid State Chem.*, **178**, 2420-2429(2005).
7. L/ E/ Kreno, K. Leong, O. K. Farha, M. Allendorf, R. P. Van Duyne, J. T. Hupp, *Chem. Rev.*, **112**, 1105-1125(2011).
8. M. D. Allendorf, A Schwartzberg, V. Stavila, A. A. Talin, *Chem.-Eur. J.* **17**, 11372-11388(2011).
9. M. D. Allendorf, R. J. T. Houk, L. Andruszkiewicz, A. A. Talin, J. Pikarsky,A. Choudhury, K. A. Gall, P. J. Hesketh, *J. American Chemical Society*, **130**, 14404-14405(2008).
10. S. Bordiga, L. Regli, F. Bonino, E. Groppo, C. Lamberti, B. Xiao, P.S. Wheatley, R. E. Morris, A. Zecchina, *Physical Chemistry Chemical Physics*, **9**, 2676-2685 (2007)
11. J. H. Lee, R. J. T. Houk, J. A. Greathouse, M. D. Allendorf, P. J. Hesketh, *Hilton Head Workshop 2010: A Solid-State Sensors, Actuators and Microsystems Workshop*, Hilton Head, SC, June (2010).
12. D. Farruseng, C. Daniel, C. Gaudillere, U. Ravon, Y. Schurmann, C. Mirodatos, D. Dubbeldam, H. Frost, R. Q. Snurr, *Langmuir*, **25**(13), 7383(2009)
13. A. Venkatasubramanian, J. H. Lee, R. J. T. Houk, M. D. Allendorf, S. Nair, P. J. Hesketh, *Fall Meeting of the ECS*, Las Vegas,October (2010).
14. V. Stavila, J. Volponi, A.M. Katzenmeyer, M. C. Dixon, M. D. Allendorf, *Chemical Science, 3*, 1531–1540(2012).
15. I. Ellern, A. Venkatasubramanian, J. H. Lee, P. J. Hesketh, M. D. Allendorf, *Fall Meeting of the ECS*, Boston, MA, October (2011).
16. J. E. ten Elshof, C.R. Abadal, J. Sekulic, S. R. Chowdhury, D. H. A. Blank, *Microporous and Mesoporous Materials*, **65**, 197-208 (2003).
17. Q. Zhu, J. N. Kondo, S. Inagaki, T. Tatsumi, *Topics in Catalysis*, **52**, 1272-1280 (2009)
18. X. Qiao, T. S. Chung, W. F. Guo, T. Matsuura, M. M. Teoh, *Journal of Membrane Science, 252*, 37-49 (2005)
19. B. E. Poling, J. M. Prasznitz, J. P. O'Connell, *The Properties of Gases and Liquids*, p. A.20-A.34, McGraw-Hill, New York (2001).
20. H. Zhou, Y. Su, X. Chen, Y. Wan, *Separation and Purification Technology*, **79**, 375-384 (2011).

ECS Transactions, 50 (12) 477-485 (2012)
©The Electrochemical Society

Manipulation of Micro Condensed Matter by Direct Peeling Method by using Atomic Force Microscope Tip

A. Kawai

Department of Electrical Engineering, Nagaoka University of Technology, Nagaoka, Niigata, 940-2188 Japan

Understanding manipulation properties of polymer nanoscale pattern is of crucial importance for development of not only MEMS/NEMS devices but other functional devices in the nanometer scale. Quantitative analysis of peel and manipulate properties of an ArF dot resist pattern ranging from 141 to 405 nm diameter and 360 nm height is demonstrated experimentally. By directly applying a certain load to top corner of resist pattern with a micro cantilever tip, a resist dot pattern can be peeled easily from a substrate accompanying slight residue formation. The load required for pattern peel decreases with decreasing the pattern diameter. In combination with the analysis of internal stress distribution in the resist pattern, an optimum condition for successful manipulation condition can be obtained. The rearrangement of 141nm diameter pattern can be demonstrated by the tip manipulation technique.

1. Introduction

In recent years, with the increasing degree of miniaturization in microelectronics, we unquestionably become confronted with structures of condensed matter on nanometer scale. Moreover, with increasing integration density of functional MEMS/NEMS devices, condensation enhancement of polymer patterns has been recognized as an important problem to be solved because of pattern collapse during development process. In this regard, in lithography process, photoresist is regarded as an important mask material for dry or wet etching processes. (1,2) In these days, various studies on peeling phenomena of photoresist material have been accomplished by several researchers.(3-5) A great deal of effort has been made on monitoring the peeling property. What seems to be lacking, however, is improvement of control accuracy. Meanwhile, with the invention of the atomic force microscope (AFM)(6), we have been equipped in good time with the appropriate tools to manipulate a condensed matter on a nanometer scale.(7-10)

In this regard, the present author has already proposed the novel principle for direct analysis method for resist pattern adhesion, that is, direct peeling with AFM tip (DPAT).(11-14) By this method, the dependency of the load for peel of the KrF excimer resist micropattern on linewidth and pattern shape are shown.(13,14) In this paper, peel and fracture mechanism of the dot shape patterns ranging from 141 to 405 nm in diameter are characterized by the DPAT method. Particularly, re-arrangement of these patterns are demonstrated experimentally under the optimized manipulation conditions.

2. Experiment

2.1 Sample preparation

An ArF chemically amplified positive resist consisting of acryl resin as a base polymer was used. The resist film 360 nm in thickness was coated onto a bottom antireflective coat (BARC) layer by a spinning method. The pre-baking treatment was carried out at 90 ℃ for 60 s on a hot plate. The silane-coupling treatment with hexamethyldisilazane (HMDS) as an adhesion promoter was performed at 90 ℃ for 60 s. The dot shapes ranging from 141 to 405 nm in diameter were imaged to the resist film by using an ArF excimer laser stepper with irradiation energy of 35 mJ/cm^2 and then were developed by dipping into tetramethylammoniumhydroxide (TMAH) 2.38 % aqueous solution for 60 s. Subsequently, the resist patterns were rinsed in the deionized water, and dried by the spin method. In order to observe the resist patterns, a scanning electron microscope (SEM; S-5000 made by HITACHI Ltd.) was used. Figure 1 shows SEM photographs of ArF resist patterns fabricated on the BARC layer. Aspect ratio of resist pattern was approximately ranging from 2.55 to 0.89. It seems that the pattern top shows slightly T-shaped and the pattern bottom at which taper shape due to chemical reaction with BARC layer can be observed.

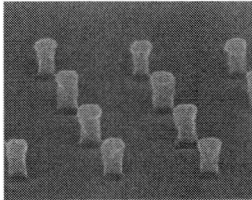 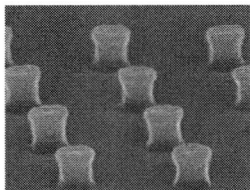

141nm diameter   241nm diameter

Fig.1 SEM images of the dot patterns of ArF excimer resist of 360 nm height.

2.2 Pattern peel investigation

An AFM, a commercially available version, integrated with a micro tip was used for the peel investigation. Figure 2 shows a SEM photograph of the tip apex. The curvature radius of tip apex is approximately 8 nm. The tip was made of Si single crystal by lithography technique. The torsion spring constant was 2200 N/m. The cantilever torsion by contacting to a resist pattern was detected experimentally using a laser reflection system. The forces in torsional direction are determined by multiplying the measured cantilever displacement by the torsional spring constant of the cantilever. Figure 3 shows the schematic explanation of DPAT concept.

- (a): Prior to the peel experiment, the resist dot-pattern was imaged in the non-contact mode by which no pattern peel occurs. Subsequently, the tip traversed the dot-pattern under the vector scan control.
- (b): By directly applying load to the top corner of resist pattern with a micro cantilever tip, a resist dot pattern adhering on a substrate can be peeled easily. In this study, the diameter dependency of dot pattern on peel property is focused.
- (c): The load required for peeling the pattern can be regarded as the meaningful value which applies the manipulation of the resist pattern from the substrate.

In the manipulation investigation, the applied torsion force on the Si cantilever had to be set to a value above 15 μN for successful peel of the resist pattern, thereby directly indicating the limit of the sample cohesion which depends on the tip geometry as well as the elastic and plastic properties of the resist pattern.

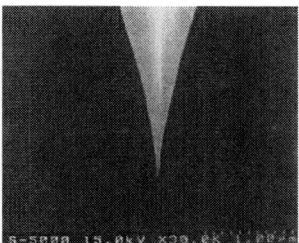

Fig.2 SEM image of the tip apex (magnification of 30000). The tetrahedral shape tip is made by lithography.

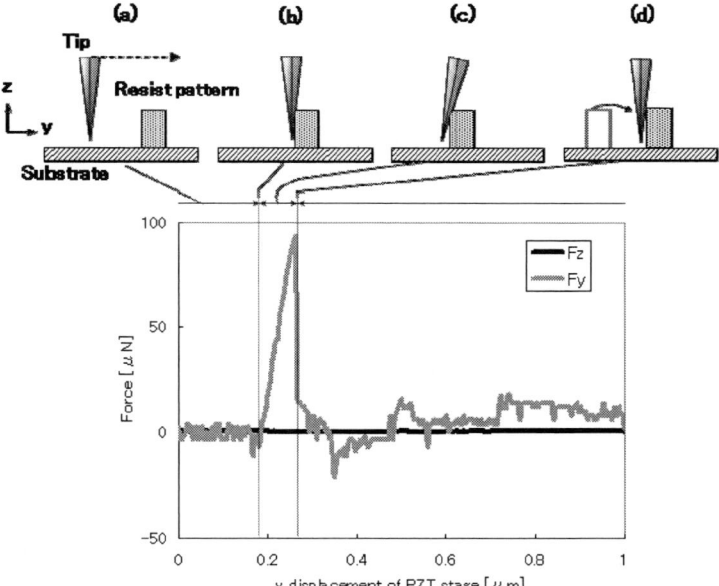

Fig.3 Schematic explanation of DPAT process. Tip movement was controlled in vector scan program and monitored simultaneously.

The pattern peel and residue formation can be reconfirmed by imaging the sample in the noncontact mode. Neither the AFM tip nor the dot resist pattern appeared to be damaged as a result of the surface scanning procedure. Five pieces of dot patterns were used for peel investigation for each size. All investigations were carried out in a dry atmosphere (23 °C, 4 %RH), because water vapor is known to coat surfaces with several monolayers after a brief exposure to ambient conditions. In the vapor ambient, the tip-sample adhesion force can lead to the formation of a thin continuous meniscus neck by slightly retracting the tip from the sample surface. Details of DPAT method and the instrumentation are described elsewhere.(13,14)

3. Results and Discussion

3.1 Diameter dependency of pattern peel

Figure 4 shows the experimental result with regard to the peel property for these differently diameter pattern. The load for peel decreases gradually as pattern diameter decreases. The standard deviation of peel load is considerably low. No peel occurs for the dot pattern of diameter over 350 nm. These peel property is almost same to those for KrF line resist pattern adhering on a Si substrate in our previous report.(13) These peel phenomena can be analyzed in detail on the point of residue formation, in the following section.

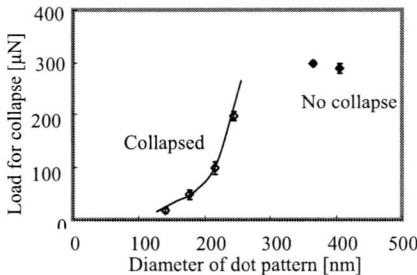

Fig.4 Dependence of load for peel on diameter of dot pattern.

3.2 Mechanism of residue formation

Figures 5(a) to 5(d) show the AFM images of resist residue formed on the substrate surface after pattern peel. In Fig.5(a) to 5(c), a considerable residue of resist material, approximately 100nm thickness, can be observed. Therefore, in the case of pattern diameter from 141 to 216 nm, one can safely state that the pattern peel occurs in the cause of insufficient cohesion of the resist material. On the contrary, slight residue around 20nm thickness can be observed for large pattern in Fig.5(d). In this case, the residue thickness at center is closely zero, therefore, it can be considered that the pattern destruction occurs in mixture of interface and cohesion destruction. In this way, we can assume from residue observation that the destruction mechanism at the interface should be different with pattern diameter variation. Usually, some information concerning the initiation of the fracture can be obtained from the direct investigation of the fractured surfaces. In a micro condensed material, a fracture is initiated, practically without exception, on the surface, usually at some type of surface defect which can serve as a precursor flaw. However, as seen in Fig.5, the fractured surface of the dot pattern is relatively smooth. This is in agreement with the general theory of crack propagation, which puts forth that a low crack propagation velocity caused by a low fracture limit and a correspondingly low acceleration of the crack propagation front results in a smooth fracture. The destruction mechanism for micro-dot pattern is discussed based on stress concentration in the next section.

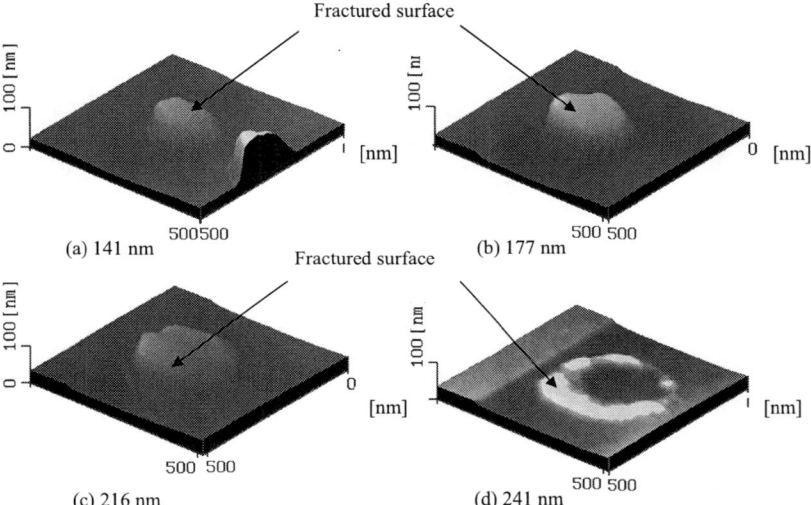

Fig.5 AFM images of resist residue formed on the substrate surface after collapsing the pattern.

3.3 Stress distribution in dot pattern

The diameter dependency of dot pattern peel can be analyzed on the point of stress concentration induced by exerting a certain load on the pattern. The internal stress distributed in resist pattern can be estimated by three dimensional finite element method (FEM). Figures 6(a) and 6(b) show the pattern model for FEM analysis. The resist dot pattern is divided into about 960 pieces of elements. By symmetry, the stresses in the resist element are isotropic. These are the basic assumptions made in performing FEM analysis as follows. (i): The physical properties of material are linearly related to external loads. This is indicative of elastic behavior. (ii): The resist/substrate interface is defined as fixed points, that is, no deformation of the substrate. Young's modulus and poisson ratio of resist material used for the calculation are 1 GPa and 0.33, respectively.(13) The loading position of the AFM tip is indicated by an arrow.

The amount of the internal stress can be obtained based on von-Mises stress S defined as a following equation.

$$S = [\ 1/2*\{(P_1 - P_2)^2 + (P_1 - P_3)^2 + (P_2 - P_3)^2\}]^{1/2}, \tag{1}$$

where the symbols P_1, P_2, P_3 represent principal stress in each element.

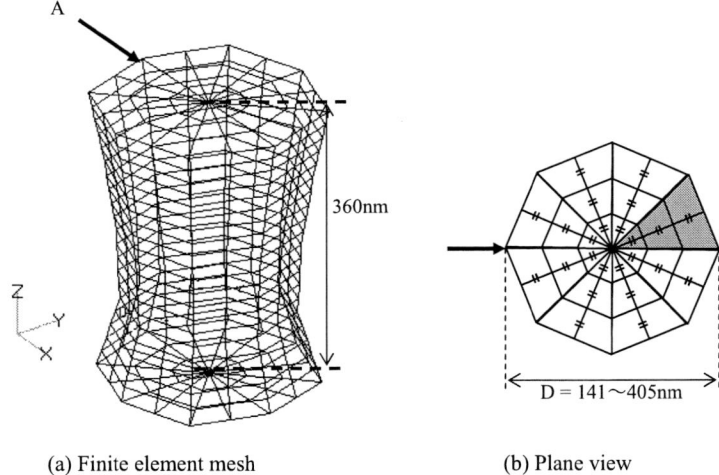

(a) Finite element mesh   (b) Plane view

Fig.6 Analysis of stress distribution in the dot pattern by FEM.

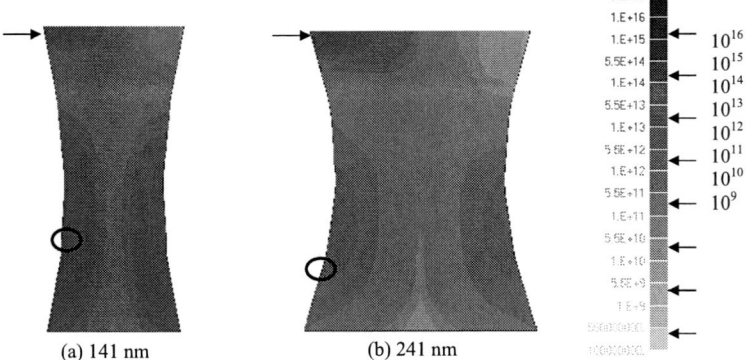

(a) 141 nm   (b) 241 nm

Fig.7 Cross sectional image of stress distribution for each different pattern diameter.

Figure 7 shows stress distribution in sectional view for each different pattern diameter. In all cases, maximum stress concentrates mainly at the position corresponding to each loading position. In the model of diameter 141nm, it was shown that the internal stress in the pattern concentrates at the upper region (indicated by open circle) slightly away from the resist-substrate interface. This is because that the considerable residue is formed on the substrate as seen in Figs. 5(a) to 5(c). However, in the model of diameter greater than 241nm, stress around the pattern bottom (indicated by open circle) is relatively released. Therefore, it can be explained that a relatively slight residue can be formed for wider pattern as seen in Fig.5(d). These simulation results are in good agreement with the peel properties of the dot patterns. In this regard, Fig.8 shows the von Mises stress concentrated at pattern bottom depending on pattern diameter. As decreasing the pattern diameter, the stress concentration is considerably large. The peel phenomenon of dot pattern as seen in Fig.4 reflects strongly the stress concentration property.

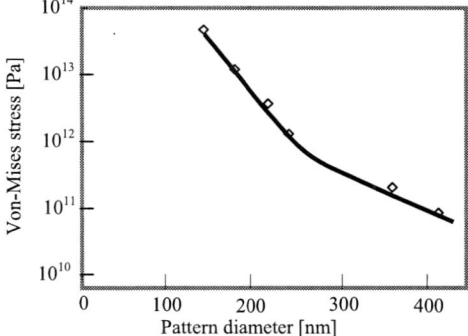

Fig.8 Dependency of von-Mises stress concentrated at pattern bottom on pattern diameter.

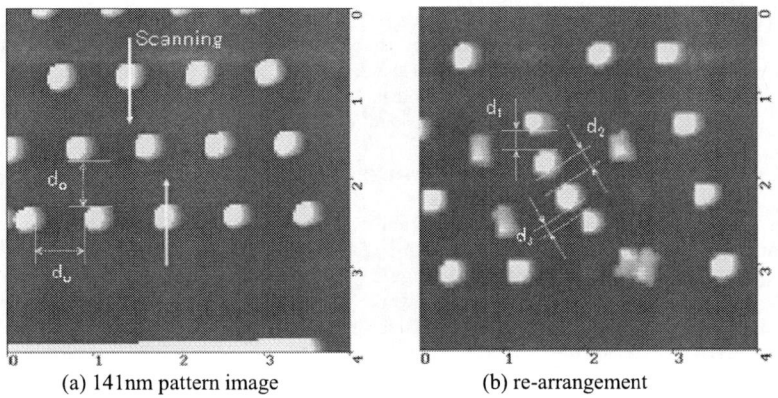

(a) 141nm pattern image   (b) re-arrangement

Fig.9 Re-arrangement of polymer pattern after pattern manipulation

In this way, the stress concentration occurring in the vicinity of resist-substrate interface can be regarded as one of major factors explaining peel behavior. Moreover, the stress concentration effect of applying load is expected to have some influence on the manipulation limit. In order to improve the measurement accuracy of the DPAT method, the following issues should be taken into consideration: (i) Plastic deformation and friction resulting from the contact between the cantilever tip and the resist pattern surface should be clarified. (ii) The actual contact area between the tip and the resist pattern surface should be estimated quantitatively. (iii) The possibility that the presence of a contamination layer between the cantilever tip and the resist pattern reduces the interaction.

Figure 9 shows the re-arranged dot pattern image of micro dot patterns of 141nm diameter after manipulation procedure. The peeling condition above mentioned was employed for the manipulation. The certain resist residue can be confirmed. In this way, by the direct peeling method with AFM tip, the nanoscale pattern displacement can be designed flexibly. Furthermore, this method gives meaningful information to material design for condensation enhancement. We believe that the DPAT method in this study can be applied to analysis of a nanoscale polymer pattern condensation less than 15nm width in near future.

Conclusion

The novel technique, DPAT method, for analyzing resist pattern peel is discussed and the pattern manipulation is demonstrated. The validity and reliability of this technology can be confirmed. In particular, present authors focused on the diameter dependency of the resist pattern peel on the 141 to 405 nm scale. The followings are obtained as the major conclusion.
1. The diameter dependency of peel behavior can be detected by this method quantitatively. The load for peeling pattern decreases considerably as decreasing the pattern diameter.
2. The resist pattern peel is dominated by the stress concentration in resist pattern. The diameter dependency of pattern peel can be explained clearly by analyzing the stress distribution in the dot pattern.
3. The residue formation after pattern peel can be also explained clearly on the point of stress concentration in the vicinity of pattern bottom.
This technique will give useful information to MEMS/NEMS devices construction and structural design for quantum effect devices and so on.

Acknowledgements

This work was partially supported from the Grant-in-Aid for Scientific Research from the Ministry of Education, Science, Sports, Culture and Technology, Japan, and by Program for High Reliable Materials Design and Manufacturing in Nagaoka University of Technology. The present work was partially supported by Grant-in-Aid for Scientific Research, Grant-in-Aid for Scientific Research (JSPS KAKENHI B:23360150) and Grant-in-Aid for challenging Exploratory Research (MEXT KAKENHI 23656235).

References

1. K. Deguchi, K. Miyoshi, T. Ishii, and T. Matsuda, *Jpn. J. Appl. Phys.* **31**, 2954 (1992).
2. T. Tanaka, M. Morigami, and N. Atoda, *Jpn. J. Appl. Phys.* **32**, 6059 (1993).
3. J. Y. Yu, et al., Proc. SPIE's 23rd Inter. Symp. *Microlithography*, Santa Clara, vol.3333 p880 (1998).
4. K. L. Mittal, *Polym. Eng. & Sci.*, **No.7**, 467, (1977).
5. A. Kawai, H. Nagata and M. Takata, *Jpn. J. Appl. Phys.* **31**, 3725 (1992).

6. G. Binnig, C. F. Quate and Ch. Gerber, *Phys. Rev. Lett.*, **56**, 930 (1986).
7. S. Johansson, J. Schweitz, L Tenerz, and J Tiren, *J. Appl. Phys.* **63**, 4799 (1988).
8. C. J. Wilson, A. Ormeggi, and M. Narbutovskih, *J. Appl. Phys.* **79**, 2386 (1996).
9. M. J. Matthewson, *Appl. Phys. Lett.* **49**, 1426 (1986).
10. D. Tomanek, G. Overney, H. Miyazaki, S. D. Mahanti, and H. J. Guntherodt, *Phys. Rev. Lett.* **63**, 876 (1989).
11. A. Kawai, and Y. Kawakami, Ext. Abstr. 44th Spring Meet. Japan Society of Applied Physics and Related Societies, Chiba, March, p.576 (1997)[in Japanese].
12. A. Kawai, *J. Vac. Sci. Techol.* **B17**, 1090 (1999).
13. A. Kawai, Proc. SPIE's 24th Inter. Symp. *Microlithography*, Santa Clara, vol.3677 p565 (1999).
14. A. Kawai, *Jpn. J. Appl. Phys.* **39**, 1426 (2000).

486

A MEMS-based Platform for Multi-physics Characterization of Ultra-thin Freestanding Films

M. A. Haque, S. Kumar and M. T. Alam

Department of Mechanical & Nuclear Engineering, The Pennsylvania State University, University Park, Pennsylvania 16802, USA

Micro-electro-mechanical systems (MEMS) based techniques can outperform conventional materials characterization tools in terms of specimen size and resolution. In addition, multi-functioning as well as access to analytical microscopy is also feasible for complete multi-physics characterization. We present the design and fabrication of a versatile tool that can perform mechanical, electrical and thermal characterization of nanoscale freestanding thin films. The tool is smaller than 3mm x 5mm and is compatible with virtually all types of analytical chambers. This feature allows 'in-situ' studies inside electron microscopes for real time acquisition of composition, microstructure and defect evolution and dynamics data. The unique advantage of such simultaneous acquisition of quantitative and qualitative data can be realized through accurate and quick 'observation-based' modeling of materials behavior. We present preliminary studies on multi-physics, as well as single domain characterization to demonstrate the novel experimental technique.

Introduction

Thin films are prevalently used in micro and nano electronics as well as sensors, actuators and energy conversion devices. The literature shows that their physical properties are thickness dependent and below a certain length-scale, even the intrinsic properties (such as Young's modulus, melting temperature) exhibit breakdown from the classical laws. For example, the primary mechanical deformation mechanism in bulk metals is switched from dislocations to diffusion at the nanoscale (1, 2), which gives rise to unpredictable fracture and fatigue behavior. Enhanced diffusion is further influenced by the large surface area to volume ratio, which results in predominance of surface and interfacial effects even at room temperature. In addition, phase transformation (conventionally known to take place in alloys only) can be driven by surface energy at the nanoscale in pure materials. This is particularly interesting since not only the mechanical properties change, the new phases imply significant changes in electrical and thermal conductivity, which are not seen or predicted in bulk materials. Similarly, when the specimen size is comparable to the electron and phonon mean free path, not only the electrical and thermal properties change, they are also influenced by the diffusion controlled mechanical deformation. Or in other words, the predominance of 2D defects and micro-structural features over conventional 1D counterpart makes physical properties of nanoscale materials sensitive to mechanical deformation. This gives rise to the concept of 'strain engineering'(3, 4), which has been applied to electron mobility, thermal conductivity, phase change etc. physical phenomena.

With the current trend in miniaturization, fundamental understanding in such multi-physics coupling is essential for the design of future devices. Even though mechanical strain influences defects, interfaces and grain boundaries as well as the lattice dynamics, *essentially all existing studies* on heat or current flow involve solids under either intrinsic strain or with *static* micro-structural (grain boundaries, dislocations and voids) features (5-7). The motivation for this research comes from the observation that only a few studies consider this, primarily computationally (8-13). Experimental efforts are appearing in the literature only very recently (14-16). It is very important to note that most of these studies use compressive deformation, which 'closes' up the defects rather than 'opening' (as in tensile mode) them. Compressive deformation is thus only expected to show effects of elastic constants or sound speed (requires unrealistic pressure) to see any appreciable changes (10, 15, 17). Moreover, the mechanics of deformation is not explicitly considered in classical models, which results in an inordinate difference between theory and experiments.

The current paradigm in thin films characterization remains to be essentially single domain, which makes multi-physics studies difficult. Therefore a paradigm shift from single to multi-domain (coupling) studies is needed. Even that is not sufficient because simultaneous or 'in-situ' characterization of materials composition, microstructure, defect evolution, surface and interfacial dynamics, crystalline phases is also necessary to capture, quantify and model the new physics behind the hypothesized strain-thermal conductivity coupling. A strong candidate for this purpose is the transmission electron microscope (TEM), where a probing electron beam transmits through the electron transparent specimen, creating differential contrasts for imaging microstructural (defects, phases, precipitates and interfaces) with up to atomic resolution, identify crystalline phases through electron diffraction and microanalysis of materials electron spectroscopy (18). Developments in the electron detectors, spectrometers and digital imaging and electron scattering theories have made the modern TEM is a *complete analytical tool* for materials research (19).

Ideally, one would like to perform in-situ tests in TEM, which would provide both quantitative (stress, strain, lattice parameter, elemental composition, defect density, thermal conductivity) and qualitative (direct visualization of microstructures, interfaces, defects) information at the same time. Because one can see through the specimen with atomic resolution while measuring properties, the underlying physics can be modeled accurately and quickly. Therefore, no other analytical microscope is as effective as the TEM for bridging nanoscale multi-physics experiments with theory. However, the TEM chamber is extremely small for conventional characterization tools. The standard TEM specimen is supported by a 3mm diameter grid. The absence of mechanical, electrical or thermal probing tools on a TEM holder makes conventional microscopy essentially a 'post-mortem' study. Drastic miniaturization is therefore needed to integrate freestanding ultra-thin specimens with sensors and actuators (for mechanical stress, strain and thermal conductivity) in a 3mm x 3mm x 0.5mm envelope.

Design and Fabrication of a MEMS Multi-physics Tool

Figure 1 shows the scanning electron micrographs with superimposed schematic diagrams to illustrate the novel MEMS-based multi-physics characterization concept, where uniaxial tension is applied and measured on the specimen, while four integrated micro-probes interrogate it thermally (using the 3-ω technique). The specimens are electron transparent, which allows us to measure (and/or visualize) virtually all aspects of microstructures and defects through three (imaging, diffraction and energy loss spectroscopy) operating modes of the TEM (20). Such 'in-situ multi-physics' tool does not exist yet in the literature.

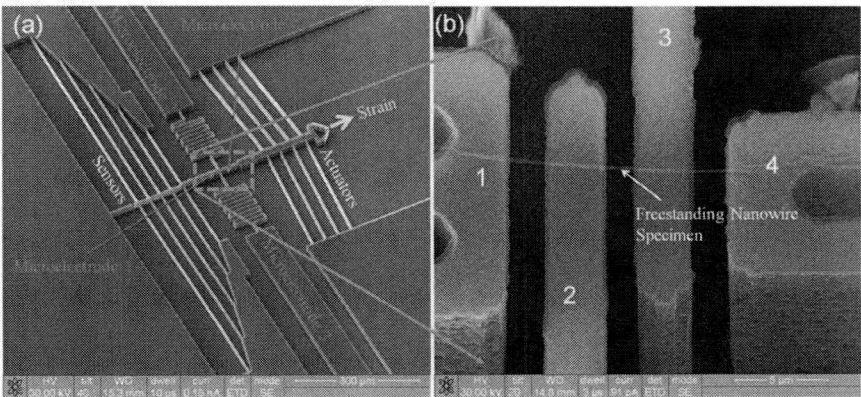

Figure 1. (a) Scanning electron micrograph of a MEMS-based multi-physics tool (b) zoomed view of a single nanowire manipulated on the four microelectrodes.

Multi-physics Characterization

In the present design, a bent-beam type thermal actuator was employed to generate the tensile forces on the specimen. The beam angle was taken to be 1° for maximum displacement. The design is optimized to generate the required specimen displacement (around 1 micron for 10 micron long specimens) at minimum possible actuator current. In addition, a modified design integrates heat sink structures with the thermal actuators, which is shown to bring the specimen end of the device at room temperature (21). The applied displacement is transmitted through the specimen, which also deflects the single crystal silicon microbeams labeled as sensors in Figure 1a. Each of these silicon beams are effectively springs with spring constant, k, given by,

$$k = \left(\frac{24EI}{L^3} \right)$$

[1]

where, L is half the total length of the force sensor beam, E is the elastic modulus of the material, I is the moment of inertia obtained from the beam cross-section. Since the beam x-x' is made from single crystal silicon with known crystal direction, the value of E is known accurately. More accurate value of the spring constant can be found by calibrating the force sensor beam with a Nano-indenter. The force (F) on the specimen is evaluated as F = kδ, where k is the spring constant of the force sensor beam, and δ is the beam deflection, measured from direct SEM or TEM observation. The displacement or strain on the specimen can be obtained by subtracting sensor beam displacement from that of the actuators. This can be performed inside the microscope with very high resolution.

The freestanding specimen is also integrated with four microelectrodes made of heavily doped single crystal silicon. This is shown in Figure 1b, where two of the electrodes are designed to be very soft springs to avoid errors in mechanical characterization. Conventional four-point electrical probing can be performed on the specimens using these microelectrodes. Here, electrical current of the form $I_0 \sin(\omega t)$ is passed through the structure (electrodes 1 and 4 in Figure 1b). At the same time, voltage across the specimen is measured (electrodes 2 and 3). The first and third harmonic of the voltage contain the electrical and thermal conductivity information respectively. Since our specimens are free from the substrate, we can use the following model for in-plane thermal conductivity and specific heat. The rms value of the 3-ω voltage ($V_{3\omega}$) is given by (22),

$$V_{3\omega} \approx \frac{4I^3 \rho \rho'}{\pi^4 \kappa \sqrt{1+(2\omega\gamma)^2}} \left(\frac{L}{S}\right)^3 \qquad [2]$$

Here, I is the rms current, κ the thermal conductivity and ρ is the electrical resistivity at background temperature T_0. Also, $\rho' = d\rho/dT$, L is the length of the specimen between voltage contacts, S the cross section and γ is the characteristic thermal time constant of the specimen. Since the overall experimental setup size is very small, it fits virtually all forms of analytical microscope chambers. In the next section, we present some experimental results obtained inside a TEM.

Nanofabrication Process Steps

Figure 2 shows the nanofabrication process for aluminum as the specimen material. Virtually any material that can be deposited and patterned using nanofabrication can be studied with this technique. The process starts with deposition of the thin film (aluminum) on a silicon on insulator (SOI) wafer. TEM compatibility requires that the device has a through-the-wafer hole under the electron transparent specimens. Therefore, the next step is to perform a back-side lithography to pattern the back side hole. The handle layer of the SOI wafer is then etched with deep reactive ion etching (DRIE). After that, the front side is patterned with the device design described above. The specimen is first patterned using a back-side alignment process, so that the specimen segment is aligned with the hole in the handle layer. The device layer of the wafer is then etched with DRIE using a carrier wafer. At this point, both device and handle layers are duely processed and are separated by the oxide layer. The wafer is then removed from the carrier and the oxide layer is etched away using a vapor-phase hydrofluoric acid etch. This separates the movable structures (actuator and force sensing beams) from the substrate (handle layer). The next step is to remove the device layer silicon beneath the specimen to make it freestanding. This is performed by another DRIE step performed from the back side. The remaining photoresist is the removed and the device is packaged by wire bonding process.

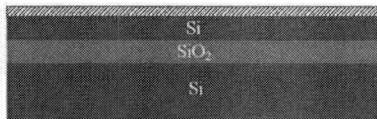

A. Aluminum is deposited of on top of the wafer

E. Aluminum is etched away on the top

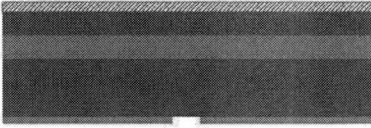

B. TEM holes are patterned on the backside of the wafer with photo-resist

F. Front-side DRIE through device layer is done

C. Backside DRIE through handle layer is done

G. Oxide is etched by HF vapor to release the device while still protecting the metallic layer

D. Backside photo-resist is removed and device is patterned on the top by photolithography

H. Silicon is etched from back side to get the free standing metal thin film; and photo-resist on top is removed by oxygen plasma

Figure 2. Nanofabrication process steps for the MEMS multi-physics characterization device.

Experimental Results and Discussion

Figure 3a shows the experimental setup mounted on a custom made TEM specimen holder that has six electrical feed through lines. We performed strain-conductivity (electrical and thermal) for 99.99% pure evaporated 100 nm thick, 30 nm average grain size aluminum as the specimen. The motivation for studying aluminum is that it has no bandgap – and hence is not expected to show any strain sensitivity towards electron or phonon transport. Most of the existing studies investigate crystalline or polycrystalline *semiconductors*, where mechanical deformation is expected to influence phonon dispersion (11, 23, 24). Therefore, stringent validation of the concept of strain-based tuning of thermal transport should involve metals or amorphous dielectrics materials, where such coupling is least expected.

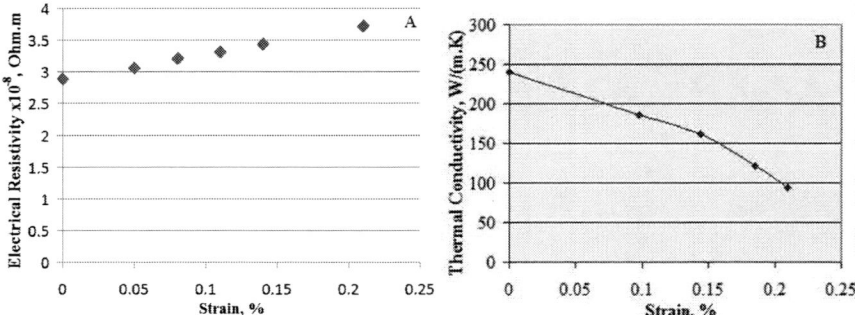

Figure 3. Strain effects on (a) electrical and (b) thermal conductivity of 100 nm thick freestanding aluminum.

Conventional deformation mechanism in metals is dislocation dominated. However, there is a critical size below which the dislocation-based plastic deformation mechanism breaks down and surface/grain boundary deformation becomes dominant (25, 26). Grain boundaries are the strongest electron and phonon scattering agents in materials and behave without any dependence on lattice thermal vibrations (or temperature). In addition, not only does the surface area to volume ratio increase non-linearly (L^{-1}) with the grain size L, in our specimens this length-scale will also be comparable to the electron and phonon mean free paths. Such a compounding effect therefore manifests itself through a more pronounced coupling between the mechanical and physical behaviors. In bulk materials, the strain is predominantly accommodated by dislocations inside the large grains so that these materials do not exhibit such a strong strain-thermal conductivity coupling as compared to the nanoscale.

Another unique way mechanical deformation can influence electron and phonon transport is through phase transformation. Size effects can induce phase transformation that cannot be predicted by classical laws. For example, diffusionless phase transformation requires unrealistically high temperature and pressure to switch between crystallographic configurations and hence is rarely observed in pure metals. Similarly, crystalline to amorphous phase transformation in metals is also improbable. In addition, these transformations are also very difficult to capture with 'post-mortem' type materials analysis. However, the small length-scale of the metal grains and the dominant role of surfaces and interfaces may bring about such transformation (27-29), which can be unambiguously captured through in-situ electron diffraction by our experimental setup. Figure 4a shows an example of face center cubic to hexagonal omega transformation in pure aluminum. Another example is shown in Figures 4b and 4c for amorphous to nanocrystalline carbon. The novelty of these transformations is that they are purely mechanical deformation induced.

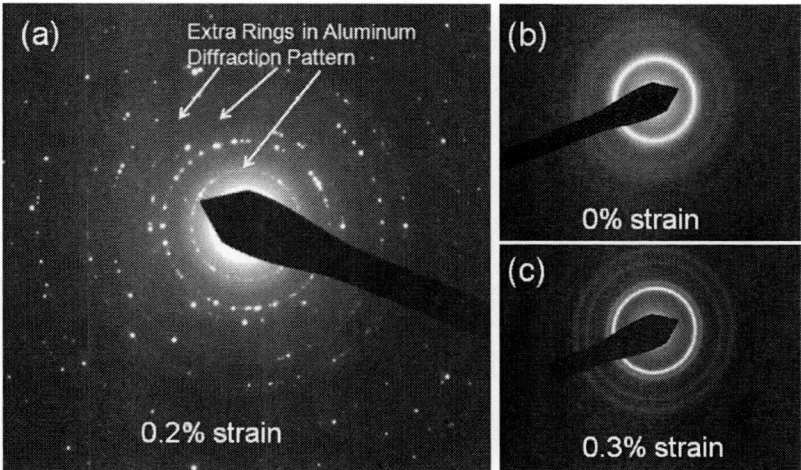

Figure 4. Mechanical deformation induced phase transformation in (a) pure face centered cubic to hexagonal omega aluminum, (b and c) amorphous to nanocrystalline carbon.

Conclusion

We present the design and fabrication of a MEMS-based chip capable of performing mechanical, electrical and thermal characterization of ultra-thin films of any material that can be deposited on a substrate. The 3mm x 3mm size of the chip results in the unique capability of in-situ testing in analytical chambers such as the transmission electron microscope. The basic concept is to 'see' the micro-mechanisms while 'measuring' the deformation and transport properties of materials and interfaces. The demonstrated multi-physics characterization is expected to be crucial for specimens with dimension comparable to the mean free paths of electron and phonons. Here, the classical mechanisms for mechanical deformation break down and also become strongly coupled with electron and phonon transport. Also, new mechanisms such as phase transformation may be induced due to size effect, which may significantly impact electrical and thermal transport. The novelty of this paper is the concept of 'size induced coupling' among various physical domains.

Acknowledgments

We gratefully acknowledges the support from the Center for Nanoscale Mechatronics & Manufacturing of the Korea Institute of Machinery & Materials and the National Science Foundation, USA (ECCS: 1028521). We also thank Dr. Benedict Samuel for his help device fabrication.

References

1. B. A. Samuel and M. A. Haque, *Journal of Micromechanics and Microengineering*, **16**, 929 (2006).
2. M. A. Haque and M. T. A. Saif, *Thin Solid Films*, **484**, 364 (2005).

3. M. T. Currie, *Strained silicon: Engineered substrates and device integration*, p. 261 (2004).
4. A. Maiti, *Nature Materials*, **2**, 440 (2003).
5. A. Bulusu and D. G. Walker, in, p. 29, American Society of Mechanical Engineers, Electronic and Photonic Packaging, EPP, New York, NY 10016-5990, United States (2004).
6. D. G. Cahill, W. K. Ford, K. E. Goodson, G. D. Mahan, A. Majumdar, H. J. Maris, R. Merlin and S. R. Phillpot, *Journal of Applied Physics*, **93**, 793 (2003).
7. S. T. Huxtable, A. Majumdar, E. T. Croke, A. R. Abramson, A. Shakouri and C. C. Ahn, in, p. 19, American Society of Mechanical Engineers, Micro-Electromechanical Systems Division Publication (MEMS) (2002).
8. R. C. Picu, T. Borca-Tasciuc and M. C. Pavel, *Journal of Applied Physics*, **93**, 3535 (2003).
9. S. Bhowmick and V. B. Shenoy, *The Journal of Chemical Physics*, **125**, 164513 (2006).
10. Z. Linli and Z. Xiaojing, *EPL (Europhysics Letters)*, **88**, 36003 (2009).
11. X. Li, K. Maute, M. L. Dunn and R. Yang, *Physical Review B*, **81**, 245318 (2010).
12. A. J. Kulkarni and M. Zhou, *Nanotechnology*, **18**, 435706 (2007).
13. J. Liu and R. Yang, *Physical Review B*, **81**, 174122 (2010).
14. S. Shen, A. Henry, J. Tong, R. Zheng and G. Chen, *Nat Nano*, **5**, 251 (2010).
15. W.-P. Hsieh, M. D. Losego, P. V. Braun, S. Shenogin, P. Keblinski and D. G. Cahill, *Physical Review B*, **83**, 174205 (2011).
16. M. Manoharan, H. Lee, R. Rajagopalan, H. Foley and M. Haque, *Nanoscale Research Letters*, **5**, 14 (2010).
17. W.-P. Hsieh, B. Chen, J. Li, P. Keblinski and D. G. Cahill, *Physical Review B*, **80**, 180302 (2009).
18. W. Bollmann, *Phys. Rev.*, **103**, 1588 (1956).
19. D. E. Newbury and D. B. Williams, *Acta Materialia*, **48**, 323 (2000).
20. S. A. Muller, U. Aebi and A. Engel, *Journal of Structural Biology*, **163**, 235 (2008).
21. S. Kumar, D. E. Wolfe and M. A. Haque, *International Journal of Plasticity*, **27**, 739 (2011).
22. L. Lu, W. Yi and D. L. Zhang, *Review of Scientific Instruments*, **72**, 2996 (2001).
23. Y. Xu and G. Li, *Journal of Applied Physics*, **106**, 114302 (2009).
24. L. Zhu and X. Zheng, *EPL (Europhysics Letters)*, **88**, 360003 (1 (2009).
25. H. Li, H. Choo, Y. Ren, T. A. Saleh, U. Lienert, P. K. Liaw and F. Ebrahimi, *Physical Review Letters*, **101**, 015502 (2008).
26. C. C. Koch, I. A. Ovidko, S. Seal and S. Veprek, *Structural Nanocrystalline Materials: Fundamentals and Applications*, Cambridge University Press (2007).
27. H. S. Park, *Nano Letters*, **6**, 958 (2006).
28. K. Gall, J. Diao, M. L. Dunn, M. Haftel, N. Bernstein and M. J. Mehl, *Journal of Engineering Materials and Technology*, **127**, 417 (2005).
29. F. Delogu, *The Journal of Physical Chemistry C*, **114**, 3364 (2010).

Effects of Adsorbate Surface Diffusion in Focused-Electron-Beam-Induced-Deposition

A. Szkudlarek[a,b], M.Gabureac[a], I. Utke[a]

[a] Laboratory for Mechanics of Materials and Nanostructures, Empa, 3602 Thun, Switzerland
[b] Department of Solid State Physics, AGH University, 30-059 Krakow, Poland

> The time-dependent continuum model is applied to simulate a growth of structures with a non-rotational symmetry. The surface diffusion of adsorbates, depending on the beam parameters and the deposit size, may lead to the different growth rate along the deposited structures. Here, a few conventional patterning strategies have been studied: *spiral inwards, spiral outwards* and *zigzag*.

Introduction

Focused Electron Beam Induced Deposition (FEBID) is a novel maskless nanolithography technique, in which continuously supplied, physisorbed gas molecules are decomposed upon electronic excitation by the impinging electron beam and the emitted secondary electron flux. As a result a local deposit with nanoscale dimensions is formed on the area, where the beam was scanning over substrate (see fig.1). Depending on the gas precursor used, various functional nanodevices can be fabricated, e.g.: magnetic sensors, strain sensors, scanning probe tips, photonic crystals, to name a few (1,2). Several theoretical approaches were applied to study the process and determine the shape (3), the growth rate (4,5) and the deposit composition (6). The continuum model, based on the local concentration of adsorbates allows for the small aspect ratio structures to calculate the local growth rate, thus the deposit profile. This can be particularly useful to predict the C-content in Hall magnetic sensors, produced by FEBID. The Hall sensors of a high lateral resolution are composed of Co-nanograins, embed in a carbonous matrix with the sensitivity strongly dependent on the carbon content (7).

Theory

For the small aspect ratio structures, the process can be described by an adsorption rate model, where four key contributions are taken into account: adsorption, desorption, electron stimulated decomposition, and surface diffusion (2):

$$\partial n/\partial t = sJ(1- n/n_0) - n/\tau - \delta fn + D(\partial^2 n/\partial x^2 +\partial^2 n/\partial y^2), \qquad [1]$$

where $n(x,y,t)$ – is the local concentration of adsorbates, s – sticking probability, J – molecular flux, n_0 – number of molecules in a complete monolayer, τ – residence time of adsorbate, δ – effective cross section, $f(x,y,t)$ – electron flux, D – diffusion coefficient.

The numerical solution of adsorption equations allows obtaining the local growth rate, which is proportional to the number of adsorbed molecules, the volume of the deposited molecule fragment - V, the local electron flux and the cross section for molecule dissociation (3):

$$R = nVf\delta. \qquad [2]$$

The contributions of diffusion and depletion can be easily separated with dimensionless variables and the following set of dimensionless parameters, introduced in (9):

$$\tau^{dep} = 1 + f_0\delta(sJ/n_0 + 1/\tau)^{-1}, \; \rho^{out} = 2(D\,\tau)^{-1/2}/FWHM, \qquad [3]$$

where τ^{dep}, ρ^{out} characterize the depletion of molecules at the center of the beam, and the surface diffusion of adsorbates with respect to the full width at half maximum of the beam size - *FWHM*, respectively (f_0 – electron flux at the center of the beam).

Based on the relation between τ^{dep}, ρ^{out} three generic FEBID regimes can be distinguished: (i)adsorbate-limited, the surface diffusion of adsorbates is small and does not influence the deposit shape, (ii)diffusion-enhanced, the surface diffusion leads to edge effects around the deposit, (iii)electron-limited, the diffusive replenishment is large enough to transfer sufficient adsorbates to the beam center for electron dissociation (9). By varying the size of exposed area and the electron beam current we can control the process regime.

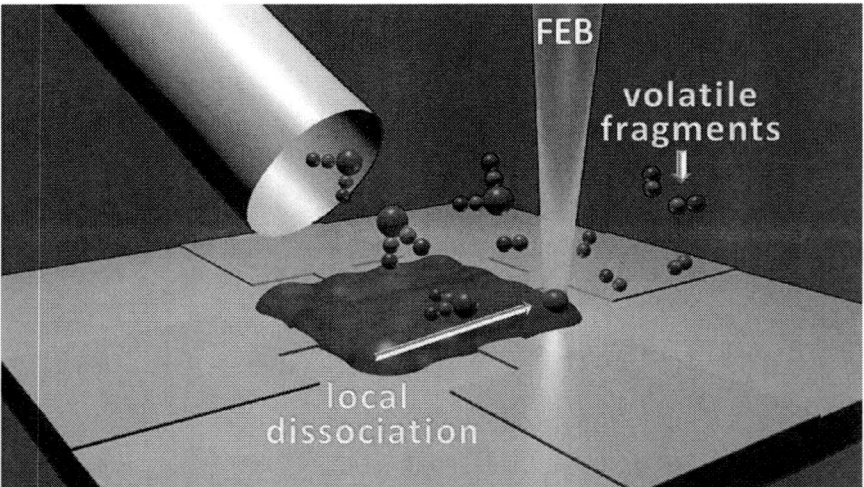

Figure 1. Principle of FEBID: the volatile molecules are supplied through the gas injection system into the SEM chamber and physisorbed on the substrate. They are dissociated upon the irradiation with the focused electron beam (FEB), being scanned over the surface. The volatile reaction products and desorbed molecules are pumped.

Simulations

We solved the equation 1 in the dimensionless form, for $\tau^{dep}=50$, $\rho^{out}=10$. These values correspond to the diffusion-enhanced regime, which means that the concentration of adsorbates at the beam center is 1/10 of adsorbates outside the irradiated area, and that the molecule, before being desorbed, can travel over the distance equal to *5FWHM*.

Figure 2 presents the simulated growth rate function in arbitrary units, for three common scanning modes: (a)*spiral inwards*, (b)*spiral outwards*, (c)*zigzag*. The arrows show how the beam is scanned. The distance between the neighboring pixels is kept equal *FWHM*. All the remaining process parameters are the same in both cases (exposure time per pixel, refresh time, number of pattern repetition, beam size, beam current). The most uniform growth is observed for the spiral outwards and zigzag modes. The difference in growth rates at the central part of the deposit can be well explained by the diffusive contribution of adsorbates, coming from a non-irradiated area. The model reproduces well known effect of the rim formation at deposit periphery.

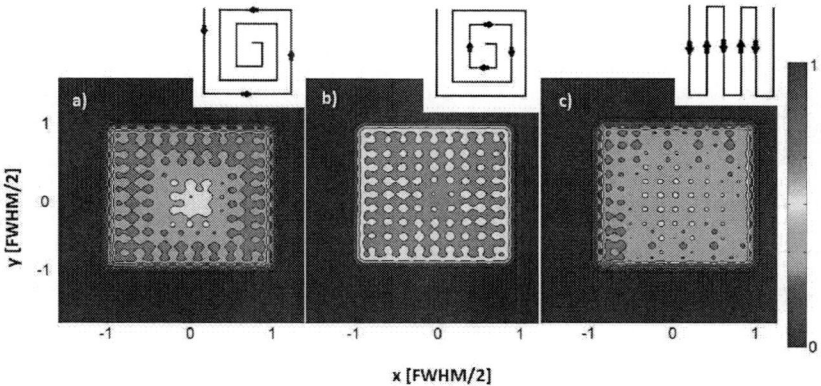

Figure 2. Numerical solution of the FEBID profiles for one-adsorbate situation for $\tau^{dep}=50$, $\rho^{out}=10$: (a)*spiral inwards*, (b)*spiral outwards* (c)*zigzag* scan, showing the change in deposit profile. The value *1* corresponds to the highest value of growth rate function. The size of the scanned area is 10 pixels x 10 pixels. Arrows indicate the scanning direction.

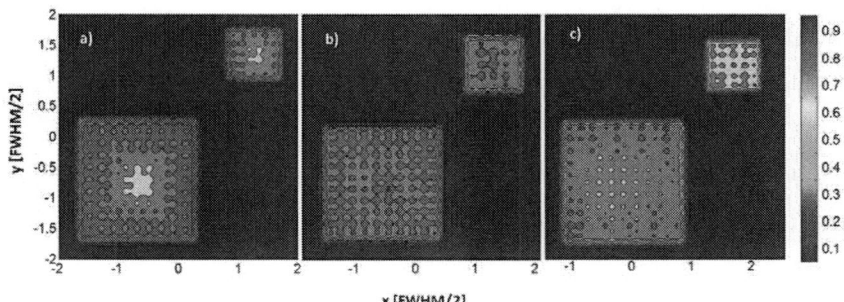

Figure 3. Numerical solution of the FEBID profiles for one-adsorbate situation for $\tau^{dep}=50$, $\rho^{out}=10$: (a)*spiral inwards*, (b)*spiral outwards* (c)*zigzag* scan and two different size of scanning area: bottom structures – 10 pixels x 10 pixels, upper ones: 5 pixels x 5 pixels

We also made a comparison between the same scanning mode and different area of being scanned (Figure 2). The small deposits do not show significant variations in growth calculated profiles, whereas for the bigger squares the differences are clearly visible. This proves that the local growth rate is influenced by the scanning mode and is a function of the deposit size.

Conclusions

We showed that surface diffusion of adsorbates plays an important role in the shape of the deposited structure. As it was already shown the composition of magnetic structures fabricated by FEBID, can vary, being dependent just on the size of the deposited structure, without changing the rest of the process parameters (10). The model presented here explains how the experiments performed in diffusion-enhanced and electron-limited regimes for different lateral deposit geometries can lead to different lateral composition, important for magnetic sensors in case of two molecule species (10).

Acknowledgments

We would like to thank **SCIEX-NMS[ch]** for the financial support of this work.

References

1. I. Utke and A.Gölzhäuser, *Angew. Chem. Int. Ed.,* **49**,9328 (2010)
2. I. Utke, P. Hoffmann and J. Melngailis, *J.Vac. Sci. Technol. B*, **26**, 4, p. 1197-1276, (2008)
3. J.Fowlkes and P. D. Rack, *ACS Nano* **4**, 3 (2010).
4. C. J. Lobo, M. Toth, R. Wagner, B. L. Thiel, and M. Lysaght,*Nanotechnolgy* **19**, 025303 (2008).
5. H. Plank, D. A. Smith, T. Haber, P. D. Rack, and F. Hofer, *ACS Nano,* **6** (2012) 286,
6. L. Bernau, M. Gabureac, R. Erni, and I. Utke, *Angew. Chem. Int. Ed.* **49**, 8880 (2010)
7. M. Gabureac, L. Bernau, I. Utke, *Nanotechnology,* **21**, 11(2010)
8. I. Utke, V. Friedli, M. Purrucker, J. Michler, *J. Vac. Sci. Technol. B*, **25**, 6, p. 2219-2223, (2007).
9. A. Szkudlarek, M. Gabureac, I. Utke,*J. Nanosci. Nanotech.,* **11**, p.8074–8078, (2011).
10. M. Gabureac, L. Bernau, I. Utke, *J. Nanosci. Nanotech*, 11, 7982–7987, (2011).

Electroplating of Micropatterned Nickel Phase Gratings for X-Ray Phase Contrast Tomography

M. Amberger[a], K. Bade[a], J. Meiser[a], D. Kunka[a], J.Mohr[a]

[a] Institute of Microstructure Technology (IMT)
Karlsruhe Institute of Technology (KIT)
76131 Karlsruhe, Germany

We report on a new approach of monitoring Nickel-Sulfamate based electroplating baths to obtain a more precise control of the height of x-ray phase contrast gratings. The influence of the Ti/TiO$_x$ film substrates on electroplating start and behavior is investigated. A four phase electroplating sequence was recorded and utilized to adaptively control the height of electroplated samples.

Introduction

X-Ray Phase Contrast Imaging is an imaging technique that allows the acquisition of both absorption and phase information of an irradiated material (1). This improved imaging technique can provide benefits for a wide range of applications, including medical (2) and material science (3) imaging. Up to now different approaches to detect phase differences in x-ray imaging have been proposed and proven to work: In-Line phase contrast imaging using Fresnel Diffraction (4), Diffraction Enhanced imaging utilizing crystal monochromators and analyzer crystals (5), Interferometer based imaging (6) and Differential Phase Contrast Imaging (DPC) (7). The latter has lately been seen as on one of the most promising approaches to detect phase information of x-rays, especially since the introduction of the Talbot-Laue Interferometry that allows phase retrieval also for polychromatic and weakly coherent x-ray beams found in x-ray tubes of clinical devices like Computer Tomography or Radiography (8).

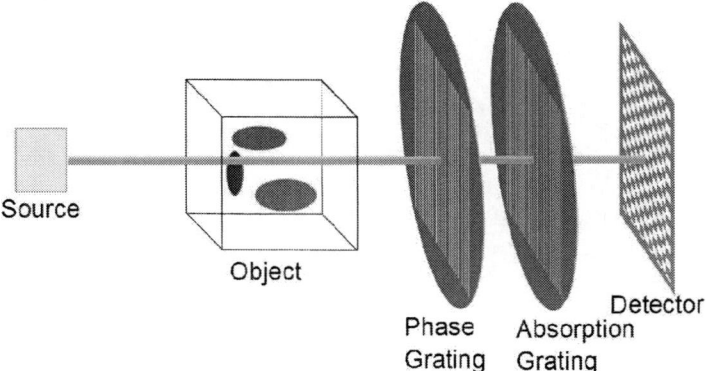

Fig. 1: X-Ray Phase Contrast Imaging Setup for Coherent Radiation

This approach consists of a setup of two (for coherent synchrotron radiation) or three (x-ray tubes) micro fabricated gratings arranged in accurately defined distances from the x-ray source and each other to make use of the Talbot Effect (Fig. 1). One crucial part of this setup is a nickel grating that creates a phase shift between the x-rays passing through the solid nickel phase on the one side and the undisturbed and undelayed x-rays on the other side.

This results in a self repeating interference pattern also called Talbot Pattern. For those phase gratings height control of the metallic structures is of crucial importance, as the resulting phase shift determines the geometric dimensions of the interferometer setup. In a typical setup with an x-ray energy of 25 keV and a phase shift of $\pi/2$, a nickel grating of 4.2 µm height is needed to achieve the best visibility for this setup. In this work we present our measures to achieve the best possible control of nickel height through electroplating to maximize the imaging performance of an x-ray phase contrast setup.

Phase Grating Manufacturing

Those phase shift gratings are produced via the LIGA process (10) by making high Aspect Ratio grating microstructures in MRL-resist. For this purpose resist is spin coated onto Ti/TiO_x coated Si substrates followed by x-ray lithography. Ti/TiO_x is used both as a plating base for the electrodeposition as well as a surface that shows very good adhesion with the resist due to its nanostructured surface. The Ti/TiO_x thin film is produced by sputter depositing 2µm thick Ti layers onto 525µm Si Wafers and subsequently oxidizing the sputtered layer in $NaOH/H_2O_2$ solution. Developed resist templates are finally filled with nickel via electrodeposition (see Fig. 2). As the resist material is of no use afterwards in the DPC setup and the nickel structures do not need any mechanical support, the MRL resist is stripped. A part of the final grating can be seen in Fig. 2.

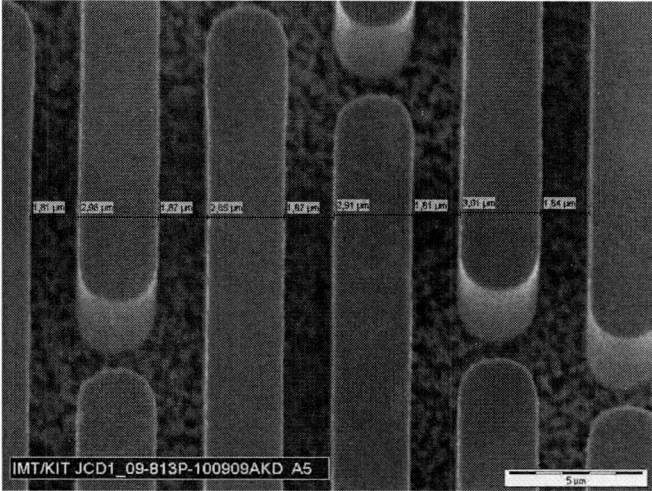

Fig. 2: Example of a Ni- Phase Grating

Electroplating was done in a commercial bright nickel-sulfamate bath. The bath contains 76 g/l Nickel, 40 g/l boric acid and 180 mg/l perflourinated tenside. Throughout the electroplating experiments current density was kept at 3mA/cm² with a plating bath temperature of 52 °C and a pH of 3.8. Current efficiency was measured by gravimetry and the height of the final structures was determined using SEM measurements on tilted samples. From those conditions the plating rate was calculated to be a growth of 3.6 μm in Ni-height per hour of electroplating, assuming 95 % current efficiency, which was derived from previous experiments.

But throughout grating manufacturing serious discrepancies between measured and initial design height were experienced (see Fig. 5). For those depositions, plating times were calculated assuming a current efficiency of 95 %. For the production of phase gratings, if the tolerance exceeds 15% of the design height the grating was regarded worthless and had to be redone. Looking at factors that could influence plating performance, we identified the TiO_x layer as a critical part in the fabrication process (9) that strongly affects the starting behavior of the nickel electrodeposition. Another factor that influences plating height is the plating area uncertainty. This uncertainty results from deviations in the gratings duty cycle (ratio of metallic structure width to grating period) as well as mask errors. As grating area errors can only be corrected to a certain degree, one way to adjust the plating height is to modify the plating process during the electroplating according to the TiO_x layers properties.

Results and Discussion

To vary morphology and chemical composition of the TiO_x layer, oxidation times in the range from 30 s to 2.5 min were used. Nickel morphology, current efficiency and cell potential were recorded for each oxidation time.

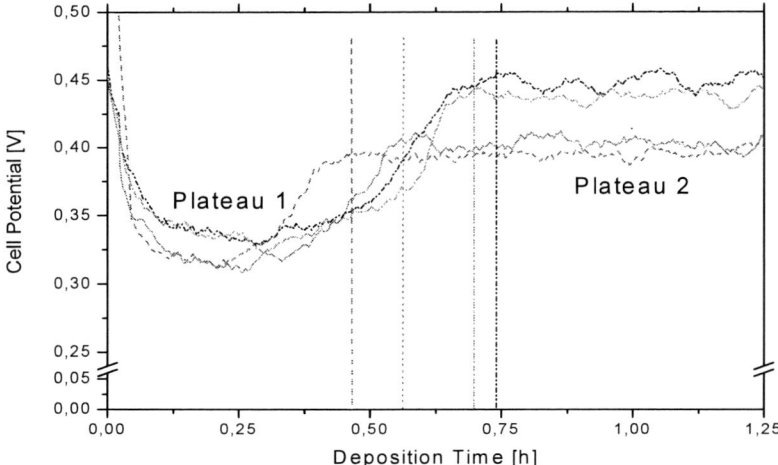

Fig. 3: Cell Voltage vs. Deposition Time for 4 random samples showing differences in Plateau length

By observing the electroplating cells potential we were able to distinguish three different plating regimes: During the first minute of electroplating, cell voltage typically rises steeply within a few seconds and then decays until a first plateau is reached. The length of this first plateau phase can vary between 5 to 60 min (see Fig. 3) and is followed by an increase in cell potential of roughly 100 mV to a second plateau that lasts till the end of deposition.

While monitoring cell voltage during the phase grating fabrication, a clear correlation between the length of plateau one and the absolute deviation in height was experienced. To investigate the origin of this two-plateau plating sequence SEM pictures of non micro patterned electroplated wafers were taken (Fig 4).

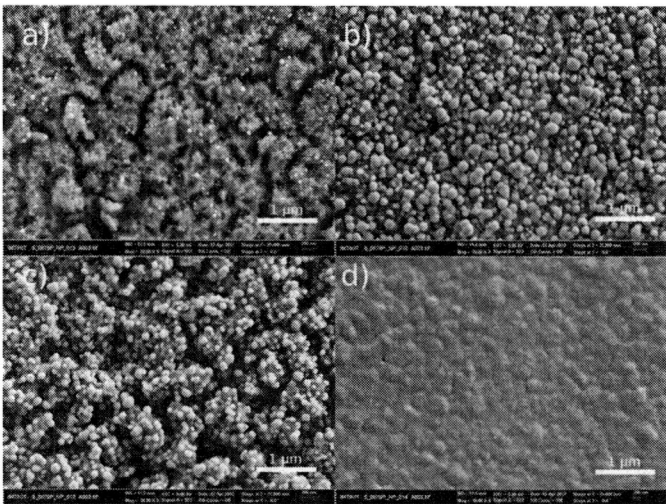

Fig. 4: SEM Pictures showing the electroplated surface in different sections of the phases: a) right after initial voltage drop, plateau 1; b) & c) at the beginning (b)at the end (c) of the voltage increase, between plateau 1 and 2; d) start of plateau 2

Within this group of samples, electroplating was stopped in each of the phases so that differences in morphology could be examined. The results showed that the transition time marked the point where a 2-phase cathode surface consisting of both Ni and TiO_x being present is replaced by a single Ni interface. Figure 4 shows four different surface samples collected in phases 2 (Fig. 4a, lower plateau), 3 (Fig. 4b and 4c increase in cell voltage) and 4 (Fig. 4d, plateau 2). With increasing time, the entire surface is finally getting covered and therefore the two phase surface system is reduced to just one remaining nickel-electrolyte interface. It was hence concluded that during the first plateau phase current is split between Ni and TiO_x area and favoring the Ti/TiO_x system. This would lead to a significant drop in current efficiency of Nickel plating during phases one and two. Current ratio between Ni and Ti/TiO_x and therefore also Ni current efficiency was influenced by the oxidation time of the TiO_x, although no clear pattern emerged. Very short oxidation times (shorter than 30 s) lead to short plateau phases, whereas very long oxidation times of 2.5 min lead to a hindered Ni-nucleation and longer plateau one phases. All intermediate oxidation times did not show any correlation, showing both long and short plateau one phases. This also correlates to Ni electroplating of phase gratings, as those also did not show any recognizable plateau length pattern. Therefore the characteristic voltage pattern itself was used to monitor and adjust the electroplating procedure.

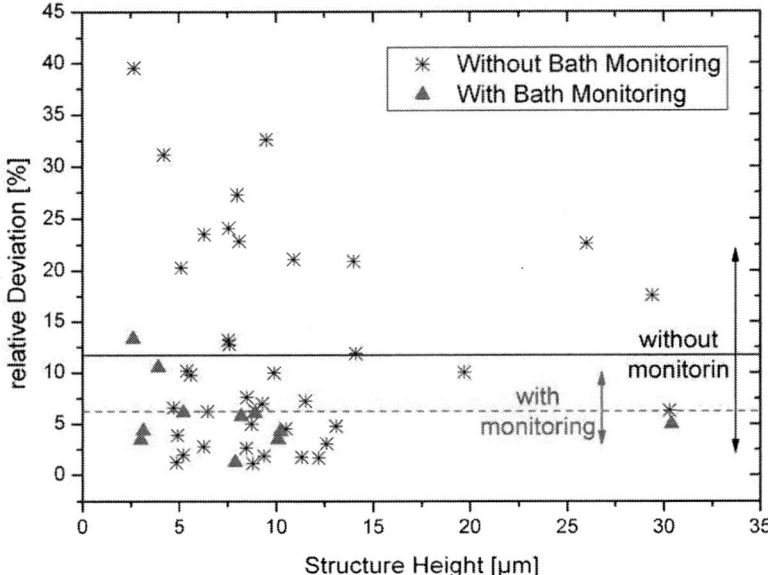

Fig. 5: Relative Deviation in height of grating structures compared to absolute structure height, lines indicating mean values and the arrows giving standard deviation

By using cell monitoring in grating fabrication, i.e. recording and displaying cell voltage "live" during electroplating, we were able to half the average height deviation from 11.8 % to 5.8 % and, even more importantly, reduce standard deviation from 10 % to less than 3.8 %, as can be seen in Fig. 5. This was done by adjusting plating time according to plateau one length, giving longer plating times for samples with a prolonged plateau one. Considering also the other error sources, especially the duty cycle variation, a plating height error of 5 % or less is the minimum that can be expected, as even small differences of just 0.01 in Duty Cycle amount to 2 % change in plating area and therefore also roughly 2 % change in grating height.

Conclusion

In conclusion, we were able to show that cell voltage monitoring can be an easy to implement way to control the height of Ni LIGA-microstructures in situ. Plating regimes were addressed to Ni-coverage of the cathode surface and TiO_x oxidation times were showed to affect overall current efficiency and plateau length.

Acknowledgments

This work was partly carried out with the support of the Karlsruhe Nano Micro Facility (KNMF, www.kit.edu/knmf), a Helmholtz Research Infrastructure at Karlsruhe Institute of Technology (KIT, ww.kit.edu).

References

1. S. W. Wilkins, T. E. Gureyev, D. Gao, A. Pogany, A. W. Stevenson, *Nature*, **384** (6607), 335-338 (1996).
2. S. Schleede, M. Bech, K. Achterhold, G. Potdevin, M. Gifford, R. Loewen, C. Limborg, R. Ruth, F. Pfeiffer, *Journal of Synchrotron Radiation*, **19** (4), 525-529, (2012).
3. I. Jerjen, V. Revol, C. Kottler, R. Kaufmann, C. Urban, T. Luethi, U. Sennhauser, *International Symposium on Digital Industrial Radiology and Computed Tomography*, Berlin (2011).
4. A. Pogany, D. Gao, S. W. Wilkins, *Rev. Sci. Instrum.*, **68** (7), 2774-2782, (1997).
5. T. J. Davis, D. Gao, T. E. Gureyev, A. W. Stevenson, S. W. Wilkins, *Nature*, **373**(6515), 595-598, (1995).
6. A. Momose, T. Takeda, Y. Itai, K. Hirano, *Nature Medicine*, **2** (4), 473-475 (1996).
7. C. David, B. Noehammer, H. H. Solak, E. Ziegler, *Applied Physics Letters*, **8** (17), 3287-3289, (2002).
8. F. Pfeiffer, T. Weitkamp, O. Bunk, C. David, *Nature Physics*, **2**(4), 258-261, (2006).
9. H. K. Chang, B.-H. Choe, J. K. Lee, *Materials Science and Engineering*: A, **409**(1-2), 317-328, (2005).
10. W. Menz, J. Mohr, O. Paul, *Microsystem Technology,* Wiley-VCH, (2001), ISBN 978-3527613007.

506

ECS Transactions, 50 (12) 507-511 (2012)
©The Electrochemical Society

Impact of Donor Dopant on Acceptor Solubility in TlBr

S. R. Bishop[a,b] and H. L. Tuller[a]

[a] Department of Materials Science and Engineering, Massachusetts Institute of Technology, Cambridge, MA, 02139, USA
[b] International Institute for Carbon Neutral Energy Research (WPI-I2CNER), Kyushu University, Nishi-ku Fukuoka, 819-0395, Japan

> The predicted dependence of TlBr room temperature ionic conductivity on divalent donor and acceptor dopant concentration is presented, taking into account acceptor dopant exsolution and dopant-defect association. Donor-acceptor compensation is found to play a key role in determining the ultimate room temperature conductivity with donors increasing the solubility of acceptors from 0.04 ppm (0 donor) to 0.84 ppm (1 ppm donor) to 8.4 ppm (10 ppm donor).

Introduction

Thallium bromide (TlBr) is a candidate material for high energy (e.g. γ-ray) radiation detection due to its relatively large energy band gap 2.68 eV, large molecular weight and ease of crystal growth (1, 2). Though dark currents are low enough for achieving high radiation detection sensitivity, ionic defect migration is believed to play a role in long-term degradation of detection properties (3-5). To clarify the source and nature of ionic conduction in TlBr, a systematic examination of the roles of dopants and temperature in controlling defect generation and transport was undertaken by our group (6-8). Using electrical conductivity measurements, significant exsolution of acceptor dopants was observed. As a consequence, a revised prediction of the dependence of the room temperature ionic conductivity on donor and acceptor dopant concentration is presented and the role of donor dopant concentration on acceptor solubility is investigated.

Theory

The total electrical conductivity (σ_{total}) is given by

$$\sigma_{total} = \sum_i |z_i| q[i] \mu_i \qquad [1]$$

where z_i, q, [i], and μ_i is the charge number, elementary charge, concentration, and mobility of charge carrier i, respectively. Since Tl ($V_{Tl}^{/}$) and Br ($V_{Br}^{\bullet}$) vacancies are the primary ionic charge carriers in TlBr (7), the total conductivity at fixed temperature (e.g. room temperature) is given by

$$\sigma_{total} = q[V_{Tl}^{/}] \mu_{Tl} + q[V_{Br}^{\bullet}] \mu_{Br} \qquad [2]$$

By doping the material with divalent acceptor ($A_{Br}^{/}$) or donor ($D_{Tl}^{\bullet}$) dopants (e.g. Se, or Pb, respectively), the concentration of Br and Tl vacancies can be modified through the charge neutrality equation, below.

507

$$[V_{Br}^{\bullet}] + [D_{Tl}^{\bullet}] = [V_{Tl}^{/}] + [A_{Br}^{/}] \qquad [3]$$

Additionally, the intrinsic equilibria for vacancies (thermally generated) occurs via the Schottky defect reaction (followed by the corresponding equilibrium equation),

$$nil \leftrightarrow V_{Tl}^{/} + V_{Br}^{\bullet} \qquad [4]$$

$$[V_{Tl}^{/}][V_{Br}^{\bullet}] = K_s \qquad [5]$$

where K_s is an equilibrium constant. Acceptor dopant exsolution leads to an upper limit on the amount of extrinsic $V_{Br}^{\bullet}$, given by the following reaction and equilibrium equation

$$A(precip.) \leftrightarrow (V_{Br}^{\bullet} - A_{Br}^{/})^X + V_{Br}^{\bullet} + A_{Br}^{/} \qquad [6]$$

$$[(C - [A_{Br}^{/}])][A_{Br}^{/}][V_{Br}^{\bullet}] = K_{sol} \qquad [7]$$

where $(V_{Br}^{\bullet} - A_{Br}^{/})^X$ is an associated defect arising from electrostatic attraction and/or strain relaxation (see Ref. 6 for a detailed discussion). C represents the total amount of soluble acceptor dopant (including free and associated acceptors) and K_{sol} is an equilibrium constant with the activity of precipitated acceptor ($A(precip.)$) equal to unity. With Br and Tl mobilities and equilibrium constants K_s and K_{sol} previously derived, the conductivity can be predicted using the above equations (6).

Results and Discussion

In Figure 1, the predicted ionic conductivity for TlBr is plotted as a function of either divalent acceptor [A] (on a Br site) or donor [D] (on a Tl site) concentration at room temperature. In the donor doped region, for donor concentrations greater than ~10 ppm, the ionic conductivity is dominated by Tl vacancy migration (σ_{Tl}). In this region, $[V_{Tl}^{/}] \approx [D_{Tl}^{\bullet}]$, which can be substituted into Eq. 5 to yield.

$$[V_{Br}^{\bullet}] = \frac{K_s}{[D_{Tl}^{\bullet}]} \qquad [8]$$

$[V_{Br}^{\bullet}]$ is thus largely suppressed by the high donor concentration. σ_{Tl} decreases while σ_{Br} increases with decreasing donor concentration, with the two becoming equal at ~ 3 ppm donor. Here, $[V_{Tl}^{/}] >> [V_{Br}^{\bullet}]$, reflecting the orders of magnitude larger mobility of Br versus Tl vacancies (see Eq. 2) (7). With further decrease in donor concentration, and subsequent increase in acceptor concentration, σ_{Br} increases, ultimately reaching a constant value given by the maximum room temperature solubility of acceptor dopant (0.04 ppm).

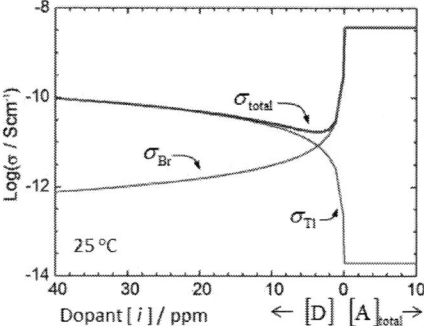

Figure 1. Total and partial ionic conductivities in TlBr. The conductivity plateau observed in the acceptor controlled region is a result of limited acceptor solubility.

In the above discussion, the background impurity content was assumed to be zero, i.e. assuming no simultaneous presence of donors and acceptors. In the following, the impact of constant donor concentration on acceptor solubility is discussed.

Figure 2 shows the ionic conductivity as a function of acceptor dopant concentration for two different donor concentrations. The first point to note is the plateau in conductivity continues to exist despite donor concentrations much larger than the aforementioned solubility limit of acceptors in figure 1, and that the onset of the plateau, where maximum acceptor solubility is reached, occurs at higher acceptor concentrations with increasing donor level. Second, the conductivity minimum shifts from ~3 ppm donor in figure 1 to ~7 ppm acceptor in figure 2B, an expected result for acceptor dopant compensation of donors. Third, the conductivity of the plateau region decreases for increasing donor concentration. For an explanation of this behavior, we turn to the defect equilibria.

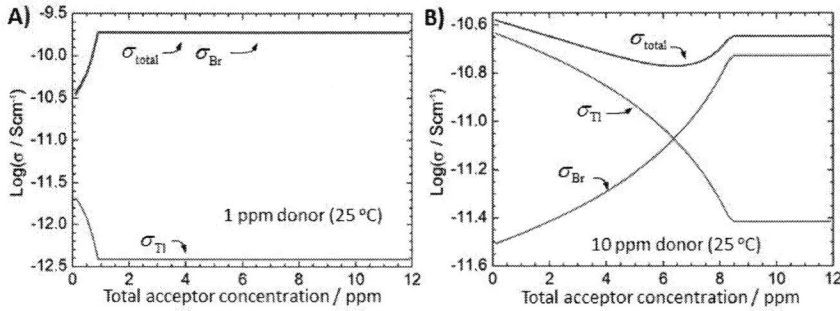

Figure 2. Ionic conductivity of TlBr as a function of acceptor content for two different donor concentrations.

As discussed above, an increased donor concentration results in a suppression of the Br vacancy concentration; see Eq. 8. Additionally, according to Eq. 7, a decrease in Br vacancy concentration is accompanied by an increase in acceptor solubility, from 0.04 ppm (0 donor) to 0.84 ppm (1 ppm donor) to 8.4 ppm (10 ppm donor), thereby explaining the shift of conductivity minimum and compensation of increasing concentrations of donors in Fig. 2.

The decrease in conductivity in the plateau region of Fig. 2 with increasing donor content arises from an increased driving force for defect association through the following defect reaction and corresponding equilibrium equation.

$$V_{Br}^{\bullet} + A_{Br}^{/} \leftrightarrow \left(V_{Br}^{\bullet} - A_{Br}^{/}\right)^{\times} \qquad [9]$$

$$\frac{\left[V_{Br}^{\bullet}\right]\left[A_{Br}^{/}\right]}{\left[\left(V_{Br}^{\bullet} - A_{Br}^{/}\right)^{\times}\right]} = K_a \qquad [10]$$

As $\left[A_{Br}^{/}\right]$ increases, $\left[\left(V_{Br}^{\bullet} - A_{Br}^{/}\right)^{\times}\right]$ must also increase for a $\left[V_{Br}^{\bullet}\right]$ largely suppressed by $\left[D_{Tl}^{\bullet}\right]$ in Eq. 8. Figure 3 shows the concentrations of defects in TlBr corresponding to the conductivity plots in Fig. 2. It is clear in Fig. 3, that for larger donor concentration, the ratio of associated defects to Br vacancies increases, leading to a corresponding decrease in σ_{Br}. Figs. 2 and 3 also show that despite $\left[V_{Tl}^{/}\right] > \left[V_{Br}^{\bullet}\right]$ for high acceptor content, $\sigma_{Tl} < \sigma_{Br}$, again highlighting the large disparity in mobilities between the two defects (7).

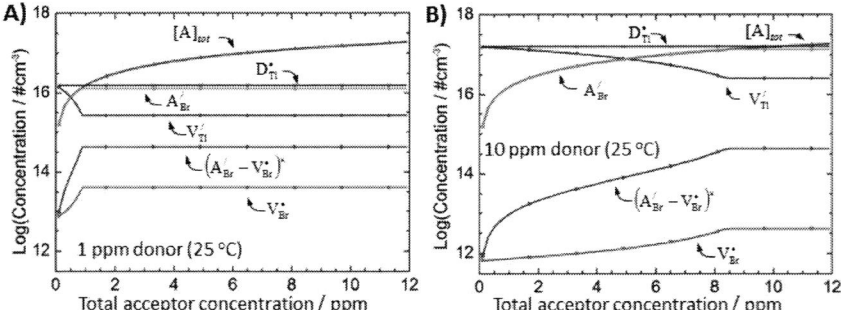

Figure 3. Concentration of defects in TlBr as a function of acceptor content for the two different donor concentrations shown in Fig. 2, where $\left[A_{Br}^{/}\right]_{tot}$ represents the total acceptor content (soluble and insoluble).

Conclusions

The ionic conductivity of TlBr was analyzed as a function of divalent donor and acceptor dopant concentration, taking into account the limited solubility of acceptor dopants. The presence of donor dopants is found to increase the solubility of acceptor dopants from 0.04 ppm (0 donor) to 0.84 ppm (1 ppm donor) to 8.4 ppm (10 ppm donor) resulting in an overall decrease in dark ionic conductivity relative to that without acceptor additions.

Acknowledgments

This work was supported by the Domestic Nuclear Detection Office under contract nos. HSHQDC-07-C-0039 and HSHQDC-10-C-00210. S.R.B. and H.L.T. thank RMD for their support. The material presented in this paper represents the positions of the authors and not necessarily that of the department of Homeland Security.

References

1. A. V. Churilov, G. Ciampi, H. Kim, L. J. Cirignano, W. M. Higgins, F. Olschner and K. S. Shah, *IEEE Trans.Nucl.Sci.,* **56**, 4 (2009).
2. R. Hofstadter, *Phys. Rev.,* **72**, 11 (1947).
3. J. Vaitkus, J. Banys, V. Gostilo, S. Zatoloka, A. Mekys, J. Storasta and A. Zindulis, *Nucl.Instrum.Methods Phys.Res.Sect.A-Accel.Spectrom.Dect.Assoc.Equip.,* **546**, 1-2 (2005).
4. J. Vaitkus, V. Gostilo, R. Jasinskaite, A. Mekys, A. Owens, S. Zatoloka and A. Zindulis, *Nucl.Instrum.Methods Phys.Res.Sect.A-Accel. Spectrom. Dect. Assoc. Equip.,* **531**, 1-2 (2004).
5. K. Hitomi, Y. Kikuchi, T. Shoji and K. Ishii, *IEEE Trans.Nucl.Sci.,* **56**, 4 (2009).
6. S. R. Bishop, H. L. Tuller, G. Ciampi, W. Higgins, J. Engel, A. Churilov and K. Shah, *Phys. Chem. Chem. Phys.,* **14** (2012).
7. S. R. Bishop, W. Higgins, G. Ciampi, A. Churilov, K. S. Shah and H. L. Tuller, *J. Electrochem. Soc.,* **158**, 2 (2011).
8. S. R. Bishop, H. L. Tuller, W. M. Higgins, G. Ciampi, A. Churilov, H. Kim, L. J. Cirignano, F. Olschner, V. Biteman, J. Tower and K. S. Shah, *ECS Trans.,* **28**, 11 (2010).

Microfabricated Systems to Measure Marine Variables

S. Aravamudhan[a] and S. Bhansali[b]

[a] Department of Nanoengineering, Joint School of Nanoscience and Nanoengineering, North Carolina A&T State University, Greensboro, North Carolina 27401, USA
[b] Department of Electrical and Computer Engineering, Florida International University, Miami, Florida 33174, USA

> The goal of this paper is to provide a basis for MEMS marine-specific sensors and their associated harsh environmental packaging. Specifically, this paper will discuss our efforts in the development of two microfabricated systems to measure marine variables – (a) MEMS-based electrochemical systems to measure marine nutrients and (b) reinforced MEMS pressure sensor to measure spatial depths. The electrochemical detection-on-chip module was fabricated using assembled polypyrrole (PPy) nanowires as the sensing element. The measurement of nitrate ions in a flow through environment is presented here. For the depth measurements, we developed the double diaphragm or reinforced configuration to measure pressure/depth at both higher sensitivity (27% higher) and wider linear operating range (> 400 psi, from combination of inner and outer diaphragms) compared to the conventional single diaphragm design.

Introduction

Oceans, lakes, rivers, and groundwater are complex, dynamic environments in which physical, chemical, and biological processes occur over varying temporal and spatial scales (e.g., eddies, nutrient fluxes, microbial organisms, algae blooms and pollution). In particular, the coastal marine environment is strongly influenced by a wide diversity of anthropogenic perturbations and natural substances, organisms and processes that may have adverse implications on human health, economy and the ecosystem. Evaluating the scale and consequences of changes in the marine variables due to natural and human activities is important. The objective of this work is to provide a basis for MEMS based marine-specific sensors and their associated harsh environmental packaging. Specifically, this paper will discuss our efforts in the development of MEMS-based electrochemical systems to measure marine nutrients across varying spatial depths including a reinforced MEMS pressure sensor to measure physical phenomena.

Background

The motivation for this work stems from the historical need for inexpensive yet highly accurate and reliable in-situ sensors and systems to continuously obtain measurements over varying spatio-temporal scales. Current accepted methods for measuring marine parameters usually involve research vessels such as Autonomous Underwater Vehicles (AUV) or Remote Operated Vehicles (ROV), which can only sense local environmental variables at a single point in space-time. The use of multiple vehicles improves measurement quality. However, the gain from higher spatial sampling frequency is directly related to the number of additional vessels used. The increased use of support vessels, whether AUVs or ships, will add to cost. While in-situ buoy systems have provided part of the solution, they are known to have limitations.

The United States Commission on Ocean Policy report, "AN OCEAN BLUEPRINT FOR THE 21ST CENTURY," states an urgent need to enact major changes in the ocean resources monitoring and management (1). According to this report "Human ingenuity and ever-improving technologies have exploited and significantly altered – the ocean's bounty to meet the society's escalating needs. Pollution runs off the land, degrading coastal waters and harming marine life. The introduction of non-native species and increase in nonpoint source pollution were found as the primary causes of nutrient enrichment, hypoxia, harmful algae blooms, toxic contamination and other problems that plague coastal waters." In these situations, an unprecedented need for marine monitoring systems is recognized. One such initiative to monitor marine variables using microfabricated systems forms the basis for this paper.

Objectives

The central objective of this paper is to develop MEMS based marine sensing elements and packaging techniques to measure marine variables. Specifically, in this research, MEMS and nanotechnology enabled sensors were developed to measure physical phenomenon such as pressure/depth variations and chemical markers such as nitrate (nutrient) fluxes. This paper will form the foundation for the miniature integrated physical, chemical and biological sensor based marine systems, thus making them ubiquitous for in-situ ocean measurements. Such miniature systems could also house electronics, control systems, data storage and communications modules into one package.

Microfabricated System Design and Fabrication

In this work, two MEMS-based sub-systems and their associated packages were developed to measure marine variables. (A) Electrochemical detection-on-chip (EoC) to measure nitrate ions in a flow through environment, and (B) reinforced diaphragm micromachined pressure sensor array to accurately measure the corresponding spatial depth for every nitrate measurement made at a particular depth.

Electrochemical Detection-on-Chip (EoC) Design and Fabrication

This module describes the development of a nitrate-selective electrochemical sensor using doped polypyrrole (PPy) nanowires on a microfluidic platform. Determination and estimation of nitrate ion is important for a variety of applications in different fields, including environmental engineering, marine cycles, food analysis, water management, biology and medicine (2). Firstly, the sensor element used in this case is polypyrrole (PPy), one of the widely studied conducting polymers because of its high conductivity, ease of preparation, flexibility, and stability as ion-selective electrode material. Polypyrrole-doped nanowires show high selectivity and rapid reactivity towards the dopant ion and integration ability as sensor elements. In its oxidized state, polypyrrole exists as a polyradical cation and at the oxidation stage; nitrate anions are attracted electrostatically into the PPy matrix as dopants or counter ions. PPy nanowires were synthesized using chemical template method by simply immersing the membrane into a solution of the desired monomer and its oxidizing agent. The diameter of nanowire is determined by the pore diameter of template (3). Next, selective chemical recognition for nitrate by electrochemical polymerization (doping) of an appropriate monomer (pyrrole) under controlled conditions was done for nitrate ions. This was followed by assembly of the PPy-based electrochemical module in a MEMS/microfluidic configuration, as shown in Figure 1.

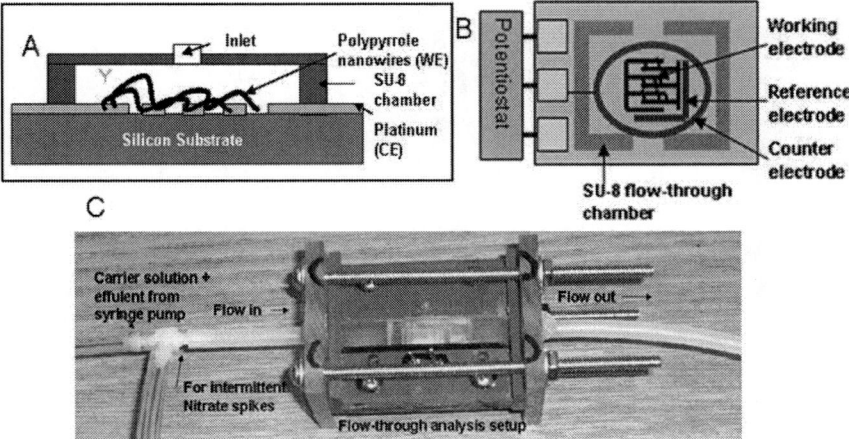

Figure 1. Electrochemical nitrate sensor module: A) Illustrative view, B) Illustration of the flow through electrochemical test cell, C) Picture of through flow-cell shown with inlet/outlet and Pogo™ pin connections.

The construction of the microfluidic platform includes the fabrication of electrochemical sensor chip, assembly of PPy-NWs on the chip, and integration of micromachined fluidic paths and microfluidic components (Figure 1). The assembly of PPy-NWs nanowires on the Pt lines achieved using dielectrophoresis (DEP) technique (3). The interdigitated assembly lines and the counter electrode were fabricated in Pt, with Ti as an adhesion layer on 7740 Pyrex substrate (2.5 x 2.5 mm). The microfluidic flow paths and reagent chamber were then patterned using 100 μm thick SU-8 film. Next, flow through analysis under both static and dynamic conditions were conducted to understand the marine processes.

Micromachined Reinforced Depth Sensor Design, Fabrication and Package

High resolution depth/pressure measurements are important for accurate leveling of marine instruments and for applications such as wave/tide detection. In this work, in order to attain wider operating range (> 400 m) without compensating on sensitivity (better than 0.1%), we developed the double or reinforced diaphragm configuration (instead of a conventional single diaphragm), with the outer diaphragm acting both as reinforcement and sensing structure along with the inner sensing diaphragm and temperature compensation bridge (4). When a conventional thin diaphragm is subjected to high pressures, it may result in large deflections, which in turn will induce non-linearities in the measurement (5). To overcome this limitation, in this research, a double or reinforced diaphragm is used (Figure 2A). A thinner inner diaphragm (5-25 μm) and thicker outer diaphragm (25-40 μm), with piezoresistors embedded in both of them are used to transfer the peak stresses from inner to outer diaphragm (Figure 2B). Even though, large deflections and non-linearities are avoided in this current design, sensitivity and operating range are not compromised. After detailed design analysis, analytical and Finite-Element-Method (FEM) performance analysis, we fabricated the reinforced diaphragm design using the masked-maskless process (6).

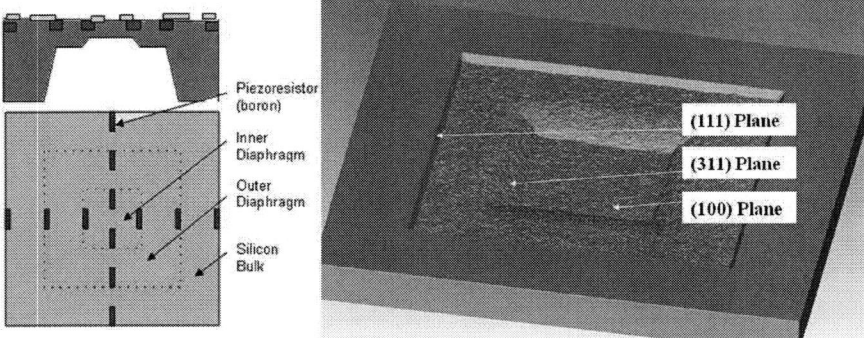

Figure 2. (A) Schematic of the double or reinforced diaphragm configuration with piezoresistors. (B) Etching of the reinforced configuration, figure illustrates a typical etching in masked-maskless process.

The reinforced fabrication process followed three variations. In the first approach, referred to as Generation I, the double diaphragm was fabricated without accounting for the masked-maskless etching process, while in the second approach; referred to as Generation II, this was taken into account at the stage of mask design. The third approach (called as Generation III) incorporated a square central boss structure on the inner diaphragm. In this paper, only Generation II and its associated packaging will be discussed. Readers are referred to Ref. (4) for detailed design and fabrication steps. The packaging process also followed two variations. In the first approach (Pack I), the zeroth level package was accomplished by routing electrical connections outside the glass reference cavity. This was later system packaged by integration through Anisotropic Conductive Film (ACF) to flexible printed circuit board (Figure 3A). In the second approach (Pack II), the zeroth-level package involved electrical interconnections drawn through the reference pressure cavity. This was accomplished by drilling through-holes through the glass substrate and later filling it with conductive epoxy paste. Then, system packaging was realized by integrating printed circuit boards with signal conditioning circuit on the pressure sensor package (Figure 3B). Figures 3A and 3B show the two packaging approaches followed in the packaging of reinforced pressure sensors. The sensing side (bottom side) in both cases was first coated with protective coatings and then attached to bottom boards either Plexiglas or glass substrates with inlet and outlet ports.

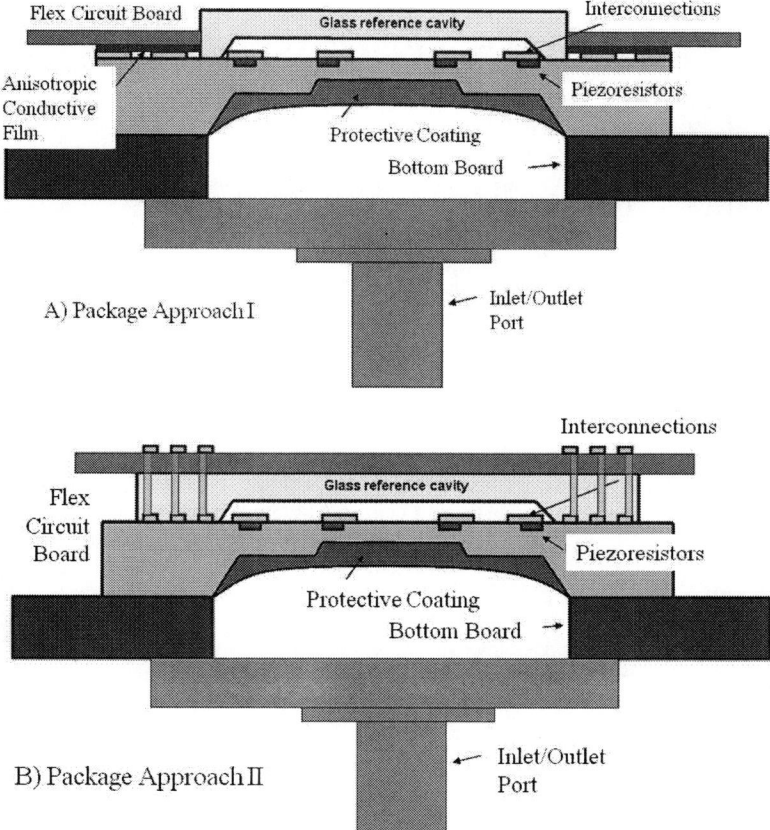

Figure 3. (A) Schematic of system-level packaging of reinforced pressure sensors following packaging approach I (Pack I). (B) Schematic of system-level packaging of reinforced pressure sensors following packaging approach II (Pack II).

Results and Discussion

The two-subsystems were characterized and measured individually in simulated marine conditions. First, the nano-enabled electrochemical sensor on a MEMS/microfluidic platform was measured for nitrate fluxes. Next, the high resolution reinforced diaphragm MEMS pressure (depth) sensor with package was tested for up to pressure of 430 psi (approximately 300 m of water) in simulated ocean conditions.

Nitrate Measurements

Amperometric measurements were performed with Autolab PFSTAT30 potentiostat/galvanostat under flow-through conditions. The measurements were performed in the standard three-electrode format with doped PPy-NWs, Ag/AgCl double junction electrode and on-chip deposited platinum electrode as the working, the reference and the counter electrode, respectively. Initially, the electrochemical response of the doped PPy-NWs to varying levels of nitrate (10-1000 μM) was recorded to obtain the calibration plot. Two sets of effluent carrier solutions were used in these measurements. In the first set of experiments referred to as DI-cal, de-ionized water was used to obtain baseline calibration and nitrate (10-200 μM) was progressively added to the baseline solution. In the second set of experiments, referred to as Ocean-cal, IAPSO standard seawater (salinity 34.996) was used to obtain the baseline calibration and nitrate (20-1000 μM) was progressively added to the standard solution. Then, sensor response was measured under flow through conditions. The flow rate was set between 1-5 mL/min. At the first potential, optimized at 0.65 V, the PPy-NWs were preoxidized, and the second potential step at -0.15 V was to reduce the nitrate ions. These conditions were determined after series of cyclic voltammograms of doped PPy-NWs which indicated the nitrate reduction peaks (3). All experiments were carried under identical conditions at room temperature.

After successful calibration testing of PPy-NWs electrodes under static conditions, the electrodes were tested in the integrated microfluidic platform. The microfluidic platform represented a more realistic and controlled environment mimicking conditions in water systems. As in the case of static measurements, constant-potential technique is used for flow analysis under dynamic conditions. In the first set of experiments, the carrier effluent contained only IAPSO standard seawater to give baseline stability. Numerous experiments were carried out to optimize the carrier flow rate, and pulsed voltammograms as the response intensity depended on the carrier conditions. An optimal flow rate was chosen between 1-5 mL/min, and was used in all further experiments. Figure 4A shows a typical peak current response at regular intervals of time for successive higher nitrate flow into the system. It can be observed that for successive higher nitrate levels, the peak current also increased linearly, maintaining a stable current response. In the next set of experiments, the baseline carrier effluent contained IAPSO standard seawater and 50 μM nitrate ions with intermittent nitrate spikes injected into the system at certain time intervals (Figure 4B).

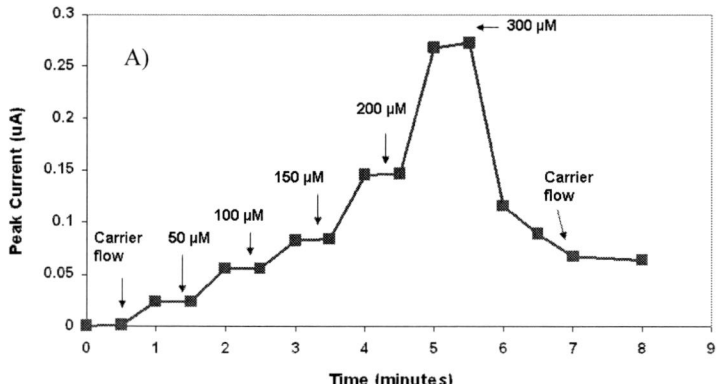

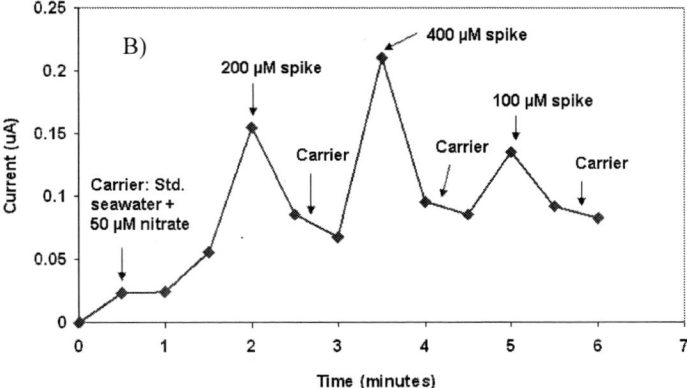

Figure 4. (A) Flow through analysis plot of peak current response at regular time intervals for successive nitrate increments (50-300 μM); (B) Flow through analysis plot of peak current response for intermittent nitrate spikes. Baseline carrier solution was IAPSO standard seawater and 50 μM nitrate effluent.

Next, the selectiveness of the doped PPy-NWs electrodes and the effect of interfering ions on current response were investigated for various anions. When the effluent solution contains interfering ions, these ions can compete for the adsorption sites with nitrate ions, and the determination of nitrate may be affected. However, we find that for up to 0.2 mM of chloride, sulfate, phosphate and perchlorate ions in baseline IAPSO seawater, no significant change in the current response was seen. In addition, the peak current response for higher levels of interfering ions decreased by only about 15-20% (7). This clearly demonstrated that the PPy-NWs electrodes have significant deviation from the Hofmeister series with regards to selectivity towards nitrate ions (8).

Pressure/Depth Measurements

The performance of the reinforced diaphragm configuration including temperature effects and package performance was measured and compared with the modeling (simulation) results. We only present results of Generation II devices under simulated ocean conditions. In the Generation II devices, the masked-maskless etching process was taken into account during the stage of mask design. Two representative Generation II reinforced diaphragm pressure sensors, referred to as II-D5 (inner L/H: 500/15 and outer L/H: 3000/45, all dimensions in μm), II-D6 (inner L/H: 1000/10 and outer L/H: 4000/30, all dimensions in μm) were analyzed for this paper. Figure 5A shows the output response of II-D5 versus applied input pressure. The outer bridge exhibited a linear response up to 220 psi with a sensitivity of 0.018 mV/V/psi. The response of the thin inner diaphragm can be broken into two parts: a) linear region up to 120 psi, with a sensitivity of 0.029 mV/V/psi, and b) saturation region, where the inner bridge is in a state of uniaxial tension. Further, the pressure hysteresis was within ±4-5% of the measured output. Figure 5B shows the response of II-D6 versus input pressure. As in the case of II-D5, the outer bridge exhibited a quasi-linear response up to 350 psi, with a sensitivity of 0.044 mV/V/psi. While, the inner bridge exhibited two regions: a) linear region up to 260 psi, with a high sensitivity of 0.067 mV/V/psi, and b) saturation region up to 350 psi. It is worth noting that even though the inner diaphragm is inoperable after 260 psi, it doesn't break or buckle because of transfer of peak stresses to the outer diaphragm after the transition point. These results showed clear evidence of the effect of reinforced edge thickness. It can be seen that outer diaphragms with same length and different thicknesses, affected both outer bridge sensitivity and inner bridge transition point. Outer diaphragms of lower thickness deflected more easily and hence had higher sensitivity. In case of the transition point, however, lower outer thickness caused lower transition point as inner diaphragm reached large deflection regime much earlier.

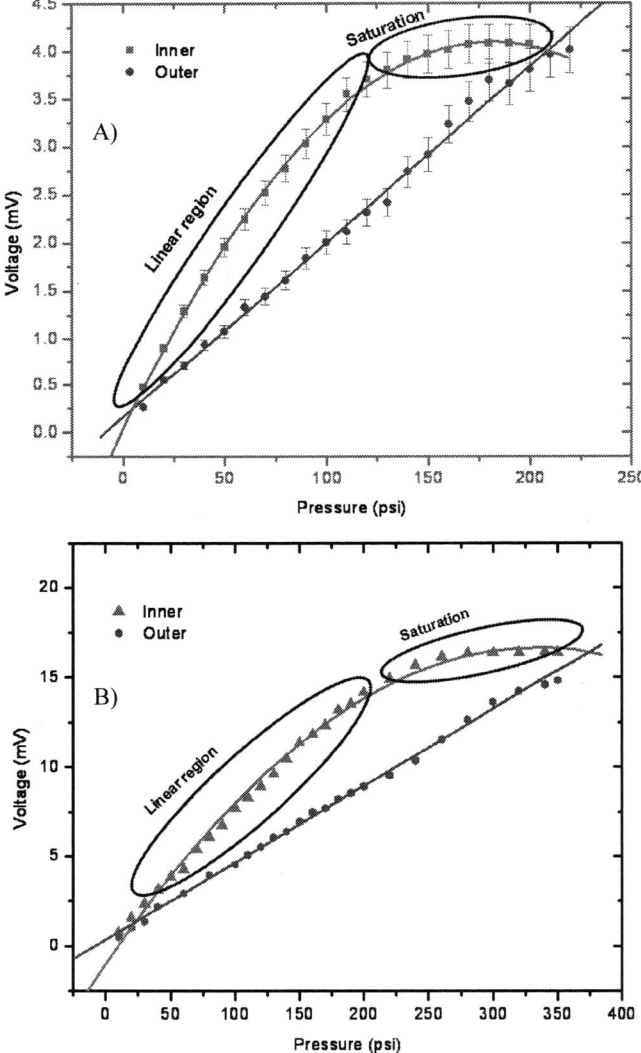

Figure 5. (A) Variation of output response to input pressure variations for sensor II-D5; (B) Variation of output response to input pressure variations for sensor II-D6.

In summary, this paper described the implementation of microfabricated and nanosensor elements and their associated packages for measurement of physical, chemical and biological variables in the marine environment. Using two sub-systems, we illustrated the basis for this measurement. Firstly, we showed the development of nanowire-based Electrochemical detection-on-Chip (EoC) to measure chemical markers, especially nitrate in the marine environment. Next, a novel reinforced diaphragm piezoresistive pressure sensor with package was developed to overcome the current limitation in operating range and sensitivity of conventional single diaphragm designs.

Acknowledgments

The authors acknowledge funding from Larry Langebrake and Center for Ocean Technology, Tampa, Florida and NSF-NER award 0403800.

References

1. J. D. Watkins et al., *U.S. Commission on Ocean Policy*, Washington, D.C. (2004).
2. M. J. Moorcroft, J. Davis and R. G. Compton, *Talanta* **54** (5), 785-803 (2001).
3. S. Aravamudhan and S. Bhansali, *Sensor Actuat B-Chem* **132** (2), 623-630 (2008).
4. S. Aravamudhan and S. Bhansali, *Sensor Actuat A-Phys* **142** (1), 111-117 (2008).
5. L. W. Lin, H. C. Chu and Y. W. Lu, *J Microelectromech S* **8** (4), 514-522 (1999).
6. A. Gotz, F. Campabadal and C. Cane, Sensor Actuat A-Phys **67** (1-3), 138-141 (1998).
7. S. Aravamudhan, S. Ketkar and S. Bhansali, *IEEE Sensors*, 205-208, 28-31 Oct. 2007.
8. F. Hofmeister, *Arch. Exp. Pathol. Pharmacol.* **24** 247-260 (1888).

522

CHAPTER 6

MEMS/NEMS POSTER SESSION

Micropatternable, Electrically Conducting Polyaniline Photoresist Blends For MEMS Applications.

C. V. Patel[1], A. Khosla[2] and S. Kassegne[1*]

[1]MEMS Research Lab, Department of Mechanical Engineering,
San Diego State University, San Diego, CA 92182
*Email: kassegne@mail.sdsu.edu

[2]School of Engineering Science, Simon Fraser University, Burnaby,
BC, Canada. V5A 1S6

We present preparation and electrical characterization of Polyaniline (PANi) SU8 micropatternable photoresist blend for MEMS applications. The blend was prepared by shear mixing of PANi and SU8-2010 at an rpm of 1000 for 15 hours. The composite was spin coated on a silicon wafer at an 850 rpm in order to achieve a thickness of 50µm, followed by soft baking at 70 °C for 35 minutes and cooling to room temperature. The desired structures were patterned using masked UV exposure for 60 seconds. Full cross linking of PANi SU-8 blend was achieved by a post-exposure bake at a temperature of 90°C for 25 minutes followed by cooling to room temperature. The desired electrode structures and trace lines were then developed in SU-8 developer (Microchem™) for 10 minutes by manual agitation. The fabricated structures were analyzed under SEM and Electron Dispersion X-ray Spectroscopy (EDS) demonstrating that an electrically conductive path is formed by PANI in SU-8 polymer matrix. It was also observed that resistivity of 4×10^3 Ω-m is achieved at 8.6 weight percentage of PANi in SU-8 polymer matrix.

Introduction

The introduction of conducting polymers has opened up significant new potential in microelectronics and micro-electro-mechanical systems (1-3). Conducting polymers offer exceptional combination of optical, electrical, and mechanical properties that make them a good candidate for certain materials currently used in microelectronics. The major applications of conductive polymers are in microelectronics, rechargeable batteries, power equipments, etc, where they could be used in metallization, interconnects, diodes, transistors, photovoltaic cells, corrosion protection of metals, and lithography (4). Polymers with a regular alternation of single and double bonds along the polymer chain form the basis of organic conducting polymers. Such an electronic structure results in a semiconductor-like band where band-gap depends on the extension of the conjugation. These conjugated polymers are typically made conductive by reacting the conjugated semiconducting polymer with an oxidizing agent, a reducing agent, or a protonic acid resulting in conductive structures (5). The conductivity of such polymers varies widely from 10^{-7} S/cm to 10^2 S/cm.

Conductive polymers can typically be differentiated into two categories: filled polymers and inherent conductive polymers (ICP). ICP contains some of polymers with good conductivity like Polyaniline (PANi), Polypyrrole, Polyvinylene and Polythiophene (2). Polyaniline that is used extensively in industry for various applications such as LED, rechargeable batteries, printed circuit boards and surface treatments for corrosion protection is the most stable ICP. PANI is generally classified into emeraldine base and emeraldine salt (6). Emeraldine base is the non-conductive form of polyaniline and is generally doped with acids like hydrochloric acid to obtain a conductive form called polyaniline emeraldine salt. However, despite their increased use, conductive polymers such as polyaniline, possess some undesirable properties such as instability in air and moisture, lack of mechanical strength, difficulty for processibility using conventional techniques and micropatterning. This study, therefore, investigates micropatterning techniques for PANi using PANi and photoresist blends.

Materials

Polyaniline (dispersed 2-5wt% in xylene) was bought from Sigma Aldrich. SU8-2010, which is a negative tone photoresist was purchased from Microchem™ USA.

Preparation of PANi SU8-2010 blend

Three different methods were employed for preparing PANi SU8-2010 blend. These are:

1) *Manual stirring:* In this method PANi is manually sired with SU-8 2010 using a glass rod for 15 minutes. This is straight-forward but highly inefficient approach that resulted in non-uniform blend of PANi and SU8-2010.

2) *Magnetic shear mixing:* This method employs magnetic stirrers where PANi and SU8-2010 are magnetically stirred at an rpm of 1000 for 15 hours. This approach is observed to reduce the viscosity of the blend resulting in better uniformity. It is also observed that stirring times longer than 15 hours do not necessarily result in a better homogeneous blend.

3) *Ultrasonic agitation:* In this method PANi and SU8-2010 blend is first manually stirred for 15 minutes and then placed in an ultrasonic bath operating at a frequency of 24 kHz for 1 hour. Ultrasonic agitation produces heat; hence, it is critical to maintain the temperature of the ultrasonic bath at 25^0 C in order to avoid detrimental effect on SU-8. It is observed that a fairly uniform blend is achieved with no air bubbles.

Based on these observations, a combination of all these three methods was employed to get a uniform and homogenous blend of PANi and SU8-2010 with no air bubbles. A blend of PANi and SU8-2010 were first manually stirred using a glass rod for 15 minutes, followed by magnetic shear mixing for 12 hours and ultrasonic agitation of the PANi SU8-2010 blend for 30 minutes. It was observed that all the air bubbles were effectively removed from the blend. It should be noted that it is critical to maintain the temperature of ultrasonic bath at 25^0 C.

Micropatterning of SU8-2010 Blend

Micropatterning of PANi SU8-2010 blend is an 6-step process. Each of these steps is discussed in this section.

Substrate Cleaning

The substrate was cleaned in 100% Micro 90 detergent using ultrasonic agitation for 5 minutes. The wafer was then rinsed with de-ionized (DI) water, acetone, isopropyl alcohol (IPA), and DI water. The wafer is then blow-dried using nitrogen followed by dehydration baking for 20 minutes at 150°C in a convection oven, followed by cooling down at room temperature (7-8).

Spin-coating of PANi-SU8-2010 blend

PANi SU8-2010 was then spun-coated on the cleaned silicon wafer. Spin speed versus film thickness of PANi -SU8-2010 photoresist blend is characterized. Figure 1 shows spin coating curve for PANi SU8-2010 blend. This curve is typical for 8.6 weight percentage of PANi in SU8-2010 polymer matrix, with varying speeds at a constant time of 95 seconds. Typically, after spin-coating, the substrate is kept on flat surface for ten minutes in order to allow the blend to settle down properly on the substrate. If an edge bead is observed, it is important to remove the edge bead as it results in unwanted gaps between the substrate and the mask, which may affect the resolution of the feature during the UV exposure.

Pre-Baking

After Spin-coating of PANi-SU8-2010 blend a soft-bake process is carried out on a hotplate in order to evaporate the solvents. Solvent evaporation rate is influenced by the rate of heat transfer and ventilation. Therefore, bake times should be optimized for proximity. Lower initial bake temperatures allow the solvent to evaporate out of the film at a more controlled rate, which ultimately results in better coating with reduced edge bead and better resist-to-substrate adhesion. If wrinkles are observed during the soft baking process, it is advisable to carry out soft bake for a longer time at 75^0 C instead of 90 ^{0}C. After soft baking is done, the substrate is allowed to cool down to room temperature. Table 1, summarizes the soft-baking parameters for PANi SU-8 blend (8.6 wt% of PANi in SU8-2010 polymer matrix).

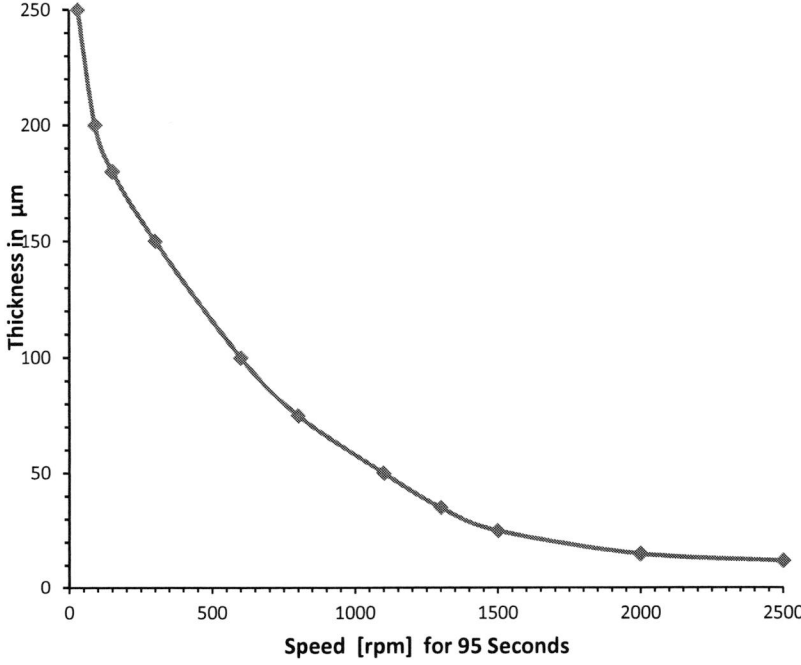

Figure 1. Spin coating curve for PANi SU8-2010 blend. This curve is typical to 8.6 weight percentage of PANi in SU8-2010 polymer matrix

Table 1. Soft-Baking parameters PANi SU-8 blend. (8.6 wt% of PANi in SU8-2010 polymer matrix)

Thickness (μm)	Soft Bake		
	At 65⁰C	At 75⁰C	90⁰C
20	5 min	10 min	10 min
45	8 min	12 min	15 min
75	15 min	15 min	20 min

UV Exposure

Due to the obstruction offered by PANi in the SU-8 matrix, UV exposure time and intensity selection play a significant role in the appropriate lithography. Absorption, reflection and diffraction are the main concepts that affect the photoresist polymerization. Diffraction of UV light results in low resolution and damaged features. Exposure intensity and time are key factors in reducing this disadvantage. More than 100 seconds of exposure might result in a larger diffraction effect. Due to reflection between PANi particles, deeper polymerization are typically attained. However, the reflected light also causes some portion of the masked region to polymerize, which perhaps reduces the feature resolution. The drawback of PANi particles is their absorbance effect. The light that is absorbed by PANi particles will be impaired from going deeper and initiating polymerization. The polymerized thickness will be limited due to the absence of light. As a result, the thickness of the film should be selected wisely. By performing various experiments, the data showed that 70 seconds of exposure under 15 mW/cm^2 intensity gave better results than longer exposure or higher intensity. The negative effect of long exposure times is shown in Figure 2 below which is imaged by Scanning Electron Microscope (SEM). Overexposure sometimes damages the features as shown in Figure 2 where the patterned features peeled-off from the substrate.

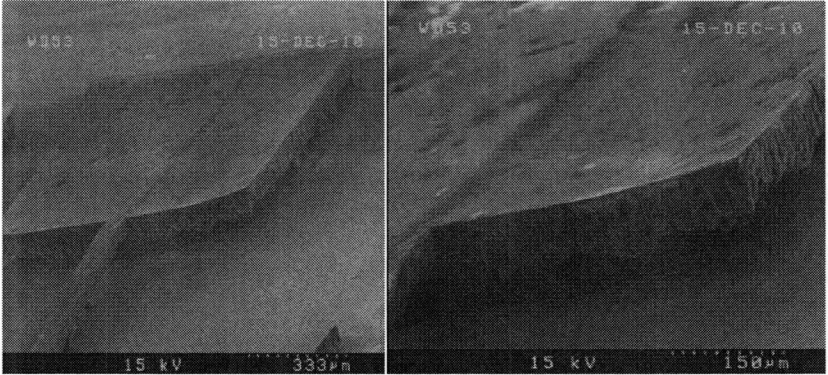

Figure 2. SEM image of over exposed PANi SU-8 structures. Undercutting and peeling-off is observed.

Post Baking

The Post-exposure baking parameters of PANi SU8-2010 blend were characterized and are summarized in Table 2. As the thermal co-efficient of expansion is different for PANi and SU-8, lower temperatures are preferred for post-exposure bake in comparison to the temperatures used for pure SU-8. In order to lower the rate of thermal expansion difference the baking temperature ramped up at the rate of 250^{0}C/hr from 65^{0}C to 75^{0}C and then to 90° C. After post baking the substrate is again allowed to cool down for 5 to 10 minutes.

Table 2. Post exposure baking parameters of PANi SU8-2010 blend (8.6 wt% of PANi in SU8-2010 polymer matrix)

Thickness (µm)	Post Baking Time		
	At 65°C	At 75°C	90°C
15-20µm	1 min	1 min	5 min
25-45µm	1 min	1 min	5 min
80-140µm	5 min	8 min	10 min
140µm	5 min	10 min	20 min

Development

After post-baking, the chip is allowed to cool down and later slowly immersed in the developer. It is then manually agitated for about 10 minutes. Longer times of development may result in loss of features. After development, the chip is rinsed with IPA for cleaning. For under-developed features, it was often observed that milky white patches form during IPA rinse. Later, the chip is air blow dried by nitrogen gun. The fabricated structures were analyzed under SEM to perform optical characterization. SEM analysis showed good patternability. The patterned structures shown in Figure 3 represent uniform stretch of polyaniline on the substrate. The features obtained were clear without any agglomeration and damaged features.

Figure 3. SEM micrographs of developed PANi SU8 blend

EDS (Electron Dispersion Spectroscopy) Characterization

A resistivity value of 4×10^3 Ω-m is achieved at 8.6 weight percentage of PANi in SU-8 2010 by using a four-point probe station. Subsequent to this, Electron Dispersion X-ray Spectroscopy (EDS) analysis was performed to establish what we call 'conductive-path', i.e., a line along where PANi is patterned. Figure 4 shows a bump pad feature marked with 5 points. At point 1, which is outside the feature (bump pad), elements like carbon, oxygen and silicon are expected. This is demonstrated in Figure 5a where a large proportion of carbon, oxygen and silicon were observed. Polyaniline comprises of various key elements like nitrogen, hydrogen, chlorine. The chlorine comes from HCl which is used to dope Polyaniline-EB to form Polyaniline-ES which is a conductive form of polyaniline. At the selected points (i.e., points 2, 3 and 4 in Figure 4), we use the presence of nitrogen as a confirmation of the presence of polyaniline on the feature. This is maintly driven by the argument that elements like hydrogen, chlorine, carbon are part of the other reagents and chemicals such as SU-8, photoinitiator, developer solution, cleaning reagents, etc whereas nitrogen is the only element which is exclusively a component of polyaniline. At point 5 which is outside the feature but on the substrate, we expect to have carbon, oxygen and silicon. Figure 5e confirms this.

In general, EDS seems to provide ample evidence that nitrogen and chlorine which are related to doped polyaniline are clearly present on the electrode features as desired offering what we consider is a 'conductive-path'. EDS analysis also shows that chlorine is also detected at points outside the feature albeit in a very negligible amount (0.01-0.03 wt. %). This is because complete removal of chlorine that could originate from the other lithography reagents is typically challenging.

Therefore, based on SEM and EDS characterizations, the following arguments support our observation that our lithographic patterning of PANi and SU-8 blend has resulted in a 'conductive path'.

1) Polyaniline ES is already conductive form of polymer. It is transformed from non-conducting (EB) to conducting (ES) form via a mechanism called doping. Hence, the presence of N^+ ions present on the features indicated conductance.

2) The majority of nitrogen is present on the microelectrode and bump-pad features.

3) Most importantly, the success of SEM on the sample (Figure 3) without coating it with a conductive materials suggest that the chip sample itself contains enough conductivity emanating from the PANi/SU-8 blend.

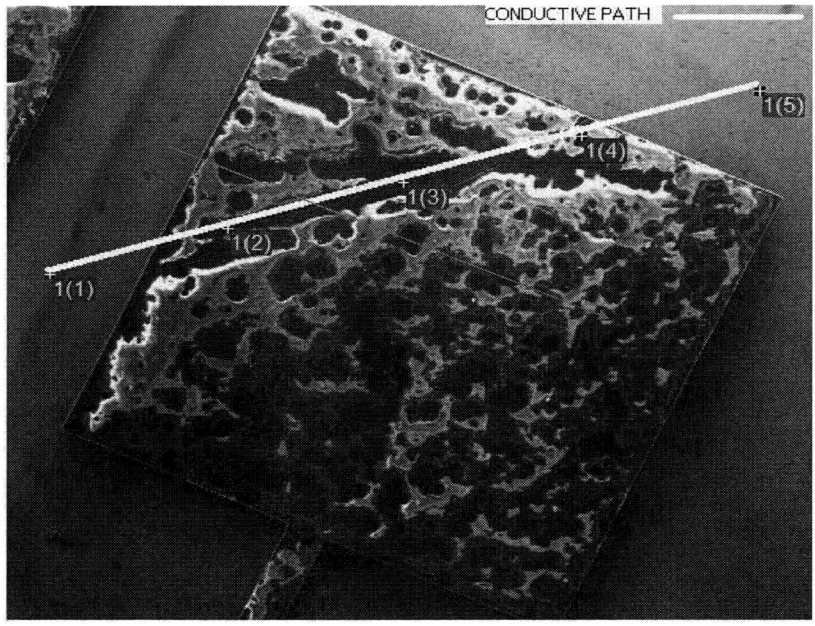

Figure 4. EDS characterization on a particular area of sample at 5 different points

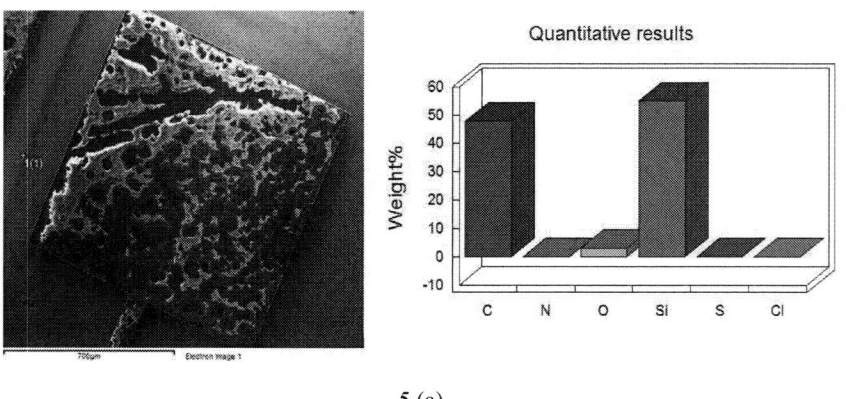

5 (a)

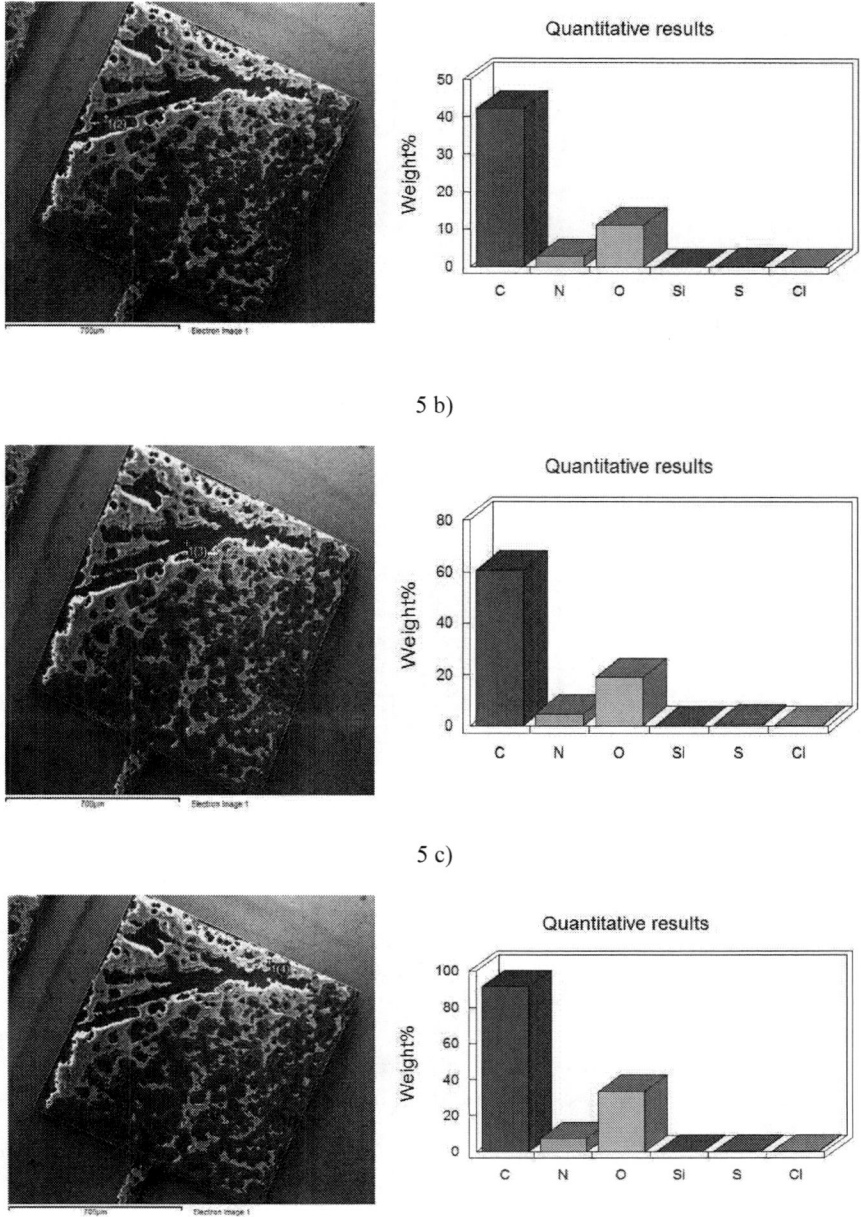

5 b)

5 c)

Figure 5 d)

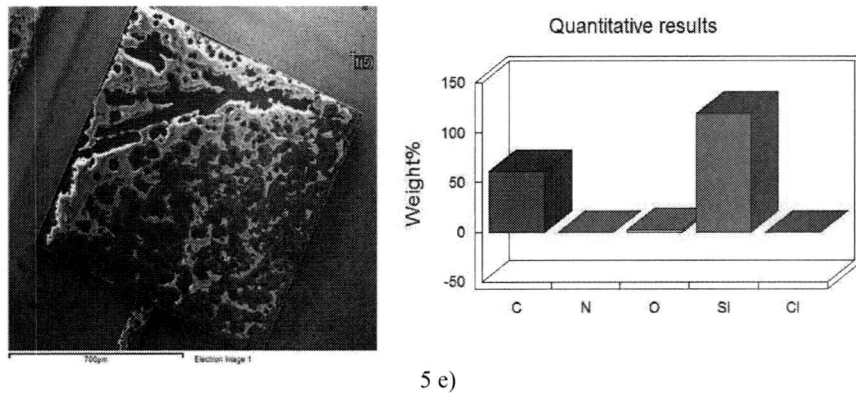

5 e)

Figure 5. EDS characterization of Bump-pad at individual points

Table 4. Data obtained through EDS characterization of Bump-pad

Bump-pad	Carbon wt %	Nitrogen wt %	Oxygen wt %	Silicon wt%	Chlorine wt%
Point 1	47.92	0.00	3.02	55	0.01
Point 2	42.21	2.72	10.99	0.00	0.09
Point 3	60.65	4.61	18.89	0.00	0.19
Point 4	91.46	7.13	33.32	0.00	0.15
Point 5	60.53	0.00	2.33	119.39	0.01

Conclusion

We have successfully developed an electrically conductive PANi SU8-2010 photoresist blend which can be micro-patterned using conventional photolithography techniques. It is observed that in order to fabricate a uniform homogenous blend of PANi and SU8-2010 with no air bubbles, both of them were first manually stirred using a glass rod for 15 minutes, followed up with magnetic shear mixing for 12 hours. Spin coating curves as well baking times have been characterized. A resistivity value of 4×10^3 Ω-m is achieved at 8.6 weight percentage of PANi in SU-8. EDS analysis confirm setting up of percolation path by PANi particle in SU8-2010 polymer matrix.

Acknowledgments

Authors would like to acknowledge San Diego State University Electron Microscope facilities and Dr. Steve Barlow for allowing us access to the SEM for our EDS analysis

References

1. A.J. Heeger, S. Kivelson, J.R. Schrieffer, W.P. Su, *Reviews of Modern Physics*, **60** (3), 781-850 (1988)
2. N. S. Sariciftci, L. Smilowitz, A. J. Heeger and F. Wudl, *Science*, **258** (5087), 1474-1476 (1992)
3. J.H. Burroughes, D.D.C. Bradley, A.R. Brown, et al., *Nature*, **347** (6293), 539–54 (1990)
4. B. Michel, A. Bernard, A. Bietsch, E. Delamarche, et al., *IBM Journal of Research & Development* , **45**, 697-719 (2001)
5. Y. Xia, X.M. Zhao, G.M. Whitesides, *Microelectronic Engineering*, **32**, 255-268 (1996)
6. J. Stejskal and R. G. Gilbert, *Pure Applied Chemistary*, **74** (5), 857-867(2002)
7. A. Khosla and B.L. Gray, *Materials Letters*, **63** (13-14), 1203-1206 (2009)
8. A. Khosla and B.L. Gray, *ECS Transactions*, **45** (3), 477-494 (2012)

536

Author Index

Ajito, K.	109	Dupont, T.	413
Akatsuka, T.	279		
Akbar, S. A.	119	Edagawa, K.	83, 401
Akiyama, T.	465	El-Kady, I.	449
Alam, M.	487	Ellern, I.	469
Alam, T. M.	423	Enachescu, M.	61
Allendorf, M. D.	469	Erdmann, C. A.	393
Amberger, M.	499	Eychmüller, A.	255
Anggraini, S. A.	179		
Aravamudhan, S.	513	Findlay, M. W.	211
Arimoto, S.	339	Finnegan, P. S.	423
		Fujii, E.	273
Bade, K.	499		
Bagheri, M.	459	Gabureac, M.	495
Balijepalli, A.	13	Gaponik, N.	255
Barsan, N.	221	Garcia, J. R.	393
Beechem, T. E.	423	Garzón, F. H.	307
Bhansali, S.	513	Gautsch, S.	465
Bishop, S. R.	507	Ghosh, N.	93
Blecke, J.	423	Goettler, D.	449
Boisen, A.	77	Górski, L.	333
Breedon, M.	129, 179	Goto, T.	267
Brosha, E. L.	307	Grossmann, K.	221
Brumbach, M. T.	423	Grutter, P.	77
Bychkov, E.	357		
		Hadvary, P.	89
Carter, M. T.	211	Hance, B. G.	423
Chai, Y.	43, 53, 69	Haque, M.	487
Chao, K.	301	Harada, S.	231
Chen, C.	413	Hasegawa, T.	279
Chen, S.	35	Hayashi, A.	363
Chin, B. A.	43, 53, 69	Heiskanen, A.	77
Cohan, B.	13	Hesketh, P. J.	469
Cui, M.	237	Hirabayashi, M.	325
		Hiremath, N.	69
Davis, D.	423	Honda, K.	377
de Rooij, N. F.	89, 465	Horikawa, S.	43, 53, 69
Devadas, B.	35	Hsieh, K. H.	301

Hu, C.	147	Lambert, T.	423
Hu, G.	441	Lee, D.	459
Huang, F.	237	Lee, F. P.	301
Hughes, M. K.	321	Lee, J.	469
Hyodo, T.	171, 267, 273	Leseman, Z. C.	449
		Li, M.	441
Iida, Y.	385	Li, S.	43, 53, 69
Imanaka, N.	189	Li, W.	307
Inaba, J.	393	Li, Y.	413
Ino, K.	205	Li, Y.	35
Ishida, H.	259	Lien, C. H.	147
Ishii, H.	435	Limwikrant, W.	109
		Lin, Y.	409
Janyasupab, M.	195	Liu, C.	441
Jeon, S.	459	Liu, C.	195
Jiang, X.	441	Liu, C.	195
Jin, Y.	237	Lunsford, S. K.	321
Jin, Y.	435		
Jorge Dulanto, J.	77	Major, T. C.	13
		Makishita, T.	259
Kamei, A.	339	Malachowski, K.	413
Kanazawa, S.	279	Malinowska, E.	333
Kaneyasu, K.	267, 315	Mani, V.	35
Kang, W.	93	Markin, V. S.	3, 23
Kanno, Y.	205	Maru, Y.	289
Kassegne, S.	325, 525	Masuda, H.	249
Kassem, M.	357	Matsue, T.	139, 205
Kawai, A.	477	Matsui, T.	231
Khosla, A.	325, 525	Matsuo, K.	273
Kim, B.	449	Matsuoka, R.	369
Kim, J.	109	Mehta, B.	325
Kim, S.	459	Meiser, J.	499
Kirsch, J. S.	345	Meyerhoff, M. E.	13
Komaba, S.	279	Milochova, M.	357
Kondo, T.	369	Miura, N.	129, 179
Kondo, T.	249	Miyado, T.	165
Kreller, C. R.	307	Mizukami, T.	165
Kukutsu, N.	109	Mizutani, F.	139
Kumar, S.	487	Mizutani, T.	289
Kunka, D.	499	Mohr, J.	499
Kuwabara, K.	435	Moribe, K.	109
		Mu, D.	237
Labuda, A.	77	Mukundan, R.	307

Musa, S.	413	Shimanoe, K.	221
		Shimazu, K.	165
Nagai, T.	189	Shimizu, Y.	171, 267, 273
Nakahara, A.	377	Shimouchi, A.	165
Nakanishi, H.	83	Shinnishi, K.	315
Naragino, H.	377	Simonian, A. L.	345
Nguyen, J.	449	Soltis, R. E.	295
Nguyen, P. K.	321	Song, H.	109
Nishijo, T.	205	Stakenborg, T.	413
Nishio, K.	249	Stavilla, V.	469
Nishiyama, T.	231	Stefan - van Staden, R. I.	61
Nose, K.	165	Stetter, J. R.	211
		Strong, J. M.	423
Okazaki, S.	289	Su, M.	449
Olsson III, R. H.	449	Sun, J.	441
Ono, K.	435	Sun, Y.	77
Ou, K.	301	Suzuki, C.	279
		Suzuki, T.	315
Palanisamy, P.	307	Szkudlarek, A.	495
Park, M.	43, 53, 69		
Patel, C. V.	525	Takada, K.	363
Patel, V.	211	Takagahara, K.	435
Pavelko, R. G.	221	Takahashi, K.	259
Peng, B.	13	Takahashi, M.	339
Pessoa, C. A.	393	Takemura, R.	259
Petrenko, V. A.	69	Takeuchi, M.	315
		Tamechika, E.	109
Quan, X.	77	Tamura, S.	189
		Tenje, M.	77
Raina, S.	93	Thundat, R.	459
Reinke, C.	449	Thundat, T.	459
Robinson, A. L.	469	Tsai, Y.	147
		Tschirky, H.	89
Sabuncuoglu Tezcan, D.	413	Tsubokawa, N.	231
Sai, M.	315	Tsunashima, K.	231
Sakata, T.	435	Tuller, H. L.	507
Sato, S.	83		
Sato, T.	129	Ueno, Y.	109
Sato, Y.	435	Utke, I.	495
Satoh, I.	385		
Sekhar, P. K.	307	van der Wal, P. D.	89, 465
Shibutani, Y.	165	Van Dorpe, P.	413
Shiku, H.	139, 205	Venkatasubramanian, A.	469

Verbeeck, R.	413
Viana, A. G.	393
Vodyanoy, V.	43
Volkov, A. G.	3, 23
Volkova, M. I.	3
Waite, A. J.	23
Wakida, S.	165
Wang, D.	147
Washburn, C. M.	423
Weerakoon, K.	69
Wen, L.	237
Wheeler, D. R.	423
Wikle III, H.	53
Wohnrath, K.	393
Wolff, A.	77
Wooten, J. D.	23
Wu, Y.	465
Yabuuchi, N.	279
Yamamoto, K.	109
Yamamura, N.	363
Yamashita, T.	171
Yamauchi, T.	231
Yan, Q.	13
Yanagi, H.	267
Yang, L.	101
Yang, X.	345
Yasui, T.	363
Yasukawa, T.	139
Yasuzawa, M.	83, 401
Yeh, C.	409
Yoshinaga, K.	377
Yoshioka, T.	339
Yuan, J.	255
Yuasa, M.	369
Yuchi, A.	363
Zhang, Y.	195
Ziaei-Moayyed, M.	449
Ziółkowski, R.	333